VOLUME

2

IMDG CODE

INTERNATIONAL MARITIME DANGEROUS GOODS CODE

2012 EDITION

INCORPORATING AMENDMENT 36-12

IMO INTERNATIONAL MARITIME ORGANIZATION

London, 2012

Published in 2012
by the INTERNATIONAL MARITIME ORGANIZATION
4 Albert Embankment, London SE1 7SR
www.imo.org

Printed in the United Kingdom by CPI Books Limited, Reading RG1 8EX

PEFC
PEFC/06-37-03

ISBN: 978-92-801-1561-1

IMO PUBLICATION
Sales number: II200E

This publication has been prepared from official documents of IMO, and every effort has been made to eliminate errors and reproduce the original text(s) faithfully. Readers should be aware that, in case of inconsistency, the official IMO text will prevail.

J1923E

Contents

Note

The following symbols placed against an item indicate changes from the previous edition, in accordance with Amendment 36-12 to the IMDG Code:

- ■ Insertion of an item
- ⊗ Deletion of an item
- △ Change to an item

For detailed information on changes, please visit http://gisis.imo.org and navigate to the IMDG Code module (free, instant registration required).

PART 3

DANGEROUS GOODS LIST, SPECIAL PROVISIONS AND EXCEPTIONS

Chapter 3.1

General

3.1.1 Scope and general provisions

3.1.1.1 The Dangerous Goods List in chapter 3.2 lists many of the dangerous goods most commonly transported. The list includes entries for specific chemical substances and articles and generic or "not otherwise specified" entries. Since it is not practical to include a separate entry for every chemical substance or article of commercial importance specifically by name, especially names for mixtures and solutions of various chemical constituents and concentrations, the Dangerous Goods List also includes generic or "not otherwise specified" names (e.g., EXTRACTS, FLAVOURING, LIQUID, UN 1197 or FLAMMABLE LIQUID, N.O.S., UN 1993). On this basis, the Dangerous Goods List is intended to include an appropriate name or entry for any dangerous good which may be transported.

3.1.1.2 Where a dangerous good is specifically listed by name in the Dangerous Goods List, it shall be transported in accordance with the provisions in the List which are appropriate for that dangerous good. A generic or "not otherwise specified" entry may be used to permit the transport of substances, materials or articles which do not appear specifically by name in the Dangerous Goods List. Such a dangerous good may be transported only after its dangerous properties have been determined. Dangerous goods shall be classified according to the class definitions, tests and criteria. The name which most appropriately describes the dangerous goods shall be used. Only when the specific name of the dangerous goods does not appear in the Dangerous Goods List or the associated primary or subsidiary hazards assigned to it are not appropriate may a generic or "not otherwise specified" name be used. The classification shall be made by the shipper/consignor or by the appropriate competent authority where so specified in the Code. Once the class of the dangerous good has been so established, all conditions for transport, as provided in this Code, shall be met. Any dangerous good having or suspected of having explosive characteristics shall first be considered for inclusion in class 1. Some collective entries may be of the generic or "not otherwise specified" type provided that the Code contains provisions ensuring safety, both by excluding extremely dangerous goods from normal transport and by covering all subsidiary risks inherent in some goods.

3.1.1.3 Inherent instability in goods may take different dangerous forms, for example explosion, polymerization with intense evolution of heat or emission of flammable, toxic, corrosive or asphyxiant gases. The Dangerous Goods List indicates that certain dangerous goods, or dangerous goods in a specific form, concentration or state, are prohibited for transport by sea. This means that the goods specified are not suitable for transport by sea under normal conditions of transport This does not mean that such goods may not be transported under any circumstances. For most goods, such inherent instability can be controlled by suitable packaging, dilution, stabilization, addition of an inhibitor, temperature control or other measures.

3.1.1.4 Where precautionary measures are laid down in the Dangerous Goods List in respect of a given dangerous good (such as that it shall be "stabilized" or "with *x*% water or phlegmatizer"), such dangerous good may not normally be transported when these measures have not been taken, unless the item in question is listed elsewhere (such as class 1) without any indication of, or with different, precautionary measures.

3.1.1.5 Certain substances, by the nature of their chemical composition, tend to polymerize or otherwise react in a dangerous manner under certain conditions of temperature or in contact with a catalyst. Mitigation of this tendency can be carried out either by requiring special transport conditions or by adding adequate amounts of chemical inhibitors or stabilizers to the product. These products shall be sufficiently stabilized to prevent any dangerous reaction during the intended voyage. If this cannot be ensured, the transport of such products is prohibited.

3.1.1.6 Where the contents of a portable tank is to be transported heated, the transport temperature is to be maintained during the intended voyage unless it is established that crystallization or solidification on cooling would not result in instability, which can occur with some stabilized or inhibited products.

3.1.2 Proper Shipping Names

Note 1: The Proper Shipping Names of the dangerous goods are those listed in chapter 3.2, Dangerous Goods List. Synonyms, secondary names, initials, abbreviations of names, etc. have been included in the Index to facilitate the search for the Proper Shipping Name (see part 5, Consignment Procedures).

Note 2: For Proper Shipping Names to be used for transport of samples, see 2.0.4. For Proper Shipping Names to be used for transport of wastes, see 5.4.1.4.3.3.

3.1.2.1 The Proper Shipping Name is that portion of the entry most accurately describing the goods in the Dangerous Goods List, which is shown in upper-case characters (plus any numbers, Greek letters, '*sec*', '*tert*', and the letters *m*, *n*, *o*, *p*, which form an integral part of the name). An alternative Proper Shipping Name may be shown in brackets following the main Proper Shipping Name (such as ETHANOL (ETHYL ALCOHOL)). Portions of an entry appearing in lower case need not be considered as part of the Proper Shipping Name but may be used.

3.1.2.2 When conjunctions such as "and" or "or" are in lower case or when segments of the name are punctuated by commas, the entire name of the entry need not necessarily be shown in the transport document or package markings. This is the case particularly when a combination of several distinct entries are listed under a single UN Number. Examples illustrating the selection of the Proper Shipping Name for such entries are:

.1 UN 1057 LIGHTERS or LIGHTER REFILLS – The Proper Shipping Name is the most appropriate of the following possible combinations:

LIGHTERS

LIGHTER REFILLS;

.2 UN 2583 ALKYLSULPHONIC ACIDS, SOLID or ARYLSULPHONIC ACIDS, SOLID with more than 5% free sulphuric acid – The Proper Shipping Name is the most appropriate of the following:

ALKYLSULPHONIC ACIDS, SOLID

ARYLSULPHONIC ACIDS, SOLID;

.3 UN 2793 FERROUS METAL BORINGS, SHAVINGS, TURNINGS or CUTTINGS in a form liable to self-heating. The Proper Shipping Name is the most appropriate of the following combinations:

FERROUS METAL BORINGS

FERROUS METAL SHAVINGS

FERROUS METAL TURNINGS

FERROUS METAL CUTTINGS.

3.1.2.3 Proper Shipping Names may be used in the singular or plural as appropriate. In addition, when qualifying words are used as part of the Proper Shipping Name, their sequence on documentation or packages is optional. Commercial or military names for goods of class 1, which contain the Proper Shipping Name supplemented by additional text, may be used.

3.1.2.4 Many substances have an entry for both the liquid and solid state (see definitions for *liquids* and *solids* in 1.2.1), or for the solid and solution. These are allocated separate UN Numbers which are not necessarily adjacent to each other. Details are provided in the alphabetical index, e.g.:

NITROXYLENES, LIQUID	–	6.1	1665
NITROXYLENES, SOLID	–	6.1	3447.

3.1.2.5 Where it is not already included, the qualifying word "MOLTEN" shall be added to the Proper Shipping Name when a substance which is solid in accordance with the definition in 1.2.1 is offered for transport in the molten state (such as ALKYLPHENOL, SOLID, N.O.S., MOLTEN). For elevated temperature substances, see 5.4.1.4.3.4.

△ **3.1.2.6** Except for self-reactive substances and organic peroxides and unless it is already included in capital letters in the name indicated in the Dangerous Goods List, the word STABILIZED shall be added as part of the Proper Shipping Name of the substance which without stabilization would be forbidden from transport in accordance with 1.1.3 due to it being liable to dangerously react under conditions normally encountered in transport (such as TOXIC LIQUID, ORGANIC, N.O.S., STABILIZED). When temperature control is used to stabilize such substances to prevent the development of any dangerous excess pressure, then:

.1 For liquids: where the SADT is less than or equal to 50°C, the provisions of 7.3.7.5 shall apply;

.2 For gases: the conditions of transport shall be approved by the competent authority.

3.1.2.7 Hydrates may be transported under the Proper Shipping Name for the anhydrous substance.

3.1.2.8 **Generic or "not otherwise specified" (N.O.S.) entries**

3.1.2.8.1 Generic and "not otherwise specified" Proper Shipping Names that are assigned to special provision 274 or 318 in column 6 of the Dangerous Goods List shall be supplemented with the technical or chemical group names unless a national law or international convention prohibits its disclosure if it is a controlled substance. For explosives of class 1, the dangerous goods description may be supplemented by additional descriptive text to indicate commercial or military names. Technical and chemical group names shall be entered in brackets immediately following the Proper Shipping Name. An appropriate modifier, such as "contains" or "containing" or other qualifying words such as "mixture", "solution", etc., and the percentage of the technical constituent may also be used. For example: "UN 1993 Flammable liquid, n.o.s. (contains xylene and benzene), 3, PG II".

3.1.2.8.1.1 The technical name shall be a recognized chemical or biological name or other name currently used in scientific and technical handbooks, journals and texts. Trade names shall not be used for this purpose. In the case of pesticides, only ISO common name(s), other name(s) in the WHO Recommended Classification of Pesticides by Hazard and Guidelines to Classification, or the name(s) of the active substance(s) may be used.

3.1.2.8.1.2 When a mixture of dangerous goods is described by one of the "N.O.S." or "generic" entries to which special provision 274 has been allocated in the Dangerous Goods List, not more than the two constituents which most predominantly contribute to the hazard or hazards of a mixture need to be shown, excluding controlled substances when their disclosure is prohibited by national law or international convention. If a package containing a mixture is labelled with any subsidiary risk label, one of the two technical names shown in brackets shall be the name of the constituent which compels the use of the subsidiary risk label.

3.1.2.8.1.3 Examples illustrating the selection of the Proper Shipping Name supplemented with the technical name of goods for such N.O.S. entries are:

UN 2902 PESTICIDE, LIQUID, TOXIC, N.O.S. (drazoxolon)

UN 3394 ORGANOMETALLIC SUBSTANCE, LIQUID, PYROPHORIC, WATER-REACTIVE (trimethylgallium).

3.1.2.9 **Marine pollutants**

3.1.2.9.1 For generic or "not otherwise specified" (N.O.S.) entries, the Proper Shipping Name shall be supplemented with the recognized chemical name of the marine pollutant.

3.1.2.9.2 Examples illustrating the selection of the Proper Shipping Name supplemented with the recognized technical name of goods for such entries are indicated below:

UN 1993 FLAMMABLE LIQUID, N.O.S. (propyl acetate, di-n-butyltin di-2-ethylhexanoate) class 3 PG III (50°C c.c.) MARINE POLLUTANT

UN 1263 PAINT (triethylbenzene) class 3 PG III (27°C c.c.) MARINE POLLUTANT

3.1.3 Mixtures or solutions

Note: Where a substance is specifically listed by name in the Dangerous Goods List, it shall be identified in transport by the Proper Shipping Name in the Dangerous Goods List. Such substances may contain technical impurities (for example those deriving from the production process) or additives for stability or other purposes that do not affect their classification. However, a substance listed by name containing technical impurities or additives for stability or other purposes affecting its classification shall be considered a mixture or solution (see 2.0.2.2 and 2.0.2.5).

3.1.3.1 A mixture or solution is not subject to the provisions of this Code if the characteristics, properties, form or physical state of the mixture or solution are such that it does not meet the criteria, including human experience criteria, for inclusion in any class.

△ 3.1.3.2 A mixture or solution meeting the classification criteria of this Code composed of a single predominant substance identified by name in the Dangerous Goods List and one or more substances not subject to the provisions of this Code and/or traces of one or more substances identified by name in the Dangerous Goods List, shall be assigned the UN Number and Proper Shipping Name of the predominant substance named in the Dangerous Goods List unless:

.1 the mixture or solution is identified by name in the Dangerous Goods List;

.2 the name and description of the substance named in the Dangerous Goods List specifically indicate that they apply only to the pure substance;

.3 the hazard class or division, subsidiary risk(s), packing group, or physical state of the mixture or solution is different from that of the substance named in the Dangerous Goods List; or

.4 the hazard characteristics and properties of the mixture or solution necessitate emergency response measures that are different from those required for the substance identified by name in the Dangerous Goods List.

3.1.3.3 Qualifying words such as "MIXTURE" or "SOLUTION", as appropriate, shall be added as part of the Proper Shipping Name, for example, "ACETONE SOLUTION". In addition, the concentration of the mixture or solution may also be indicated after the basic description of the mixture or solution, for example, "ACETONE 75% SOLUTION".

3.1.3.4 A mixture or solution meeting the classification criteria of this Code that is not identified by name in the Dangerous Goods List and that is composed of two or more dangerous goods shall be assigned to an entry that has the Proper Shipping Name, description, hazard class or division, subsidiary risk(s) and packing group that most precisely describe the mixture or solution.

3.1.4 Segregation groups

△ 3.1.4.1 For the purpose of segregation, dangerous goods having certain similar chemical properties have been grouped together in segregation groups, see 7.2.5. Where, in the Dangerous Goods List entry in column 16 (stowage and segregation) a particular segregation requirement refers to a group of substances, the particular segregation requirement applies to the goods allocated to the respective segregation group.

△ 3.1.4.2 It is recognized that not all substances, mixtures, solutions or preparations falling within a segregation group are listed in the IMDG Code by name. These are shipped under N.O.S. entries. Although these N.O.S. entries are not themselves listed in the segregation groups (see 3.1.4.4), the consignor shall decide whether inclusion under the segregation group is appropriate and, if so, shall mention that fact in the transport document (see 5.4.1.5.11).

△ 3.1.4.3 The segregation groups in this Code do not cover substances which fall outside the classification criteria of the Code. It is recognized that some non-hazardous substances have similar chemical properties as substances listed in the segregation groups. A consignor or the person responsible for packing the goods into a cargo transport unit who does have knowledge of the chemical properties of such non-dangerous goods may decide to implement the segregation provisions of a related segregation group on a voluntary basis.

3.1.4.4 The following segregation groups are identified.

1 **Acids**

1052 Hydrogen fluoride, anhydrous*
1182 Ethyl chloroformate
1183 Ethyldichlorosilane
1238 Methyl chloroformate
1242 Methyldichlorosilane
1250 Methyltrichlorosilane
1295 Trichlorosilane
1298 Trimethylchlorosilane
1305 Vinyltrichlorosilane
1572 Cacodylic acid
1595 Dimethyl sulphate
1715 Acetic anhydride
1716 Acetyl bromide
1717 Acetyl chloride
1718 Butyl acid phosphate
1722 Allyl chloroformate
1723 Allyl iodide
1724 Allyltrichlorosilane, stabilized
1725 Aluminium bromide, anhydrous
1726 Aluminium chloride, anhydrous
1727 Ammonium hydrogendifluoride, solid
1728 Amyltrichlorosilane
1729 Anisoyl chloride

1730	Antimony pentachloride, liquid
1731	Antimony pentachloride solution
1732	Antimony pentafluoride
1733	Antimony trichloride
1736	Benzoyl chloride
1737	Benzyl bromide
1738	Benzyl chloride
1739	Benzyl chloroformate
1740	Hydrogendifluorides, n.o.s.
1742	Boron trifluoride acetic acid complex, liquid
1743	Boron trifluoride propionic acid complex, liquid
1744	Bromine or bromine solution
1745	Bromine pentafluoride
1746	Bromine trifluoride
1747	Butyltrichlorosilane
1750	Chloroacetic acid solution
1751	Chloroacetic acid, solid
1752	Chloroacetyl chloride
1753	Chlorophenyltrichlorosilane
1754	Chlorosulphonic acid (with or without sulphur trioxide)
1755	Chromic acid solution
1756	Chromic fluoride, solid
1757	Chromic fluoride solution
1758	Chromium oxychloride
1762	Cyclohexenyltrichlorosilane
1763	Cyclohexyltrichlorosilane
1764	Dichloroacetic acid
1765	Dichloroacetyl chloride
1766	Dichlorophenyltrichlorosilane
1767	Diethyldichlorosilane
1768	Difluorophosphoric acid, anhydrous
1769	Diphenyldichlorosilane
1770	Diphenylmethyl bromide
1771	Dodecyltrichlorosilane
1773	Ferric chloride, anhydrous
1775	Fluoroboric acid
1776	Fluorophosphoric acid, anhydrous
1777	Fluorosulphonic acid*
1778	Fluorosilicic acid
1779	Formic acid with more than 85% acid by mass
1780	Fumaryl chloride
1781	Hexadecyltrichlorosilane
1782	Hexafluorophosphoric acid
1784	Hexyltrichlorosilane
1786	Hydrofluoric acid and sulphuric acid mixture*
1787	Hydriodic acid*
1788	Hydrobromic acid*
1789	Hydrochloric acid*
1790	Hydrofluoric acid*
△ 1792	Iodine monochloride, solid

1793	Isopropyl acid phosphate
1794	Lead sulphate with more than 3% free acid
1796	Nitrating acid mixture*
1798	Nitrohydrochloric acid*
1799	Nonyltrichlorosilane
1800	Octadecyltrichlorosilane
1801	Octyltrichlorosilane
1802	Perchloric acid with not more than 50% acid, by mass*
1803	Phenolsulphonic acid, liquid
1804	Phenyltrichlorosilane
1805	Phosphoric acid, solution
1806	Phosphorus pentachloride
1807	Phosphorus pentoxide
1808	Phosphorus tribromide
1809	Phosphorus trichloride
1810	Phosphorus oxychloride
1811	Potassium hydrogendifluoride, solid
1815	Propionyl chloride
1816	Propyltrichlorosilane
1817	Pyrosulphuryl chloride
1818	Silicon tetrachloride
1826	Nitrating acid mixture, spent*
1827	Stannic chloride, anhydrous
1828	Sulphur chlorides
1829	Sulphur trioxide, inhibited or sulphur trioxide, stabilized
1830	Sulphuric acid with more than 51% acid*
1831	Sulphuric acid, fuming*
1832	Sulphuric acid, spent*
1833	Sulphurous acid
1834	Sulphuryl chloride
1836	Thionyl chloride
1837	Thiophosphoryl chloride
1838	Titanium tetrachloride
1839	Trichloroacetic acid
1840	Zinc chloride solution
1848	Propionic acid with not less than 10% and less than 90% acid, by mass
1873	Perchloric acid with more than 50% but not more than 72% acid, by mass*
1898	Acetyl iodide
1902	Diisooctyl acid phosphate
1905	Selenic acid
1906	Sludge acid*
1938	Bromoacetic acid solution
1939	Phosphorus oxybromide
1940	Thioglycolic acid
2031	Nitric acid, other than red fuming*
2032	Nitric acid, red fuming*
2214	Phthalic anhydride with more than 0.05% of maleic anhydride
2215	Maleic anhydride
2218	Acrylic acid, inhibited
2225	Benzenesulphonyl chloride

2226	Benzotrichloride
2240	Chromosulphuric acid*
2262	Dimethylcarbamoyl chloride
2267	Dimethyl thiophosphoryl chloride
2305	Nitrobenzenesulphonic acid
2308	Nitrosylsulphuric acid, liquid*
2331	Zinc chloride, anhydrous
2353	Butyryl chloride
2395	Isobutyryl chloride
2407	Isopropyl chloroformate
2434	Dibenzyldichlorosilane
2435	Ethylphenyldichlorosilane
2437	Methylphenyldichlorosilane
2438	Trimethylacetyl chloride
2439	Sodium hydrogendifluoride
2440	Stannic chloride pentahydrate
2442	Trichloroacetyl chloride
2443	Vanadium oxytrichloride
2444	Vanadium tetrachloride
2475	Vanadium trichloride
2495	Iodine pentafluoride
2496	Propionic anhydride
2502	Valeryl chloride
2503	Zirconium tetrachloride
2506	Ammonium hydrogen sulphate
2507	Chloroplatinic acid, solid
2508	Molybdenum pentachloride
2509	Potassium hydrogen sulphate
2511	2-Chloropropionic acid
2513	Bromoacetyl bromide
2531	Methacrylic acid, stabilized
2564	Trichloroacetic acid solution
2571	Alkylsulphuric acids
2576	Phosphorus oxybromide, molten
2577	Phenylacetyl chloride
2578	Phosphorus trioxide
2580	Aluminium bromide solution
2581	Aluminium chloride solution
2582	Ferric chloride solution
2583	Alkylsulphonic acids, solid or arylsulphonic acids, solid with more than 5% free sulphuric acid
2584	Alkylsulphonic acids, liquid or arylsulphonic acids, liquid with more than 5% free sulphuric acid
2585	Alkylsulphonic acids, solid or arylsulphonic acids, solid with not more than 5% free sulphuric acid
2586	Alkylsulphonic acids, liquid or arylsulphonic acids, liquid with not more than 5% free sulphuric acid
2604	Boron trifluoride diethyl etherate
2626	Chloric acid, aqueous solution with not more than 10% chloric acid
2642	Fluoroacetic acid
2670	Cyanuric chloride
2691	Phosphorus pentabromide
2692	Boron tribromide

2698	Tetrahydrophthalic anhydrides with more than 0.05% maleic anhydride
2699	Trifluoroacetic acid
2739	Butyric anhydride
2740	Propyl chloroformate
2742	Chloroformates, toxic, corrosive, flammable, n.o.s.
2743	*n*-Butyl chloroformate
2744	Cyclobutyl chloroformate
2745	Chloromethyl chloroformate
2746	Phenyl chloroformate
2748	2-Ethylhexyl chloroformate
2751	Diethylthiophosphoryl chloride
2789	Acetic acid, glacial or acetic acid solution, more than 80% acid, by mass
2790	Acetic acid solution, more than 10% but not more than 80% acid, by mass
2794	Batteries, wet, filled with acid electric storage
2796	Sulphuric acid with not more than 51% acid or battery fluid, acid*
2798	Phenylphosphorus dichloride
2799	Phenylphosphorus thiodichloride
2802	Copper chloride
2817	Ammonium hydrogendifluoride solution
2819	Amyl acid phosphate
2820	Butyric acid
2823	Crotonic acid, solid
2826	Ethyl chlorothioformate
2829	Caproic acid
2834	Phosphorous acid
2851	Boron trifluoride dihydrate
2865	Hydroxylamine sulphate
2869	Titanium trichloride mixture
2879	Selenium oxychloride
2967	Sulphamic acid
2985	Chlorosilanes, flammable, corrosive, n.o.s.
2986	Chlorosilanes, corrosive, flammable, n.o.s.
2987	Chlorosilanes, corrosive, n.o.s.
2988	Chlorosilanes, water-reactive, flammable, corrosive, n.o.s.
3246	Methanesulphonyl chloride
3250	Chloroacetic acid, molten
3260	Corrosive solid, acidic, inorganic, n.o.s.
3261	Corrosive solid, acidic, organic, n.o.s.
3264	Corrosive liquid, acidic, inorganic, n.o.s.
3265	Corrosive liquid, acidic, organic, n.o.s.
3277	Chloroformates, toxic, corrosive, n.o.s.
3361	Chlorosilanes, toxic, corrosive, n.o.s.
3362	Chlorosilanes, toxic, corrosive, flammable, n.o.s.
3412	Formic acid with not less than 10% but not more than 85% acid by mass
3412	Formic acid with not less than 5% but not more than 10% acid by mass
3419	Boron trifluoride acetic acid complex, solid
3420	Boron trifluoride propionic acid complex, solid
3421	Potassium hydrogendifluoride solution
3425	Bromoacetic acid, solid
3453	Phosphoric acid, solid

3456 Nitrosylsulphuric acid, solid

3463 Propionic acid with not less than 90% acid by mass

3472 Crotonic acid, liquid

■ 3498 Iodine monochloride, liquid

* identifies strong acids

2 **Ammonium compounds**

0004 Ammonium picrate dry or wetted with less than 10% water, by mass

0222 Ammonium nitrate, with more than 0.2% combustible substances

0402 Ammonium perchlorate

1310 Ammonium picrate, wetted with not less than 10% water, by mass

1439 Ammonium dichromate

1442 Ammonium perchlorate

1444 Ammonium persulphate

1512 Zinc ammonium nitrite

1546 Ammonium arsenate

1630 Mercury ammonium chloride

1727 Ammonium hydrogendifluoride, solid

1835 Tetramethylammonium hydroxide solution

1843 Ammonium dinitro-*o*-cresolate, solid

1942 Ammonium nitrate with not more than 0.2% combustible substances

2067 Ammonium nitrate based fertilizer

2071 Ammonium nitrate based fertilizer

2073 Ammonia solution, relative density less than 0.880 at 15°C in water, with more than 35% but not more than 50% ammonia

2426 Ammonium nitrate, liquid (hot concentrated solution)

2505 Ammonium fluoride

2506 Ammonium hydrogen sulphate

2683 Ammonium sulphide solution

2687 Dicyclohexylammonium nitrite

2817 Ammonium hydrogendifluoride solution

2818 Ammonium polysulphide solution

2854 Ammonium fluorosilicate

2859 Ammonium metavanadate

2861 Ammonium polyvanadate

2863 Sodium ammonium vanadate

3375 Ammonium nitrate emulsion or suspension or gel intermediate for blasting explosives

3423 Tetramethylammonium hydroxide, solid

3424 Ammonium dinitro-*o*-cresolate solution

3 **Bromates**

1450 Bromates, inorganic, n.o.s.

1473 Magnesium bromate

1484 Potassium bromate

1494 Sodium bromate

2469 Zinc bromate

2719 Barium bromate

3213 Ammonium bromate

3213 Bromates, inorganic, aqueous solution, n.o.s.

4 **Chlorates**

1445 Barium chlorate, solid

1452 Calcium chlorate

1458	Chlorate and borate mixture
1459	Chlorate and magnesium chloride mixture, solid
1461	Chlorates, inorganic, n.o.s.
1485	Potassium chlorate
1495	Sodium chlorate
1506	Strontium chlorate
1513	Zinc chlorate
2427	Potassium chlorate, aqueous solution
2428	Sodium chlorate, aqueous solution
2429	Calcium chlorate, aqueous solution
2573	Thallium chlorate
2721	Copper chlorate
2723	Magnesium chlorate
3405	Barium chlorate solution
3407	Chlorate and magnesium chloride mixture solution

5 **Chlorites**

1453	Calcium chlorite
1462	Chlorites, inorganic, n.o.s.
1496	Sodium chlorite
1908	Chlorite solution

6 **Cyanides**

1541	Acetone cyanhydrin, stabilized
1565	Barium cyanide
1575	Calcium cyanide
1587	Copper cyanide
1588	Cyanides, inorganic, solid, n.o.s.
1620	Lead cyanide
1626	Mercuric potassium cyanide
1636	Mercury cyanide
1642	Mercury oxycyanide, desensitized
1653	Nickel cyanide
1679	Potassium cuprocyanide
1680	Potassium cyanide, solid
1684	Silver cyanide
1689	Sodium cyanide, solid
1694	Bromobenzyl cyanides, liquid
1713	Zinc cyanide
1889	Cyanogen bromide
1935	Cyanide solution, n.o.s.
2205	Adiponitrile
2316	Sodium cuprocyanide, solid
2317	Sodium cuprocyanide solution
3413	Potassium cyanide solution
3414	Sodium cyanide solution
3449	Bromobenzyl cyanides, solid

7 **Heavy metals and their salts (including their organometallic compounds)**

0129	Lead azide, wetted, with not less than 20% water, or mixture of alcohol and water, by mass
0130	Lead styphnate (lead trinitroresorcinate), wetted with not less than 20% water, or mixture of alcohol and water, by mass

0135	Mercury fulminate, wetted with not less than 20% water, or mixture of alcohol and water, by mass
1347	Silver picrate, wetted with not less than 30% water, by mass
1366	Diethylzinc
1370	Dimethylzinc
1389	Alkali metal amalgam, liquid
1392	Alkaline earth metal amalgam, liquid
1435	Zinc ashes
1436	Zinc dust or zinc powder
1469	Lead nitrate
1470	Lead perchlorate, solid
1493	Silver nitrate
1512	Zinc ammonium nitrite
1513	Zinc chlorate
1514	Zinc nitrate
1515	Zinc permanganate
1516	Zinc peroxide
1587	Copper cyanide
1616	Lead acetate
1617	Lead arsenates
1618	Lead arsenites
1620	Lead cyanide
1623	Mercuric arsenate
1624	Mercuric chloride
1625	Mercuric nitrate
1626	Mercuric potassium cyanide
1627	Mercurous nitrate
1629	Mercury acetate
1630	Mercury ammonium chloride
1631	Mercury benzoate
1634	Mercury bromides
1636	Mercury cyanide
1637	Mercury gluconate
1638	Mercury iodide
1639	Mercury nucleate
1640	Mercury oleate
1641	Mercury oxide
1642	Mercury oxycyanide, desensitized
1643	Mercury potassium iodide
1644	Mercury salicylate
1645	Mercury sulphate
1646	Mercury thiocyanate
1649	Motor fuel anti-knock mixture
1653	Nickel cyanide
1674	Phenylmercuric acetate
1683	Silver arsenite
1684	Silver cyanide
1712	Zinc arsenate and zinc arsenite mixture
1713	Zinc cyanide
1714	Zinc phosphide
1794	Lead sulphate with more than 3% free acid

1838	Titanium tetrachloride
1840	Zinc chloride solution
1872	Lead dioxide
1894	Phenylmercuric hydroxide
1895	Phenylmercuric nitrate
1931	Zinc hydrosulphite
1931	Zinc dithionite
2024	Mercury compound, liquid, n.o.s.
2025	Mercury compound, solid, n.o.s.
2026	Phenylmercuric compound, n.o.s.
2291	Lead compound, soluble, n.o.s.
2331	Zinc chloride, anhydrous
2441	Titanium trichloride, pyrophoric or titanium trichloride mixture, pyrophoric
2469	Zinc bromate
2546	Titanium powder, dry
2714	Zinc resinate
2777	Mercury based pesticide, solid, toxic
2778	Mercury based pesticide, liquid, flammable, toxic
2809	Mercury
2855	Zinc fluorosilicate
2869	Titanium trichloride mixture
2878	Titanium, sponge granules or titanium, sponge powders
2881	Metal catalyst, dry
2989	Lead phosphite, dibasic
3011	Mercury based pesticide, liquid, toxic, flammable
3012	Mercury based pesticide, liquid, toxic
3089	Metal powder, flammable, n.o.s.
3174	Titanium disulphide
3181	Metal salts of organic compounds, flammable, n.o.s.
3189	Metal powder, self-heating, n.o.s.
3401	Alkali metal amalgam, solid
3402	Alkaline earth metal amalgam, solid
3408	Lead perchlorate solution
3483	Motor fuel anti-knock mixture, flammable

8 **Hypochlorites**

1471	Lithium hypochlorite
1748	Calcium hypochlorite mixture
1791	Hypochlorite solution
2208	Calcium hypochlorite mixture, dry with more than 10% but not more than 39% available chlorine
2741	Barium hypochlorite with more than 22% available chlorine
2880	Calcium hypochlorite, hydrated or calcium hypochlorite, hydrated mixture with not less than 5.5% but not more than 16% water
3212	Hypochlorites, inorganic, n.o.s.
3255	*tert*-Butyl hypochlorite
3485	Calcium hypochlorite, dry, corrosive or calcium hypochlorite mixture, dry, corrosive with more than 39% available chlorine (8.8% available oxygen)
3486	Calcium hypochlorite mixture, dry, corrosive with more than 10% but not more than 39% available chlorine
3487	Calcium hypochlorite, hydrated, corrosive or calcium hypochlorite, hydrated mixture, corrosive, with not less than 5.5% but not more than 16% water

3

9 Lead and its compounds

0129	Lead azide, wetted with not less than 20% water, or mixture of alcohol and water, by mass
0130	Lead styphnate, wetted with not less than 20% water, or mixture of alcohol and water, by mass
0130	Lead trinitroresorcinate, wetted with not less than 20% water, or mixture of alcohol and water, by mass
1469	Lead nitrate
1470	Lead perchlorate, solid
1616	Lead acetate
1617	Lead arsenates
1618	Lead arsenites
1620	Lead cyanide
1649	Motor fuel anti-knock mixture
1794	Lead sulphate with more than 3% free acid
1872	Lead dioxide
2291	Lead compound, soluble, n.o.s.
2989	Lead phosphide, dibasic
3408	Lead perchlorate solution
3483	Motor fuel anti-knock mixture, flammable

10 Liquid halogenated hydrocarbons

1099	Allyl bromide
1100	Allyl chloride
1107	Amyl chloride
1126	1-Bromobutane
1127	Chlorobutanes
1134	Chlorobenzene
1150	1,2-Dichloroethylene
1152	Dichloropentanes
1184	Ethylene dichloride
1278	1-Chloropropane
1279	1,2-Dichloropropane
1303	Vinylidene chloride, stabilized
1591	*o*-Dichlorobenzene
1593	Dichloromethane
1605	Ethylene dibromide
1647	Methyl bromide and ethylene dibromide mixture, liquid
1669	Pentachloroethane
1701	Xylyl bromide
1702	1,1,2,2-Tetrachloroethane
1710	Trichloroethylene
1723	Allyl iodide
1737	Benzyl bromide
1738	Benzyl chloride
1846	Carbon tetrachloride
1887	Bromochloromethane
1888	Chloroform
1891	Ethyl bromide
1897	Tetrachloroethylene
1991	Chloroprene, stabilized
2234	Chlorobenzotrifluorides
2238	Chlorotoluenes

2279 Hexachlorobutadiene
2321 Trichlorobenzenes, liquid
2322 Trichlorobutene
2339 2-Bromobutane
2341 1-Bromo-3-methylbutane
2342 Bromomethylpropanes
2343 2-Bromopentane
2344 Bromopropanes
2356 2-Chloropropane
2362 1,1-Dichloroethane
2387 Fluorobenzene
2388 Fluorotoluenes
2390 2-Iodobutane
2391 Iodomethylpropanes
2392 Iodopropanes
2456 2-Chloropropene
2504 Tetrabromoethane
2515 Bromoform
2554 Methylallyl chloride
2644 Methyl iodide
2646 Hexachlorocyclopentadiene
2664 Dibromomethane
2688 1-Bromo-3-chloropropane
2831 1,1,1-Trichloroethane
2872 Dibromochloropropanes

11 Mercury and mercury compounds

0135 Mercury fulminate, wetted with not less than 20% water
1389 Alkali metal amalgam, liquid
1392 Alkaline earth metal amalgam, liquid
1623 Mercuric arsenate
1624 Mercuric chloride
1625 Mercuric nitrate
1626 Mercuric potassium cyanide
1627 Mercurous nitrate
1629 Mercury acetate
1630 Mercury ammonium chloride
1631 Mercury benzoate
1634 Mercury bromides
1636 Mercury cyanide
1637 Mercury gluconate
1638 Mercury iodide
1639 Mercury nucleate
1640 Mercury oleate
1641 Mercury oxide
1642 Mercury oxycyanide, desensitized
1643 Mercury potassium iodide
1644 Mercury salicylate
1645 Mercury sulphate
1646 Mercury thiocyanate
1894 Phenylmercuric hydroxide

1895 Phenylmercuric nitrate
2024 Mercury compound, liquid, n.o.s.
2025 Mercury compound, solid, n.o.s
2026 Phenylmercuric compound, n.o.s.
2777 Mercury based pesticide, solid, toxic
2778 Mercury based pesticide, liquid, flammable, toxic
2809 Mercury
3011 Mercury based pesticide, liquid, toxic, flammable
3012 Mercury based pesticide, liquid, toxic
3401 Alkali metal amalgam, solid
3402 Alkaline earth metal amalgam, solid

12 Nitrites and their mixtures

1487 Potassium nitrate and sodium nitrite mixture
1488 Potassium nitrite
1500 Sodium nitrite
1512 Zinc ammonium nitrite
2627 Nitrites, inorganic, n.o.s.
2726 Nickel nitrite
3219 Nitrites, inorganic, aqueous solution, n.o.s

13 Perchlorates

1442 Ammonium perchlorate
1447 Barium perchlorate, solid
1455 Calcium perchlorate
1470 Lead perchlorate, solid
1475 Magnesium perchlorate
1481 Perchlorates, inorganic, n.o.s.
1489 Potassium perchlorate
1502 Sodium perchlorate
1508 Strontium perchlorate
3211 Perchlorates, inorganic, aqueous solution, n.o.s.
3406 Barium perchlorate solution
3408 Lead perchlorate solution

14 Permanganates

1448 Barium permanganate
1456 Calcium permanganate
1482 Permanganates, inorganic, n.o.s.
1490 Potassium permanganate
1503 Sodium permanganate
1515 Zinc permanganate
3214 Permanganates, inorganic, aqueous solution, n.o.s.

15 Powdered metals

1309 Aluminium powder, coated
1326 Hafnium powder, wetted with not less than 25% water
1352 Titanium powder, wetted with not less than 25% water
1358 Zirconium powder, wetted with not less than 25% water
1383 Pyrophoric alloy or pyrophoric metal, n.o.s.
1396 Aluminium powder, uncoated
1398 Aluminium silicon powder, uncoated
1418 Magnesium powder
1435 Zinc ashes

1436 Zinc dust or zinc powder
1854 Barium alloys, pyrophoric
2008 Zirconium powder, dry
2009 Zirconium, dry, sheets, strip or coiled wire
2545 Hafnium powder, dry
2546 Titanium powder, dry
2878 Titanium sponge powders
2881 Metal catalyst, dry
2950 Magnesium granules, coated, particle size not less than 149 microns
3078 Cerium, turnings or gritty powder
3089 Metal powder, flammable, n.o.s.
3170 Aluminium smelting by-products
3189 Metal powder, self-heating, n.o.s.

16 Peroxides

1449 Barium peroxide
1457 Calcium peroxide
1472 Lithium peroxide
1476 Magnesium peroxide
1483 Peroxides, inorganic, n.o.s.
1491 Potassium peroxide
1504 Sodium peroxide
1509 Strontium peroxide
1516 Zinc peroxide
2014 Hydrogen peroxide, aqueous solution, 20–60%
2015 Hydrogen peroxide, aqueous solution, stabilized
2466 Potassium superoxide
2547 Sodium superoxide
3149 Hydrogen peroxide and peroxyacetic acid mixture
3377 Sodium perborate monohydrate
3378 Sodium carbonate peroxyhydrate

17 Azides

0129 Lead azide, wetted
0224 Barium azide, dry
1571 Barium azide, wetted
1687 Sodium azide

18 Alkalis

1005 Ammonia, anhydrous
1160 Dimethylamine, aqueous solution
1163 Dimethylhydrazine, unsymmetrical
1235 Methylamine, aqueous solution
1244 Methylhydrazine
1382 Potassium sulphide, anhydrous or potassium sulphide with less than 30% water of crystallization
1385 Sodium sulphide, anhydrous or sodium sulphide with less than 30% water of crystallization
1604 Ethylenediamine
1719 Caustic alkali liquid, n.o.s.
1813 Potassium hydroxide, solid
1814 Potassium hydroxide solution
1819 Sodium aluminate solution
1823 Sodium hydroxide, solid
1824 Sodium hydroxide solution

1825 Sodium monoxide

1835 Tetramethylammonium hydroxide solution

1847 Potassium sulphide, hydrated with not less than 30% water of crystallization

1849 Sodium sulphide, hydrated with not less than 30% water

1907 Soda lime with more than 4% sodium hydroxide

1922 Pyrrolidine

2029 Hydrazine, anhydrous

2030 Hydrazine, aqueous solution with more than 37% hydrazine, by mass

2033 Potassium monoxide

2073 Ammonia solution relative density less than 0.880 at 15°C in water, with more than 35% but not more than 50% ammonia

2079 Diethylenetriamine

2259 Triethylenetetramine

2270 Ethylamine, aqueous solution, with not less than 50% but not more than 70% ethylamine

2318 Sodium hydrosulphide with less than 25% water of crystallization

2320 Tetraethylenepentamine

2379 1,3-Dimethylbutylamine

2382 Dimethylhydrazine, symmetrical

2386 1-Ethylpiperidine

2399 1-Methylpiperidine

2401 Piperidine

2491 Ethanolamine or ethanolamine solution

2579 Piperazine

2671 Aminopyridines

2672 Ammonia solution relative density between 0.880 and 0.957 at 15°C in water, with more than 10% but not more than 35% ammonia, by mass

2677 Rubidium hydroxide solution

2678 Rubidium hydroxide, solid

2679 Lithium hydroxide solution

2680 Lithium hydroxide

2681 Caesium hydroxide solution

2682 Caesium hydroxide

2683 Ammonium sulphide solution

2733 Amines, flammable, corrosive, n.o.s. or polyamines, flammable, corrosive, n.o.s.

2734 Amines, liquid, corrosive, flammable, n.o.s. or polyamines, liquid, corrosive, flammable, n.o.s.

2735 Amines, liquid, corrosive, n.o.s. or polyamines, liquid, corrosive, n.o.s.

2795 Batteries, wet, filled with alkali electric storage

2797 Battery fluid, alkali

2818 Ammonium polysulphide solution

2949 Sodium hydrosulphide, solid with not less than 25% water of crystallization

3028 Batteries, dry, containing potassium hydroxide, solid electric storage

3073 Vinylpyridines, stabilized

3253 Disodium trioxosilicate

3259 Amines, solid, corrosive, n.o.s. or polyamines, solid, corrosive, n.o.s.

3262 Corrosive solid, basic, inorganic, n.o.s.

3263 Corrosive solid, basic, organic, n.o.s.

3266 Corrosive liquid, basic, inorganic, n.o.s.

3267 Corrosive liquid, basic, organic, n.o.s.

3293 Hydrazine, aqueous solution with not more than 37% hydrazine, by mass

3318 Ammonia solution relative density less than 0.880 at 15°C in water, with more than 50% ammonia

3320	Sodium borohydride and sodium hydroxide solution with not more than 12% sodium borohydride and not more than 40% sodium hydroxide, by mass
3423	Tetramethylammonium hydroxide, solid
3484	Hydrazine aqueous solution, flammable, with more than 37% hydrazine, by mass

Chapter 3.2

Dangerous Goods List

3.2.1 Structure of the Dangerous Goods List

The Dangerous Goods List is divided into 18 columns as follows:

Column 1 **UN No.** – this column contains the United Nations Number assigned to a dangerous good by the United Nations Sub-Committee of Experts on the Transport of Dangerous Goods (UN List).

Column 2 **Proper Shipping Name (PSN)** – this column contains the Proper Shipping Names in upper-case characters, which may have to be followed by additional descriptive text in lower-case characters (see 3.1.2). Proper Shipping Names may be shown in plural where isomers of similar classification exist. Hydrates may be included under the Proper Shipping Name for the anhydrous substances. Unless otherwise indicated for an entry in the Dangerous Goods List, the word "SOLUTION" in a Proper Shipping Name means one or more named dangerous goods dissolved in a liquid that is not otherwise subject to this Code. When a flashpoint is mentioned in this column, the data is based on closed-cup (c.c.) methods.

Column 3 **Class or division** – this column contains the class and, in the case of class 1, the division and the compatibility group assigned to the substance or article according to the classification system described in part 2, chapter 2.1.

Column 4 **Subsidiary risk(s)** – this column contains the class number(s) of any subsidiary risk(s) which have been identified by applying the classification system described in part 2. This column also identifies a dangerous good as a marine pollutant as follows:

P – Marine pollutant: a non-exhaustive list of known marine pollutants, based on previous criteria and assignment

Column 5 **Packing group** – this column contains the packing group number (i.e., I, II or III) where assigned to the substance or article. If more than one packing group is indicated for the entry, the packing group of the substance or formulation to be transported shall be determined, based on its properties, through application of the hazard grouping criteria as provided in part 2.

Column 6 **Special provisions** – this column contains a number referring to any special provision(s) indicated in chapter 3.3 that is relevant to the substance or article. Special provisions apply to all packing groups permitted for a particular substance or article unless the wording makes it otherwise apparent. The special provision numbers specific to the sea mode start from 900.

Note: When a special provision is no longer needed, this special provision is deleted but the special provision number is not allocated again, in order not to confuse the users of this Code. For this reason, some of the numbers are missing.

Column 7a **Limited quantities** – this column provides the maximum quantity per inner packaging or article for transporting dangerous goods as limited quantities in accordance with chapter 3.4.

Column 7b **Excepted quantities** – this column provides an alpha-numeric code described in sub-section 3.5.1.2 which indicates the maximum quantity per inner and outer packaging for transporting dangerous goods as excepted quantities in accordance with chapter 3.5.

Column 8 **Packing instructions** – this column contains alpha-numeric codes which refer to the relevant packing instruction(s) in 4.1.4. The packing instructions indicate the packagings (including large packagings) which may be used for the transport of substances and articles.

A code including the letter "P" refers to packing instructions for the use of packagings described in chapter 6.1, 6.2 or 6.3.

A code including the letters "LP" refers to packing instructions for the use of large packagings described in chapter 6.6.

When a code including the letter(s) "P" or "LP" is not provided, it means that the substance is not allowed in that type of packaging.

Column 9 **Special packing provisions** – this column contains alpha-numeric codes which refer to the relevant special packing provisions specified in 4.1.4. The special packing provisions indicate the packagings (including large packagings).

A special packing provision including the letters "PP" refers to a special packing provision applicable to the use of a packing instruction bearing the Code "P" in 4.1.4.1.

A special packing provision including the letter "L" refers to a special packing provision applicable to a packing instruction bearing the code "LP" in 4.1.4.3.

Column 10 **IBC packing instructions** – this column contains alpha-numeric codes that refer to the relevant IBC instruction, which indicates the type of IBC that shall be used for the transport of the substance under reference. A code including the letters "IBC" refers to packing instructions for the use of IBCs described in chapter 6.5. When a code is not provided, it means the substance is not authorized in IBC.

Column 11 **IBC special provisions** – this column contains an alpha-numeric code, including the letter "B", which refers to special packing provisions applicable to the use of packing instructions bearing the code "IBC" in 4.1.4.2.

Column 12 [Reserved]

△ Column 13 **Tank and bulk container instructions** – this column contains T codes (see 4.2.5.2.6) applicable to the transport of dangerous goods in portable tanks and road tank vehicles.

When a T code is not provided in this column, it means that the dangerous goods are not authorized for transport in tanks unless specifically approved by the competent authority.

A code including the letters "BK" refers to the type of bulk containers used for the transport of bulk goods described in chapters 4.3 and 6.9.

The gases authorized for transport in MEGCs are indicated in the column "MEGC" in tables 1 and 2 of packing instruction P200 in 4.1.4.1.

Column 14 **Tank special provisions** – this column contains TP notes (see 4.2.5.3) applicable to the transport of dangerous goods in portable tanks and road tank vehicles. The TP notes specified in this column apply to the portable tanks specified in both columns 12 and 13.

Column 15 **EmS** – this column refers to the relevant emergency schedules for FIRE and SPILLAGE in "The EmS Guide – Emergency Response Procedures for Ships Carrying Dangerous Goods".

The first EmS code refers to the relevant Fire Schedule (e.g., Fire Schedule Alfa "F-A" General Fire Schedule).

The second EmS code refers to the relevant Spillage Schedule (e.g., Spillage Schedule Alfa "S-A" Toxic Substances).

Underlined EmS codes (special cases) indicate a substance, material or article for which additional advice is given in the emergency response procedures.

For dangerous goods offered for transport under N.O.S. entries or other generic entries, the most relevant emergency response procedures may vary with the properties of the hazardous constituents. As a consequence, shippers may have to declare different EmS codes from those indicated, if, to their knowledge, such codes are more appropriate.

The provisions in this column are not mandatory.

Column 16 **Stowage and segregation** – this column contains the stowage and segregation provisions as prescribed in part 7.

Column 17 **Properties and observations** – this column contains properties of and observations on the dangerous goods listed. The provisions in this column are not mandatory.

Properties of most gases include an indication of its density in relation to air. The figures in brackets give the density relative to air.

.1 "lighter than air" when the vapour density is down to half that of air;

.2 "much lighter than air" when the vapour density is less than half that of air;

.3 "heavier than air" when the vapour density is up to twice that of air; and

.4 "much heavier than air" when the vapour density is more than twice that of air.

When explosive limits are given, these refer to the volume percentage of the vapour of the substance when mixed with air.

The ease and extent to which different liquids mix with water varies greatly and most entries have included an indication of miscibility. In these cases "miscible with water" normally means capable of being mixed with water in all proportions to form a completely homogeneous liquid.

Column 18 **UN No.** – see column 1.

3.2.2 Abbreviations and symbols

The following abbreviations and symbols are used in the Dangerous Goods List and have the meanings shown:

Abbreviation/symbol	Column	Meaning
N.O.S.	2	Not otherwise specified
P	4	Marine pollutant

3

Dangerous Goods List

UN No.	Proper Shipping Name (PSN)	Class or division	Subsidiary risk(s)	Packing group	Special provisions	Limited and excepted quantity provisions		Packing		IBC	
						Limited quantities	Excepted quantities	Instructions	Provisions	Instructions	Provisions
(1)	(2) 3.1.2	(3) 2.0	(4) 2.0	(5) 2.0.1.3	(6) 3.3	(7a) 3.4	(7b) 3.5	(8) 4.1.4	(9) 4.1.4	(10) 4.1.4	(11) 4.1.4
△ 0004	AMMONIUM PICRATE dry or wetted with less than 10% water, by mass	1.1D	–	–	–	0	E0	P112 (a), (b) or (c)	PP26	–	–
△ 0005	CARTRIDGES FOR WEAPONS with bursting charge	1.1F	–	–	–	0	E0	P130	–	–	–
△ 0006	CARTRIDGES FOR WEAPONS with bursting charge	1.1E	–	–	–	0	E0	P130 LP101	PP67 L1	–	–
△ 0007	CARTRIDGES FOR WEAPONS with bursting charge	1.2F	–	–	–	0	E0	P130	–	–	–
△ 0009	AMMUNITION, INCENDIARY with or without burster, expelling charge or propelling charge	1.2G	–	–	–	0	E0	P130 LP101	PP67 L1	–	–
△ 0010	AMMUNITION, INCENDIARY with or without burster, expelling charge or propelling charge	1.3G	–	–	–	0	E0	P130 LP101	PP67 L1	–	–
△ 0012	CARTRIDGES FOR WEAPONS, INERT PROJECTILE or CARTRIDGES, SMALL ARMS	1.4S	–	–	364	5 kg	E0	P130	–	–	–
△ 0014	CARTRIDGES FOR WEAPONS, BLANK or CARTRIDGES, SMALL ARMS, BLANK or CARTRIDGES FOR TOOLS, BLANK	1.4S	–	–	364	5 kg	E0	P130	–	–	–
△ 0015	AMMUNITION, SMOKE with or without burster, expelling charge or propelling charge	1.2G	See SP204	–	204	0	E0	P130 LP101	PP67 L1	–	–
△ 0016	AMMUNITION, SMOKE with or without burster, expelling charge or propelling charge	1.3G	See SP204	–	204	0	E0	P130 LP101	PP67 L1	–	–
△ 0018	AMMUNITION, TEAR-PRODUCING with burster, expelling charge or propelling charge	1.2G	6.1/8	–	–	0	E0	P130 LP101	PP67 L1	–	–
△ 0019	AMMUNITION, TEAR-PRODUCING with burster, expelling charge or propelling charge	1.3G	6.1/8	–	–	0	E0	P130 LP101	PP67 L1	–	–
△ 0020	AMMUNITION, TOXIC with burster, expelling charge or propelling charge	1.2K	6.1	–	274	0	E0	P101	–	–	–
△ 0021	AMMUNITION, TOXIC with burster, expelling charge or propelling charge	1.3K	6.1	–	274	0	E0	P101	–	–	–
△ 0027	BLACK POWDER (GUNPOWDER) granular, or as a meal	1.1D	–	–	–	0	E0	P113	PP50	–	–
△ 0028	BLACK POWDER (GUNPOWDER), COMPRESSED or BLACK POWDER (GUNPOWDER) IN PELLETS	1.1D	–	–	–	0	E0	P113	PP51	–	–
△ 0029	DETONATORS, NON-ELECTRIC for blasting	1.1B	–	–	–	0	E0	P131	PP68	–	–
△ 0030	DETONATORS, ELECTRIC for blasting	1.1B	–	–	–	0	E0	P131	–	–	–
△ 0033	BOMBS with bursting charge	1.1F	–	–	–	0	E0	P130	–	–	–

Portable tanks and bulk containers			EmS	Stowage and segregation	Properties and observations	UN No.
	Tank instructions	Provisions				
(12)	(13) 4.2.5 4.3	(14) 4.2.5	(15) 5.4.3.2 7.3	(16) 7.1 7.2	(17)	(18)
–	–	–	F-B, S-Y	Category 04. "Away from" explosives containing chlorates or perchlorates. "Away from" lead and its compounds. Protected from sources of heat.	Substance.	0004 △
–	–	–	F-B, S-X	Category 05. Protected from sources of heat.	See glossary of terms in appendix B.	0005 △
–	–	–	F-B, S-X	Category 04. Protected from sources of heat.	See glossary of terms in appendix B.	0006 △
–	–	–	F-B, S-X	Category 05. Protected from sources of heat.	See glossary of terms in appendix B.	0007 △
–	–	–	F-B, S-X	Category 03. Protected from sources of heat.	See glossary of terms in appendix B.	0009 △
–	–	–	F-B, S-X	Category 03. Protected from sources of heat.	See glossary of terms in appendix B.	0010 △
–	–	–	F-B, S-X	Category 01. Protected from sources of heat.	See glossary of terms in appendix B.	0012 △
–	–	–	F-B, S-X	Category 01. Protected from sources of heat.	See glossary of terms in appendix B.	0014 △
–	–	–	F-B, S-X	Category 03. Protected from sources of heat.	See glossary of terms in appendix B.	0015 △
–	–	–	F-B, S-X	Category 03. Protected from sources of heat.	See glossary of terms in appendix B.	0016 △
–	–	–	<u>F-B</u>, S-Z	Category 03. Segregation as for class 1.2G. Protected from sources of heat.	See glossary of terms in appendix B.	0018 △
–	–	–	<u>F-B</u>, S-Z	Category 03. Segregation as for Class 1.3G. Protected from sources of heat.	See glossary of terms in appendix B.	0019 △
–	–	–	<u>F-B</u>, S-Z	Category 05. Protected from sources of heat.	See glossary of terms in appendix B.	0020 △
–	–	–	<u>F-B</u>, S-Z	Category 05. Protected from sources of heat.	See glossary of terms in appendix B.	0021 △
–	–	–	F-B, S-Y	Category 04. Protected from sources of heat.	See glossary of terms in appendix B.	0027 △
–	–	–	F-B, S-Y	Category 04. Protected from sources of heat.	See glossary of terms in appendix B.	0028 △
–	–	–	F-B, S-X	Category 05. Protected from sources of heat.	See glossary of terms in appendix B.	0029 △
–	–	–	F-B, S-X	Category 05. Protected from sources of heat.	See glossary of terms in appendix B.	0030 △
–	–	–	F-B, S-X	Category 05. Protected from sources of heat.	See glossary of terms in appendix B.	0033 △

UN No.	Proper Shipping Name (PSN)	Class or division	Subsidiary risk(s)	Packing group	Special provisions	Limited and excepted quantity provisions		Packing		IBC	
						Limited quantities	Excepted quantities	Instructions	Provisions	Instructions	Provisions
(1)	(2) 3.1.2	(3) 2.0	(4) 2.0	(5) 2.0.1.3	(6) 3.3	(7a) 3.4	(7b) 3.5	(8) 4.1.4	(9) 4.1.4	(10) 4.1.4	(11) 4.1.4
△ 0034	BOMBS with bursting charge	1.1D	–	–	–	0	E0	P130 LP101	PP67 L1	–	–
△ 0035	BOMBS with bursting charge	1.2D	–	–	–	0	E0	P130 LP101	PP67 L1	–	–
△ 0037	BOMBS, PHOTO-FLASH	1.1F	–	–	–	0	E0	P130	–	–	–
△ 0038	BOMBS, PHOTO-FLASH	1.1D	–	–	–	0	E0	P130 LP101	PP67 L1	–	–
△ 0039	BOMBS, PHOTO-FLASH	1.2G	–	–	–	0	E0	P130 LP101	PP67 L1	–	–
△ 0042	BOOSTERS without detonator	1.1D	–	–	–	0	E0	P132 (a) or (b)	–	–	–
△ 0043	BURSTERS explosive	1.1D	–	–	–	0	E0	P133	PP69	–	–
△ 0044	PRIMERS, CAP TYPE	1.4S	–	–	–	0	E0	P133	–	–	–
△ 0048	CHARGES, DEMOLITION	1.1D	–	–	–	0	E0	P130 LP101	PP67 L1	–	–
△ 0049	CARTRIDGES, FLASH	1.1G	–	–	–	0	E0	P135	–	–	–
△ 0050	CARTRIDGES, FLASH	1.3G	–	–	–	0	E0	P135	–	–	–
△ 0054	CARTRIDGES, SIGNAL	1.3G	–	–	–	0	E0	P135	–	–	–
△ 0055	CASES, CARTRIDGE, EMPTY, WITH PRIMER	1.4S	–	–	364	5 kg	E0	P136	–	–	–
△ 0056	CHARGES, DEPTH	1.1D	–	–	–	0	E0	P130 LP101	PP67 L1	–	–
△ 0059	CHARGES, SHAPED without detonator	1.1D	–	–	–	0	E0	P137	PP70	–	–
△ 0060	CHARGES, SUPPLEMENTARY, EXPLOSIVE	1.1D	–	–	–	0	E0	P132 (a) or (b)	–	–	–
△ 0065	CORD, DETONATING flexible	1.1D	–	–	–	0	E0	P139	PP71 PP72	–	–
△ 0066	CORD, IGNITER	1.4G	–	–	–	0	E0	P140	–	–	–
△ 0070	CUTTERS, CABLE, EXPLOSIVE	1.4S	–	–	–	0	E0	P134 LP102	–	–	–
△ 0072	CYCLOTRIMETHYLENE-TRINITRAMINE (CYCLONITE), (RDX), (HEXOGEN), WETTED with not less than 15% water, by mass	1.1D	–	–	266	0	E0	P112 (a)	PP45	–	–
△ 0073	DETONATORS FOR AMMUNITION	1.1B	–	–	–	0	E0	P133	–	–	–
△ 0074	DIAZODINITROPHENOL, WETTED with not less than 40% water or mixture of alcohol and water, by mass	1.1A	–	–	266	0	E0	P110 (a) or (b)	PP42	–	–
△ 0075	DIETHYLENEGLYCOL DINITRATE, DESENSITIZED with not less than 25% non-volatile water-insoluble phlegmatizer, by mass	1.1D	–	–	266	0	E0	P115	PP53 PP54 PP57 PP58	–	–

Portable tanks and bulk containers			EmS	Stowage and segregation	Properties and observations	UN No.	
	Tank instructions	Provisions					
(12)	(13) 4.2.5 4.3	(14) 4 2.5	(15) 5.4.3.2 7.3	(16) 7.1 7.2	(17)	(18)	
–	–	–	F-B, S-X	Category 04. Protected from sources of heat.	See glossary of terms in appendix B.	0034	△
–	–	–	F-B, S-X	Category 04. Protected from sources of heat.	See glossary of terms in appendix B.	0035	△
–	–	–	F-B, S-X	Category 05. Protected from sources of heat.	See glossary of terms in appendix B.	0037	△
–	–	–	F-B, S-X	Category 04. Protected from sources of heat.	See glossary of terms in appendix B.	0038	△
–	–	–	F-B, S-X	Category 03. Protected from sources of heat.	See glossary of terms in appendix B.	0039	△
–	–	–	F-B, S-X	Category 04. Protected from sources of heat.	See glossary of terms in appendix B.	0042	△
–	–	–	F-B, S-X	Category 04. Protected from sources of heat.	See glossary of terms in appendix B.	0043	△
–	–	–	F-B, S-X	Category 01. Protected from sources of heat.	See glossary of terms in appendix B.	0044	△
–	–	–	F-B, S-X	Category 04. Protected from sources of heat.	See glossary of terms in appendix B.	0048	△
–	–	–	F-B, S-X	Category 03. Protected from sources of heat.	See glossary of terms in appendix B.	0049	△
–	–	–	F-B, S-X	Category 03. Protected from sources of heat.	See glossary of terms in appendix B.	0050	△
–	–	–	F-B, S-X	Category 03. Protected from sources of heat.	See glossary of terms in appendix B.	0054	△
–	–	–	F-B, S-X	Category 01. Protected from sources of heat.	See glossary of terms in appendix B.	0055	△
–	–	–	F-B, S-X	Category 04. Protected from sources of heat.	See glossary of terms in appendix B.	0056	△
–	–	–	F-B, S-X	Category 04. Protected from sources of heat.	See glossary of terms in appendix B.	0059	△
–	–	–	F-B, S-X	Category 04. Protected from sources of heat.	See glossary of terms in appendix B.	0060	△
–	–	–	F-B, S-X	Category 04. Protected from sources of heat.	See glossary of terms in appendix B.	0065	△
–	–	–	F-B, S-X	Category 02. Protected from sources of heat.	See glossary of terms in appendix B.	0066	△
–	–	–	F-B, S-X	Category 01. Protected from sources of heat.	See glossary of terms in appendix B.	0070	△
–	–	–	F-B, S-Y	Category 04. Protected from sources of heat.	Mass detonating explosive which becomes more sensitive if the wetting agent is lost. This substance, when containing less alcohol, water or phlegmatizer than specified, shall not be transported, unless specifically authorized by the competent authority.	0072	△
–	–	–	F-B, S-X	Category 05. Protected from sources of heat.	See glossary of terms in appendix B.	0073	△
–	–	–	F-B, S-Y	Category 05. Protected from sources of heat.	Sensitive substance used in detonators, which becomes extremely sensitive if the wetting agents are lost. This substance, when containing less alcohol, water or phlegmatizer than specified, shall not be transported, unless specifically authorized by the competent authority.	0074	△
–	–	–	F-B, S-Y	Category 04. Protected from sources of heat.	This substance, when containing less alcohol, water or phlegmatizer than specified, shall not be transported, unless specifically authorized by the competent authority.	0075	△

	UN No.	Proper Shipping Name (PSN)	Class or division	Subsidiary risk(s)	Packing group	Special provisions	Limited and excepted quantity provisions		Packing		IBC	
							Limited quantities	Excepted quantities	Instructions	Provisions	Instructions	Provisions
	(1)	(2) 3.1.2	(3) 2.0	(4) 2.0	(5) 2.0.1.3	(6) 3.3	(7a) 3.4	(7b) 3.5	(8) 4.1.4	(9) 4.1.4	(10) 4.1.4	(11) 4.1.4
△	0076	DINITROPHENOL dry or wetted with less than 15% water, by mass	1.1D	6.1 P	–	–	0	E0	P112 (a), (b) or (c)	PP26	–	–
△	0077	DINITROPHENOLATES alkali metals, dry or wetted with less than 15% water, by mass	1.3C	6.1 P	–	–	0	E0	P114 (a) or (b)	PP26	–	–
△	0078	DINITRORESORCINOL dry or wetted with less than 15% water, by mass	1.1D	–	–	–	0	E0	P112 (a), (b) or (c)	PP26	–	–
△	0079	HEXANITRODIPHENYLAMINE (DIPICRYLAMINE), (HEXYL)	1.1D	–	–	–	0	E0	P112 (b) or (c)	–	–	–
△	0081	EXPLOSIVE, BLASTING, TYPE A	1.1D	–	–	–	0	E0	P116	PP63 PP66	–	–
△	0082	EXPLOSIVE, BLASTING, TYPE B	1.1D	–	–	–	0	E0	P116	PP61 PP62 PP65	IBC100	B9
△	0083	EXPLOSIVE, BLASTING, TYPE C	1.1D	–	–	267	0	E0	P116	–	–	–
△	0084	EXPLOSIVE, BLASTING, TYPE D	1.1D	–	–	–	0	E0	P116	–	–	–
△	0092	FLARES, SURFACE	1.3G	–	–	–	0	E0	P135	–	–	–
△	0093	FLARES, AERIAL	1.3G	–	–	–	0	E0	P135	–	–	–
△	0094	FLASH POWDER	1.1G	–	–	–	0	E0	P113	PP49	–	–
△	0099	FRACTURING DEVICES, EXPLOSIVE for oil wells, without detonator	1.1D	–	–	–	0	E0	P134 LP102	–	–	–
△	0101	FUSE, NON-DETONATING	1.3G	–	–	–	0	E0	P140	PP74 PP75	–	–
△	0102	CORD (FUSE), DETONATING metal-clad	1.2D	–	–	–	0	E0	P139	PP71	–	–
△	0103	FUSE, IGNITER tubular, metal-clad	1.4G	–	–	–	0	E0	P140	–	–	–
△	0104	CORD (FUSE), DETONATING, MILD EFFECT metal-clad	1.4D	–	–	–	0	E0	P139	PP71	–	–
△	0105	FUSE, SAFETY	1.4S	–	–	–	0	E0	P140	PP73	–	–
△	0106	FUZES, DETONATING	1.1B	–	–	–	0	E0	P141	–	–	–
△	0107	FUZES, DETONATING	1.2B	–	–	–	0	E0	P141	–	–	–
△	0110	GRENADES, PRACTICE hand or rifle	1.4S	–	–	–	0	E0	P141	–	–	–

3 DGL

Portable tanks and bulk containers			EmS	Stowage and segregation	Properties and observations	UN No.	
	Tank instructions	Provisions					
(12)	(13) 4.2.5 4.3	(14) 4.2.5	(15) 5.4.3.2 7.3	(16) 7.1 7.2	(17)	(18)	
–	–	–	F-B, S-Z	Category 04. "Away from" lead and its compounds. Protected from sources of heat.	Substance.	0076	△
–	–	–	F-B, S-Z	Category 04. "Away from" lead and its compounds. Protected from sources of heat.	Substance.	0077	△
–	–	–	F-B, S-Y	Category 04. "Away from" lead and its compounds. Protected from sources of heat.	Substance.	0078	△
–	–	–	F-B, S-Y	Category 04. Protected from sources of heat.	Substance.	0079	△
–	–	–	F-B, S-Y	Category 04. When containing ammonium compounds, "away from" chlorates or perchlorates and explosives containing chlorates or perchlorates. Protected from sources of heat.	Substance. See glossary of terms in appendix B.	0081	△
–	–	–	F-B, S-Y	Category 04. When containing ammonium compounds, "away from" chlorates or perchlorates and explosives containing chlorates or perchlorates. Protected from sources of heat.	Substance. See glossary of terms in appendix B.	0082	△
–	–	–	F-B, S-Y	Category 04. "Away from" ammonium compounds and explosives containing ammonium compounds or salts. Protected from sources of heat.	Substance. See glossary of terms in appendix B.	0083	△
–	–	–	F-B, S-Y	Category 04. Protected from sources of heat.	Substance. See glossary of terms in appendix B.	0084	△
–	–	–	F-B, S-X	Category 03. Protected from sources of heat.	See glossary of terms in appendix B.	0092	△
–	–	–	F-B, S-X	Category 03. Protected from sources of heat.	See glossary of terms in appendix B.	0093	△
–	–	–	F-B, S-Y	Category 03. Protected from sources of heat.	See glossary of terms in appendix B.	0094	△
–	–	–	F-B, S-X	Category 04. Protected from sources of heat.	See glossary of terms in appendix B.	0099	△
–	–	–	F-B, S-X	Category 03. Protected from sources of heat.	See glossary of terms in appendix B.	0101	△
–	–	–	F-B, S-X	Category 04. Protected from sources of heat.	See glossary of terms in appendix B.	0102	△
–	–	–	F-B, S-X	Category 02. Protected from sources of heat.	See glossary of terms in appendix B.	0103	△
–	–	–	F-B, S-X	Category 02. Protected from sources of heat.	See glossary of terms in appendix B.	0104	△
–	–	–	F-B, S-X	Category 01. Protected from sources of heat.	See glossary of terms in appendix B.	0105	△
–	–	–	F-B, S-X	Category 05. Protected from sources of heat.	See glossary of terms in appendix B.	0106	△
–	–	–	F-B, S-X	Category 05. Protected from sources of heat.	See glossary of terms in appendix B.	0107	△
–	–	–	F-B, S-X	Category 01. Protected from sources of heat.	See glossary of terms in appendix B.	0110	△

UN No.	Proper Shipping Name (PSN)	Class or division	Subsidiary risk(s)	Packing group	Special provisions	Limited and excepted quantity provisions		Packing		IBC	
						Limited quantities	Excepted quantities	Instructions	Provisions	Instructions	Provisions
(1)	(2) 3.1.2	(3) 2.0	(4) 2.0	(5) 2.0.1.3	(6) 3.3	(7a) 3.4	(7b) 3.5	(8) 4.1.4	(9) 4.1.4	(10) 4.1.4	(11) 4.1.4
△ 0113	GUANYL NITROSAMINO-GUANYLIDENEHYDRAZINE, WETTED with not less than 30% water, by mass	1.1A	–	–	266	0	E0	P110 (a) or (b)	PP42	–	–
△ 0114	GUANYL NITROSAMINO-GUANYLTETRAZENE (TETRAZENE), WETTED with not less than 30% water, or mixture of alcohol and water, by mass	1.1A	–	–	266	0	E0	P110 (a) or (b)	PP42	–	–
△ 0118	HEXOLITE (HEXOTOL) dry or wetted with less than 15% water, by mass	1.1D	–	–	–	0	E0	P112 (a), (b) or (c)	–	–	–
△ 0121	IGNITERS	1.1G	–	–	–	0	E0	P142	–	–	–
△ 0124	JET PERFORATING GUNS, CHARGED oil well, without detonator	1.1D	–	–	–	0	E0	P101	–	–	–
△ 0129	LEAD AZIDE, WETTED with not less than 20% water, or mixture of alcohol and water, by mass	1.1A	–	–	266	0	E0	P110 (a) or (b)	PP42	–	–
△ 0130	LEAD STYPHNATE (LEAD TRINITRORESORCINATE), WETTED with not less than 20% water, or mixture of alcohol and water, by mass	1.1A	–	–	266	0	E0	P110 (a) or (b)	PP42	–	–
△ 0131	LIGHTERS, FUSE	1.4S	–	–	–	0	E0	P142	–	–	–
△ 0132	DEFLAGRATING METAL SALTS OF AROMATIC NITRO-DERIVATIVES, N.O.S.	1.3C	–	–	–	0	E0	P114 (b)	PP26	–	–
△ 0133	MANNITOL HEXANITRATE (NITROMANNITE), WETTED with not less than 40% water, or mixture of alcohol and water, by mass	1.1D	–	–	266	0	E0	P112 (a)	–	–	–
△ 0135	MERCURY FULMINATE, WETTED with not less than 20% water, or mixture of alcohol and water, by mass	1.1A	–	–	266	0	E0	P110 (a) or (b)	PP42	–	–
△ 0136	MINES with bursting charge	1.1F	–	–	–	0	E0	P130	–	–	–
△ 0137	MINES with bursting charge	1.1D	–	–	–	0	E0	P130 LP101	PP67 L1	–	–
△ 0138	MINES with bursting charge	1.2D	–	–	–	0	E0	P130 LP101	PP67 L1	–	–
△ 0143	NITROGLYCERIN, DESENSITIZED with not less than 40% non-volatile water-insoluble phlegmatizer, by mass	1.1D	See SP271	–	266 271 272	0	E0	P115	PP53 PP54 PP57 PP58	–	–

Portable tanks and bulk containers			EmS	Stowage and segregation	Properties and observations	UN No.	
	Tank instructions	Provisions					
(12)	(13) 4.2.5 4.3	(14) 4.2.5	(15) 5.4.3.2 7.3	(16) 7.1 7.2	(17)	(18)	
–	–	–	F-B, S-Y	Category 05. Protected from sources of heat.	Sensitive substance used in detonators, which becomes extremely sensitive if the wetting agents are lost. This substance, when containing less alcohol, water or phlegmatizer than specified, shall not be transported, unless specifically authorized by the competent authority.	**0113**	△
–	–	–	F-B, S-Y	Category 05. Protected from sources of heat.	Sensitive substance used in detonators, which becomes extremely sensitive if the wetting agents are lost. This substance, when containing less alcohol, water or phlegmatizer than specified, shall not be transported, unless specifically authorized by the competent authority.	**0114**	△
–	–	–	F-B, S-Y	Category 04. Protected from sources of heat.	Substance. Mixtures of mass detonating explosives.	**0118**	△
–	–	–	F-B, S-X	Category 03. Protected from sources of heat.	See glossary of terms in appendix B.	**0121**	△
–	–	–	F-B, S-X	Category 04. Protected from sources of heat.	See glossary of terms in appendix B.	**0124**	△
–	–	–	F-B, S-Y	Category 05. Protected from sources of heat.	Sensitive substance used in detonators, which becomes extremely sensitive if the wetting agents are lost. This substance, when containing less alcohol, water or phlegmatizer than specified, shall not be transported, unless specifically authorized by the competent authority.	**0129**	△
–	–	–	F-B, S-Y	Category 05. Protected from sources of heat.	Sensitive substance used in detonators, which becomes extremely sensitive if the wetting agents are lost. This substance, when containing less alcohol, water or phlegmatizer than specified, shall not be transported, unless specifically authorized by the competent authority.	**0130**	△
–	–	–	F-B, S-X	Category 01. Protected from sources of heat.	See glossary of terms in appendix B.	**0131**	△
–	–	–	F-B, S-Y	Category 04. "Away from" lead and its compounds. Protected from sources of heat.	Substance.	**0132**	△
–	–	–	F-B, S-Y	Category 04. Protected from sources of heat.	This substance, when containing less alcohol, water or phlegmatizer than specified, shall not be transported, unless specifically authorized by the competent authority.	**0133**	△
–	–	–	F-B, S-Y	Category 05. Protected from sources of heat.	Sensitive substance used in detonators which will become extremely sensitive if it loses its wetting or desensitizing agent. This substance, when containing less alcohol, water or phlegmatizer than specified, shall not be transported, unless specifically authorized by the competent authority.	**0135**	△
–	–	–	F-B, S-X	Category 05. Protected from sources of heat.	See glossary of terms in appendix B.	**0136**	△
–	–	–	F-B, S-X	Category 04. Protected from sources of heat.	See glossary of terms in appendix B.	**0137**	△
–	–	–	F-B, S-X	Category 04. Protected from sources of heat.	See glossary of terms in appendix B.	**0138**	△
–	–	–	F-B, S-Z	Category 04. Protected from sources of heat.	Substance. This substance, when containing less alcohol, water or phlegmatizer than specified, shall not be transported, unless specifically authorized by the competent authority.	**0143**	△

UN No. (1)	Proper Shipping Name (PSN) (2) 3.1.2	Class or division (3) 2.0	Subsidiary risk(s) (4) 2.0	Packing group (5) 2.0.1.3	Special provisions (6) 3.3	Limited and excepted quantity provisions: Limited quantities (7a) 3.4	Limited and excepted quantity provisions: Excepted quantities (7b) 3.5	Packing: Instructions (8) 4.1.4	Packing: Provisions (9) 4.1.4	IBC: Instructions (10) 4.1.4	IBC: Provisions (11) 4.1.4
△ 0144	NITROGLYCERIN SOLUTION IN ALCOHOL with more than 1% but not more than 10% nitroglycerin	1.1D	–	–	358	0	E0	P115	PP45 PP55 PP56 PP59 PP60	–	–
△ 0146	NITROSTARCH dry or wetted, with less than 20% water, by mass	1.1D	–	–	–	0	E0	P112 (a), (b) or (c)	–	–	–
△ 0147	NITRO UREA	1.1D	–	–	–	0	E0	P112 (b)	–	–	–
△ 0150	PENTAERYTHRITE TETRANITRATE (PENTAERYTHRITOL TETRANITRATE; PETN), WETTED with not less than 25% water, by mass or PENTAERYTHRITE TETRANITRATE (PENTAERYTHRITOL TETRANITRATE; PETN), DESENSITIZED with not less than 15% phlegmatizer, by mass	1.1D	–	–	266	0	E0	P112 (a) or (b)	–	–	–
△ 0151	PENTOLITE dry or wetted with less than 15% water, by mass	1.1D	–	–	–	0	E0	P112 (a), (b) or (c)	–	–	–
△ 0153	TRINITROANILINE (PICRAMIDE)	1.1D	–	–	–	0	E0	P112 (b) or (c)	–	–	–
△ 0154	TRINITROPHENOL (PICRIC ACID) dry or wetted with less than 30% water, by mass	1.1D	–	–	–	0	E0	P112 (a), (b) or (c)	PP26	–	–
△ 0155	TRINITROCHLOROBENZENE (PICRYL CHLORIDE)	1.1D	–	–	–	0	E0	P112 (b) or (c)	–	–	–
△ 0159	POWDER CAKE (POWDER PASTE), WETTED with not less than 25% water, by mass	1.3C	–	–	266	0	E0	P111	PP43	–	–
△ 0160	POWDER, SMOKELESS	1.1C	–	–	–	0	E0	P114 (b)	PP50 PP52	–	–
△ 0161	POWDER, SMOKELESS	1.3C	–	–	–	0	E0	P114 (b)	PP50 PP52	–	–
△ 0167	PROJECTILES with bursting charge	1.1F	–	–	–	0	E0	P130	–	–	–
△ 0168	PROJECTILES with bursting charge	1.1D	–	–	–	0	E0	P130 LP101	PP67 L1	–	–
△ 0169	PROJECTILES with bursting charge	1.2D	–	–	–	0	E0	P130 LP101	PP67 L1	–	–
△ 0171	AMMUNITION, ILLUMINATING with or without burster, expelling charge or propelling charge	1.2G	–	–	–	0	E0	P130 LP101	PP67 L1	–	–
△ 0173	RELEASE DEVICES, EXPLOSIVE	1.4S	–	–	–	0	E0	P134 LP102	–	–	–
△ 0174	RIVETS, EXPLOSIVE	1.4S	–	–	–	0	E0	P134 LP102	–	–	–
△ 0180	ROCKETS with bursting charge	1.1F	–	–	–	0	E0	P130	–	–	–

Portable tanks and bulk containers			EmS	Stowage and segregation	Properties and observations	UN No.	
	Tank instructions	Provisions					
(12)	(13) 4.2.5 4.3	(14) 4.2.5	(15) 5.4.3.2 7.3	(16) 7.1 7.2	(17)	(18)	
–	–	–	F-B, S-Y	Category 04. Protected from sources of heat.	Substance.	0144	△
–	–	–	F-B, S-Y	Category 04. Protected from sources of heat.	Substance.	0146	△
–	–	–	F-B, S-Y	Category 04. Protected from sources of heat.	Substance.	0147	△
–	–	–	F-B, S-Y	Category 04. Protected from sources of heat.	Substance. Mass detonating explosive which will become more sensitive if it loses its wetting or desensitizing agent. This substance, when containing less alcohol, water or phlegmatizer than specified, shall not be transported, unless specifically authorized by the competent authority.	0150	△
–	–	–	F-B, S-Y	Category 04. Protected from sources of heat.	Substance. Mixtures of mass detonating explosive substances.	0151	△
–	–	–	F-B, S-Y	Category 04. Protected from sources of heat.	Substance.	0153	△
–	–	–	F-B, S-Y	Category 04. Protected from sources of heat. "Away from" lead and its compounds.	Substance.	0154	△
–	–	–	F-B, S-Y	Category 04. Protected from sources of heat.	Substance.	0155	△
–	–	–	F-B, S-Y	Category 04. Protected from sources of heat.	Substance consisting of nitrocellulose impregnated with not more than 60% of nitroglycerin or other liquid organic nitrates or a mixture of these. This substance, when containing less alcohol, water or phlegmatizer than specified, shall not be transported, unless specifically authorized by the competent authority.	0159	△
–	–	–	F-B, S-Y	Category 04. Protected from sources of heat.	Substances based on nitrocellulose used as propellant. Sensitive to sparks, friction, pressure and electrostatic discharge.	0160	△
–	–	–	F-B, S-Y	Category 04. Protected from sources of heat.	Substances based on nitrocellulose used as propellant. Sensitive to sparks, friction, pressure and electrostatic discharge.	0161	△
–	–	–	F-B, S-X	Category 05. Protected from sources of heat.	See glossary of terms in appendix B.	0167	△
–	–	–	F-B, S-X	Category 04. Protected from sources of heat.	See glossary of terms in appendix B.	0168	△
–	–	–	F-B, S-X	Category 04. Protected from sources of heat.	See glossary of terms in appendix B.	0169	△
–	–	–	F-B, S-X	Category 03. Protected from sources of heat.	See glossary of terms in appendix B.	0171	△
–	–	–	F-B, S-X	Category 01. Protected from sources of heat.	See glossary of terms in appendix B.	0173	△
–	–	–	F-B, S-X	Category 01. Protected from sources of heat.	See glossary of terms in appendix B.	0174	△
–	–	–	F-B, S-X	Category 05. Protected from sources of heat.	See glossary of terms in appendix B.	0180	△

	UN No.	Proper Shipping Name (PSN)	Class or division	Subsidiary risk(s)	Packing group	Special provisions	Limited and excepted quantity provisions		Packing		IBC	
							Limited quantities	Excepted quantities	Instructions	Provisions	Instructions	Provisions
	(1)	(2) 3.1.2	(3) 2.0	(4) 2.0	(5) 2.0.1.3	(6) 3.3	(7a) 3.4	(7b) 3.5	(8) 4.1.4	(9) 4.1.4	(10) 4.1.4	(11) 4.1.4
△	0181	ROCKETS with bursting charge	1.1E	–	–	–	0	E0	P130 LP101	PP67 L1	–	–
△	0182	ROCKETS with bursting charge	1.2E	–	–	–	0	E0	P130 LP101	PP67 L1	–	–
△	0183	ROCKETS with inert head	1.3C	–	–	–	0	E0	P130 LP101	PP67 L1	–	–
△	0186	ROCKET MOTORS	1.3C	–	–	–	0	E0	P130 LP101	PP67 L1	–	–
△	0190	SAMPLES, EXPLOSIVE other than initiating explosive	1	–	–	16 274	0	E0	P101	–	–	–
△	0191	SIGNAL DEVICES, HAND	1.4G	–	–	–	0	E0	P135	–	–	–
△	0192	SIGNALS, RAILWAY TRACK, EXPLOSIVE	1.1G	–	–	–	0	E0	P135	–	–	–
△	0193	SIGNALS, RAILWAY TRACK, EXPLOSIVE	1.4S	–	–	–	0	E0	P135	–	–	–
△	0194	SIGNALS, DISTRESS, ship	1.1G	–	–	–	0	E0	P135	–	–	–
△	0195	SIGNALS, DISTRESS, ship	1.3G	–	–	–	0	E0	P135	–	–	–
△	0196	SIGNALS, SMOKE	1.1G	–	–	–	0	E0	P135	–	–	–
△	0197	SIGNALS, SMOKE	1.4G	–	–	–	0	E0	P135	–	–	–
△	0204	SOUNDING DEVICES, EXPLOSIVE	1.2F	–	–	–	0	E0	P134 LP102	–	–	–
△	0207	TETRANITROANILINE	1.1D	–	–	–	0	E0	P112 (b) or (c)	–	–	–
△	0208	TRINITROPHENYLMETHYL-NITRAMINE (TETRYL)	1.1D	–	–	–	0	E0	P112 (b) or (c)	–	–	–
△	0209	TRINITROTOLUENE (TNT) dry or wetted with less than 30% water, by mass	1.1D	–	–	–	0	E0	P112 (a), (b) or (c)	PP46	–	–
△	0212	TRACERS FOR AMMUNITION	1.3G	–	–	–	0	E0	P133	PP69	–	–
△	0213	TRINITROANISOLE	1.1D	–	–	–	0	E0	P112 (b) or (c)	–	–	–
△	0214	TRINITROBENZENE dry or wetted with less than 30% water, by mass	1.1D	–	–	–	0	E0	P112 (a), (b) or (c)	–	–	–
△	0215	TRINITROBENZOIC ACID dry or wetted with less than 30% water, by mass	1.1D	–	–	–	0	E0	P112 (a), (b) or (c)	–	–	–
△	0216	TRINITRO-*m*-CRESOL	1.1D	–	–	–	0	E0	P112 (b) or (c)	PP26	–	–
△	0217	TRINITRONAPHTHALENE	1.1D	–	–	–	0	E0	P112 (b) or (c)	–	–	–
△	0218	TRINITROPHENETOLE	1.1D	–	–	–	0	E0	P112 (b) or (c)	–	–	–

Portable tanks and bulk containers			EmS	Stowage and segregation	Properties and observations	UN No.	
	Tank instructions	Provisions					
(12)	(13) 4.2.5 4.3	(14) 4.2.5	(15) 5.4.3.2 7.3	(16) 7.1 7.2	(17)	(18)	
–	–	–	F-B, S-X	Category 04. Protected from sources of heat.	See glossary of terms in appendix B.	0181	△
–	–	–	F-B, S-X	Category 04. Protected from sources of heat.	See glossary of terms in appendix B.	0182	△
–	–	–	F-B, S-X	Category 04. Protected from sources of heat.	See glossary of terms in appendix B.	0183	△
–	–	–	F-B, S-X	Category 04. Protected from sources of heat.	See glossary of terms in appendix B.	0186	△
–	–	–	F-B, S-X	Category 05. Protected from sources of heat.	See glossary of terms in appendix B.	0190	△
–	–	–	F-B, S-X	Category 02. Protected from sources of heat.	See glossary of terms in appendix B.	0191	△
–	–	–	F-B, S-X	Category 03. Protected from sources of heat.	See glossary of terms in appendix B.	0192	△
–	–	–	F-B, S-X	Category 01. Protected from sources of heat.	See glossary of terms in appendix B.	0193	△
–	–	–	F-B, S-X	Category 03. Protected from sources of heat.	See glossary of terms in appendix B.	0194	△
–	–	–	F-B, S-X	Category 03. Protected from sources of heat.	See glossary of terms in appendix B.	0195	△
–	–	–	F-B, S-X	Category 03. Protected from sources of heat.	See glossary of terms in appendix B.	0196	△
–	–	–	F-B, S-X	Category 02. Protected from sources of heat.	See glossary of terms in appendix B.	0197	△
–	–	–	F-B, S-X	Category 05. Protected from sources of heat.	See glossary of terms in appendix B.	0204	△
–	–	–	F-B, S-Y	Category 04. Protected from sources of heat.	Substance.	0207	△
–	–	–	F-B, S-Y	Category 04. Protected from sources of heat.	Substance. Mass detonating explosive.	0208	△
–	–	–	F-B, S-Y	Category 04. Protected from sources of heat.	Substance. Tritonal is a substance consisting of trinitrotoluene (TNT) mixed with aluminium.	0209	△
–	–	–	F-B, S-X	Category 03. Protected from sources of heat.	See glossary of terms in appendix B.	0212	△
–	–	–	F-B, S-Y	Category 04. Protected from sources of heat.	Substance.	0213	△
–	–	–	F-B, S-Y	Category 04. Protected from sources of heat.	Substance.	0214	△
–	–	–	F-B, S-Y	Category 04. Protected from sources of heat.	Substance.	0215	△
–	–	–	F-B, S-Y	Category 04. "Away from" lead and its compounds. Protected from sources of heat.	Substance.	0216	△
–	–	–	F-B, S-Y	Category 04. Protected from sources of heat.	Substance.	0217	△
–	–	–	F-B, S-Y	Category 04. Protected from sources of heat.	Substance.	0218	△

UN No.	Proper Shipping Name (PSN)	Class or division	Subsidiary risk(s)	Packing group	Special provisions	Limited and excepted quantity provisions		Packing		IBC	
						Limited quantities	Excepted quantities	Instructions	Provisions	Instructions	Provisions
(1)	(2) 3.1.2	(3) 2.0	(4) 2.0	(5) 2.0.1.3	(6) 3.3	(7a) 3.4	(7b) 3.5	(8) 4.1.4	(9) 4.1.4	(10) 4.1.4	(11) 4.1.4
△ 0219	TRINITRORESORCINOL (STYPHNIC ACID) dry or wetted with less than 20% water, or mixture of alcohol and water, by mass	1.1D	–	–	–	0	E0	P112 (a), (b) or (c)	PP26	–	–
△ 0220	UREA NITRATE dry or wetted with less than 20% water, by mass	1.1D	–	–	–	0	E0	P112 (a), (b) or (c)	–	–	–
△ 0221	WARHEADS, TORPEDO with bursting charge	1.1D	–	–	–	0	E0	P130 LP101	PP67 L1	–	–
△ 0222	AMMONIUM NITRATE with more than 0.2% combustible substances, including any organic substance calculated as carbon, to the exclusion of any other added substance	1.1D	–	–	–	0	E0	P112 (b) or (c)	PP47	–	–
△ 0224	BARIUM AZIDE, dry or wetted with less than 50% water, by mass	1.1A	6.1	–	–	0	E0	P110 (a) or (b)	PP42	–	–
△ 0225	BOOSTERS WITH DETONATOR	1.1B	–	–	–	0	E0	P133	PP69	–	–
△ 0226	CYCLOTETRAMETHYLENE-TETRANITRAMINE (HMX; OCTOGEN), WETTED with not less than 15% water, by mass	1.1D	–	–	266	0	E0	P112 (a)	PP45	–	–
△ 0234	SODIUM DINITRO-o-CRESOLATE dry or wetted with less than 15% water, by mass	1.3C	6.1 P	–	–	0	E0	P114 (a) or (b)	PP26	–	–
△ 0235	SODIUM PICRAMATE dry or wetted with less than 20% water, by mass	1.3C	–	–	–	0	E0	P114 (a) or (b)	PP26	–	–
△ 0236	ZIRCONIUM PICRAMATE dry or wetted with less than 20% water, by mass	1.3C	–	–	–	0	E0	P114 (a) or (b)	PP26	–	–
△ 0237	CHARGES, SHAPED, FLEXIBLE, LINEAR	1.4D	–	–	–	0	E0	P138	–	–	–
△ 0238	ROCKETS, LINE-THROWING	1.2G	–	–	–	0	E0	P130	–	–	–
△ 0240	ROCKETS, LINE-THROWING	1.3G	–	–	–	0	E0	P130	–	–	–
△ 0241	EXPLOSIVE, BLASTING, TYPE E	1.1D	–	–	–	0	E0	P116	PP61 PP62 PP65	IBC100	B10
△ 0242	CHARGES, PROPELLING, FOR CANNON	1.3C	–	–	–	0	E0	P130	–	–	–
△ 0243	AMMUNITION, INCENDIARY, WHITE PHOSPHORUS with burster, expelling charge or propelling charge	1.2H	–	–	–	0	E0	P130 LP101	PP67 L1	–	–
△ 0244	AMMUNITION, INCENDIARY, WHITE PHOSPHORUS with burster, expelling charge or propelling charge	1.3H	–	–	–	0	E0	P130 LP101	PP67 L1	–	–

Portable tanks and bulk containers			EmS	Stowage and segregation	Properties and observations	UN No.	
	Tank instructions	Provisions					
(12)	(13) 4.2.5 4.3	(14) 4.2.5	(15) 5.4.3.2 7.3	(16) 7.1 7.2	(17)	(18)	
–	–	–	F-B, S-Y	Category 04. "Away from" explosives containing chlorates or perchlorates. Protected from sources of heat.	Substance.	0219	△
–	–	–	F-B, S-Y	Category 04. Protected from sources of heat.	Substance.	0220	△
–	–	–	F-B, S-X	Category 04. Protected from sources of heat.	See glossary of terms in appendix B.	0221	△
–	–	–	F-B, S-Y	Category 04. "Away from" explosives containing chlorates or perchlorates. Protected from sources of heat.	Substance.	0222	△
–	–	–	F-B, S-Z	Category 05. Protected from sources of heat.	See glossary of terms in appendix B.	0224	△
–	–	–	F-B, S-X	Category 05. Protected from sources of heat.	See glossary of terms in appendix B.	0225	△
–	–	–	F-B, S-Y	Category 04. Protected from sources of heat.	Substance. Mass detonating explosive which will become more sensitive if the wetting or desensitizing agent is lost. This substance, when containing less alcohol, water or phlegmatizer than specified, shall not be transported, unless specifically authorized by the competent authority.	0226	△
–	–	–	F-B, S-Z	Category 04. "Away from" lead and its compounds. Protected from sources of heat.	Substance.	0234	△
–	–	–	F-B, S-Y	Category 04. "Away from" lead and its compounds. Protected from sources of heat.	Substance.	0235	△
–	–	–	F-B, S-Y	Category 04. "Away from" lead and its compounds. Protected from sources of heat.	Substance.	0236	△
–	–	–	F-B, S-X	Category 02. Protected from sources of heat.	See glossary of terms in appendix B.	0237	△
–	–	–	F-B, S-X	Category 03. Protected from sources of heat.	See glossary of terms in appendix B.	0238	△
–	–	–	F-B, S-X	Category 03. Protected from sources of heat.	See glossary of terms in appendix B.	0240	△
–	–	–	F-B, S-X	Category 04. When containing ammonium compounds, "away from" chlorates or perchlorates and explosives containing chlorates or perchlorates. Protected from sources of heat.	See glossary of terms in appendix B.	0241	△
–	–	–	F-B, S-X	Category 04. Protected from sources of heat.	See glossary of terms in appendix B.	0242	△
–	–	–	F-B, S-X	Category 05. Protected from sources of heat.	See glossary of terms in appendix B.	0243	△
–	–	–	F-B, S-X	Category 05. Protected from sources of heat.	See glossary of terms in appendix B.	0244	△

UN No. (1)	Proper Shipping Name (PSN) (2) 3.1.2	Class or division (3) 2.0	Subsidiary risk(s) (4) 2.0	Packing group (5) 2.0.1.3	Special provisions (6) 3.3	Limited and excepted quantity provisions: Limited quantities (7a) 3.4	Limited and excepted quantity provisions: Excepted quantities (7b) 3.5	Packing: Instructions (8) 4.1.4	Packing: Provisions (9) 4.1.4	IBC: Instructions (10) 4.1.4	IBC: Provisions (11) 4.1.4
△ 0245	AMMUNITION, SMOKE, WHITE PHOSPHORUS with burster, expelling charge or propelling charge	1.2H	–	–	–	0	E0	P130 LP101	PP67 L1	–	–
△ 0246	AMMUNITION, SMOKE, WHITE PHOSPHORUS with burster, expelling charge or propelling charge	1.3H	–	–	–	0	E0	P130 LP101	PP67 L1	–	–
△ 0247	AMMUNITION, INCENDIARY liquid or gel, with burster, expelling charge or propelling charge	1.3J	–	–	–	0	E0	P101	–	–	–
△ 0248	CONTRIVANCES, WATER-ACTIVATED with burster, expelling charge or propelling charge	1.2L	4.3	–	274	0	E0	P144	PP77	–	–
△ 0249	CONTRIVANCES, WATER-ACTIVATED with burster, expelling charge or propelling charge	1.3L	4.3	–	274	0	E0	P144	PP77	–	–
△ 0250	ROCKET MOTORS WITH HYPERGOLIC LIQUIDS with or without expelling charge	1.3L	–	–	–	0	E0	P101	–	–	–
△ 0254	AMMUNITION, ILLUMINATING with or without burster, expelling charge or propelling charge	1.3G	–	–	–	0	E0	P130 LP101	PP67 L1	–	–
△ 0255	DETONATORS, ELECTRIC for blasting	1.4B	–	–	–	0	E0	P131	–	–	–
△ 0257	FUZES, DETONATING	1.4B	–	–	–	0	E0	P141	–	–	–
△ 0266	OCTOLITE (OCTOL) dry or wetted with less than 15% water, by mass	1.1D	–	–	–	0	E0	P112 (a), (b) or (c)	–	–	–
△ 0267	DETONATORS, NON-ELECTRIC for blasting	1.4B	–	–	–	0	E0	P131	PP68	–	–
△ 0268	BOOSTERS WITH DETONATOR	1.2B	–	–	–	0	E0	P133	PP69	–	–
△ 0271	CHARGES, PROPELLING	1.1C	–	–	–	0	E0	P143	PP76	–	–
△ 0272	CHARGES, PROPELLING	1.3C	–	–	–	0	E0	P143	PP76	–	–
△ 0275	CARTRIDGES, POWER DEVICE	1.3C	–	–	–	0	E0	P134 LP102	–	–	–
△ 0276	CARTRIDGES, POWER DEVICE	1.4C	–	–	–	0	E0	P134 LP102	–	–	–
△ 0277	CARTRIDGES, OIL WELL	1.3C	–	–	–	0	E0	P134 LP102	–	–	–
△ 0278	CARTRIDGES, OIL WELL	1.4C	–	–	–	0	E0	P134 LP102	–	–	–
△ 0279	CHARGES, PROPELLING, FOR CANNON	1.1C	–	–	–	0	E0	P130	–	–	–
△ 0280	ROCKET MOTORS	1.1C	–	–	–	0	E0	P130 LP101	PP67 L1	–	–
△ 0281	ROCKET MOTORS	1.2C	–	–	–	0	E0	P130 LP101	PP67 L1	–	–

Portable tanks and bulk containers			EmS	Stowage and segregation	Properties and observations	UN No.	
	Tank instructions	Provisions					
(12)	(13) 4.2.5 4.3	(14) 4.2.5	(15) 5.4.3.2 7.3	(16) 7.1 7.2	(17)	(18)	
–	–	–	F-B, S-X	Category 05. Protected from sources of heat.	See glossary of terms in appendix B.	0245	△
–	–	–	F-B, S-X	Category 05. Protected from sources of heat.	See glossary of terms in appendix B.	0246	△
–	–	–	F-B, S-X	Category 05. Protected from sources of heat.	See glossary of terms in appendix B.	0247	△
–	–	–	F-B, S-Y	Category 05. Protected from sources of heat.	See glossary of terms in appendix B.	0248	△
–	–	–	F-B, S-Y	Category 05. Protected from sources of heat.	See glossary of terms in appendix B.	0249	△
–	–	–	F-B, S-X	Category 05. Protected from sources of heat.	See glossary of terms in appendix B.	0250	△
–	–	–	F-B, S-X	Category 03. Protected from sources of heat.	See glossary of terms in appendix B.	0254	△
–	–	–	F-B, S-X	Category 05. Protected from sources of heat.	See glossary of terms in appendix B.	0255	△
–	–	–	F-B, S-X	Category 05. Protected from sources of heat.	See glossary of terms in appendix B.	0257	△
–	–	–	F-B, S-Y	Category 04. Protected from sources of heat.	Substance. Mixtures of mass detonating explosives.	0266	△
–	–	–	F-B, S-X	Category 05. Protected from sources of heat.	See glossary of terms in appendix B.	0267	△
–	–	–	F-B, S-X	Category 05. Protected from sources of heat.	See glossary of terms in appendix B.	0268	△
–	–	–	F-B, S-X	Category 04. Protected from sources of heat.	See glossary of terms in appendix B.	0271	△
–	–	–	F-B, S-X	Category 04. Protected from sources of heat.	See glossary of terms in appendix B.	0272	△
–	–	–	F-B, S-X	Category 04. Protected from sources of heat.	See glossary of terms in appendix B.	0275	△
–	–	–	F-B, S-X	Category 02. Protected from sources of heat.	See glossary of terms in appendix B.	0276	△
–	–	–	F-B, S-X	Category 04. Protected from sources of heat.	See glossary of terms in appendix B.	0277	△
–	–	–	F-B, S-X	Category 02. Protected from sources of heat.	See glossary of terms in appendix B.	0278	△
–	–	–	F-B, S-X	Category 04. Protected from sources of heat.	See glossary of terms in appendix B.	0279	△
–	–	–	F-B, S-X	Category 04. Protected from sources of heat.	See glossary of terms in appendix B.	0280	△
–	–	–	F-B, S-X	Category 04. Protected from sources of heat.	See glossary of terms in appendix B.	0281	△

UN No.	Proper Shipping Name (PSN)	Class or division	Subsidiary risk(s)	Packing group	Special provisions	Limited and excepted quantity provisions		Packing		IBC	
						Limited quantities	Excepted quantities	Instructions	Provisions	Instructions	Provisions
(1)	(2) 3.1.2	(3) 2.0	(4) 2.0	(5) 2.0.1.3	(6) 3.3	(7a) 3.4	(7b) 3.5	(8) 4.1.4	(9) 4.1.4	(10) 4.1.4	(11) 4.1.4
△ 0282	NITROGUANIDINE (PICRITE) dry or wetted with less than 20% water, by mass	1.1D	–	–	–	0	E0	P112 (a), (b) or (c)	–	–	–
△ 0283	BOOSTERS without detonator	1.2D	–	–	–	0	E0	P132 (a) or (b)	–	–	–
△ 0284	GRENADES hand or rifle, with bursting charge	1.1D	–	–	–	0	E0	P141	–	–	–
△ 0285	GRENADES hand or rifle, with bursting charge	1.2D	–	–	–	0	E0	P141	–	–	–
△ 0286	WARHEADS, ROCKET with bursting charge	1.1D	–	–	–	0	E0	P130 LP101	PP67 L1	–	–
△ 0287	WARHEADS, ROCKET with bursting charge	1.2D	–	–	–	0	E0	P130 LP101	PP67 L1	–	–
△ 0288	CHARGES, SHAPED, FLEXIBLE, LINEAR	1.1D	–	–	–	0	E0	P138	–	–	–
△ 0289	CORD, DETONATING flexible	1.4D	–	–	–	0	E0	P139	PP71 PP72	–	–
△ 0290	CORD (FUSE), DETONATING metal-clad	1.1D	–	–	–	0	E0	P139	PP71	–	–
△ 0291	BOMBS with bursting charge	1.2F	–	–	–	0	E0	P130	–	–	–
△ 0292	GRENADES hand or rifle, with bursting charge	1.1F	–	–	–	0	E0	P141	–	–	–
△ 0293	GRENADES hand or rifle, with bursting charge	1.2F	–	–	–	0	E0	P141	–	–	–
△ 0294	MINES with bursting charge	1.2F	–	–	–	0	E0	P130	–	–	–
△ 0295	ROCKETS with bursting charge	1.2F	–	–	–	0	E0	P130	–	–	–
△ 0296	SOUNDING DEVICES, EXPLOSIVE	1.1F	–	–	–	0	E0	P134 LP102	–	–	–
△ 0297	AMMUNITION, ILLUMINATING with or without burster, expelling charge or propelling charge	1.4G	–	–	–	0	E0	P130 LP101	PP67 L1	–	–
△ 0299	BOMBS, PHOTO-FLASH	1.3G	–	–	–	0	E0	P130 LP101	PP67 L1	–	–
△ 0300	AMMUNITION, INCENDIARY with or without burster, expelling charge or propelling charge	1.4G	–	–	–	0	E0	P130 LP101	PP67 L1	–	–
△ 0301	AMMUNITION, TEAR-PRODUCING with burster, expelling charge or propelling charge	1.4G	6.1/8	–	–	0	E0	P130 LP101	PP67 L1	–	–
△ 0303	AMMUNITION, SMOKE with or without burster, expelling charge or propelling charge	1.4G	See SP204	–	204	0	E0	P130 LP101	PP67 L1	–	–
△ 0305	FLASH POWDER	1.3G	–	–	–	0	E0	P113	PP49	–	–
△ 0306	TRACERS FOR AMMUNITION	1.4G	–	–	–	0	E0	P133	PP69	–	–
△ 0312	CARTRIDGES, SIGNAL	1.4G	–	–	–	0	E0	P135	–	–	–
△ 0313	SIGNALS, SMOKE	1.2G	–	–	–	0	E0	P135	–	–	–

Portable tanks and bulk containers			EmS	Stowage and segregation	Properties and observations	UN No.
	Tank instructions	Provisions				
(12)	(13) 4.2.5 4.3	(14) 4.2.5	(15) 5.4.3.2 7.3	(16) 7.1 7.2	(17)	(18)
–	–	–	F-B, S-Y	Category 04. Protected from sources of heat.	Substance.	0282 △
–	–	–	F-B, S-X	Category 04. Protected from sources of heat.	See glossary of terms in appendix B.	0283 △
–	–	–	F-B, S-X	Category 04. Protected from sources of heat.	See glossary of terms in appendix B.	0284 △
–	–	–	F-B, S-Y	Category 04. Protected from sources of heat.	See glossary of terms in appendix B.	0285 △
–	–	–	F-B, S-X	Category 04. Protected from sources of heat.	See glossary of terms in appendix B.	0286 △
	–	–	F-B, S-X	Category 04. Protected from sources of heat.	See glossary of terms in appendix B.	0287 △
–	–	–	F-B, S-X	Category 04. Protected from sources of heat.	See glossary of terms in appendix B.	0288 △
–	–	–	F-B, S-X	Category 02. Protected from sources of heat.	See glossary of terms in appendix B.	0289 △
–	–	–	F-B, S-X	Category 04. Protected from sources of heat.	See glossary of terms in appendix B.	0290 △
–	–	–	F-B, S-X	Category 05. Protected from sources of heat.	See glossary of terms in appendix B.	0291 △
–	–	–	F-B, S-X	Category 05. Protected from sources of heat.	See glossary of terms in appendix B.	0292 △
–	–	–	F-B, S-X	Category 05. Protected from sources of heat.	See glossary of terms in appendix B.	0293 △
–	–	–	F-B, S-X	Category 05. Protected from sources of heat.	See glossary of terms in appendix B.	0294 △
–	–	–	F-B, S-X	Category 05. Protected from sources of heat.	See glossary of terms in appendix B.	0295 △
–	–	–	F-B, S-X	Category 05. Protected from sources of heat.	See glossary of terms in appendix B.	0296 △
–	–	–	F-B, S-X	Category 02. Protected from sources of heat.	See glossary of terms in appendix B.	0297 △
–	–	–	F-B, S-X	Category 03. Protected from sources of heat.	See glossary of terms in appendix B.	0299 △
–	–	–	F-B, S-X	Category 02. Protected from sources of heat.	See glossary of terms in appendix B.	0300 △
–	–	–	F-B, S-Z	Category 02. Segregation as for Class 1.4G. Protected from sources of heat.	See glossary of terms in appendix B.	0301 △
–	–	–	F-B, S-X	Category 02. Protected from sources of heat.	See glossary of terms in appendix B.	0303 △
–	–	–	F-B, S-Y	Category 03. Protected from sources of heat.	See glossary of terms in appendix B.	0305 △
–	–	–	F-B, S-X	Category 02. Protected from sources of heat.	See glossary of terms in appendix B.	0306 △
–	–	–	F-B, S-X	Category 02. Protected from sources of heat.	See glossary of terms in appendix B.	0312 △
–	–	–	F-B, S-X	Category 03. Protected from sources of heat.	See glossary of terms in appendix B.	0313 △

	UN No.	Proper Shipping Name (PSN)	Class or division	Subsidiary risk(s)	Packing group	Special provisions	Limited and excepted quantity provisions		Packing		IBC	
							Limited quantities	Excepted quantities	Instructions	Provisions	Instructions	Provisions
	(1)	(2) 3.1.2	(3) 2.0	(4) 2.0	(5) 2.0.1.3	(6) 3.3	(7a) 3.4	(7b) 3.5	(8) 4.1.4	(9) 4.1.4	(10) 4.1.4	(11) 4.1.4
△	0314	IGNITERS	1.2G	–	–	–	0	E0	P142	–	–	–
△	0315	IGNITERS	1.3G	–	–	–	0	E0	P142	–	–	–
△	0316	FUZES, IGNITING	1.3G	–	–	–	0	E0	P141	–	–	–
△	0317	FUZES, IGNITING	1.4G	–	–	–	0	E0	P141	–	–	–
△	0318	GRENADES, PRACTICE hand or rifle	1.3G	–	–	–	0	E0	P141	–	–	–
△	0319	PRIMERS, TUBULAR	1.3G	–	–	–	0	E0	P133	–	–	–
△	0320	PRIMERS, TUBULAR	1.4G	–	–	–	0	E0	P133	–	–	–
△	0321	CARTRIDGES FOR WEAPONS with bursting charge	1.2E	–	–	–	0	E0	P130 LP101	PP67 L1	–	–
△	0322	ROCKET MOTORS WITH HYPERGOLIC LIQUIDS with or without expelling charge	1.2L	–	–	–	0	E0	P101	–	–	–
△	0323	CARTRIDGES, POWER DEVICE	1.4S	–	–	347	0	E0	P134 LP102	–	–	–
△	0324	PROJECTILES with bursting charge	1.2F	–	–	–	0	E0	P130	–	–	–
△	0325	IGNITERS	1.4G	–	–	–	0	E0	P142	–	–	–
△	0326	CARTRIDGES FOR WEAPONS, BLANK	1.1C	–	–	–	0	E0	P130	–	–	–
△	0327	CARTRIDGES FOR WEAPONS, BLANK or CARTRIDGES, SMALL ARMS, BLANK	1.3C	–	–	–	0	E0	P130	–	–	–
△	0328	CARTRIDGES FOR WEAPONS, INERT PROJECTILE	1.2C	–	–	–	0	E0	P130 LP101	PP67 L1	–	–
△	0329	TORPEDOES with bursting charge	1.1E	–	–	–	0	E0	P130 LP101	PP67 L1	–	–
△	0330	TORPEDOES with bursting charge	1.1F	–	–	–	0	E0	P130	–	–	–
△	0331	EXPLOSIVE, BLASTING, TYPE B (AGENT, BLASTING, TYPE B)	1.5D	–	–	–	0	E0	P116	PP61 PP62 PP64 PP65	IBC100	–
△	0332	EXPLOSIVE, BLASTING, TYPE E (AGENT, BLASTING, TYPE E)	1.5D	–	–	–	0	E0	P116	PP61 PP62 PP65	IBC100	–
△	0333	FIREWORKS	1.1G	–	–	–	0	E0	P135	–	–	–
△	0334	FIREWORKS	1.2G	–	–	–	0	E0	P135	–	–	–
△	0335	FIREWORKS	1.3G	–	–	–	0	E0	P135	–	–	–

Portable tanks and bulk containers			EmS	Stowage and segregation	Properties and observations	UN No.
	Tank instructions	Provisions				
(12)	(13) 4.2.5 4.3	(14) 4.2.5	(15) 5.4.3.2 7.3	(16) 7.1 7.2	(17)	(18)
–	–	–	F-B, S-X	Category 03. Protected from sources of heat.	See glossary of terms in appendix B.	0314 △
–	–	–	F-B, S-X	Category 03. Protected from sources of heat.	See glossary of terms in appendix B.	0315 △
–	–	–	F-B, S-X	Category 03. Protected from sources of heat.	See glossary of terms in appendix B.	0316 △
–	–	–	F-B, S-X	Category 02. Protected from sources of heat.	See glossary of terms in appendix B.	0317 △
–	–	–	F-B, S-X	Category 03. Protected from sources of heat.	See glossary of terms in appendix B.	0318 △
–	–	–	F-B, S-X	Category 03. Protected from sources of heat.	See glossary of terms in appendix B.	0319 △
–	–	–	F-B, S-X	Category 02. Protected from sources of heat.	See glossary of terms in appendix B.	0320 △
–	–	–	F-B, S-X	Category 04. Protected from sources of heat.	See glossary of terms in appendix B.	0321 △
–	–	–	F-B, S-X	Category 05. Protected from sources of heat.	See glossary of terms in appendix B.	0322 △
–	–	–	F-B, S-X	Category 01. Protected from sources of heat.	See glossary of terms in appendix B.	0323 △
–	–	–	F-B, S-X	Category 05. Protected from sources of heat.	See glossary of terms in appendix B.	0324 △
–	–	–	F-B, S-X	Category 02. Protected from sources of heat.	See glossary of terms in appendix B.	0325 △
–	–	–	F-B, S-X	Category 04. Protected from sources of heat.	See glossary of terms in appendix B.	0326 △
–	–	–	F-B, S-X	Category 04. Protected from sources of heat.	See glossary of terms in appendix B.	0327 △
–	–	–	F-B, S-X	Category 04. Protected from sources of heat.	See glossary of terms in appendix B.	0328 △
–	–	–	F-B, S-X	Category 04. Protected from sources of heat.	See glossary of terms in appendix B.	0329 △
–	–	–	F-B, S-X	Category 05. Protected from sources of heat.	See glossary of terms in appendix B.	0330 △
–	T1	TP1 TP17 TP32	F-B, S-Y	Category 03. When containing ammonium compounds, "away from" chlorates or perchlorates and explosives containing chlorates or perchlorates. Protected from sources of heat.	See glossary of terms in appendix B.	0331 △
–	T1	TP1 TP17 TP32	F-B, S-Y	Category 03. When containing ammonium compounds, "away from" chlorates or perchlorates and explosives containing chlorates or perchlorates. Protected from sources of heat.	See glossary of terms in appendix B.	0332 △
–	–	–	F-B, S-X	Category 03. Protected from sources of heat.	See glossary of terms in appendix B.	0333 △
–	–	–	F-B, S-X	Category 03. Protected from sources of heat.	See glossary of terms in appendix B.	0334 △
–	–	–	F-B, S-X	Category 03. Protected from sources of heat.	See glossary of terms in appendix B.	0335 △

3

DGL

	UN No.	Proper Shipping Name (PSN)	Class or division	Subsidiary risk(s)	Packing group	Special provisions	Limited and excepted quantity provisions		Packing		IBC	
							Limited quantities	Excepted quantities	Instructions	Provisions	Instructions	Provisions
	(1)	(2) 3.1.2	(3) 2.0	(4) 2.0	(5) 2.0.1.3	(6) 3.3	(7a) 3.4	(7b) 3.5	(8) 4.1.4	(9) 4.1.4	(10) 4.1.4	(11) 4.1.4
△	0336	FIREWORKS	1.4G	–	–	–	0	E0	P135	–	–	–
△	0337	FIREWORKS	1.4S	–	–	–	0	E0	P135	–	–	–
△	0338	CARTRIDGES FOR WEAPONS, BLANK or CARTRIDGES, SMALL ARMS, BLANK	1.4C	–	–	–	0	E0	P130	–	–	–
△	0339	CARTRIDGES FOR WEAPONS, INERT PROJECTILE or CARTRIDGES, SMALL ARMS	1.4C	–	–	–	0	E0	P130	–	–	–
△	0340	NITROCELLULOSE dry or wetted with less than 25% water (or alcohol), by mass	1.1D	–	–	–	0	E0	P112 (a) or (b)	–	–	–
△	0341	NITROCELLULOSE unmodified or plasticized with less than 18% plasticizing substance, by mass	1.1D	–	–	–	0	E0	P112 (b)	–	–	–
△	0342	NITROCELLULOSE, WETTED with not less than 25% alcohol, by mass	1.3C	–	–	105	0	E0	P114 (a)	PP43	–	–
△	0343	NITROCELLULOSE, PLASTICIZED with not less than 18% plasticizing substance, by mass	1.3C	–	–	105	0	E0	P111	–	–	–
△	0344	PROJECTILES with bursting charge	1.4D	–	–	–	0	E0	P130 LP101	PP67 L1	–	–
△	0345	PROJECTILES inert, with tracer	1.4S	–	–	–	0	E0	P130 LP101	PP67 L1	–	–
△	0346	PROJECTILES with burster or expelling charge	1.2D	–	–	–	0	E0	P130 LP101	PP67 L1	–	–
△	0347	PROJECTILES with burster or expelling charge	1.4D	–	–	–	0	E0	P130 LP101	PP67 L1	–	–
△	0348	CARTRIDGES FOR WEAPONS with bursting charge	1.4F	–	–	–	0	E0	P130	–	–	–
△	0349	ARTICLES, EXPLOSIVE, N.O.S.	1.4S	–	–	178 274	0	E0	P101	–	–	–
△	0350	ARTICLES, EXPLOSIVE, N.O.S.	1.4B	–	–	178 274	0	E0	P101	–	–	–
△	0351	ARTICLES, EXPLOSIVE, N.O.S.	1.4C	–	–	178 274	0	E0	P101	–	–	–
△	0352	ARTICLES, EXPLOSIVE, N.O.S.	1.4D	–	–	178 274	0	E0	P101	–	–	–
△	0353	ARTICLES, EXPLOSIVE, N.O.S.	1.4G	–	–	178 274	0	E0	P101	–	–	–
△	0354	ARTICLES, EXPLOSIVE, N.O.S.	1.1L	See SP943	–	178 274	0	E0	P101	–	–	–
△	0355	ARTICLES, EXPLOSIVE, N.O.S.	1.2L	See SP943	–	178 274	0	E0	P101	–	–	–
△	0356	ARTICLES, EXPLOSIVE, N.O.S.	1.3L	See SP943	–	178 274	0	E0	P101	–	–	–
△	0357	SUBSTANCES, EXPLOSIVE, N.O.S.	1.1L	–	–	178 274	0	E0	P101	–	–	–
△	0358	SUBSTANCES, EXPLOSIVE, N.O.S.	1.2L	–	–	178 274	0	E0	P101	–	–	–

Portable tanks and bulk containers			EmS	Stowage and segregation	Properties and observations	UN No.	
	Tank instructions	Provisions					
(12)	(13) 4.2.5 4.3	(14) 4.2.5	(15) 5.4.3.2 7.3	(16) 7.1 7.2	(17)	(18)	
–	–	–	F-B, S-X	Category 02. Protected from sources of heat.	See glossary of terms in appendix B.	0336	△
–	–	–	F-B, S-X	Category 01. Protected from sources of heat.	See glossary of terms in appendix B.	0337	△
–	–	–	F-B, S-X	Category 02. Protected from sources of heat.	See glossary of terms in appendix B.	0338	△
–	–	–	F-B, S-X	Category 02. Protected from sources of heat.	See glossary of terms in appendix B.	0339	△
–	–	–	F-B, S-Y	Category 04. Protected from sources of heat.	Substance.	0340	△
–	–	–	F-B, S-Y	Category 04. Protected from sources of heat.	Substance.	0341	△
–	–	–	F-B, S-Y	Category 04. Protected from sources of heat.	Substance.	0342	△
–	–	–	F-B, S-Y	Category 04. Protected from sources of heat.	Substance.	0343	△
–	–	–	F-B, S-X	Category 02. Protected from sources of heat.	See glossary of terms in appendix B.	0344	△
–	–	–	F-B, S-X	Category 01. Protected from sources of heat.	See glossary of terms in appendix B.	0345	△
–	–	–	F-B, S-X	Category 04. Protected from sources of heat.	See glossary of terms in appendix B.	0346	△
–	–	–	F-B, S-X	Category 02. Protected from sources of heat.	See glossary of terms in appendix B.	0347	△
–	–	–	F-B, S-X	Category 05. Protected from sources of heat.	See glossary of terms in appendix B.	0348	△
–	–	–	F-B, S-X	Category 01. Protected from sources of heat.	–	0349	△
–	–	–	F-B, S-X	Category 05. Protected from sources of heat.	–	0350	△
–	–	–	F-B, S-X	Category 02. Protected from sources of heat.	–	0351	△
–	–	–	F-B, S-X	Category 02. Protected from sources of heat.	–	0352	△
–	–	–	F-B, S-X	Category 02. Protected from sources of heat.	–	0353	△
–	–	–	F-B, S-X	Category 05. Protected from sources of heat.	–	0354	△
–	–	–	F-B, S-X	Category 05. Protected from sources of heat.	–	0355	△
–	–	–	F-B, S-X	Category 05. Protected from sources of heat.	–	0356	△
–	–	–	F-B, S-Y	Category 05. Protected from sources of heat.	–	0357	△
–	–	–	F-B, S-Y	Category 05. Protected from sources of heat.	–	0358	△

	UN No.	Proper Shipping Name (PSN)	Class or division	Subsidiary risk(s)	Packing group	Special provisions	Limited and excepted quantity provisions		Packing		IBC	
							Limited quantities	Excepted quantities	Instructions	Provisions	Instructions	Provisions
	(1)	(2) 3.1.2	(3) 2.0	(4) 2.0	(5) 2.0.1.3	(6) 3.3	(7a) 3.4	(7b) 3.5	(8) 4.1.4	(9) 4.1.4	(10) 4.1.4	(11) 4.1.4
△	0359	SUBSTANCES, EXPLOSIVE, N.O.S.	1.3L	–	–	178 274	0	E0	P101	–	–	–
△	0360	DETONATOR ASSEMBLIES, NON-ELECTRIC for blasting	1.1B	–	–	–	0	E0	P131	–	–	–
△	0361	DETONATOR ASSEMBLIES, NON-ELECTRIC for blasting	1.4B	–	–	–	0	E0	P131	–	–	–
△	0362	AMMUNITION, PRACTICE	1.4G	–	–	–	0	E0	P130 LP101	PP67 L1	–	–
△	0363	AMMUNITION, PROOF	1.4G	–	–	–	0	E0	P130 LP101	PP67 L1	–	–
△	0364	DETONATORS FOR AMMUNITION	1.2B	–	–	–	0	E0	P133	–	–	–
△	0365	DETONATORS FOR AMMUNITION	1.4B	–	–	–	0	E0	P133	–	–	–
△	0366	DETONATORS FOR AMMUNITION	1.4S	–	–	347	0	E0	P133	–	–	–
△	0367	FUZES, DETONATING	1.4S	–	–	–	0	E0	P141	–	–	–
△	0368	FUZES, IGNITING	1.4S	–	–	–	0	E0	P141	–	–	–
△	0369	WARHEADS, ROCKET with bursting charge	1.1F	–	–	–	0	E0	P130	–	–	–
△	0370	WARHEADS, ROCKET with burster or expelling charge	1.4D	–	–	–	0	E0	P130 LP101	PP67 L1	–	–
△	0371	WARHEADS, ROCKET with burster or expelling charge	1.4F	–	–	–	0	E0	P130	–	–	–
△	0372	GRENADES, PRACTICE hand or rifle	1.2G	–	–	–	0	E0	P141	–	–	–
△	0373	SIGNAL DEVICES, HAND	1.4S	–	–	–	0	E0	P135	–	–	–
△	0374	SOUNDING DEVICES, EXPLOSIVE	1.1D	–	–	–	0	E0	P134 LP102	–	–	–
△	0375	SOUNDING DEVICES, EXPLOSIVE	1.2D	–	–	–	0	E0	P134 LP102	–	–	–
△	0376	PRIMERS, TUBULAR	1.4S	–	–	–	0	E0	P133	–	–	–
△	0377	PRIMERS, CAP TYPE	1.1B	–	–	–	0	E0	P133	–	–	–
△	0378	PRIMERS, CAP TYPE	1.4B	–	–	–	0	E0	P133	–	–	–
△	0379	CASES, CARTRIDGE, EMPTY, WITH PRIMER	1.4C	–	–	–	0	E0	P136	–	–	–
△	0380	ARTICLES, PYROPHORIC	1.2L	–	–	–	0	E0	P101	–	–	–
△	0381	CARTRIDGES, POWER DEVICE	1.2C	–	–	–	0	E0	P134 LP102	–	–	–
△	0382	COMPONENTS, EXPLOSIVE TRAIN, N.O.S.	1.2B	–	–	178 274	0	E0	P101	–	–	–
△	0383	COMPONENTS, EXPLOSIVE TRAIN, N.O.S.	1.4B	–	–	178 274	0	E0	P101	–	–	–
△	0384	COMPONENTS, EXPLOSIVE TRAIN, N.O.S.	1.4S	–	–	178 274	0	E0	P101	–	–	–

Portable tanks and bulk containers			EmS	Stowage and segregation	Properties and observations	UN No.	
	Tank instructions	Provisions					
(12)	(13) 4.2.5 4.3	(14) 4.2.5	(15) 5.4.3.2 7.3	(16) 7.1 7.2	(17)	(18)	
–	–	–	F-B, S-Y	Category 05. Protected from sources of heat.	–	0359	△
–	–	–	F-B, S-X	Category 05. Protected from sources of heat.	See glossary of terms in appendix B.	0360	△
–	–	–	F-B, S-X	Category 05. Protected from sources of heat.	See glossary of terms in appendix B.	0361	△
–	–	–	F-B, S-X	Category 02. Protected from sources of heat.	See glossary of terms in appendix B.	0362	△
–	–	–	F-B, S-X	Category 02. Protected from sources of heat.	See glossary of terms in appendix B.	0363	△
–	–	–	F-B, S-X	Category 05. Protected from sources of heat.	See glossary of terms in appendix B.	0364	△
–	–	–	F-B, S-X	Category 05. Protected from sources of heat.	See glossary of terms in appendix B.	0365	△
–	–	–	F-B, S-X	Category 01. Protected from sources of heat.	See glossary of terms in appendix B.	0366	△
–	–	–	F-B, S-X	Category 01. Protected from sources of heat.	See glossary of terms in appendix B.	0367	△
–	–	–	F-B, S-X	Category 01. Protected from sources of heat.	See glossary of terms in appendix B.	0368	△
–	–	–	F-B, S-X	Category 05. Protected from sources of heat.	See glossary of terms in appendix B.	0369	△
–	–	–	F-B, S-X	Category 02. Protected from sources of heat.	See glossary of terms in appendix B.	0370	△
–	–	–	F-B, S-X	Category 05. Protected from sources of heat.	See glossary of terms in appendix B.	0371	△
–	–	–	F-B, S-X	Category 03. Protected from sources of heat.	See glossary of terms in appendix B.	0372	△
–	–	–	F-B, S-X	Category 01. Protected from sources of heat.	See glossary of terms in appendix B.	0373	△
–	–	–	F-B, S-X	Category 04. Protected from sources of heat.	See glossary of terms in appendix B.	0374	△
–	–	–	F-B, S-X	Category 04. Protected from sources of heat.	See glossary of terms in appendix B.	0375	△
–	–	–	F-B, S-X	Category 01. Protected from sources of heat.	See glossary of terms in appendix B.	0376	△
–	–	–	F-B, S-X	Category 05. Protected from sources of heat.	See glossary of terms in appendix B.	0377	△
–	–	–	F-B, S-X	Category 05. Protected from sources of heat.	See glossary of terms in appendix B.	0378	△
–	–	–	F-B, S-X	Category 02. Protected from sources of heat.	See glossary of terms in appendix B.	0379	△
–	–	–	F-B, S-X	Category 05. Protected from sources of heat.	See glossary of terms in appendix B.	0380	△
–	–	–	F-B, S-X	Category 04. Protected from sources of heat.	See glossary of terms in appendix B.	0381	△
–	–	–	F-B, S-X	Category 05. Protected from sources of heat.	See glossary of terms in appendix B.	0382	△
–	–	–	F-B, S-X	Category 05. Protected from sources of heat.	See glossary of terms in appendix B.	0383	△
–	–	–	F-B, S-X	Category 01. Protected from sources of heat.	See glossary of terms in appendix B.	0384	△

3

DGL

	UN No.	Proper Shipping Name (PSN)	Class or division	Subsidiary risk(s)	Packing group	Special provisions	Limited and excepted quantity provisions		Packing		IBC	
							Limited quantities	Excepted quantities	Instructions	Provisions	Instructions	Provisions
	(1)	(2) 3.1.2	(3) 2.0	(4) 2.0	(5) 2.0.1.3	(6) 3.3	(7a) 3.4	(7b) 3.5	(8) 4.1.4	(9) 4.1.4	(10) 4.1.4	(11) 4.1.4
△	0385	5-NITROBENZOTRIAZOL	1.1D	–	–	–	0	E0	P112 (b) or (c)	–	–	–
△	0386	TRINITROBENZENE-SULPHONIC ACID	1.1D	–	–	–	0	E0	P112 (b) or (c)	PP26	–	–
△	0387	TRINITROFLUORENONE	1.1D	–	–	–	0	E0	P112 (b) or (c)	–	–	–
△	0388	TRINITROTOLUENE (TNT) AND TRINITROBENZENE MIXTURE or TRINITROTOLUENE (TNT) AND HEXANITROSTILBENE MIXTURE	1.1D	–	–	–	0	E0	P112 (b) or (c)	–	–	–
△	0389	TRINITROTOLUENE (TNT) MIXTURE CONTAINING TRINITROBENZENE AND HEXANITROSTILBENE	1.1D	–	–	–	0	E0	P112 (b) or (c)	–	–	–
△	0390	TRITONAL	1.1D	–	–	–	0	E0	P112 (b) or (c)	–	–	–
△	0391	CYCLOTRIMETHYLENE-TRINITRAMINE (CYCLONITE; HEXOGEN; RDX) AND CYCLOTETRAMETHYLENE-TETRANITRAMINE (HMX; OCTOGEN) MIXTURE, WETTED with not less than 15% water, by mass or CYCLOTRIMETHYLENE-TRINITRAMINE (CYCLONITE; HEXOGEN; RDX) AND CYCLOTETRAMETHYLENE-TETRANITRAMINE (HMX; OCTOGEN) MIXTURE, DESENSITIZED with not less than 10% phlegmatizer, by mass	1.1D	–	–	266	0	E0	P112 (a) or (b)	–	–	–
△	0392	HEXANITROSTILBENE	1.1D	–	–	–	0	E0	P112 (b) or (c)	–	–	–
△	0393	HEXOTONAL	1.1D	–	–	–	0	E0	P112 (b)	–	–	–
△	0394	TRINITRORESORCINOL (STYPHNIC ACID), WETTED with not less than 20% water, or mixture of alcohol and water, by mass	1.1D	–	–	–	0	E0	P112 (a)	PP26	–	–
△	0395	ROCKET MOTORS, LIQUID FUELLED	1.2J	–	–	–	0	E0	P101	–	–	–
△	0396	ROCKET MOTORS, LIQUID FUELLED	1.3J	–	–	–	0	E0	P101	–	–	–

Portable tanks and bulk containers			EmS	Stowage and segregation	Properties and observations	UN No.	
	Tank instructions	Provisions					
(12)	(13) 4.2.5 4.3	(14) 4.2.5	(15) 5.4.3.2 7.3	(16) 7.1 7.2	(17)	(18)	
–	–	–	F-B, S-Y	Category 04. Protected from sources of heat.	Substance.	0385	△
–	–	–	F-B, S-Y	Category 04. "Away from" lead and its compounds. Protected from sources of heat.	Substance.	0386	△
–	–	–	F-B, S-Y	Category 04. Protected from sources of heat.	Substance.	0387	△
–	–	–	F-B, S-Y	Category 04. Protected from sources of heat.	Substance.	0388	△
–	–	–	F-B, S-Y	Category 04. Protected from sources of heat.	Substance.	0389	△
–	–	–	F-B, S-Y	Category 04. Protected from sources of heat.	Tritonal is a substance consisting of trinitrotoluene (TNT) mixed with aluminium.	0390	△
–	–	–	F-B, S-Y	Category 04. Protected from sources of heat.	Substance. Mass detonating explosive which will become more sensitive if the wetting or desensitizing agents are lost. This substance, when containing less alcohol, water or phlegmatizer than specified, shall not be transported, unless specifically authorized by the competent authority.	0391	△
–	–	–	F-B, S-Y	Category 04. Protected from sources of heat.	Substance. Mass detonating explosive.	0392	△
–	–	–	F-B, S-Y	Category 04. Protected from sources of heat.	Substance. Mass detonating explosive.	0393	△
–	–	–	F-B, S-Y	Category 04. "Away from" lead and its compounds. Protected from sources of heat.	Substance. Mass detonating explosive.	0394	△
–	–	–	F-B, S-X	Category 05. "Separated from" division 1.4 and "separated longitudinally by an intervening complete compartment or hold from" divisions 1.1, 1.2, 1.3, 1.5 and 1.6 except from explosives of compatibility group J. Protected from sources of heat.	See glossary of terms in appendix B.	0395	△
–	–	–	F-B, S-X	Category 05. "Separated from" division 1.4 and "separated longitudinally by an intervening complete compartment or hold from" divisions 1.1, 1.2, 1.3, 1.5 and 1.6 except from explosives of compatibility group J. Protected from sources of heat.	See glossary of terms in appendix B.	0396	△

UN No. (1)	Proper Shipping Name (PSN) (2) 3.1.2	Class or division (3) 2.0	Subsidiary risk(s) (4) 2.0	Packing group (5) 2.0.1.3	Special provisions (6) 3.3	Limited and excepted quantity provisions: Limited quantities (7a) 3.4	Limited and excepted quantity provisions: Excepted quantities (7b) 3.5	Packing: Instructions (8) 4.1.4	Packing: Provisions (9) 4.1.4	IBC: Instructions (10) 4.1.4	IBC: Provisions (11) 4.1.4
△ 0397	ROCKETS, LIQUID FUELLED with bursting charge	1.1J	–	–	–	0	E0	P101	–	–	–
△ 0398	ROCKETS, LIQUID FUELLED with bursting charge	1.2J	–	–	–	0	E0	P101	–	–	–
△ 0399	BOMBS WITH FLAMMABLE LIQUID with bursting charge	1.1J	–	–	–	0	E0	P101	–	–	–
△ 0400	BOMBS WITH FLAMMABLE LIQUID with bursting charge	1.2J	–	–	–	0	E0	P101	–	–	–
△ 0401	DIPICRYL SULPHIDE dry or wetted with less than 10% water, by mass	1.1D	–	–	–	0	E0	P112 (a), (b) or (c)	–	–	–
△ 0402	AMMONIUM PERCHLORATE	1.1D	–	–	152	0	E0	P112 (b) or (c)	–	–	–
△ 0403	FLARES, AERIAL	1.4G	–	–	–	0	E0	P135	–	–	–
△ 0404	FLARES, AERIAL	1.4S	–	–	–	0	E0	P135	–	–	–
△ 0405	CARTRIDGES, SIGNAL	1.4S	–	–	–	0	E0	P135	–	–	–
△ 0406	DINITROSOBENZENE	1.3C	–	–	–	0	E0	P114 (b)	–	–	–
△ 0407	TETRAZOL-1-ACETIC ACID	1.4C	–	–	–	0	E0	P114 (b)	–	–	–
△ 0408	FUZES, DETONATING with protective features	1.1D	–	–	–	0	E0	P141	–	–	–
△ 0409	FUZES, DETONATING with protective features	1.2D	–	–	–	0	E0	P141	–	–	–
△ 0410	FUZES, DETONATING with protective features	1.4D	–	–	–	0	E0	P141	–	–	–

Portable tanks and bulk containers			EmS	Stowage and segregation	Properties and observations	UN No.	
	Tank instructions	Provisions					
(12)	(13) 4.2.5 4.3	(14) 4.2.5	(15) 5.4.3.2 7.3	(16) 7.1 7.2	(17)	(18)	
–	–	–	F-B, S-X	Category 05. "Separated from" division 1.4 and "separated longitudinally by an intervening complete compartment or hold from" divisions 1.1, 1.2, 1.3, 1.5 and 1.6 except from explosives of compatibility group J. Protected from sources of heat.	See glossary of terms in appendix B.	0397	△
–	–	–	F-B, S-X	Category 05. "Separated from" division 1.4 and "separated longitudinally by an intervening complete compartment or hold from" divisions 1.1, 1.2, 1.3, 1.5 and 1.6 except from explosives of compatibility group J. Protected from sources of heat.	See glossary of terms in appendix B.	0398	△
–	–	–	F-B, S-X	Category 05. "Separated from" division 1.4 and "separated longitudinally by an intervening complete compartment or hold from" divisions 1.1, 1.2, 1.3, 1.5 and 1.6 except from explosives of compatibility group J. Protected from sources of heat.	See glossary of terms in appendix B.	0399	△
–	–	–	F-B, S-X	Category 05. "Separated from" division 1.4 and "separated longitudinally by an intervening complete compartment or hold from" divisions 1.1, 1.2, 1.3, 1.5 and 1.6 except from explosives of compatibility group J. Protected from sources of heat.	See glossary of terms in appendix B.	0400	△
–	–	–	F-B, S-Y	Category 04. Protected from sources of heat.	Substance.	0401	△
–	–	–	F-B, S-Y	Category 04. "Away from" explosives containing chlorates or perchlorates. Protected from sources of heat.	Substance.	0402	△
–	–	–	F-B, S-X	Category 02. Protected from sources of heat.	See glossary of terms in appendix B.	0403	△
–	–	–	F-B, S-X	Category 01. Protected from sources of heat.	See glossary of terms in appendix B.	0404	△
–	–	–	F-B, S-X	Category 01. Protected from sources of heat.	See glossary of terms in appendix B.	0405	△
–	–	–	F-B, S-Y	Category 04. Protected from sources of heat.	Substance.	0406	△
–	–	–	F-B, S-Y	Category 02. Protected from sources of heat.	Substance.	0407	△
–	–	–	F-B, S-X	Category 04. Protected from sources of heat.	See glossary of terms in appendix B.	0408	△
–	–	–	F-B, S-X	Category 04. Protected from sources of heat.	See glossary of terms in appendix B.	0409	△
–	–	–	F-B, S-X	Category 02. Protected from sources of heat.	See glossary of terms in appendix B.	0410	△

UN No.	Proper Shipping Name (PSN)	Class or division	Subsidiary risk(s)	Packing group	Special provisions	Limited and excepted quantity provisions		Packing		IBC	
						Limited quantities	Excepted quantities	Instructions	Provisions	Instructions	Provisions
(1)	(2) 3.1.2	(3) 2.0	(4) 2.0	(5) 2.0.1.3	(6) 3.3	(7a) 3.4	(7b) 3.5	(8) 4.1.4	(9) 4.1.4	(10) 4.1.4	(11) 4.1.4
△ 0411	PENTAERYTHRITE TETRANITRATE (PENTAERYTHRITOL TETRANITRATE; PETN) with not less than 7% wax, by mass	1.1D	–	–	131	0	E0	P112 (b) or (c)	–	–	–
△ 0412	CARTRIDGES FOR WEAPONS with bursting charge	1.4E	–	–	–	0	E0	P130 LP101	PP67 L1	–	–
△ 0413	CARTRIDGES FOR WEAPONS, BLANK	1.2C	–	–	–	0	E0	P130	–	–	–
△ 0414	CHARGES, PROPELLING, FOR CANNON	1.2C	–	–	–	0	E0	P130	–	–	–
△ 0415	CHARGES, PROPELLING	1.2C	–	–	–	0	E0	P143	PP76	–	–
△ 0417	CARTRIDGES FOR WEAPONS, INERT PROJECTILE or CARTRIDGES, SMALL ARMS	1.3C	–	–	–	0	E0	P130	–	–	–
△ 0418	FLARES, SURFACE	1.1G	–	–	–	0	E0	P135	–	–	–
△ 0419	FLARES, SURFACE	1.2G	–	–	–	0	E0	P135	–	–	–
△ 0420	FLARES, AERIAL	1.1G	–	–	–	0	E0	P135	–	–	–
△ 0421	FLARES, AERIAL	1.2G	–	–	–	0	E0	P135	–	–	–
△ 0424	PROJECTILES inert, with tracer	1.3G	–	–	–	0	E0	P130 LP101	PP67 L1	–	–
△ 0425	PROJECTILES inert, with tracer	1.4G	–	–	–	0	E0	P130 LP101	PP67 L1	–	–
△ 0426	PROJECTILES with burster or expelling charge	1.2F	–	–	–	0	E0	P130	–	–	–
△ 0427	PROJECTILES with burster or expelling charge	1.4F	–	–	–	0	E0	P130	–	–	–
△ 0428	ARTICLES, PYROTECHNIC for technical purposes	1.1G	–	–	–	0	E0	P135	–	–	–
△ 0429	ARTICLES, PYROTECHNIC for technical purposes	1.2G	–	–	–	0	E0	P135	–	–	–
△ 0430	ARTICLES, PYROTECHNIC for technical purposes	1.3G	–	–	–	0	E0	P135	–	–	–
△ 0431	ARTICLES, PYROTECHNIC for technical purposes	1.4G	–	–	–	0	E0	P135	–	–	–
△ 0432	ARTICLES, PYROTECHNIC for technical purposes	1.4S	–	–	–	0	E0	P135	–	–	–
△ 0433	POWDER CAKE (POWDER PASTE), WETTED with not less than 17% alcohol, by mass	1.1C	–	–	266	0	E0	P111	–	–	–
△ 0434	PROJECTILES with burster or expelling charge	1.2G	–	–	–	0	E0	P130 LP101	PP67 L1	–	–
△ 0435	PROJECTILES with burster or expelling charge	1.4G	–	–	–	0	E0	P130 LP101	PP67 L1	–	–
△ 0436	ROCKETS with expelling charge	1.2C	–	–	–	0	E0	P130 LP101	PP67 L1	–	–
△ 0437	ROCKETS with expelling charge	1.3C	–	–	–	0	E0	P130 LP101	PP67 L1	–	–

Portable tanks and bulk containers			EmS	Stowage and segregation	Properties and observations	UN No.
	Tank instructions	Provisions				
(12)	(13) 4.2.5 4.3	(14) 4.2.5	(15) 5.4.3.2 7.3	(16) 7.1 7.2	(17)	(18)
–	–	–	F-B, S-Y	Category 04. Protected from sources of heat.	Substance.	0411 △
–	–	–	F-B, S-X	Category 03. Protected from sources of heat.	See glossary of terms in appendix B.	0412 △
–	–	–	F-B, S-X	Category 04. Protected from sources of heat.	See glossary of terms in appendix B.	0413 △
–	–	–	F-B, S-X	Category 04. Protected from sources of heat.	See glossary of terms in appendix B.	0414 △
–	–	–	F-B, S-X	Category 04. Protected from sources of heat.	See glossary of terms in appendix B.	0415 △
–	–	–	F-B, S-X	Category 04. Protected from sources of heat.	See glossary of terms in appendix B.	0417 △
–	–	–	F-B, S-X	Category 03. Protected from sources of heat.	See glossary of terms in appendix B.	0418 △
–	–	–	F-B, S-X	Category 03. Protected from sources of heat.	See glossary of terms in appendix B.	0419 △
–	–	–	F-B, S-X	Category 03. Protected from sources of heat.	See glossary of terms in appendix B.	0420 △
–	–	–	F-B, S-X	Category 03. Protected from sources of heat.	See glossary of terms in appendix B.	0421 △
–	–	–	F-B, S-X	Category 03. Protected from sources of heat.	See glossary of terms in appendix B.	0424 △
–	–	–	F-B, S-X	Category 02. Protected from sources of heat.	See glossary of terms in appendix B.	0425 △
–	–	–	F-B, S-X	Category 05. Protected from sources of heat.	See glossary of terms in appendix B.	0426 △
–	–	–	F-B, S-X	Category 05. Protected from sources of heat.	See glossary of terms in appendix B.	0427 △
–	–	–	F-B, S-X	Category 03. Protected from sources of heat.	See glossary of terms in appendix B.	0428 △
–	–	–	F-B, S-X	Category 03. Protected from sources of heat.	See glossary of terms in appendix B.	0429 △
–	–	–	F-B, S-X	Category 03. Protected from sources of heat.	See glossary of terms in appendix B.	0430 △
–	–	–	F-B, S-X	Category 02. Protected from sources of heat.	See glossary of terms in appendix B.	0431 △
–	–	–	F-B, S-X	Category 01. Protected from sources of heat.	See glossary of terms in appendix B.	0432 △
–	–	–	F-B, S-Y	Category 04. Protected from sources of heat.	See glossary of terms in appendix B.	0433 △
–	–	–	F-B, S-X	Category 03. Protected from sources of heat.	See glossary of terms in appendix B.	0434 △
–	–	–	F-B, S-X	Category 02. Protected from sources of heat.	See glossary of terms in appendix B.	0435 △
–	–	–	F-B, S-X	Category 04. Protected from sources of heat.	See glossary of terms in appendix B.	0436 △
–	–	–	F-B, S-X	Category 04. Protected from sources of heat.	See glossary of terms in appendix B.	0437 △

	UN No.	Proper Shipping Name (PSN)	Class or division	Subsidiary risk(s)	Packing group	Special provisions	Limited and excepted quantity provisions		Packing		IBC	
							Limited quantities	Excepted quantities	Instructions	Provisions	Instructions	Provisions
	(1)	(2) 3.1.2	(3) 2.0	(4) 2.0	(5) 2.0.1.3	(6) 3.3	(7a) 3.4	(7b) 3.5	(8) 4.1.4	(9) 4.1.4	(10) 4.1.4	(11) 4.1.4
△	0438	ROCKETS with expelling charge	1.4C	–	–	–	0	E0	P130 LP101	PP67 L1	–	–
△	0439	CHARGES, SHAPED without detonator	1.2D	–	–	–	0	E0	P137	PP70	–	–
△	0440	CHARGES, SHAPED without detonator	1.4D	–	–	–	0	E0	P137	PP70	–	–
△	0441	CHARGES, SHAPED without detonator	1.4S	–	–	347	0	E0	P137	PP70	–	–
△	0442	CHARGES, EXPLOSIVE, COMMERCIAL without detonator	1.1D	–	–	–	0	E0	P137	–	–	–
△	0443	CHARGES, EXPLOSIVE, COMMERCIAL without detonator	1.2D	–	–	–	0	E0	P137	–	–	–
△	0444	CHARGES, EXPLOSIVE, COMMERCIAL without detonator	1.4D	–	–	–	0	E0	P137	–	–	–
△	0445	CHARGES, EXPLOSIVE, COMMERCIAL without detonator	1.4S	–	–	347	0	E0	P137	–	–	–
△	0446	CASES, COMBUSTIBLE, EMPTY, WITHOUT PRIMER	1.4C	–	–	–	0	E0	P136	–	–	–
△	0447	CASES, COMBUSTIBLE, EMPTY, WITHOUT PRIMER	1.3C	–	–	–	0	E0	P136	–	–	–
△	0448	5-MERCAPTOTETRAZOL-1-ACETIC ACID	1.4C	–	–	–	0	E0	P114 (b)	–	–	–
△	0449	TORPEDOES, LIQUID-FUELLED with or without bursting charge	1.1J	–	–	–	0	E0	P101	–	–	–
△	0450	TORPEDOES, LIQUID-FUELLED with inert head	1.3J	–	–	–	0	E0	P101	–	–	–
△	0451	TORPEDOES with bursting charge	1.1D	–	–	–	0	E0	P130 LP101	PP67 L1	–	–
△	0452	GRENADES, PRACTICE hand or rifle	1.4G	–	–	–	0	E0	P141	–	–	–
△	0453	ROCKETS, LINE-THROWING	1.4G	–	–	–	0	E0	P130	–	–	–
△	0454	IGNITERS	1.4S	–	–	–	0	E0	P142	–	–	–
△	0455	DETONATORS, NON-ELECTRIC for blasting	1.4S	–	–	347	0	E0	P131	PP68	–	–
△	0456	DETONATORS, ELECTRIC for blasting	1.4S	–	–	347	0	E0	P131	–	–	–
△	0457	CHARGES, BURSTING, PLASTICS BONDED	1.1D	–	–	–	0	E0	P130	–	–	–
△	0458	CHARGES, BURSTING, PLASTICS BONDED	1.2D	–	–	–	0	E0	P130	–	–	–

Portable tanks and bulk containers			EmS	Stowage and segregation	Properties and observations	UN No.	
	Tank instructions	Provisions					
(12)	(13) 4.2.5 4.3	(14) 4.2.5	(15) 5.4.3.2 7.3	(16) 7.1 7.2	(17)	(18)	
–	–	–	F-B, S-X	Category 02. Protected from sources of heat.	See glossary of terms in appendix B.	0438	△
–	–	–	F-B, S-X	Category 04. Protected from sources of heat.	See glossary of terms in appendix B.	0439	△
–	–	–	F-B, S-X	Category 02. Protected from sources of heat.	See glossary of terms in appendix B.	0440	△
–	–	–	F-B, S-X	Category 01. Protected from sources of heat.	See glossary of terms in appendix B.	0441	△
–	–	–	F-B, S-X	Category 04. Protected from sources of heat.	See glossary of terms in appendix B.	0442	△
–	–	–	F-B, S-X	Category 04. Protected from sources of heat.	See glossary of terms in appendix B.	0443	△
–	–	–	F-B, S-X	Category 02. Protected from sources of heat.	See glossary of terms in appendix B.	0444	△
–	–	–	F-B, S-X	Category 01. Protected from sources of heat.	See glossary of terms in appendix B.	0445	△
–	–	–	F-B, S-X	Category 02. Protected from sources of heat.	See glossary of terms in appendix B.	0446	△
–	–	–	F-B, S-X	Category 04. Protected from sources of heat.	See glossary of terms in appendix B.	0447	△
–	–	–	F-B, S-Y	Category 02. Protected from sources of heat.	Substance.	0448	△
–	–	–	F-B, S-X	Category 05. "Separated from" division 1.4 and "separated longitudinally by an intervening complete compartment or hold from" divisions 1.1, 1.2, 1.3, 1.5 and 1.6 except from explosives of compatibility group J. Protected from sources of heat.	See glossary of terms in appendix B.	0449	△
–	–	–	F-B, S-X	Category 05. "Separated from" division 1.4 and "separated longitudinally by an intervening complete compartment or hold from" divisions 1.1, 1.2, 1.3, 1.5 and 1.6 except from explosives of compatibility group J. Protected from sources of heat.	See glossary of terms in appendix B.	0450	△
–	–	–	F-B, S-X	Category 04. Protected from sources of heat.	See glossary of terms in appendix B.	0451	△
–	–	–	F-B, S-X	Category 02. Protected from sources of heat.	See glossary of terms in appendix B.	0452	△
–	–	–	F-B, S-X	Category 02. Protected from sources of heat.	See glossary of terms in appendix B.	0453	△
–	–	–	F-B, S-X	Category 01. Protected from sources of heat.	See glossary of terms in appendix B.	0454	△
–	–	–	F-B, S-X	Category 01. Protected from sources of heat.	See glossary of terms in appendix B.	0455	△
–	–	–	F-B, S-X	Category 01. Protected from sources of heat.	See glossary of terms in appendix B.	0456	△
–	–	–	F-B, S-X	Category 04. Protected from sources of heat.	See glossary of terms in appendix B.	0457	△
–	–	–	F-B, S-X	Category 04. Protected from sources of heat.	See glossary of terms in appendix B.	0458	△

UN No.	Proper Shipping Name (PSN)	Class or division	Subsidiary risk(s)	Packing group	Special provisions	Limited and excepted quantity provisions		Packing		IBC	
						Limited quantities	Excepted quantities	Instructions	Provisions	Instructions	Provisions
(1)	(2) 3.1.2	(3) 2.0	(4) 2.0	(5) 2.0.1.3	(6) 3.3	(7a) 3.4	(7b) 3.5	(8) 4.1.4	(9) 4.1.4	(10) 4.1.4	(11) 4.1.4
△ 0459	CHARGES, BURSTING, PLASTICS BONDED	1.4D	–	–	–	0	E0	P130	–	–	–
△ 0460	CHARGES, BURSTING, PLASTICS BONDED	1.4S	–	–	347	0	E0	P130	–	–	–
△ 0461	COMPONENTS, EXPLOSIVE TRAIN, N.O.S.	1.1B	–	–	178 274	0	E0	P101	–	–	–
△ 0462	ARTICLES, EXPLOSIVE, N.O.S.	1.1C	–	–	178 274	0	E0	P101	–	–	–
△ 0463	ARTICLES, EXPLOSIVE, N.O.S.	1.1D	–	–	178 274	0	E0	P101	–	–	–
△ 0464	ARTICLES, EXPLOSIVE, N.O.S.	1.1E	–	–	178 274	0	E0	P101	–	–	–
△ 0465	ARTICLES, EXPLOSIVE, N.O.S.	1.1F	–	–	178 274	0	E0	P101	–	–	–
△ 0466	ARTICLES, EXPLOSIVE, N.O.S.	1.2C	–	–	178 274	0	E0	P101	–	–	–
△ 0467	ARTICLES, EXPLOSIVE, N.O.S.	1.2D	–	–	178 274	0	E0	P101	–	–	–
△ 0468	ARTICLES, EXPLOSIVE, N.O.S.	1.2E	–	–	178 274	0	E0	P101	–	–	–
△ 0469	ARTICLES, EXPLOSIVE, N.O.S.	1.2F	–	–	178 274	0	E0	P101	–	–	–
△ 0470	ARTICLES, EXPLOSIVE, N.O.S.	1.3C	–	–	178 274	0	E0	P101	–	–	–
△ 0471	ARTICLES, EXPLOSIVE, N.O.S.	1.4E	–	–	178 274	0	E0	P101	–	–	–
△ 0472	ARTICLES, EXPLOSIVE, N.O.S.	1.4F	–	–	178 274	0	E0	P101	–	–	–
△ 0473	SUBSTANCES, EXPLOSIVE, N.O.S.	1.1A	–	–	178 274	0	E0	P101	–	–	–
△ 0474	SUBSTANCES, EXPLOSIVE, N.O.S.	1.1C	–	–	178 274	0	E0	P101	–	–	–
△ 0475	SUBSTANCES, EXPLOSIVE, N.O.S.	1.1D	–	–	178 274	0	E0	P101	–	–	–
△ 0476	SUBSTANCES, EXPLOSIVE, N.O.S.	1.1G	–	–	178 274	0	E0	P101	–	–	–
△ 0477	SUBSTANCES, EXPLOSIVE, N.O.S.	1.3C	–	–	178 274	0	E0	P101	–	–	–
△ 0478	SUBSTANCES, EXPLOSIVE, N.O.S.	1.3G	–	–	178 274	0	E0	P101	–	–	–
△ 0479	SUBSTANCES, EXPLOSIVE, N.O.S.	1.4C	–	–	178 274	0	E0	P101	–	–	–
△ 0480	SUBSTANCES, EXPLOSIVE, N.O.S.	1.4D	–	–	178 274	0	E0	P101	–	–	–
△ 0481	SUBSTANCES, EXPLOSIVE, N.O.S.	1.4S	–	–	178 274	0	E0	P101	–	–	–
△ 0482	SUBSTANCES, EXPLOSIVE, VERY INSENSITIVE (SUBSTANCES, EVI), N.O.S.	1.5D	–	–	178 274	0	E0	P101	–	–	–
△ 0483	CYCLOTRIMETHYLENE-TRINITRAMINE (CYCLONITE; HEXOGEN; RDX), DESENSITIZED	1.1D	–	–	–	0	E0	P112 (b) or (c)	–	–	–

Portable tanks and bulk containers			EmS	Stowage and segregation	Properties and observations	UN No.	
	Tank instructions	Provisions					
(12)	(13) 4.2.5 4.3	(14) 4.2.5	(15) 5.4.3.2 7.3	(16) 7.1 7.2	(17)	(18)	
–	–	–	F-B, S-X	Category 02. Protected from sources of heat.	See glossary of terms in appendix B.	0459	△
–	–	–	F-B, S-X	Category 01. Protected from sources of heat.	See glossary of terms in appendix B.	0460	△
–	–	–	F-B, S-X	Category 05. Protected from sources of heat.	See glossary of terms in appendix B.	0461	△
–	–	–	F-B, S-X	Category 04. Protected from sources of heat.	–	0462	△
–	–	–	F-B, S-X	Category 04. Protected from sources of heat.	–	0463	△
–	–	–	F-B, S-X	Category 04. Protected from sources of heat.	–	0464	△
–	–	–	F-B, S-X	Category 05. Protected from sources of heat.	–	0465	△
–	–	–	F-B, S-X	Category 04. Protected from sources of heat.	–	0466	△
–	–	–	F-B, S-X	Category 04. Protected from sources of heat.	–	0467	△
–	–	–	F-B, S-X	Category 04. Protected from sources of heat.	–	0468	△
–	–	–	F-B, S-X	Category 05. Protected from sources of heat.	–	0469	△
–	–	–	F-B, S-X	Category 04. Protected from sources of heat.	–	0470	△
–	–	–	F-B, S-X	Category 03. Protected from sources of heat.	–	0471	△
–	–	–	F-B, S-X	Category 05. Protected from sources of heat.	–	0472	△
–	–	–	F-B, S-Y	Category 05. Protected from sources of heat.	–	0473	△
–	–	–	F-B, S-Y	Category 04. Protected from sources of heat.	–	0474	△
–	–	–	F-B, S-Y	Category 04. Protected from sources of heat.	–	0475	△
–	–	–	F-B, S-Y	Category 03. Protected from sources of heat.	–	0476	△
–	–	–	F-B, S-Y	Category 04. Protected from sources of heat.	–	0477	△
–	–	–	F-B, S-Y	Category 03. Protected from sources of heat.	–	0478	△
–	–	–	F-B, S-Y	Category 02. Protected from sources of heat.	–	0479	△
–	–	–	F-B, S-Y	Category 02. Protected from sources of heat.	–	0480	△
–	–	–	F-B, S-Y	Category 01. Protected from sources of heat.	–	0481	△
–	–	–	F-B, S-Y	Category 03. Protected from sources of heat.	–	0482	△
–	–	–	F-B, S-Y	Category 04. Protected from sources of heat.	Substance. Mass detonating explosive which will become more sensitive if the wetting or desensitizing agents are lost.	0483	△

UN No.	Proper Shipping Name (PSN)	Class or division	Subsidiary risk(s)	Packing group	Special provisions	Limited and excepted quantity provisions		Packing		IBC	
						Limited quantities	Excepted quantities	Instructions	Provisions	Instructions	Provisions
(1)	(2) 3.1.2	(3) 2.0	(4) 2.0	(5) 2.0.1.3	(6) 3.3	(7a) 3.4	(7b) 3.5	(8) 4.1.4	(9) 4.1.4	(10) 4.1.4	(11) 4.1.4
△ 0484	CYCLOTETRAMETHYLENE-TETRANITRAMINE (OCTOGEN; HMX), DESENSITIZED	1.1D	–	–	–	0	E0	P112 (b) or (c)	–	–	–
△ 0485	SUBSTANCES, EXPLOSIVE, N.O.S.	1.4G	–	–	178 274	0	E0	P101	–	–	–
△ 0486	ARTICLES, EXPLOSIVE, EXTREMELY INSENSITIVE (ARTICLES, EEI)	1.6N	–	–	–	0	E0	P101	–	–	–
△ 0487	SIGNALS, SMOKE	1.3G	–	–	–	0	E0	P135	–	–	–
△ 0488	AMMUNITION, PRACTICE	1.3G	–	–	–	0	E0	P130 LP101	PP67 L1	–	–
△ 0489	DINITROGLYCOLURIL (DINGU)	1.1D	–	–	–	0	E0	P112 (b) or (c)	–	–	–
△ 0490	NITROTRIAZOLONE (NTO)	1.1D	–	–	–	0	E0	P112 (b) or (c)	–	–	–
△ 0491	CHARGES, PROPELLING	1.4C	–	–	–	0	E0	P143	PP76	–	–
△ 0492	SIGNALS, RAILWAY TRACK, EXPLOSIVE	1.3G	–	–	–	0	E0	P135	–	–	–
△ 0493	SIGNALS, RAILWAY TRACK, EXPLOSIVE	1.4G	–	–	–	0	E0	P135	–	–	–
△ 0494	JET PERFORATING GUNS, CHARGED oil well, without detonator	1.4D	–	–	–	0	E0	P101	–	–	–
△ 0495	PROPELLANT, LIQUID	1.3C	–	–	224	0	E0	P115	PP53 PP54 PP57 PP58	–	–
△ 0496	OCTONAL	1.1D	–	–	–	0	E0	P112 (b) or (c)	–	–	–
△ 0497	PROPELLANT, LIQUID	1.1C	–	–	224	0	E0	P115	PP53 PP54 PP57 PP58	–	–
△ 0498	PROPELLANT, SOLID	1.1C	–	–	–	0	E0	P114 (b)	–	–	–
△ 0499	PROPELLANT, SOLID	1.3C	–	–	–	0	E0	P114 (b)	–	–	–
△ 0500	DETONATOR ASSEMBLIES, NON-ELECTRIC for blasting	1.4S	–	–	347	0	E0	P131	–	–	–
△ 0501	PROPELLANT, SOLID	1.4C	–	–	–	0	E0	P114 (b)	–	–	–
△ 0502	ROCKETS with inert head	1.2C	–	–	–	0	E0	P130 LP101	PP67 L1	–	–
△ 0503	AIR BAG INFLATORS or AIR BAG MODULES or SEAT-BELT PRETENSIONERS	1.4G	–	–	235 289	0	E0	P135	–	–	–
△ 0504	1*H*-TETRAZOLE	1.1D	–	–	–	0	E0	P112 (c)	PP48	–	–
△ 0505	SIGNALS, DISTRESS, ship	1.4G	–	–	–	0	E0	P135	–	–	–
△ 0506	SIGNALS, DISTRESS, ship	1.4S	–	–	–	0	E0	P135	–	–	–

Portable tanks and bulk containers			EmS	Stowage and segregation	Properties and observations	UN No.	
	Tank instructions	Provisions					
(12)	(13) 4.2.5 4.3	(14) 4.2.5	(15) 5.4.3.2 7.3	(16) 7.1 7.2	(17)	(18)	
–	–	–	F-B, S-Y	Category 04. Protected from sources of heat.	Substance. Mass detonating explosive which will become more sensitive if the wetting or desensitizing agents are lost.	0484	△
–	–	–	F-B, S-Y	Category 02. Protected from sources of heat.	–	0485	△
–	–	–	F-B, S-X	Category 03. Protected from sources of heat.	See glossary of terms in appendix B.	0486	△
–	–	–	F-B, S-X	Category 03. Protected from sources of heat.	See glossary of terms in appendix B.	0487	△
–	–	–	F-B, S-X	Category 03. Protected from sources of heat.	See glossary of terms in appendix B.	0488	△
–	–	–	F-B, S-Y	Category 04. Protected from sources of heat.	Substance.	0489	△
–	–	–	F-B, S-Y	Category 04. Protected from sources of heat.	Substance.	0490	△
–	–	–	F-B, S-X	Category 02. Protected from sources of heat.	See glossary of terms in appendix B.	0491	△
–	–	–	F-B, S-X	Category 03. Protected from sources of heat.	See glossary of terms in appendix B.	0492	△
–	–	–	F-B, S-X	Category 02. Protected from sources of heat.	See glossary of terms in appendix B.	0493	△
–	–	–	F-B, S-X	Category 02. Protected from sources of heat.	See glossary of terms in appendix B.	0494	△
–	–	–	F-B, S-Y	Category 04. Protected from sources of heat.	See glossary of terms in appendix B.	0495	△
–	–	–	F-B, S-Y	Category 04. Protected from sources of heat.	Substance. Mixtures of mass detonating explosives.	0496	△
–	–	–	F-B, S-Y	Category 04. Protected from sources of heat.	See glossary of terms in appendix B.	0497	△
–	–	–	F-B, S-Y	Category 04. Protected from sources of heat.	See glossary of terms in appendix B.	0498	△
–	–	–	F-B, S-Y	Category 04. Protected from sources of heat.	See glossary of terms in appendix B.	0499	△
–	–	–	F-B, S-Y	Category 01. Protected from sources of heat.	See glossary of terms in appendix B.	0500	△
–	–	–	F-B, S-X	Category 02. Protected from sources of heat.	See glossary of terms in appendix B.	0501	△
–	–	–	F-B, S-X	Category 04. Protected from sources of heat.	See glossary of terms in appendix B.	0502	△
–	–	–	F-B, S-X	Category 02. Protected from sources of heat.	See glossary of terms in appendix B.	0503	△
–	–	–	F-B, S-Y	Category 04. Protected from sources of heat.	Substance.	0504	△
–	–	–	F-B, S-X	Category 02. Protected from sources of heat.	See glossary of terms in appendix B.	0505	△
–	–	–	F-B, S-X	Category 01. Protected from sources of heat.	See glossary of terms in appendix B.	0506	△

UN No.	Proper Shipping Name (PSN)	Class or division	Subsidiary risk(s)	Packing group	Special provisions	Limited and excepted quantity provisions		Packing		IBC	
						Limited quantities	Excepted quantities	Instructions	Provisions	Instructions	Provisions
(1)	(2) 3.1.2	(3) 2.0	(4) 2.0	(5) 2.0.1.3	(6) 3.3	(7a) 3.4	(7b) 3.5	(8) 4.1.4	(9) 4.1.4	(10) 4.1.4	(11) 4.1.4
△ 0507	SIGNALS, SMOKE	1.4S	–	–	–	0	E0	P135	–	–	–
△ 0508	1-HYDROXYBENZOTRIAZOLE, ANHYDROUS, dry or wetted with less than 20% water, by mass	1.3C	–	–	–	0	E0	P114 (b)	PP48 PP50	–	–
△ 0509	POWDER, SMOKELESS	1.4C	–	–	–	0	E0	P114(b)	PP48	–	–
△ 1001	ACETYLENE, DISSOLVED	2.1	–	–	–	0	E0	P200	–	–	–
1002	AIR, COMPRESSED	2.2	–	–	–	120 mL	E1	P200	–	–	–
1003	AIR, REFRIGERATED LIQUID	2.2	5.1	–	–	0	E0	P203	–	–	–
1005	AMMONIA, ANHYDROUS	2.3	8	–	23	0	E0	P200	–	–	–
1006	ARGON, COMPRESSED	2.2	–	–	–	120 mL	E1	P200	–	–	–
1008	BORON TRIFLUORIDE	2.3	8	–	–	0	E0	P200	–	–	–
1009	BROMOTRIFLUOROMETHANE (REFRIGERANT GAS R 13B1)	2.2	–	–	–	120 mL	E1	P200	–	–	–
△ 1010	BUTADIENES, STABILIZED or BUTADIENES AND HYDROCARBON MIXTURE, STABILIZED, containing more than 40% butadienes	2.1	–	–	–	0	E0	P200	–	–	–
1011	BUTANE	2.1	–	–	–	0	E0	P200	–	–	–
1012	BUTYLENE	2.1	–	–	–	0	E0	P200	–	–	–
1013	CARBON DIOXIDE	2.2	–	–	–	120 mL	E1	P200	–	–	–
1016	CARBON MONOXIDE, COMPRESSED	2.3	2.1	–	–	0	E0	P200	–	–	–
1017	CHLORINE	2.3	5.1/8 P	–	–	0	E0	P200	–	–	–
1018	CHLORODIFLUOROMETHANE (REFRIGERANT GAS R 22)	2.2	–	–	–	120 mL	E1	P200	–	–	–
1020	CHLOROPENTAFLUORO-ETHANE (REFRIGERANT GAS R 115)	2.2	–	–	–	120 mL	E1	P200	–	–	–

Portable tanks and bulk containers			EmS	Stowage and segregation	Properties and observations	UN No.	
	Tank instructions	Provisions					
(12)	(13) 4.2.5 4.3	(14) 4.2.5	(15) 5.4.3.2 7.3	(16) 7.1 7.2	(17)	(18)	
–	–	–	F-B, S-X	Category 01. Protected from sources of heat.	See glossary of terms in appendix B.	0507	△
–	–	–	F-B, S-Y	Category 04. Protected from sources of heat.	Substance.	0508	△
–	–	–	F-B, S-Y	Category 02. Protected from sources of heat.	See glossary of terms in appendix B.	0509	△
–	–	–	F-D, S-U	Category D. Protected from sources of heat. Clear of living quarters. "Separated from" chlorine.	Flammable gas with slight odour. Explosive limits: 2.1% to 80%. Lighter than air (0.907). Rough handling and exposure to local heating should be avoided, since these conditions may result in delayed explosion. Empty cylinders should be carried with the same precautions as filled cylinders.	1001	△
–	–	–	F-C, S-V	Category A.	Non-flammable gas.	1002	
–	T75	TP5 TP22	F-C, S-W	Category D.	Liquefied, non-flammable gas. Strong oxidizing agent. Mixtures of liquid air with combustible materials or oils may explode. May ignite organic materials.	1003	
–	T50	–	F-C, S-U	Category D. Clear of living quarters. "Separated from" chlorine. "Separated from" acids.	Liquefied, non-flammable, toxic and corrosive gas with a pungent odour. Lighter than air (0.6). Suffocating in low concentrations. Even though this substance has a flammability hazard, it only exhibits such hazard under extreme fire conditions in confined areas. Reacts violently with acids. Highly irritating to skin, eyes and mucous membranes.	1005	
–	–	–	F-C, S-V	Category A.	Inert gas. Heavier than air (1.4).	1006	
–	–	–	F-C, S-U	Category D. Clear of living quarters.	Non-flammable, toxic and corrosive gas. Forms dense white corrosive fumes in moist air. Reacts violently with water, evolving hydrogen fluoride, an irritating and corrosive gas apparent as white fumes. In the presence of moisture, highly corrosive to glass and most metals. Much heavier than air (2.35). Highly irritating to skin, eyes and mucous membranes.	1008	
–	T50	–	F-C, S-V	Category A.	Liquefied, non-flammable gas with a slight odour. Much heavier than air (5.2).	1009	
–	T50	–	F-D, S-U	Category B. Clear of living quarters.	Liquefied, flammable gas with an unpleasant odour. Explosive limits: 2% to 12%. Heavier than air (1.84).	1010	△
–	T50	–	F-D, S-U	Category E. Clear of living quarters.	Flammable hydrocarbon gas. Explosive limits: 1.8% to 8.4%. Heavier than air (2.11).	1011	
–	T50	–	F-D, S-U	Category E. Clear of living quarters.	Flammable hydrocarbon gas. Explosive limits: 1.6% to 10%. Heavier than air (2.0).	1012	
–	–	–	F-C, S-V	Category A.	Liquefied, non-flammable gas. Heavier than air (1.5). Cannot remain in the liquid state above 31°C.	1013	
–	–	–	F-D, S-U	Category D. Clear of living quarters.	Flammable, toxic, odourless gas. Explosive limits: 12% to 75%. Slightly lighter than air (0.97).	1016	
–	T50	TP19	F-C, S-U	Category D. Clear of living quarters. Segregation as for class 5.1 but "separated from" class 7.	Non-flammable, toxic and corrosive yellow gas with a pungent odour. Corrosive to glass and to most metals. Much heavier than air (2.4). Highly irritating to skin, eyes and mucous membranes. Powerful oxidant which may cause fire.	1017	
–	T50	–	F-C, S-V	Category A.	Liquefied, non-flammable gas with a chloroform-like odour. Much heavier than air (3.0).	1018	
–	T50	–	F-C, S-V	Category A.	Liquefied, non-flammable gas. Much heavier than air (5.4).	1020	

3 DGL

UN No.	Proper Shipping Name (PSN)	Class or division	Subsidiary risk(s)	Packing group	Special provisions	Limited and excepted quantity provisions		Packing		IBC	
						Limited quantities	Excepted quantities	Instructions	Provisions	Instructions	Provisions
(1)	(2) 3.1.2	(3) 2.0	(4) 2.0	(5) 2.0.1.3	(6) 3.3	(7a) 3.4	(7b) 3.5	(8) 4.1.4	(9) 4.1.4	(10) 4.1.4	(11) 4.1.4
1021	1-CHLORO-1,2,2,2-TETRA-FLUOROETHANE (REFRIGERANT GAS R 124)	2.2	–	–	–	120 mL	E1	P200	–	–	–
1022	CHLOROTRIFLUOROMETHANE (REFRIGERANT GAS R 13)	2.2	–	–	–	120 mL	E1	P200	–	–	–
1023	COAL GAS, COMPRESSED	2.3	2.1	–	–	0	E0	P200	–	–	–
1026	CYANOGEN	2.3	2.1	–	–	0	E0	P200	–	–	–
1027	CYCLOPROPANE	2.1	–	–	–	0	E0	P200	–	–	–
1028	DICHLORODIFLUOROMETHANE (REFRIGERANT GAS R 12)	2.2	–	–	–	120 mL	E1	P200	–	–	–
1029	DICHLOROFLUOROMETHANE (REFRIGERANT GAS R 21)	2.2	–	–	–	120 mL	E1	P200	–	–	–
1030	1,1-DIFLUOROETHANE (REFRIGERANT GAS R 152a)	2.1	–	–	–	0	E0	P200	–	–	–
1032	DIMETHYLAMINE, ANHYDROUS	2.1	–	–	–	0	E0	P200	–	–	–
1033	DIMETHYL ETHER	2.1	–	–	–	0	E0	P200	–	–	–
1035	ETHANE	2.1	–	–	–	0	E0	P200	–	–	–
1036	ETHYLAMINE	2.1	–	–	912	0	E0	P200	–	–	–
1037	ETHYL CHLORIDE	2.1	–	–	–	0	E0	P200	–	–	–
1038	ETHYLENE, REFRIGERATED LIQUID	2.1	–	–	–	0	E0	P203	–	–	–
1039	ETHYL METHYL ETHER	2.1	–	–	–	0	E0	P200	–	–	–
1040	ETHYLENE OXIDE or ETHYLENE OXIDE WITH NITROGEN up to a total pressure of 1 MPa (10 bar) at 50°C	2.3	2.1	–	342	0	E0	P200	–	–	–
1041	ETHYLENE OXIDE AND CARBON DIOXIDE MIXTURE with more than 9% but not more than 87% ethylene oxide	2.1	–	–	–	0	E0	P200	–	–	–
1043	FERTILIZER AMMONIATING SOLUTION with free ammonia	2.2	–	–	–	120 mL	E1	P200	–	–	–
1044	FIRE EXTINGUISHERS with compressed or liquefied gas	2.2	–	–	225	120 mL	E0	P003	–	–	–
1045	FLUORINE, COMPRESSED	2.3	5.1/8	–	–	0	E0	P200	–	–	–
1046	HELIUM, COMPRESSED	2.2	–	–	–	120 mL	E1	P200	–	–	–

Portable tanks and bulk containers			EmS	Stowage and segregation	Properties and observations	UN No.
	Tank instructions	Provisions				
(12)	(13) 4.2.5 4.3	(14) 4.2.5	(15) 5.4.3.2 7.3	(16) 7.1 7.2	(17)	(18)
–	T50	–	F-C, S-V	Category A.	Liquefied, non-flammable gas. Much heavier than air (4.7).	**1021**
–	–	–	F-C, S-V	Category A.	Liquefied, non-flammable gas. Much heavier than air (3.6). Cannot remain in the liquid state above 29°C.	1022
–	–	–	F-D, S-U	Category D. Clear of living quarters.	Flammable, toxic gas. Explosive limits: 4.5% to 40%. Much lighter than air (0.4 to 0.6).	**1023**
–	–	–	F-D, S-U	Category D. Clear of living quarters.	Liquefied, flammable, toxic gas with a pungent odour. Explosive limits: 6.6% to 43%. Heavier than air (1.9).	1026
–	T50	–	F-D, S-U	Category E. Clear of living quarters.	Flammable hydrocarbon gas. Heavier than air.	**1027**
–	T50	–	F-C, S-V	Category A.	Liquefied, non-flammable gas. Much heavier than air (4.2).	1028
–	T50	–	F-C, S-V	Category A.	Liquefied, non-flammable gas with a chloroform-like odour. Much heavier than air (3.6). Boiling point: 9°C.	**1029**
–	T50	–	F-D, S-U	Category B. Clear of living quarters.	Flammable gas. Explosive limits: 5% to 17%. Much heavier than air (2.3).	1030
–	T50	–	F-D, S-U	Category D. Clear of living quarters.	Liquefied, flammable gas with an ammonia-like odour. Heavier than air (1.6). Boiling point: 7°C. Suffocating in low concentrations.	**1032**
–	T50	–	F-D, S-U	Category B. Clear of living quarters.	Flammable gas with a chloroform-like odour. Heavier than air (1.6).	1033
–	–	–	F-D, S-U	Category E. Clear of living quarters.	Flammable gas. Explosive limits: 3% to 16%. Slightly heavier than air (1.05).	**1035**
–	T50	–	F-D, S-U	Category D. Clear of living quarters.	Liquefied, flammable gas with an ammonia-like odour. Explosive limits: 3.5% to 14%. Heavier than air (1.6). Boiling point: 17°C	1036
–	T50	–	F-D, S-U	Category B. Clear of living quarters.	Liquefied, flammable gas. Explosive limits: 3.5% to 15%. Much heavier than air (2.2). Boiling point: 13°C.	**1037**
–	T75	TP5	F-D, S-U	Category D. Clear of living quarters.	Liquefied, flammable gas. Explosive limits: 3% to 34%. Lighter than air (0.98).	1038
–	–	–	F-D, S-U	Category B. Clear of living quarters.	Liquefied, flammable gas. Explosive limits: 2% to 10%. Much heavier than air (2.1). Boiling point: 11°C.	**1039**
–	T50	TP20 TP90	F-D, S-U	Category D. Clear of living quarters.	Liquefied, flammable, toxic gases with an ether-like odour. Heavier than air (1.5). Boiling point: 11°C.	1040
–	T50	–	F-D, S-U	Category B. Clear of living quarters.	Liquefied, flammable gas with an ether-like odour. Heavier than air (1.5).	**1041**
–	–	–	F-C, S-V	Category E. Clear of living quarters.	Non-flammable aqueous solution of ammonium nitrate, calcium nitrate, urea and their mixtures containing ammonia gas. Emits toxic vapours of ammonia.	1043
–	–	–	F-C, S-V	Category A.	Fire extinguishers, containing compressed or liquefied gases under pressure above 175 kPa for expelling fire-extinguishing contents.	**1044**
–	–	–	F-C, S-W	Category D. Clear of living quarters. Segregation as for class 5.1 but "separated from" class 7.	Non-flammable, toxic and corrosive pale yellowish gas with a pungent odour. Powerful oxidant which may cause fire. Reacts with water or moist air to produce toxic and corrosive fumes. Corrosive to glass and to most metals. Will explode when mixed with hydrogen. Heavier than air (1.3). Highly irritating to skin, eyes and mucous membranes.	1045
–	–	–	F-C, S-V	Category A.	Inert gas. Much lighter than air (0.14).	**1046**

UN No.	Proper Shipping Name (PSN)	Class or division	Subsidiary risk(s)	Packing group	Special provisions	Limited and excepted quantity provisions		Packing		IBC	
						Limited quantities	Excepted quantities	Instructions	Provisions	Instructions	Provisions
(1)	(2) 3.1.2	(3) 2.0	(4) 2.0	(5) 2.0.1.3	(6) 3.3	(7a) 3.4	(7b) 3.5	(8) 4.1.4	(9) 4.1.4	(10) 4.1.4	(11) 4.1.4
1048	HYDROGEN BROMIDE, ANHYDROUS	2.3	8	–	–	0	E0	P200	–	–	–
1049	HYDROGEN, COMPRESSED	2.1	–	–	–	0	E0	P200	–	–	–
1050	HYDROGEN CHLORIDE, ANHYDROUS	2.3	8	–	–	0	E0	P200	–	–	–
1051	HYDROGEN CYANIDE, STABILIZED containing less than 3% water	6.1	3 P	I	–	0	E5	P200	–	–	–
1052	HYDROGEN FLUORIDE, ANHYDROUS	8	6.1	I	–	0	E0	P200	–	–	–
1053	HYDROGEN SULPHIDE	2.3	2.1	–	–	0	E0	P200	–	–	–
1055	ISOBUTYLENE	2.1	–	–	–	0	E0	P200	–	–	–
1056	KRYPTON, COMPRESSED	2.2	–	–	–	120 mL	E1	P200	–	–	–
1057	LIGHTERS or LIGHTER REFILLS containing flammable gas	2.1	–	–	201	0	E0	P002	PP84	–	–
1058	LIQUEFIED GASES non-flammable, charged with nitrogen, carbon dioxide or air	2.2	–	–	–	120 mL	E1	P200	–	–	–
1060	METHYLACETYLENE AND PROPADIENE MIXTURE, STABILIZED	2.1	–	–	–	0	E0	P200	–	–	–
1061	METHYLAMINE, ANHYDROUS	2.1	–	–	–	0	E0	P200	–	–	–
1062	METHYL BROMIDE with not more than 2.0% chloropicrin	2.3	–	–	23	0	E0	P200	–	–	–
1063	METHYL CHLORIDE (REFRIGERANT GAS R 40)	2.1	–	–	–	0	E0	P200	–	–	–
1064	METHYL MERCAPTAN	2.3	2.1 P	–	–	0	E0	P200	–	–	–
1065	NEON, COMPRESSED	2.2	–	–	–	120 mL	E1	P200	–	–	–
1066	NITROGEN, COMPRESSED	2.2	–	–	–	120 mL	E1	P200	–	–	–
1067	DINITROGEN TETROXIDE (NITROGEN DIOXIDE)	2.3	5.1/8	–	–	0	E0	P200	–	–	–
1069	NITROSYL CHLORIDE	2.3	8	–	–	0	E0	P200	–	–	–
1070	NITROUS OXIDE	2.2	5.1	–	–	0	E0	P200	–	–	–

Portable tanks and bulk containers			EmS	Stowage and segregation	Properties and observations	UN No.
	Tank instructions	Provisions				
(12)	(13) 4.2.5 4.3	(14) 4.2.5	(15) 5.4.3.2 7.3	(16) 7.1 7.2	(17)	(18)
–	–	–	F-C, S-U	Category D. Clear of living quarters.	Non-flammable, toxic and corrosive gas with a pungent odour. Highly corrosive in the presence of water. Much heavier than air (3.6). Highly irritating to the skin, eyes and mucous membranes.	1048
–	–	–	F-D, S-U	Category E. Clear of living quarters. "Separated from" chlorine.	Flammable, odourless gas. Explosive limits: 4% to 75%. Much lighter than air (0.07).	1049
–	–	–	F-C, S-U	Category D. Clear of living quarters.	Non-flammable, toxic and corrosive colourless gas with a pungent odour. Highly corrosive in the presence of water. Heavier than air (1.3). Highly irritating to skin, eyes and mucous membranes.	1050
–	–	–	F-E, S-D	Category D. Clear of living quarters.	Very volatile, colourless flammable liquid, evolving extremely toxic flammable vapours. Boiling point: 26°C. Flashpoint: –18°C c.c. Miscible with water. Highly toxic if swallowed, by skin contact or by inhalation.	1051
–	T10	TP2	F-C, S-U	Category D. Clear of living quarters.	Colourless, fuming and highly volatile liquid with an irritating and pungent odour. Highly corrosive to metals and glass in the presence of moisture. Boiling point: 20°C. Toxic if swallowed, by skin contact or by inhalation. Causes severe burns to skin, eyes and mucous membranes.	1052
–	–	–	F-D, S-U	Category D. Clear of living quarters.	Liquefied, flammable, toxic gas with a foul odour. Heavier than air (1.2).	1053
–	T50	–	F-D, S-U	Category E. Clear of living quarters.	Flammable hydrocarbon gas. Explosive limits: 1.8% to 8.8%. May contain propane, cyclopropane, propylene, butane, butylene, etc., in varying proportions. Heavier than air (1.94).	1055
–	–	–	F-C, S-V	Category A.	Inert gas. Much heavier than air (2.9).	1056
–	–	–	F-D, S-U	Category B. Clear of living quarters.	Lighters or lighter refills containing butane or other flammable gas.	1057
–	–	–	F-C, S-V	Category A.	Non-flammable gases or mixtures of such gases which are used for filling receptacles from which the contents are to be dispersed under pressure. Vapour may be heavier than air.	1058
–	T50	–	F-D, S-U	Category B. Clear of living quarters.	Flammable gas. Explosive limits: 3% to 11%. Heavier than air (1.4).	1060
–	T50	–	F-D, S-U	Category B. Clear of living quarters.	Liquefied, flammable gas with an ammonia-like odour. Heavier than air (1.09).	1061
–	T50	–	F-C, S-U	Category D. Clear of living quarters.	Liquefied, toxic gas with a chloroform-like odour. Much heavier than air (3.3). Boiling point: 4.5°C. Even though this substance has a flammability hazard, it only exhibits such hazard under extreme fire conditions in confined areas.	1062
–	T50	–	F-D, S-U	Category D. Clear of living quarters.	Liquefied, flammable gas. Explosive limits: 8% to 20%. Heavier than air (1.8).	1063
–	T50	–	F-D, S-U	Category D. Clear of living quarters.	Liquefied, flammable, toxic gas with a foul odour. Heavier than air (1.7). Boiling point: 6°C.	1064
–	–	–	F-C, S-V	Category A.	Inert gas. Lighter than air (0.7).	1065
–	–	–	F-C, S-V	Category A.	Non-flammable, odourless gas. Lighter than air (0.97).	1066
–	T50	TP21	F-C, S-W	Category D. Clear of living quarters. Segregation as for class 5.1 but "separated from" class 7.	Liquefied, non-flammable, toxic and corrosive gas which gives off brown vapour with a pungent odour. Strong oxidizing agent. Boiling point: 21°C. Highly irritating to skin, eyes and mucous membranes. Toxic by inhalation, with delayed effect, similar to phosgene.	1067
–	–	–	F-C, S-U	Category D. Clear of living quarters.	Non-flammable, toxic yellow gas with an irritating odour. Corrosive to steel. Much heavier than air (2.3). Highly irritating to skin, eyes and mucous membranes.	1069
–	–	–	F-C, S-W	Category A. Clear of living quarters.	Non-flammable gas. Strong oxidizing agent. Heavier than air (1.5).	1070

UN No.	Proper Shipping Name (PSN)	Class or division	Subsidiary risk(s)	Packing group	Special provisions	Limited and excepted quantity provisions		Packing		IBC	
						Limited quantities	Excepted quantities	Instructions	Provisions	Instructions	Provisions
(1)	(2) 3.1.2	(3) 2.0	(4) 2.0	(5) 2.0.1.3	(6) 3.3	(7a) 3.4	(7b) 3.5	(8) 4.1.4	(9) 4.1.4	(10) 4.1.4	(11) 4.1.4
1071	OIL GAS, COMPRESSED	2.3	2.1	–	–	0	E0	P200	–	–	–
1072	OXYGEN, COMPRESSED	2.2	5.1	–	355	0	E0	P200	–	–	–
1073	OXYGEN, REFRIGERATED LIQUID	2.2	5.1	–	–	0	E0	P203	–	–	–
1075	PETROLEUM GASES, LIQUEFIED	2.1	–	–	–	0	E0	P200	–	–	–
1076	PHOSGENE	2.3	8	–	–	0	E0	P200	–	–	–
1077	PROPYLENE	2.1	–	–	–	0	E0	P200	–	–	–
1078	REFRIGERANT GAS, N.O.S.	2.2	–	–	274	120 mL	E1	P200	–	–	–
1079	SULPHUR DIOXIDE	2.3	8	–	–	0	E0	P200	–	–	–
1080	SULPHUR HEXAFLUORIDE	2.2	–	–	–	120 mL	E1	P200	–	–	–
1081	TETRAFLUOROETHYLENE, STABILIZED	2.1	–	–	–	0	E0	P200	–	–	–
1082	TRIFLUOROCHLORO-ETHYLENE, STABILIZED	2.3	2.1	–	–	0	E0	P200	–	–	–
1083	TRIMETHYLAMINE, ANHYDROUS	2.1	–	–	–	0	E0	P200	–	–	–
1085	VINYL BROMIDE, STABILIZED	2.1	–	–	–	0	E0	P200	–	–	–
1086	VINYL CHLORIDE, STABILIZED	2.1	–	–	–	0	E0	P200	–	–	–
1087	VINYL METHYL ETHER, STABILIZED	2.1	–	–	–	0	E0	P200	–	–	–
1088	ACETAL	3	–	II	–	1 L	E2	P001	–	IBC02	–
1089	ACETALDEHYDE	3	–	I	–	0	E3	P001	–	–	–
△ 1090	ACETONE	3	–	II	–	1 L	E2	P001	–	IBC02	–
1091	ACETONE OILS	3	–	II	–	1 L	E2	P001	–	IBC02	–
1092	ACROLEIN, STABILIZED	6.1	3 P	I	354	0	E0	P601	–	–	–

Portable tanks and bulk containers			EmS	Stowage and segregation	Properties and observations	UN No.
	Tank instructions	Provisions				
(12)	(13) 4.2.5 4.3	(14) 4.2.5	(15) 5.4.3.2 7.3	(16) 7.1 7.2	(17)	(18)
–	–	–	F-D, S-U	Category D. Clear of living quarters.	Flammable, toxic gas. A mixture of hydrocarbons and carbon monoxide.	1071
–	–	–	F-C, S-W	Category A.	Non-flammable, odourless gas. Strong oxidizing agent. Heavier than air (1.1).	1072
–	T75	TP5 TP22	F-C, S-W	Category D.	Liquefied, non-flammable gas. Strong oxidizing agent. Mixtures of liquid oxygen with acetylene or oils may explode.	1073
–	T50	–	F-D, S-U	Category E. Clear of living quarters.	Flammable hydrocarbon gases or mixtures obtained from natural gas or by distillation of mineral oils or coal, etc. May contain propane, cyclopropane, propylene, butane, butylene, etc., in varying proportions. Heavier than air.	1075
–	–	–	F-C, S-U	Category D. Clear of living quarters.	Liquefied, non-flammable, toxic and corrosive gas with a foul odour. Corrosive in the presence of water. Much heavier than air (3.5). Boiling point: 8°C. Highly irritating to skin, eyes and mucous membranes. This gas is particularly dangerous in that it may be inhaled without immediate effect but can cause severe damage and death after a few hours' delay.	1076
–	T50	–	F-D, S-U	Category E. Clear of living quarters.	Flammable hydrocarbon gas. Explosive limits: 2% to 11.1%. Heavier than air (1.5).	1077
–	T50	–	F-C, S-V	Category A.	Different chlorofluorohydrocarbons or other non-flammable, non-toxic gases considered as refrigerant agents.	1078
–	T50	TP19	F-C, S-U	Category D. Clear of living quarters.	Non-flammable, toxic and corrosive gas with a pungent odour. Much heavier than air (2.3). Highly irritating to skin, eyes and mucous membranes.	1079
–	–	–	F-C, S-V	Category A.	Liquefied, non-flammable, odourless gas. Much heavier than air (5.1).	1080
–	–	–	F-D, S-U	Category E. Clear of living quarters.	Liquefied, flammable gas. Explosive limits: 11% to 60%. Much heavier than air (3.5). Irritating to skin, eyes and mucous membranes.	1081
–	T50	–	F-D, S-U	Category D. Clear of living quarters.	Flammable, toxic, odourless gas. Explosive limits: 8.4% to 38.7%. Much heavier than air (4.0).	1082
–	T50	–	F-D, S-U	Category B. Clear of living quarters.	Liquefied, flammable gas with a fishy odour. Explosive limits: 2% to 12%. Much heavier than air (2.1). Boiling point: 3°C.	1083
–	T50	–	F-D, S-U	Category B. Clear of living quarters.	Liquefied, flammable gas. Much heavier than air (3.7). Boiling point: 16°C.	1085
–	T50	–	F-D, S-U	Category B. Clear of living quarters.	Liquefied, flammable gas. Explosive limits: 4% to 31%. Much heavier than air (2.2).	1086
–	T50	–	F-D, S-U	Category B. Clear of living quarters.	Liquefied, flammable gas. Explosive limits: 2.6% to 39%. Heavier than air (2.0). Boiling point: 6°C.	1087
–	T4	TP1	F-E, S-D	Category E.	Colourless, volatile liquid with an agreeable odour. Flashpoint: below –18°C c.c. Explosive limits: 1.6% to 10.4%. Miscible with water.	1088
–	T11	TP2 TP7	F-E, S-D	Category E.	Colourless liquid with a pungent, fruity odour. Flashpoint: –27°C c.c. Explosive limits: 4% to 57%. Boiling point: 21°C. Miscible with water. Harmful if swallowed or by inhalation.	1089
–	T4	TP1	F-E, S-D	Category E.	Colourless, clear liquid, with a characteristic mint-like odour. Flashpoint: –20°C to –18°C c.c. Explosive limits: 2.5% to 13%. Miscible with water.	1090 △
–	T4	TP1 TP8	F-E, S-D	Category B.	Light yellow to brownish, oily liquids. Flashpoint: –4°C to 8°C c.c. Immiscible with water.	1091
–	T22	TP2 TP7 TP13 TP35	F-E, S-D	Category D. Clear of living quarters.	Colourless or yellow liquid with a most irritating odour. Flashpoint: –26°C c.c. Explosive limits: 2.8% to 31%. Boiling point: 52°C. Miscible with water. Highly toxic if swallowed, by skin contact or by inhalation.	1092

	UN No.	Proper Shipping Name (PSN)	Class or division	Subsidiary risk(s)	Packing group	Special provisions	Limited and excepted quantity provisions		Packing		IBC	
							Limited quantities	Excepted quantities	Instructions	Provisions	Instructions	Provisions
	(1)	(2) 3.1.2	(3) 2.0	(4) 2.0	(5) 2.0.1.3	(6) 3.3	(7a) 3.4	(7b) 3.5	(8) 4.1.4	(9) 4.1.4	(10) 4.1.4	(11) 4.1.4
	1093	ACRYLONITRILE, STABILIZED	3	6.1	I	–	0	E0	P001	–	–	–
	1098	ALLYL ALCOHOL	6.1	3	I	354	0	E0	P602	–	–	–
	1099	ALLYL BROMIDE	3	6.1 P	I	–	0	E0	P001	–	–	–
	1100	ALLYL CHLORIDE	3	6.1	I	–	0	E0	P001	–	–	–
	1104	AMYL ACETATES	3	–	III	–	5 L	E1	P001 LP01	–	IBC03	–
	1105	PENTANOLS	3	–	II	–	1 L	E2	P001	–	IBC02	–
	1105	PENTANOLS	3	–	III	223	5 L	E1	P001 LP01	–	IBC03	–
△	1106	AMYLAMINE	3	8	II	–	1 L	E2	P001	–	IBC02	–
△	1106	AMYLAMINE	3	8	III	223	5 L	E1	P001	–	IBC03	–
△	1107	AMYL CHLORIDE	3	–	II	–	1 L	E2	P001	–	IBC02	–
	1108	1-PENTENE (*n*-AMYLENE)	3	–	I	–	0	E3	P001	–	–	–
	1109	AMYL FORMATES	3	–	III	–	5 L	E1	P001 LP01	–	IBC03	–
	1110	*n*-AMYL METHYL KETONE	3	–	III	–	5 L	E1	P001 LP01	–	IBC03	–
△	1111	AMYL MERCAPTAN	3	–	II	–	1 L	E2	P001	–	IBC02	–
△	1112	AMYL NITRATE	3	–	III	–	5 L	E1	P001 LP01	–	IBC03	–

Portable tanks and bulk containers			EmS	Stowage and segregation	Properties and observations	UN No.	
	Tank instructions	Provisions					
(12)	(13) 4.2.5 4.3	(14) 4.2.5	(15) 5.4.3.2 7.3	(16) 7.1 7.2	(17)	(18)	
–	T14	TP2 TP13	F-E, S-D	Category E. Clear of living quarters.	Colourless, mobile liquid with a mild pungent odour. Flashpoint: –5°C c.c. Explosive limits: 3% to 17%. Partially miscible with water. Toxic if swallowed, by skin contact or by inhalation. Practice has shown that this substance may leak from packagings that ordinarily are leakproof to other chemicals.	1093	
–	T20	TP2 TP13 TP35	F-E, S-D	Category D. Clear of living quarters.	Colourless liquid with a pungent mustard-like odour. Flashpoint: 21°C c.c. Explosive limits: 2.5% to 18%. Miscible with water. Highly toxic if swallowed, by skin contact or by inhalation.	1098	
–	T14	TP2 TP13	F-E, S-D	Category B. Clear of living quarters.	Colourless to light yellow liquid with an irritating odour. Flashpoint: –1°C c.c. Explosive limits: 4.4% to 7.3%. Immiscible with water. Highly toxic if swallowed, by skin contact or by inhalation.	1099	
–	T14	TP2 TP13	F-E, S-D	Category E. Clear of living quarters.	Colourless liquid with an unpleasant pungent odour. Flashpoint: –29°C c.c. Explosive limits: 3.3% to 11.1%. Boiling point: 44°C. Immiscible with water. Toxic if swallowed, by skin contact or by inhalation.	1100	
–	T2	TP1	F-E, S-D	Category A.	Colourless liquids with a pear- or banana-like odour. *normal*-AMYL ACETATE: flashpoint 25°C c.c. *secondary*-AMYL ACETATE: flashpoint 32°C c.c. Immiscible with water.	1104	
–	T4	TP1 TP29	F-E, S-D	Category B.	Colourless liquids with a strong odour. Immiscible with water. *tertiary*-AMYL ALCOHOL: flashpoint 19°C to 21°C c.c.	1105	
–	T2	TP1	F-E, S-D	Category A.	See entry above. Explosive limits: 1.2% to 10.5%.	1105	
–	T7	TP1	F-E, S-C	Category B.	Colourless, clear liquids. Explosive limits: 2.2% to 22%. *normal*-AMYLAMINE (1-PENTYLAMINE): flashpoint 4°C c.c. *tertiary*-AMYLAMINE (3-PENTYLAMINE): flashpoint 2°C c.c. Miscible with water. Harmful by inhalation. Cause burns to skin, eyes and mucous membranes.	1106	△
–	T4	TP1	F-E, S-C	Category A.	See entry above. However, irritating to skin, eyes and mucous membranes.	1106	△
–	T4	TP1	F-E, S-D	Category B.	Colourless or light brown liquids with an aromatic odour. *n*-AMYL CHLORIDE: flashpoint 11°C. Explosive limits: *normal*-AMYL CHLORIDE 1.4% to 8.6%. Immiscible with water.	1107	△
–	T11	TP2	F-E, S-D	Category E.	Colourless, volatile liquid with a disagreeable odour. Flashpoint: –20°C c.c. Explosive limits: 1.4% to 8.7%. Boiling point: 30°C Immiscible with water. Irritating to skin, eyes and mucous membranes. Narcotic in high concentrations.	1108	
–	T2	TP1	F-E, S-D	Category A.	Colourless liquids with a pleasant odour. *normal*-AMYL FORMATE: flashpoint 27°C c.c. ISOAMYL FORMATE: flashpoint 26°C c.c. Explosive limits: 1.7% to 10%. Immiscible with water.	1109	
–	T2	TP1	F-E, S-D	Category A.	Colourless liquid. Flashpoint: 49°C c.c. Immiscible with water.	1110	
–	T4	TP1	F-E, S-D	Category B. Segregation from foodstuffs as in 7.3.4.2, 7.6.3.1.2 or 7.7.3.6. "Separated from" odour-absorbing cargoes.	Colourless to yellow liquids with an extremely disagreeable garlic-like odour. *tertiary*-AMYL MERCAPTAN: flashpoint –7°C c.c. *normal*-AMYL MERCAPTAN: flashpoint 19°C c.c. ISOAMYL MERCAPTAN: flashpoint 18°C c.c. Immiscible with water. These substances may leak from packagings that ordinarily are leakproof to other chemicals.	1111	△
–	T2	TP1	F-E, S-D	Category A. Clear of living quarters.	Colourless liquids with an ether-like odour. *normal*-AMYL NITRATE: flashpoint 48°C c.c. ISOAMYL NITRATE: flashpoint 52°C c.c. Immiscible with water. Harmful by inhalation.	1112	△

UN No.	Proper Shipping Name (PSN)	Class or division	Subsidiary risk(s)	Packing group	Special provisions	Limited and excepted quantity provisions		Packing		IBC	
						Limited quantities	Excepted quantities	Instructions	Provisions	Instructions	Provisions
(1)	(2) 3.1.2	(3) 2.0	(4) 2.0	(5) 2.0.1.3	(6) 3.3	(7a) 3.4	(7b) 3.5	(8) 4.1.4	(9) 4.1.4	(10) 4.1.4	(11) 4.1.4
1113	AMYL NITRITE	3	–	II	–	1 L	E2	P001	–	IBC02	–
1114	BENZENE	3	–	II	–	1 L	E2	P001	–	IBC02	–
1120	BUTANOLS	3	–	II	–	1 L	E2	P001	–	IBC02	–
1120	BUTANOLS	3	–	III	223	5 L	E1	P001 LP01	–	IBC03	–
1123	BUTYL ACETATES	3	–	II	–	1 L	E2	P001	–	IBC02	–
1123	BUTYL ACETATES	3	–	III	223	5 L	E1	P001 LP01	–	IBC03	–
1125	*n*-BUTYLAMINE	3	8	II	–	1 L	E2	P001	–	IBC02	–
1126	1-BROMOBUTANE	3	–	II	–	1 L	E2	P001	–	IBC02	–
1127	CHLOROBUTANES	3	–	II	–	1 L	E2	P001	–	IBC02	–
1128	*n*-BUTYL FORMATE	3	–	II	–	1 L	E2	P001	–	IBC02	–
1129	BUTYRALDEHYDE	3	–	II	–	1 L	E2	P001	–	IBC02	–
1130	CAMPHOR OIL	3	–	III	–	5 L	E1	P001 LP01	–	IBC03	–
1131	CARBON DISULPHIDE	3	6.1	I	–	0	E0	P001	PP31	–	–
1133	ADHESIVES containing flammable liquid	3	–	I	–	500 mL	E3	P001	–	–	–
1133	ADHESIVES containing flammable liquid	3	–	II	–	5 L	E2	P001	PP1	IBC02	–
1133	ADHESIVES containing flammable liquid	3	–	III	223 955	5 L	E1	P001 LP01	PP1	IBC03	–
1134	CHLOROBENZENE	3	–	III	–	5 L	E1	P001 LP01	–	IBC03	–

Portable tanks and bulk containers			EmS	Stowage and segregation	Properties and observations	UN No.
	Tank instructions	Provisions				
(12)	(13) 4.2.5 4.3	(14) 4.2.5	(15) 5.4.3.2 7.3	(16) 7.1 7.2	(17)	(18)
–	T4	TP1	F-E, S-D	Category E. Clear of living quarters.	Yellowish, transparent, volatile liquid with a fragrant fruity odour. Flashpoint of the pure ISOAMYL NITRITE: –20°C c.c. Flashpoint of pure *normal*-AMYL NITRITE: 10°C c.c. Decomposes on exposure to air, light or water, evolving toxic nitrous fumes which are orange in colour. Immiscible with water. Harmful by inhalation.	**1113**
–	T4	TP1	F-E, S-D	Category B. Clear of living quarters.	Colourless liquid with a characteristic odour. Flashpoint: –11°C c.c. Explosive limits: 1.4% to 8%. Freezing point 5°C; flashes below its freezing point. Immiscible with water. Narcotic. Exposure to this substance may produce serious chronic effects of a toxic nature.	1114
–	T4	TP1 TP29	F-E, S-D	Category B.	Colourless liquids with a disagreeable odour. Explosive limits: *normal*-BUTANOL 1.4% to 11.2%. *secondary*-BUTANOL 1.7% to 9.8%. *tertiary*-BUTANOL 2.4% to 8%. *tertiary*-BUTANOL solidifies at about 25°C. *normal*-BUTANOL is immiscible with water. *secondary*-BUTANOL is immiscible with water. *tertiary*-BUTANOL is miscible with water. Irritating to skin, eyes and mucous membranes.	**1120**
–	T2	TP1	F-E, S-D	Category A.	See entry above.	1120
–	T4	TP1	F-E, S-D	Category B.	Colourless liquids with a pineapple-like odour. Immiscible with water. *normal*-BUTYL ACETATE: flashpoint 27°C c.c. Explosive limits: 1.5% to 15%.	**1123**
–	T2	TP1	F-E, S-D	Category A.	See entry above.	1123
–	T7	TP1	F-E, S-C	Category B. Clear of living quarters.	Flashpoint: –9°C c.c. Explosive limits: 1.7% to 10%. Colourless, volatile liquid with an ammonia-like odour. Miscible with water. Causes burns to skin, eyes and mucous membranes.	**1125**
–	T4	TP1	F-E, S-D	Category B. Clear of living quarters.	Colourless to pale straw-coloured, clear liquid. Flashpoint: 13°C c.c. Explosive limits: 2.6% to 6.6%. Immiscible with water. Narcotic.	1126
–	T4	TP1	F-E, S-D	Category B.	Colourless liquids. *tertiary*-BUTYL CHLORIDE: flashpoint –30°C c.c., boiling point 51°C. Immiscible with water.	**1127**
–	T4	TP1	F-E, S-D	Category B.	Colourless liquid. Flashpoint: 18°C c.c. Explosive limits: 1.6% to 8.3%. Immiscible with water.	1128
–	T4	TP1	F-E, S-D	Category B.	Colourless liquid with a characteristic pungent odour. Flashpoint: –7°C c.c. Explosive limits: 1.4% to 12.5%. Immiscible with water.	**1129**
–	T2	TP1	F-E, S-E	Category A.	Colourless oil with a characteristic odour. Flashpoint: 47°C c.c. Immiscible with water.	1130
–	T14	TP2 TP7 TP13	F-E, S-D	Category D. Clear of living quarters. "Separated longitudinally by an intervening complete compartment or hold from" class 1.	Colourless or faintly yellow, clear liquid, almost odourless when pure; the commercial substance has a strong disagreeable odour. Flashpoint: –30°C c.c. Explosive limits: 1% to 60%. Boiling point: 46°C. Ignition temperature: 100°C. Immiscible with water. Vapours are heavier than air, will travel a considerable distance to a source of ignition and will flash back. Vapours may be ignited by contact with an ordinary light bulb or a warm steam pipe. Toxic if swallowed, by skin contact or by inhalation.	**1131**
–	T11	TP1 TP8 TP27	F-E, S-D	Category E.	Adhesives are solutions of gums, resins, etc., usually volatile due to the solvents. Miscibility with water depends upon their composition.	1133
–	T4	TP1 TP8	F-E, S-D	Category B.	See entry above.	**1133**
–	T2	TP1	F-E, S-D	Category A.	See entry above.	1133
–	T2	TP1	F-E, S-D	Category A.	Colourless liquid with an almond-like odour. Flashpoint: 29°C c.c. Explosive limits: 1.3% to 11%. Immiscible with water.	**1134**

UN No.	Proper Shipping Name (PSN)	Class or division	Subsidiary risk(s)	Packing group	Special provisions	Limited and excepted quantity provisions		Packing		IBC	
						Limited quantities	Excepted quantities	Instructions	Provisions	Instructions	Provisions
(1)	(2) 3.1.2	(3) 2.0	(4) 2.0	(5) 2.0.1.3	(6) 3.3	(7a) 3.4	(7b) 3.5	(8) 4.1.4	(9) 4.1.4	(10) 4.1.4	(11) 4.1.4
1135	ETHYLENE CHLOROHYDRIN	6.1	3	I	354	0	E0	P602	–	–	–
1136	COAL TAR DISTILLATES, FLAMMABLE	3	–	II	–	1 L	E2	P001	–	IBC02	–
1136	COAL TAR DISTILLATES, FLAMMABLE	3	–	III	223 955	5 L	E1	P001 LP01	–	IBC03	–
△ 1139	COATING SOLUTION (includes surface treatments or coatings used for industrial or other purposes such as vehicle under-coating, drum or barrel lining)	3	–	I	–	500 mL	E3	P001	–	–	–
△ 1139	COATING SOLUTION (includes surface treatments or coatings used for industrial or other purposes such as vehicle under-coating, drum or barrel lining)	3	–	II	–	5 L	E2	P001	–	IBC02	–
△ 1139	COATING SOLUTION (includes surface treatments or coatings used for industrial or other purposes such as vehicle under-coating, drum or barrel lining)	3	–	III	955	5 L	E1	P001 LP01	–	IBC03	–
1143	CROTONALDEHYDE or CROTONALDEHYDE, STABILIZED	6.1	3 P	I	324 354	0	E0	P602	–	–	–
1144	CROTONYLENE	3	–	I	–	0	E3	P001	–	–	–
1145	CYCLOHEXANE	3	–	II	–	1 L	E2	P001	–	IBC02	–
1146	CYCLOPENTANE	3	–	II	–	1 L	E2	P001	–	IBC02	–
△ 1147	DECAHYDRONAPHTHALENE	3	–	III	–	5 L	E1	P001 LP01	–	IBC03	–
1148	DIACETONE ALCOHOL	3	–	II	–	1 L	E2	P001	–	IBC02	–
1148	DIACETONE ALCOHOL	3	–	III	223	5 L	E1	P001 LP01	–	IBC03	–
1149	DIBUTYL ETHERS	3	–	III	–	5 L	E1	P001	–	IBC03	–
1150	1,2-DICHLOROETHYLENE	3	–	II	–	1 L	E2	P001	–	IBC02	–
1152	DICHLOROPENTANES	3	–	III	–	5 L	E1	P001 LP01	–	IBC03	–
1153	ETHYLENE GLYCOL DIETHYL ETHER	3	–	II	–	1 L	E2	P001	–	IBC02	–
1153	ETHYLENE GLYCOL DIETHYL ETHER	3	–	III	–	5 L	E1	P001 LP01	–	IBC03	–

Portable tanks and bulk containers			EmS	Stowage and segregation	Properties and observations	UN No.	
	Tank instructions	Provisions					
(12)	(13) 4.2.5 4.3	(14) 4.2.5	(15) 5.4.3.2 7.3	(16) 7.1 7.2	(17)	(18)	
–	T20	TP2 TP13 TP37	F-E, S-D	Category D. Clear of living quarters.	Colourless flammable liquid with a faint, ethereal odour. Flashpoint: 60°C o c. Explosive limits: 4.9% to 15.9%. Miscible with water. When involved in a fire, evolves extremely toxic (phosgene) and corrosive (hydrogen chloride) fumes. Highly toxic if swallowed, by skin contact or by inhalation.	**1135**	
–	T4	TP1	F-E, <u>S-E</u>	Category B.	Immiscible with water. May form extremely sensitive compounds with heavy metals or their salts.	**1136**	
–	T4	TP1 TP29	F-E, <u>S-E</u>	Category A.	See entry above.	**1136**	
–	T11	TP1 TP8 TP27	F-E, <u>S-E</u>	Category E.	Miscibility with water depends upon the composition.	**1139**	△
–	T4	TP1 TP8	F-E, <u>S-E</u>	Category B.	See entry above.	**1139**	△
–	T2	TP1	F-E, <u>S-E</u>	Category A.	See entry above.	**1139**	△
–	T20	TP2 TP13 TP35	F-E, S-D	Category D. Clear of living quarters.	Colourless, mobile liquid with a pungent odour. Turns to pale yellow in contact with light and air. Miscible with water. Flashpoint: 13°C c.c. Highly toxic if swallowed, by skin contact or by inhalation. May cause lung damage.	**1143**	
–	T11	TP2	F-E, S-D	Category E.	Colourless liquid. Flashpoint: –53°C c.c. Explosive limits: 1.4% to .... Boiling point: 27°C. Immiscible with water	**1144**	
–	T4	TP1	F-E, S-D	Category E.	Colourless, mobile liquid with a sweet aromatic odour. Flashpoint: –18°C c.c. Explosive limits: 1.2% to 8.4%. Immiscible with water. Slightly irritating to skin, eyes and mucous membranes. Narcotic in high concentrations.	**1145**	
–	T7	TP1	F-E, S-D	Category E.	Colourless liquid with a pungent odour. Flashpoint: below –18°C c.c. Explosive limits: 1.4% to 8%. Boiling point: 49°C. Immiscible with water. Irritating to skin, eyes and mucous membranes. Narcotic in high concentrations.	**1146**	
–	T2	TP1	F-E, S-D	Category A.	Colourless liquids with an aromatic odour. Flashpoint: 52°C to 57°C c.c. Explosive limits: 0.7% to 4.9%. Immiscible with water. Harmful by inhalation.	**1147**	△
–	T4	TP1	F-E, S-D	Category B.	Colourless liquid. Explosive limits: 1.4% to 8%. Miscible with water.	**1148**	
–	T2	TP1	F-E, S-D	Category A.	See entry above.	**1148**	
–	T2	TP1	F-E, S-D	Category A.	Colourless liquids with a mild ether-like odour. Explosive limits: 0.9% to 8.5%. Immiscible with water. *normal*-DIBUTYL ETHER: flashpoint 25°C c.c.	**1149**	
–	T7	TP2	F-E, S-D	Category B.	Colourless liquid with a chloroform-like odour. Flashpoint: 6°C c.c. Explosive limits: 5.6% to 16%. Immiscible with water. Boiling range: 48°C to 61°C.	**1150**	
–	T2	TP1	F-E, S-D	Category A.	Light yellow liquids. 1,5-DICHLOROPENTANE: flashpoint 26°C c.c. Immiscible with water.	**1152**	
–	T4	TP1	F-E, S-D	Category A.	Colourless liquid with an ether-like odour. Flashpoint: 35°C c.c. Immiscible with water.	**1153**	
–	T2	TP1	F-E, S-D	Category A.	See entry above.	**1153**	

UN No.	Proper Shipping Name (PSN)	Class or division	Subsidiary risk(s)	Packing group	Special provisions	Limited and excepted quantity provisions		Packing		IBC	
						Limited quantities	Excepted quantities	Instructions	Provisions	Instructions	Provisions
(1)	(2) 3.1.2	(3) 2.0	(4) 2.0	(5) 2.0.1.3	(6) 3.3	(7a) 3.4	(7b) 3.5	(8) 4.1.4	(9) 4.1.4	(10) 4.1.4	(11) 4.1.4
1154	DIETHYLAMINE	3	8	II	–	1 L	E2	P001	–	IBC02	–
1155	DIETHYL ETHER (ETHYL ETHER)	3	–	I	–	0	E3	P001	–	–	–
1156	DIETHYL KETONE	3	–	II	–	1 L	E2	P001	–	IBC02	–
1157	DIISOBUTYL KETONE	3	–	III	–	5 L	E1	P001 LP01	–	IBC03	–
1158	DIISOPROPYLAMINE	3	8	II	–	1 L	E2	P001	–	IBC02	–
1159	DIISOPROPYL ETHER	3	–	II	–	1 L	E2	P001	–	IBC02	–
1160	DIMETHYLAMINE, AQUEOUS SOLUTION	3	8	II	–	1 L	E2	P001	–	IBC02	–
1161	DIMETHYL CARBONATE	3	–	II	–	1 L	E2	P001	–	IBC02	–
△ 1162	DIMETHYLDICHLOROSILANE	3	8	II	–	0	E0	P010	–	–	–
1163	DIMETHYLHYDRAZINE, UNSYMMETRICAL	6.1	3/8 P	I	354	0	E0	P602	–	–	–
1164	DIMETHYL SULPHIDE	3	–	II	–	1 L	E2	P001	–	IBC02	B8
1165	DIOXANE	3	–	II	–	1 L	E2	P001	–	IBC02	–
1166	DIOXOLANE	3	–	II	–	1 L	E2	P001	–	IBC02	–
1167	DIVINYL ETHER, STABILIZED	3	–	I	–	0	E3	P001	–	–	–
1169	EXTRACTS, AROMATIC, LIQUID	3	–	II	–	5 L	E2	P001	–	IBC02	–

Portable tanks and bulk containers			EmS	Stowage and segregation	Properties and observations	UN No.
	Tank instructions	Provisions				
(12)	(13) 4.2.5 4.3	(14) 4.2.5	(15) 5.4.3.2 7.3	(16) 7.1 7.2	(17)	(18)
–	T7	TP1	F-E, S-C	Category E. Clear of living quarters.	Colourless liquid with an ammonia-like odour. Flashpoint: −39°C c.c. Explosive limits: 1.7% to 10.1%. Boiling point: 55°C. Miscible with water. Harmful if swallowed. Causes burns to skin, eyes and mucous membranes. Higher concentrations cause dangerous lung irritation.	1154
–	T11	TP2	F-E, S-D	Category E. Clear of living quarters.	Colourless, volatile and mobile liquid with a pleasant aromatic odour. Flashpoint: −40°C c.c. Explosive limits: 1.7% to 48%. Boiling point: 34°C. Immiscible with water. In the presence of oxygen or on long standing or exposure to sunlight, unstable peroxides sometimes form; these may explode spontaneously or when heated. Strongly narcotic. Readily ignited by static electricity.	1155
–	T4	TP1	F-E, S-D	Category B.	Colourless, mobile liquid. Flashpoint: 13°C c.c. Explosive limits: 1.6% to.... Immiscible with water.	1156
–	T2	TP1	F-E, S-D	Category A.	Colourless liquid. Flashpoint: 49°C c.c. Explosive limits: 0.8% to 7.1%. Immiscible with water.	1157
–	T7	TP1	F-E, S-C	Category B.	Colourless, volatile liquid with a fishy odour. Flashpoint: −7°C c.c. Explosive limits: 1.1% to 7.1%. Partially miscible with water. Harmful by inhalation. Causes burns to skin, eyes and mucous membranes.	1158
–	T4	TP1	F-E, S-D	Category E. Clear of living quarters.	Colourless liquid with an ether-like odour. Flashpoint: −29°C c.c. Explosive limits: 1.1% to 21%. Immiscible with water. In the presence of oxygen or on long standing or exposure to sunlight, unstable peroxides sometimes form; these may explode spontaneously or when heated. Strongly narcotic. Readily ignited by static electricity.	1159
–	T7	TP1	F-E, S-C	Category B. "Separated from" acids.	Aqueous solution of a flammable gas with an ammonia-like odour. Flashpoint for 60% solution in water: −32°C c.c. Explosive limits: 2.8% to 14.4%. Boiling point for 60% solution in water: 36°C. Flashpoint for 25% solution in water: 0°C c.c. Miscible with water. Harmful by inhalation. Causes burns to skin, eyes and mucous membranes. Reacts violently with acids.	1160
–	T4	TP1	F-E, S-D	Category B.	Colourless liquid. Immiscible with water. Flashpoint: 18°C c.c.	1161
–	T10	TP2 TP7 TP13	F-E, S-C	Category B. Clear of living quarters.	Colourless liquid with a pungent odour. Flashpoint: −9°C c.c. Explosive limits: 1.4% to 9.5%. Immiscible with water. Reacts with water to form a complex mixture of dimethylsiloxanes and evolves hydrogen chloride, a toxic and corrosive gas. Harmful by inhalation. Causes burns to skin, eyes and mucous membranes.	1162
–	T20	TP2 TP13 TP35	F-E, S-C	Category D. Clear of living quarters. Segregation as for class 3 but "away from" class 4.1 and class 8. "Separated from" acids.	Colourless liquid with an ammonia-like odour. Flashpoint: −18°C c.c. Explosive limits: 2% to 95%. Miscible with water, generating heat. Reacts violently with acids. Highly toxic if swallowed, by skin contact or by inhalation. Causes burns to skin, eyes and mucous membranes. May react dangerously with oxidizing substances.	1163
–	T7	TP2	F-E, S-D	Category E. Clear of living quarters.	Colourless liquid with a disagreeable odour. Flashpoint: −37°C c.c. Explosive limits: 2.2% to 19.7%. Boiling point: 37°C. Immiscible with water. When involved in a fire, evolves toxic gases. Narcotic in high concentrations.	1164
–	T4	TP1	F-E, S-D	Category B.	Colourless liquid with an ether-like odour. Flashpoint: 12°C c.c. Explosive limits: 2% to 22%. Miscible with water. Harmful by inhalation.	1165
–	T4	TP1	F-E, S-D	Category B. Clear of living quarters.	Colourless liquid. Flashpoint: 2°C c.c. Miscible with water. Harmful by inhalation.	1166
–	T11	TP2	F-E, S-D	Category E. Clear of living quarters.	Colourless, clear liquid with a characteristic odour. Flashpoint: −30°C c.c. Explosive limits: 1.7% to 27%. Boiling point: 30°C. Immiscible with water. In the presence of oxygen or on long standing or exposure to sunlight, unstable peroxides sometimes form; these may explode spontaneously or when heated. Strongly narcotic. Readily ignited by static electricity.	1167
–	T4	TP1 TP8	F-E, S-D	Category B.	Usually consist of alcoholic solutions. Miscibility with water depends upon the composition.	1169

UN No.	Proper Shipping Name (PSN)	Class or division	Subsidiary risk(s)	Packing group	Special provisions	Limited and excepted quantity provisions		Packing		IBC	
						Limited quantities	Excepted quantities	Instructions	Provisions	Instructions	Provisions
(1)	(2) 3.1.2	(3) 2.0	(4) 2.0	(5) 2.0.1.3	(6) 3.3	(7a) 3.4	(7b) 3.5	(8) 4.1.4	(9) 4.1.4	(10) 4.1.4	(11) 4.1.4
1169	EXTRACTS, AROMATIC, LIQUID	3	–	III	223 955	5 L	E1	P001 LP01	–	IBC03	–
1170	ETHANOL (ETHYL ALCOHOL) or ETHANOL SOLUTION (ETHYL ALCOHOL SOLUTION)	3	–	II	144	1 L	E2	P001	–	IBC02	–
1170	ETHANOL (ETHYL ALCOHOL) or ETHANOL SOLUTION (ETHYL ALCOHOL SOLUTION)	3	–	III	144 223	5 L	E1	P001 LP01	–	IBC03	–
1171	ETHYLENE GLYCOL MONOETHYL ETHER	3	–	III	–	5 L	E1	P001 LP01	–	IBC03	–
1172	ETHYLENE GLYCOL MONOETHYL ETHER ACETATE	3	–	III	–	5 L	E1	P001 LP01	–	IBC03	–
1173	ETHYL ACETATE	3	–	II	–	1 L	E2	P001	–	IBC02	–
1175	ETHYLBENZENE	3	–	II	–	1 L	E2	P001	–	IBC02	–
1176	ETHYL BORATE	3	–	II	–	1 L	E2	P001	–	IBC02	–
1177	2-ETHYLBUTYL ACETATE	3	–	III	–	5 L	E1	P001 LP01	–	IBC03	–
1178	2-ETHYLBUTYRALDEHYDE	3	–	II	–	1 L	E2	P001	–	IBC02	–
1179	ETHYL BUTYL ETHER	3	–	II	–	1 L	E2	P001	–	IBC02	–
1180	ETHYL BUTYRATE	3	–	III	–	5 L	E1	P001 LP01	–	IBC03	–
1181	ETHYL CHLOROACETATE	6.1	3	II	–	100 mL	E4	P001	–	IBC02	–
1182	ETHYL CHLOROFORMATE	6.1	3/8	I	354	0	E0	P602	–	–	–
1183	ETHYLDICHLOROSILANE	4.3	3/8	I	–	0	E0	P401	PP31	–	–
1184	ETHYLENE DICHLORIDE	3	6.1	II	–	1 L	E2	P001	–	IBC02	–
1185	ETHYLENEIMINE, STABILIZED	6.1	3	I	354	0	E0	P601	–	–	–
1188	ETHYLENE GLYCOL MONOMETHYL ETHER	3	–	III	–	5 L	E1	P001 LP01	–	IBC03	–
1189	ETHYLENE GLYCOL MONOMETHYL ETHER ACETATE	3	–	III	–	5 L	E1	P001 LP01	–	IBC03	–
1190	ETHYL FORMATE	3	–	II	–	1 L	E2	P001	–	IBC02	–

Portable tanks and bulk containers			EmS	Stowage and segregation	Properties and observations	UN No.
	Tank instructions	Provisions				
(12)	(13) 4.2.5 4.3	(14) 4.2.5	(15) 5.4.3.2 7.3	(16) 7.1 7.2	(17)	(18)
–	T2	TP1	F-E, S-D	Category A.	Usually consist of alcoholic solutions. Miscibility with water depends upon the composition.	1169
–	T4	TP1	F-E, S-D	Category A.	Colourless, volatile liquids. Pure ETHANOL: flashpoint 13°C c.c. Explosive limits: 3.3% to 19%. Miscible with water.	1170
–	T2	TP1	F-E, S-D	Category A.	See entry above.	1170
–	T2	TP1	F-E, S-D	Category A.	Colourless liquid. Flashpoint: 40°C c.c. Explosive limits: 1.7% to 15.6%. Miscible with water.	1171
–	T2	TP1	F-E, S-D	Category A.	Colourless liquid. Flashpoint: 51°C c.c. Explosive limits: 1.7% to 10.1%. Partially miscible with water.	1172
–	T4	TP1	F-E, S-D	Category B.	Colourless liquid with a fragrant odour. Flashpoint: −4°C c.c. Explosive limits: 2.18% to 11.5%. Immiscible with water.	1173
–	T4	TP1	F-E, S-D	Category B.	Colourless liquid with an aromatic odour. Flashpoint: 22°C c.c. Explosive limits: 1% to 6.7%. Immiscible with water.	1175
–	T4	TP1	F-E, S-D	Category B.	Colourless liquid. Flashpoint: 11°C c.c. Immiscible with water.	1176
–	T2	TP1	F-E, S-D	Category A.	Colourless liquid. Flashpoint: 54°C o.c. Immiscible with water.	1177
–	T4	TP1	F-E, S-D	Category B.	Colourless liquid. Flashpoint: 11°C c.c. Explosive limits: 1.2% to 7.7%. Immiscible with water.	1178
–	T4	TP1	F-E, S-D	Category B.	Colourless liquid. Flashpoint: −1°C c.c. Immiscible with water.	1179
–	T2	TP1	F-E, S-D	Category A.	Colourless, volatile liquid with a pineapple-like odour. Flashpoint: 26°C c.c. Immiscible with water.	1180
–	T7	TP2	F-E, S-D	Category A.	Colourless, flammable liquid with a pungent and fruity odour. Flashpoint: 54°C c.c. Immiscible with water. When heated, evolves toxic and corrosive fumes. Toxic if swallowed, by skin contact or by inhalation.	1181
–	T20	TP2 TP13 TP37	F-E, S-C	Category D. Clear of living quarters. Segregation as for class 3 but "away from" class 4.1.	Colourless liquid. Flashpoint: 16°C c.c. Reacts and decomposes with water or heat, evolving hydrogen chloride, an irritating and corrosive gas apparent as white fumes. In the presence of moisture, highly corrosive to most metals. Highly toxic if swallowed, by skin contact or by inhalation. Causes burns to skin, eyes and mucous membranes.	1182
–	T14	TP2 TP7 TP13	F-G, S-O	Category D. Clear of living quarters. Segregation as for class 3 but "away from" classes 3, 4.1 and 8.	Colourless, very volatile liquid with a pungent odour. Flashpoint: −1°C c.c. Immiscible with water. Reacts violently with water or steam to produce heat which may lead to self-ignition; toxic and corrosive fumes will be evolved. May react vigorously in contact with oxidizing substances. Causes burns to skin, eyes and mucous membranes.	1183
–	T7	TP1	F-E, S-D	Category B. Clear of living quarters.	Colourless liquid with a chloroform-like odour. Flashpoint: 13°C c.c. Explosive limits: 6.2% to 15.9%. Immiscible with water. Toxic by inhalation. Irritating to skin, eyes and mucous membranes.	1184
–	T22	TP2 TP13	F-E, S-D	Category D. Clear of living quarters.	Colourless oily flammable liquid with a pungent ammonia-like odour. Flashpoint: −13°C c.c. Boiling point: 55°C. Explosive limits: 3.3% to 6.0%. Miscible with water. Highly toxic if swallowed, by skin contact or by inhalation.	1185
–	T2	TP1	F-E, S-D	Category A.	Colourless liquid. Flashpoint: 38°C c.c. Explosive limits: 1.3% to 20%. Miscible with water.	1188
–	T2	TP1	F-E, S-D	Category A.	Colourless liquid with a characteristic odour. Flashpoint: 44°C c.c. Explosive limits: 1.7% to 8.2%. Miscible with water.	1189
–	T4	TP1	F-E, S-D	Category E.	Colourless liquid with a pleasant aromatic odour. Flashpoint: −20°C c.c. Explosive limits: 3.5% to 16.5%. Boiling point: 54°C. Immiscible with water.	1190

UN No.	Proper Shipping Name (PSN)	Class or division	Subsidiary risk(s)	Packing group	Special provisions	Limited and excepted quantity provisions		Packing		IBC	
						Limited quantities	Excepted quantities	Instructions	Provisions	Instructions	Provisions
(1)	(2) 3.1.2	(3) 2.0	(4) 2.0	(5) 2.0.1.3	(6) 3.3	(7a) 3.4	(7b) 3.5	(8) 4.1.4	(9) 4.1.4	(10) 4.1.4	(11) 4.1.4
1191	OCTYL ALDEHYDES	3	–	III	–	5 L	E1	P001 LP01	–	IBC03	–
1192	ETHYL LACTATE	3	–	III	–	5 L	E1	P001 LP01	–	IBC03	–
1193	ETHYL METHYL KETONE (METHYL ETHYL KETONE)	3	–	II	–	1 L	E2	P001	–	IBC02	–
△ 1194	ETHYL NITRITE SOLUTION	3	6.1	I	900	0	E0	P001	–	–	–
1195	ETHYL PROPIONATE	3	–	II	–	1 L	E2	P001	–	IBC02	–
△ 1196	ETHYLTRICHLOROSILANE	3	8	II	–	0	E0	P010	–	–	–
1197	EXTRACTS, FLAVOURING, LIQUID	3	–	II	–	5 L	E2	P001	–	IBC02	–
1197	EXTRACTS, FLAVOURING, LIQUID	3	–	III	223 955	5 L	E1	P001 LP01	–	IBC03	–
1198	FORMALDEHYDE SOLUTION, FLAMMABLE	3	8	III	–	5 L	E0	P001	–	IBC03	–
1199	FURALDEHYDES	6.1	3	II	–	100 mL	E4	P001	–	IBC02	–
1201	FUSEL OIL	3	–	II	–	1 L	E2	P001	–	IBC02	–
1201	FUSEL OIL	3	–	III	223 955	5 L	E1	P001 LP01	–	IBC03	–
△ 1202	GAS OIL or DIESEL FUEL or HEATING OIL, LIGHT	3	–	III	363	5 L	E1	P001 LP01	–	IBC03	–
△ 1203	MOTOR SPIRIT or GASOLINE or PETROL	3	–	II	243 363	1 L	E2	P001	–	IBC02	–
1204	NITROGLYCERIN SOLUTION IN ALCOHOL with not more than 1% nitroglycerin	3	–	II	–	1 L	E0	P001	PP5	IBC02	–
1206	HEPTANES	3	–	II	–	1 L	E2	P001	–	IBC02	–
1207	HEXALDEHYDE	3	–	III	–	5 L	E1	P001 LP01	–	IBC03	–
1208	HEXANES	3	–	II	–	1 L	E2	P001	–	IBC02	–

Portable tanks and bulk containers			EmS	Stowage and segregation	Properties and observations	UN No.
	Tank instructions	Provisions				
(12)	(13) 4.2.5 4.3	(14) 4.2.5	(15) 5.4.3.2 7.3	(16) 7.1 7.2	(17)	(18)
–	T2	TP1	F-E, S-D	Category A.	Colourless liquids with a characteristic odour. Flashpoint: 44°C to 52°C c.c. Explosive limits: 0.9% to 7.2%. Immiscible with water.	**1191**
–	T2	TP1	F-E, S-D	Category A.	Colourless liquid. Flashpoint: 46°C c.c. Explosive limits: 1.5% to 11.4%. Miscible with water.	**1192**
–	T4	TP1	F-E, S-D	Category B.	Colourless liquid. Flashpoint: −1°C c.c. Explosive limits: 1.8% to 11.5%. Miscible with water.	**1193**
–	–	–	F-E, S-D	Category D. Clear of living quarters.	Alcoholic solution of ethyl nitrite. Extremely volatile, with an aromatic, ethereal odour. Explosive limits of the pure product: 3% to 50%. Boiling point of pure product: 17°C. Miscible or partially miscible with water. Decomposes under exposure to air, light, water or heat to evolve toxic nitrous fumes. Toxic if swallowed, by skin contact or by inhalation. Inhalation of ethyl nitrite vapours, even in small quantities, rapidly affects the heart and can be dangerous. Transport of ETHYL NITRITE pure is prohibited.	**1194** △
–	T4	TP1	F-E, S-D	Category B.	Colourless liquid with a pineapple-like odour. Flashpoint: 12°C c.c. Explosive limits: 1.8% to 11%. Immiscible with water.	**1195**
–	T10	TP2 TP7 TP13	F-E, S-C	Category B. Clear of living quarters.	Colourless liquid with a pungent odour. Flashpoint: 14°C c.c. Readily hydrolysed by moisture, evolving hydrogen chloride, an irritating and corrosive gas apparent as white fumes. Causes burns to skin and eyes. Irritating to mucous membranes.	**1196** △
–	T4	TP1 TP8	F-E, S-D	Category B.	Usually consist of alcoholic solutions. Miscibility with water depends upon the composition.	**1197**
–	T2	TP1	F-E, S-D	Category A.	See entry above.	**1197**
–	T4	TP1	F-E, S-C	Category A. Clear of living quarters.	Colourless liquids with a pungent odour. Flashpoint: 32–60°C c.c. Miscible with water. Irritating to skin, eyes and mucous membranes.	**1198**
–	T7	TP2	F-E, S-D	Category A.	Colourless or reddish-brown, mobile liquids with a pungent odour. Miscible with water. Explosive limits for 2-FURALDEHYDE: 2.1% to 19.3%. Flashpoints: 2-FURALDEHYDE 60°C c.c., 3-FURALDEHYDE 48°C c.c. Toxic if swallowed, by skin contact or by inhalation.	**1199**
–	T4	TP1	F-E, S-D	Category B.	Colourless, oily liquid with a disagreeable odour. A mixture consisting of amyl alcohols. Immiscible with water.	**1201**
–	T2	TP1	F-E, S-D	Category A.	See entry above.	**1201**
–	T2	TP1	F-E, S-E	Category A.	Immiscible with water.	**1202** △
–	T4	TP1	F-E, S-E	Category E.	Immiscible with water.	**1203** △
–	–	–	F-E, S-D	Category B.	Immiscible with water. Ignites readily. When involved in a fire, evolves toxic nitrous fumes. Not explosive in this state but damage to, or leakage from, a package may allow solvent to evaporate and thus leave the nitroglycerin in an explosive state.	**1204**
–	T4	TP1	F-E, S-D	Category B.	Colourless, volatile liquids. Explosive limits: 1.1% to 6.7%. *n*-HEPTANE: flashpoint −4°C c.c. Immiscible with water. Irritating to skin, eyes and mucous membranes.	**1206**
–	T2	TP1	F-E, S-D	Category A.	Colourless liquid with a pungent odour. Flashpoint: 32°C c.c. Immiscible with water.	**1207**
–	T4	TP1	F-E, S-D	Category E.	Colourless, volatile liquids with a faint odour. Explosive limits: 1.1% to 7.5%. *n*-HEXANE: flashpoint −22°C c.c., boiling point 69°C. NEOHEXANE: flashpoint −48°C c.c., boiling point 50°C. Immiscible with water. Slightly irritating to skin, eyes and mucous membranes.	**1208**

UN No.	Proper Shipping Name (PSN)	Class or division	Subsidiary risk(s)	Packing group	Special provisions	Limited and excepted quantity provisions		Packing		IBC	
						Limited quantities	Excepted quantities	Instructions	Provisions	Instructions	Provisions
(1)	(2) 3.1.2	(3) 2.0	(4) 2.0	(5) 2.0.1.3	(6) 3.3	(7a) 3.4	(7b) 3.5	(8) 4.1.4	(9) 4.1.4	(10) 4.1.4	(11) 4.1.4
1210	PRINTING INK flammable or PRINTING INK RELATED MATERIAL (including printing ink thinning or reducing compound), flammable	3	–	I	163	500 mL	E3	P001	–	–	–
1210	PRINTING INK flammable or PRINTING INK RELATED MATERIAL (including printing ink thinning or reducing compound), flammable	3	–	II	163	5 L	E2	P001	PP1	IBC02	–
1210	PRINTING INK flammable or PRINTING INK RELATED MATERIAL (including printing ink thinning or reducing compound), flammable	3	–	III	163 223 955	5 L	E1	P001 LP01	PP1	IBC03	–
1212	ISOBUTANOL (ISOBUTYL ALCOHOL)	3	–	III	–	5 L	E1	P001 LP01	–	IBC03	–
1213	ISOBUTYL ACETATE	3	–	II	–	1 L	E2	P001	–	IBC02	–
1214	ISOBUTYLAMINE	3	8	II	–	1 L	E2	P001	–	IBC02	–
1216	ISOOCTENES	3	–	II	–	1 L	E2	P001	–	IBC02	–
1218	ISOPRENE, STABILIZED	3	–	I	–	0	E3	P001	–	–	–
1219	ISOPROPANOL (ISOPROPYL ALCOHOL)	3	–	II	–	1 L	E2	P001	–	IBC02	–
1220	ISOPROPYL ACETATE	3	–	II	–	1 L	E2	P001	–	IBC02	–
1221	ISOPROPYLAMINE	3	8	I	–	0	E0	P001	–	–	–
1222	ISOPROPYL NITRATE	3	–	II	26	1 L	E2	P001	–	–	–
△ 1223	KEROSENE	3	–	III	363	5 L	E1	P001 LP01	–	IBC03	–
1224	KETONES, LIQUID, N.O.S.	3	–	II	274	1 L	E2	P001	–	IBC02	–
1224	KETONES, LIQUID, N.O.S.	3	–	III	223 274	5 L	E1	P001 LP01	–	IBC03	–
△ 1228	MERCAPTANS, LIQUID, FLAMMABLE, TOXIC, N.O.S. or MERCAPTAN MIXTURE, LIQUID, FLAMMABLE, TOXIC, N.O.S.	3	6.1	II	274	1 L	E2	P001	–	IBC02	–
△ 1228	MERCAPTANS, LIQUID, FLAMMABLE, TOXIC, N.O.S. or MERCAPTAN MIXTURE, LIQUID, FLAMMABLE, TOXIC, N.O.S.	3	6.1	III	223 274	5 L	E1	P001	–	IBC03	–
1229	MESITYL OXIDE	3	–	III	–	5 L	E1	P001 LP01	–	IBC03	–

Portable tanks and bulk containers			EmS	Stowage and segregation	Properties and observations	UN No.	
	Tank instructions	Provisions					
(12)	(13) 4.2.5 4.3	(14) 4.2.5	(15) 5.4.3.2 7.3	(16) 7.1 7.2	(17)	(18)	
–	T11	TP1 TP8	F-E, S-D	Category E.	Fluid or viscous liquid containing colouring matter in solution or suspension. Miscibility with water depends upon the solvent.	**1210**	
–	T4	TP1 TP8	F-E, S-D	Category B.	See entry above.	1210	
–	T2	TP1	F-E, S-D	Category A.	See entry above.	**1210**	
–	T2	TP1	F-E, S-D	Category A.	Colourless liquid with a sweet odour. Flashpoint: 28°C c.c. Explosive limits: 1.2% to 10.9%. Partially miscible with water.	1212	
–	T4	TP1	F-E, S-D	Category B.	Colourless liquid with a pineapple-like odour. Flashpoint: 18°C c.c. Explosive limits: 1.3% to 10.5%. Immiscible with water.	**1213**	
–	T7	TP1	F-E, S-C	Category B. Clear of living quarters.	Colourless liquid. Flashpoint: −9°C c.c. Explosive limits: 3.4% to 9%. Miscible with water. Harmful by inhalation. Causes burns to skin and eyes. Irritating to mucous membranes.	1214	
–	T4	TP1	F-E, S-D	Category B.	Colourless liquids. Immiscible with water.	**1216**	
–	T11	TP2	F-E, S-D	Category E.	Colourless, volatile liquid. Flashpoint: −48°C c.c. Explosive limits: 1.5% to 9.7%. Boiling point: 34°C. Immiscible with water.	1218	
–	T4	TP1	F-E, S-D	Category B.	Colourless, mobile liquid. Flashpoint: 12°C c.c. Explosive limits: 2% to 12%. Miscible with water.	**1219**	
–	T4	TP1	F-E, S-D	Category B.	Colourless liquid with an aromatic odour. Flashpoint: 11°C c.c. Explosive limits: 1.8% to 7.8%. Immiscible with water.	1220	
–	T11	TP2	F-E, S-C	Category E. Clear of living quarters.	Colourless, volatile liquid with an ammonia-like odour. Flashpoint: −37°C c.c. Explosive limits: 2.3% to 10.4%. Boiling point: 32°C. Miscible with water. Harmful if swallowed. Causes burns to skin, eyes and mucous membranes.	**1221**	
–	–	–	F-E, S-D	Category D.	Colourless liquid. Flashpoint: 12°C c.c. Explosive limits: up to 100%. Immiscible with water. May explode on heating. Harmful by inhalation.	1222	
–	T2	TP2	F-E, S-E	Category A.	Immiscible with water.	**1223**	△
–	T7	TP1 TP8 TP28	F-E, S-D	Category B.	–	1224	
–	T4	TP1 TP29	F-E, S-D	Category A.	–	**1224**	
–	T11	TP2 TP27	F-E, S-D	Category B. Clear of living quarters. Segregation from foodstuffs as in 7.3.4.2, 7.6.3.1.2 or 7.7.3.6. "Separated from" odour-absorbing cargoes.	Colourless to yellow liquids with a garlic odour. Immiscible with water. Toxic if swallowed, by skin contact or by inhalation.	1228	△
–	T7	TP1 TP28	F-E, S-D	Category B. Clear of living quarters. Segregation from foodstuffs as in 7.3.4.2, 7.6.3.1.2 or 7.7.3.6. "Separated from" odour-absorbing cargoes.	See entry above.	**1228**	△
–	T2	TP1	F-E, S-D	Category A.	Colourless, oily liquid with a sweet odour. Flashpoint: 32°C c.c. Miscible with water.	1229	

UN No.	Proper Shipping Name (PSN)	Class or division	Subsidiary risk(s)	Packing group	Special provisions	Limited and excepted quantity provisions		Packing		IBC	
						Limited quantities	Excepted quantities	Instructions	Provisions	Instructions	Provisions
(1)	(2) 3.1.2	(3) 2.0	(4) 2.0	(5) 2.0.1.3	(6) 3.3	(7a) 3.4	(7b) 3.5	(8) 4.1.4	(9) 4.1.4	(10) 4.1.4	(11) 4.1.4
1230	METHANOL	3	6.1	II	279	1 L	E2	P001	–	IBC02	–
1231	METHYL ACETATE	3	–	II	–	1 L	E2	P001	–	IBC02	–
1233	METHYLAMYL ACETATE	3	–	III	–	5 L	E1	P001 LP01	–	IBC03	–
1234	METHYLAL	3	–	II	–	1 L	E2	P001	–	IBC02	B8
1235	METHYLAMINE, AQUEOUS SOLUTION	3	8	II	–	1 L	E2	P001	–	IBC02	–
1237	METHYL BUTYRATE	3	–	II	–	1 L	E2	P001	–	IBC02	–
1238	METHYL CHLOROFORMATE	6.1	3/8	I	354	0	E0	P602	–	–	–
1239	METHYL CHLOROMETHYL ETHER	6.1	3	I	354	0	E0	P602	–	–	–
1242	METHYLDICHLOROSILANE	4.3	3/8	I	–	0	E0	P401	PP31	–	–
1243	METHYL FORMATE	3	–	I	–	0	E3	P001	–	–	–
1244	METHYLHYDRAZINE	6.1	3/8	I	354	0	E0	P602	–	–	–
1245	METHYL ISOBUTYL KETONE	3	–	II	–	1 L	E2	P001	–	IBC02	–
1246	METHYL ISOPROPENYL KETONE, STABILIZED	3	–	II	–	1 L	E2	P001	–	IBC02	–
1247	METHYL METHACRYLATE MONOMER, STABILIZED	3	–	II	–	1 L	E2	P001	–	IBC02	–
1248	METHYL PROPIONATE	3	–	II	–	1 L	E2	P001	–	IBC02	–
1249	METHYL PROPYL KETONE	3	–	II	–	1 L	E2	P001	–	IBC02	–
△ 1250	METHYLTRICHLOROSILANE	3	8	II	–	0	E0	P010	–	–	–

Portable tanks and bulk containers			EmS	Stowage and segregation	Properties and observations	UN No.
	Tank instructions	Provisions				
(12)	(13) 4.2.5 4.3	(14) 4.2.5	(15) 5.4.3.2 7.3	(16) 7.1 7.2	(17)	(18)
–	T7	TP2	F-E, S-D	Category B. Clear of living quarters.	Colourless, volatile liquid. Flashpoint: 12°C c.c. Explosive limits: 6% to 36.5%. Miscible with water. Toxic if swallowed; may cause blindness. Avoid skin contact.	**1230**
–	T4	TP1	F-E, S-D	Category B.	Colourless, volatile liquid with a fragrant odour. Flashpoint: –10°C c.c. Explosive limits: 3% to 16%. Miscible with water.	**1231**
–	T2	TP1	F-E, S-D	Category A.	Colourless liquid. Flashpoint: 43°C o.c. Immiscible with water.	**1233**
–	T7	TP2	F-E, S-D	Category E.	Colourless, volatile liquid with a chloroform-like odour. Flashpoint: –28°C c.c. Explosive limits: 3.6% to 12.6%. Boiling point: 42°C. Miscible with water. Irritating to skin, eyes and mucous membranes.	**1234**
–	T7	TP1	F-E, S-C	Category E. "Separated from" mercury and mercury compounds. "Separated from" acids.	Aqueous solution of a flammable gas having an ammonia-like odour. Explosive limits: 5% to 20.7% (pure product). Boiling point: –7°C (pure product). Commercial product is a 40% solution with: boiling point 48°C, flashpoint –13°C c.c. Miscible with water. May react explosively with mercury. Causes burns to skin, eyes and mucous membranes.	**1235**
–	T4	TP1	F-E, S-D	Category B.	Colourless liquid. Flashpoint: 14°C c.c. Immiscible with water.	**1237**
–	T22	TP2 TP13 TP35	F-E, S-C	Category D. Clear of living quarters. Segregation as for class 3 but "away from" class 4.1.	Colourless liquid. Flashpoint: 5°C c.c. Immiscible with water. Highly toxic if swallowed, by skin contact or by inhalation. Causes burns to skin, eyes and mucous membranes.	**1238**
–	T22	TP2 TP13 TP35	F-E, S-D	Category D. Clear of living quarters.	Colourless liquid. Flashpoint: below –18°C c.c. Immiscible with water. Highly toxic if swallowed, by skin contact or by inhalation.	**1239**
–	T14	TP2 TP7 TP13	F-G, S-O	Category D. Clear of living quarters. Segregation as for class 3 but "away from" classes 3, 4.1 and 8.	Colourless, very volatile liquid with a pungent odour. Flashpoint: –26°C c.c. Explosive limits: 4.5% to 70%. Boiling point: 41°C. Immiscible with water. Reacts violently with water or steam to produce heat which may lead to self-ignition; toxic and corrosive fumes will be evolved. May react vigorously in contact with oxidizing substances. Causes burns to skin, eyes and mucous membranes.	**1242**
–	T11	TP2	F-E, S-D	Category E.	Colourless liquid with an agreeable odour. Flashpoint: –32°C c.c. Explosive limits: 5% to 22.7%. Boiling point: 32°C. Miscible with water.	**1243**
–	T22	TP2 TP13 TP35	F-E, S-C	Category D. Clear of living quarters. Segregation as for class 3 but "away from" class 4.1 and class 8. "Separated from" acids.	Colourless liquid with an ammonia-like odour. Flashpoint: 20°C c.c. Explosive limits: 2.5% to 98%. Miscible with water. Reacts violently with acids. May react dangerously with oxidizing substances. Highly toxic if swallowed, by skin contact or by inhalation. Causes burns to skin, eyes and mucous membranes.	**1244**
–	T4	TP1	F-E, S-D	Category B.	Colourless liquid with a pleasant odour. Flashpoint: 14°C c.c. Explosive limits: 1.4% to 7.5%. Immiscible with water.	**1245**
–	T4	TP1	F-E, S-D	Category B.	Colourless liquid with a pleasant odour. Explosive limits: 1.8% to 9%. Immiscible with water.	**1246**
–	T4	TP1	F-E, S-D	Category B. Clear of living quarters.	Colourless, volatile liquid. Flashpoint: 8°C c.c. Explosive limits: 1.5% to 11.6%. Immiscible with water. Irritating to skin, eyes and mucous membranes.	**1247**
–	T4	TP1	F-E, S-D	Category B.	Colourless liquid. Flashpoint: –2°C c.c. Explosive limits: 2.4% to 13%. Immiscible with water.	**1248**
–	T4	TP1	F-E, S-D	Category B.	Colourless liquid. Flashpoint: 7°C c.c. Explosive limits: 1.5% to 8.2%. Immiscible with water.	**1249**
–	T10	TP2 TP7 TP13	F-E, S-C	Category B. Clear of living quarters.	Colourless liquid with a pungent odour. Flashpoint: 8°C o.c. Explosive limits: 5.1% to 20%. Immiscible with water. Readily hydrolysed by moisture, evolving hydrogen chloride, an irritating and corrosive gas apparent as white fumes. In the presence of moisture, corrosive to most metals. Causes burns to skin and eyes. Irritating to mucous membranes.	**1250** △

UN No.	Proper Shipping Name (PSN)	Class or division	Subsidiary risk(s)	Packing group	Special provisions	Limited and excepted quantity provisions		Packing		IBC	
						Limited quantities	Excepted quantities	Instructions	Provisions	Instructions	Provisions
(1)	(2) 3.1.2	(3) 2.0	(4) 2.0	(5) 2.0.1.3	(6) 3.3	(7a) 3.4	(7b) 3.5	(8) 4.1.4	(9) 4.1.4	(10) 4.1.4	(11) 4.1.4
1251	METHYL VINYL KETONE, STABILIZED	6.1	3/8	I	354	0	E0	P601	–	–	–
1259	NICKEL CARBONYL	6.1	3 P	I	–	0	E5	P601	–	–	–
1261	NITROMETHANE	3	–	II	26	1 L	E2	P001	–	–	–
1262	OCTANES	3	–	II	–	1 L	E2	P001	–	IBC02	–
△ 1263	PAINT (including paint, lacquer, enamel, stain, shellac, varnish, polish, liquid filler and liquid lacquer base) or PAINT RELATED MATERIAL (including paint thinning or reducing compound)	3	–	I	163	500 mL	E3	P001	–	–	–
△ 1263	PAINT (including paint, lacquer, enamel, stain, shellac, varnish, polish, liquid filler and liquid lacquer base) or PAINT RELATED MATERIAL (including paint thinning or reducing compound)	3	–	II	163	5 L	E2	P001	PP1	IBC02	–
△ 1263	PAINT (including paint, lacquer, enamel, stain, shellac, varnish, polish, liquid filler and liquid lacquer base) or PAINT RELATED MATERIAL (including paint thinning or reducing compound)	3	–	III	163 223 955	5 L	E1	P001 LP01	PP1	IBC03	–
1264	PARALDEHYDE	3	–	III	–	5 L	E1	P001 LP01	–	IBC03	–
1265	PENTANES, liquid	3	–	I	–	0	E3	P001	–	–	–
1265	PENTANES, liquid	3	–	II	–	1 L	E2	P001	–	IBC02	–
△ 1266	PERFUMERY PRODUCTS with flammable solvents	3	–	II	163	5 L	E2	P001	–	IBC02	–
△ 1266	PERFUMERY PRODUCTS with flammable solvents	3	–	III	163 223 904 955	5 L	E1	P001 LP01	–	IBC03	–
1267	PETROLEUM CRUDE OIL	3	–	I	357	500 mL	E3	P001	–	–	–
1267	PETROLEUM CRUDE OIL	3	–	II	357	1 L	E2	P001	–	IBC02	–
1267	PETROLEUM CRUDE OIL	3	–	III	223 357	5 L	E1	P001 LP01	–	IBC03	–

Portable tanks and bulk containers			EmS	Stowage and segregation	Properties and observations	UN No.	
	Tank instructions	Provisions					
(12)	(13) 4.2.5 4.3	(14) 4.2.5	(15) 5.4.3.2 7.3	(16) 7.1 7.2	(17)	(18)	
–	T22	TP2 TP13 TP37	F-E, S-C	Category D. Clear of living quarters. Segregation as for class 3 but "away from" class 4.1.	Colourless liquid with a pungent odour. Miscible with water. Explosive limits: 2.1% to 15.6%. Flashpoint: −7°C c.c. Highly toxic if swallowed, by skin contact or by inhalation. Causes burns to skin, eyes and mucous membranes.	**1251**	
–	–	–	F-E, S-D	Category D. Clear of living quarters. "Separated longitudinally by an intervening complete compartment or hold from" class 1.	Colourless or yellow, volatile, flammable liquid. Flashpoint: below −20°C c.c. Oxidizes in air and explodes at a temperature of 60°C. Lower explosive limit: 2.0%. Immiscible with water. Highly toxic if swallowed, by skin contact or by inhalation.	**1259**	
–	–	–	F-E, S-D	Category A.	Colourless liquid. Flashpoint: 35°C c.c. Explosive limits: 7.1% to 63%. Miscible with water. Fire and explosion hazard if package is ruptured.	**1261**	
–	T4	TP1	F-E, S-E	Category B.	Colourless liquids. Explosive limits: 1% to 6.5%. ISOOCTANE: flashpoint −12°C c.c. *n*-OCTANE: flashpoint 13°C c.c. Immiscible with water.	**1262**	
–	T11	TP1 TP8 TP27	F-E, S-E	Category E.	Miscibility with water depends upon the composition.	**1263**	△
–	T4	TP1 TP8 TP28	F-E, S-E	Category B.	See entry above.	**1263**	△
–	T2	TP1 TP29	F-E, S-E	Category A.	See entry above.	**1263**	△
–	T2	TP1	F-E, S-D	Category A.	Colourless liquid. Flashpoint: 27°C c.c. Explosive limits: 1.3% to .... Miscible with water.	**1264**	
–	T11	TP2	F-E, S-D	Category E.	Colourless liquids with a paraffin-like odour. Explosive limits: 1.4% to 8%. ISOPENTANE (2-METHYLBUTANE): boiling point 28°C. Immiscible with water. Slightly irritating to skin, eyes and mucous membranes. Narcotic in high concentrations.	**1265**	
–	T4	TP1	F-E, S-D	Category E.	See entry above. *normal*-PENTANE: boiling point 36°C.	**1265**	
–	T4	TP1 TP8	F-E, S-D	Category B.	Miscibility with water depends upon the composition.	**1266**	△
–	T2	TP1	F-E, S-D	Category A.	See entry above.	**1266**	△
–	T11	TP1 TP8	F-E, S-E	Category E.	Immiscible with water.	**1267**	
–	T4	TP1 TP8	F-E, S-E	Category B.	See entry above.	**1267**	
–	T2	TP1	F-E, S-E	Category A.	See entry above.	**1267**	

	UN No. (1)	Proper Shipping Name (PSN) (2) 3.1.2	Class or division (3) 2.0	Subsidiary risk(s) (4) 2.0	Packing group (5) 2.0.1.3	Special provisions (6) 3.3	Limited and excepted quantity provisions: Limited quantities (7a) 3.4	Excepted quantities (7b) 3.5	Packing: Instructions (8) 4.1.4	Provisions (9) 4.1.4	IBC: Instructions (10) 4.1.4	Provisions (11) 4.1.4
△	1268	PETROLEUM DISTILLATES, N.O.S. or PETROLEUM PRODUCTS, N.O.S.	3	–	I	363	500 mL	E3	P001	–	–	–
△	1268	PETROLEUM DISTILLATES, N.O.S. or PETROLEUM PRODUCTS, N.O.S.	3	–	II	363	1 L	E2	P001	–	IBC02	–
△	1268	PETROLEUM DISTILLATES, N.O.S. or PETROLEUM PRODUCTS, N.O.S.	3	–	III	223 363 955	5 L	E1	P001 LP01	–	IBC03	–
	1272	PINE OIL	3	–	III	–	5 L	E1	P001 LP01	–	IBC03	–
	1274	*n*-PROPANOL (PROPYL ALCOHOL, NORMAL)	3	–	II	–	1 L	E2	P001	–	IBC02	–
	1274	*n*-PROPANOL (PROPYL ALCOHOL, NORMAL)	3	–	III	223	5 L	E1	P001 LP01	–	IBC03	–
	1275	PROPIONALDEHYDE	3	–	II	–	1 L	E2	P001	–	IBC02	–
	1276	*n*-PROPYL ACETATE	3	–	II	–	1 L	E2	P001	–	IBC02	–
	1277	PROPYLAMINE	3	8	II	–	1 L	E2	P001	–	IBC02	–
	1278	1-CHLOROPROPANE	3	–	II	–	1 L	E2	P001	–	IBC02	B8
	1279	1,2-DICHLOROPROPANE	3	–	II	–	1 L	E2	P001	–	IBC02	–
	1280	PROPYLENE OXIDE	3	–	I	–	0	E3	P001	–	–	–
	1281	PROPYL FORMATES	3	–	II	–	1 L	E2	P001	–	IBC02	–
	1282	PYRIDINE	3	–	II	–	1 L	E2	P001	–	IBC02	–
	1286	ROSIN OIL	3	–	II	–	1 L	E2	P001	–	IBC02	–
	1286	ROSIN OIL	3	–	III	223	5 L	E1	P001 LP01	–	IBC03	–
	1287	RUBBER SOLUTION	3	–	II	–	5 L	E2	P001	–	IBC02	–
	1287	RUBBER SOLUTION	3	–	III	223 955	5 L	E1	P001 LP01	–	IBC03	–
	1288	SHALE OIL	3	–	II	–	1 L	E2	P001	–	IBC02	–
	1288	SHALE OIL	3	–	III	223	5 L	E1	P001 LP01	–	IBC03	–
	1289	SODIUM METHYLATE SOLUTION in alcohol	3	8	II	–	1 L	E2	P001	–	IBC02	–
	1289	SODIUM METHYLATE SOLUTION in alcohol	3	8	III	223	5 L	E1	P001	–	IBC03	–

3

DGL

Portable tanks and bulk containers			EmS	Stowage and segregation	Properties and observations	UN No.
	Tank instructions	Provisions				
(12)	(13) 4.2.5 4.3	(14) 4.2.5	(15) 5.4.3.2 7.3	(16) 7.1 7.2	(17)	(18)
–	T11	TP1 TP8	F-E, S-E	Category E.	Immiscible with water.	**1268** △
–	T7	TP1 TP8 TP28	F-E, S-E	Category B.	See entry above.	1268 △
–	T4	TP1 TP29	F-E, S-E	Category A.	See entry above.	**1268** △
–	T2	TP1	F-E, S-E	Category A.	Volatile oils with characteristic odours. Flashpoint: 57°C to 60°C c.c. Immiscible with water.	1272
–	T4	TP1	F-E, S-D	Category B.	Colourless liquid. Explosive limits: 2% to 12%. Flashpoint: 15°C to 23°C c.c. Miscible with water.	**1274**
–	T2	TP1	F-E, S-D	Category A.	See entry above. Flashpoint: 23°C to 26°C c.c.	1274
–	T7	TP1	F-E, S-D	Category E.	Colourless liquid with a pungent odour. Flashpoint: below –18°C c.c. Explosive limits: 2.3% to 21%. Boiling point: 49°C. Miscible with water. Irritating to skin, eyes and mucous membranes.	**1275**
–	T4	TP1	F-E, S-D	Category B.	Colourless, clear liquid with a pleasant odour. Flashpoint: 10°C c.c. Explosive limits: 1.8% to 8%. Immiscible with water.	1276
–	T7	TP1	F-E, S-C	Category E. Clear of living quarters.	Colourless liquid. Flashpoint: below –18°C c.c. Explosive limits: 2% to 10.4%. Boiling point: 48°C. Miscible with water. Harmful if swallowed. Causes burns to skin, eyes and mucous membranes.	**1277**
–	T7	TP2	F-E, S-D	Category E.	Colourless liquid with a chloroform-like odour. Flashpoint: –18°C c.c. Explosive limits: 2.6% to 10.5%. Boiling point: 47°C. Immiscible with water.	1278
–	T4	TP1	F-E, S-D	Category B.	Colourless liquid. Flashpoint: 15°C c.c. Immiscible with water. Harmful by inhalation. Irritating to skin and eyes.	**1279**
–	T11	TP2 TP7	F-E, S-D	Category E. Clear of living quarters.	Colourless, volatile liquid with an ether-like odour. Flashpoint: below –18°C c.c. Explosive limits: 2% to 22%. Boiling point: 34°C. Partially miscible with water.	1280
–	T4	TP1	F-E, S-D	Category B.	Colourless liquids with a pleasant odour. Explosive limits: 2.4% to 7.8%. Miscibility with water depends upon the composition. Irritating to skin, eyes and mucous membranes.	**1281**
–	T4	TP2	F-E, S-D	Category B. Clear of living quarters.	Colourless or slightly yellow liquid with a pungent odour. Flashpoint: 17°C c.c. Explosive limits: 1.8% to 12.4%. Miscible with water. Harmful by inhalation.	1282
–	T4	TP1	F-E, S-E	Category B.	Colourless to brown viscous liquid. Immiscible with water.	**1286**
–	T2	TP1	F-E, S-E	Category A.	See entry above.	1286
–	T4	TP1 TP8	F-E, S-D	Category B.	Miscibility with water depends upon the composition.	**1287**
–	T2	TP1	F-E, S-D	Category A.	See entry above.	1287
–	T4	TP1 TP8	F-E, S-E	Category B.	Immiscible with water.	**1288**
–	T2	TP1	F-E, S-E	Category A.	See entry above.	1288
–	T7	TP1 TP8	F-E, S-C	Category B.	Reacts violently with water. Causes burns to skin, eyes and mucous membranes.	**1289**
–	T4	TP1	F-E, S-C	Category A.	See entry above. Irritating to skin, eyes and mucous membranes.	1289

UN No.	Proper Shipping Name (PSN)	Class or division	Subsidiary risk(s)	Packing group	Special provisions	Limited and excepted quantity provisions		Packing		IBC	
						Limited quantities	Excepted quantities	Instructions	Provisions	Instructions	Provisions
(1)	(2) 3.1.2	(3) 2.0	(4) 2.0	(5) 2.0.1.3	(6) 3.3	(7a) 3.4	(7b) 3.5	(8) 4.1.4	(9) 4.1.4	(10) 4.1.4	(11) 4.1.4
1292	TETRAETHYL SILICATE	3	–	III	–	5 L	E1	P001 LP01	–	IBC03	–
1293	TINCTURES, MEDICINAL	3	–	II	–	1 L	E2	P001	–	IBC02	–
1293	TINCTURES, MEDICINAL	3	–	III	904 955	5 L	E1	P001 LP01	–	IBC03	–
1294	TOLUENE	3	–	II	–	1 L	E2	P001	–	IBC02	–
△ 1295	TRICHLOROSILANE	4.3	8/3	I	–	0	E0	P401	PP31	–	–
1296	TRIETHYLAMINE	3	8	II	–	1 L	E2	P001	–	IBC02	–
1297	TRIMETHYLAMINE, AQUEOUS SOLUTION not more than 50% trimethylamine, by mass	3	8	I	–	0	E0	P001	–	–	–
1297	TRIMETHYLAMINE, AQUEOUS SOLUTION not more than 50% trimethylamine, by mass	3	8	II	–	1 L	E2	P001	–	IBC02	–
1297	TRIMETHYLAMINE, AQUEOUS SOLUTION not more than 50% trimethylamine, by mass	3	8	III	223	5 L	E1	P001	–	IBC03	–
△ 1298	TRIMETHYLCHLOROSILANE	3	8	II	–	0	E0	P010	–	–	–
1299	TURPENTINE	3	–	III	–	5 L	E1	P001 LP01	–	IBC03	–
1300	TURPENTINE SUBSTITUTE	3	–	II	–	1 L	E2	P001	–	IBC02	–
1300	TURPENTINE SUBSTITUTE	3	–	III	223	5 L	E1	P001 LP01	–	IBC03	–
1301	VINYL ACETATE, STABILIZED	3	–	II	–	1 L	E2	P001	–	IBC02	–
1302	VINYL ETHYL ETHER, STABILIZED	3	–	I	–	0	E3	P001	–	–	–
1303	VINYLIDENE CHLORIDE, STABILIZED	3	– P	I	–	0	E3	P001	–	–	–
1304	VINYL ISOBUTYL ETHER, STABILIZED	3	–	II	–	1 L	E2	P001	–	IBC02	–
△ 1305	VINYLTRICHLOROSILANE	3	8	II	–	0	E0	P010	–	–	–

Portable tanks and bulk containers			EmS	Stowage and segregation	Properties and observations	UN No.
	Tank instructions	Provisions				
(12)	(13) 4.2.5 4.3	(14) 4.2.5	(15) 5.4.3.2 7.3	(16) 7.1 7.2	(17)	(18)
–	T2	TP1	F-E, S-D	Category A.	Colourless liquid. Flashpoint: 37°C c.c. Explosive limits: 1.3% to 23%. Immiscible with water.	1292
–	T4	TP1 TP8	F-E, S-D	Category B.	Miscibility with water depends upon the composition.	1293
–	T2	TP1	F-E, S-D	Category A.	See entry above.	1293
–	T4	TP1	F-E, S-D	Category B.	Colourless liquid with a benzene-like odour. Flashpoint: 7°C c.c. Explosive limits: 1.27% to 7%. Immiscible with water.	1294
–	T14	TP2 TP7 TP13	F-G, S-O	Category D. Clear of living quarters. Segregation as for class 3 but "away from" classes 3, 4.1 and 8. See 7.2.6.3.2.	Colourless, very volatile, flammable and corrosive liquid. Flashpoint: below −50°C. Explosive limits: 1.2% to 90.5%. Boiling point: 32°C. Reacts with water or steam to produce heat, which may lead to self-ignition; toxic and corrosive fumes will be evolved. May react vigorously in contact with oxidizing substances. Causes burns to skin, eyes and mucous membranes.	1295 △
–	T7	TP1	F-E, S-C	Category B. Clear of living quarters.	Colourless liquid with a strong ammonia-like odour. Flashpoint: −11°C c.c. Explosive limits: 1.2% to 8%. Miscible with water. Harmful by inhalation. Causes burns to skin and eyes. Irritating to mucous membranes.	1296
–	T11	TP1	F-E, S-C	Category D. Clear of living quarters. "Separated from" mercury and mercury compounds.	Aqueous solution of a flammable gas with an ammonia-like odour. Flashpoint depending on percentage of dissolved gas. May react explosively with mercury. Miscible with water. An aqueous solution of 45% TRIMETHYLAMINE, by mass, has a flashpoint of −45°C c.c. and a boiling point of 30°C (applicable to PG I only). Harmful by inhalation. Causes burns to skin, eyes and mucous membranes.	1297
–	T7	TP1	F-E, S-C	Category B. Clear of living quarters. "Separated from" mercury and mercury compounds.	See entry above.	1297
–	T7	TP1	F-E, S-C	Category A. Clear of living quarters. "Separated from" mercury and mercury compounds.	See entry above. Irritating to skin, eyes and mucous membranes.	1297
–	T10	TP2 TP7 TP13	F-E, S-C	Category E. Clear of living quarters.	Colourless liquid. Flashpoint: below −18°C c.c. Explosive limits: 1.8% to 6%. Boiling point: 57°C. Immiscible with water. Readily hydrolysed by moisture, evolving hydrogen chloride, a toxic and corrosive gas. Causes burns to skin, eyes and mucous membranes.	1298 △
–	T2	TP1	F-E, S-E	Category A.	Colourless liquid. Flashpoint: 35°C c.c. Mixture of resin and volatile oils. Immiscible with water.	1299
–	T4	TP1	F-E, S-E	Category B.	Immiscible with water.	1300
–	T2	TP1	F-E, S-E	Category A.	See entry above.	1300
–	T4	TP1	F-E, S-D	Category B.	Colourless to light yellow liquid. Flashpoint: −8°C c.c. Explosive limits: 2.6% to 14%. Immiscible with water.	1301
–	T11	TP2	F-E, S-D	Category D.	Colourless liquid. Flashpoint: below −18°C c.c. Explosive limits: 1.7% to 28%. Boiling point: 33°C. Immiscible with water. Extremely reactive; may polymerize.	1302
–	T12	TP2 TP7	F-E, S-D	Category E. Clear of living quarters.	Colourless to straw-coloured, volatile liquid with a sweet odour. Flashpoint: −28°C c.c. Explosive limits: 6.5% to 15.5%. Boiling point: 32°C. Immiscible with water.	1303
–	T4	TP1	F-E, S-D	Category B.	Colourless liquid. Flashpoint: −9°C o.c. Immiscible with water.	1304
–	T10	TP2 TP7 TP13	F-E, S-C	Category B. Clear of living quarters.	Colourless, pale yellow or pink liquid with a pungent odour. Flashpoint: 11°C c.c. Explosive limits: 3% to.... Readily hydrolysed by moisture, evolving hydrogen chloride, an irritating and corrosive gas apparent as white fumes. Immiscible with water. In the presence of moisture, corrosive to most metals.	1305 △

UN No. (1)	Proper Shipping Name (PSN) (2) 3.1.2	Class or division (3) 2.0	Subsidiary risk(s) (4) 2.0	Packing group (5) 2.0.1.3	Special provisions (6) 3.3	Limited and excepted quantity provisions: Limited quantities (7a) 3.4	Excepted quantities (7b) 3.5	Packing: Instructions (8) 4.1.4	Packing: Provisions (9) 4.1.4	IBC: Instructions (10) 4.1.4	IBC: Provisions (11) 4.1.4
1306	WOOD PRESERVATIVES, LIQUID	3	–	II	–	5 L	E2	P001	–	IBC02	–
1306	WOOD PRESERVATIVES, LIQUID	3	–	III	223 955	5 L	E1	P001 LP01	–	IBC03	–
1307	XYLENES	3	–	II	–	1 L	E2	P001	–	IBC02	–
1307	XYLENES	3	–	III	223	5 L	E1	P001 LP01	–	IBC03	–
1308	ZIRCONIUM, SUSPENDED IN A FLAMMABLE LIQUID	3	–	I	–	0	E3	P001	PP33	–	–
1308	ZIRCONIUM, SUSPENDED IN A FLAMMABLE LIQUID	3	–	II	–	1 L	E2	P001	PP33	–	–
1308	ZIRCONIUM, SUSPENDED IN A FLAMMABLE LIQUID	3	–	III	223	5 L	E1	P001	–	–	–
1309	ALUMINIUM POWDER, COATED	4.1	–	II	–	1 kg	E2	P002	PP38	IBC08	B2 B4
1309	ALUMINIUM POWDER, COATED	4.1	–	III	223	5 kg	E1	P002 LP02	PP11 PP38	IBC08	B3
1310	AMMONIUM PICRATE, WETTED with not less than 10% water, by mass	4.1	–	I	28	0	E0	P406	PP26 PP31	–	–
1312	BORNEOL	4.1	–	III	–	5 kg	E1	P002 LP02	–	IBC08	B3
1313	CALCIUM RESINATE	4.1	–	III	–	5 kg	E1	P002	–	IBC06	–
1314	CALCIUM RESINATE, FUSED	4.1	–	III	–	5 kg	E1	P002	–	IBC04	–
1318	COBALT RESINATE, PRECIPITATED	4.1	–	III	–	5 kg	E1	P002	–	IBC06	–
1320	DINITROPHENOL, WETTED with not less than 15% water, by mass	4.1	6.1 P	I	28	0	E0	P406	PP26 PP31	–	–
1321	DINITROPHENOLATES, WETTED with not less than 15% water, by mass	4.1	6.1 P	I	28	0	E0	P406	PP26 PP31	–	–
1322	DINITRORESORCINOL, WETTED with not less than 15% water, by mass	4.1	–	I	28	0	E0	P406	PP26 PP31	–	–
1323	FERROCERIUM	4.1	–	II	249	1 kg	E2	P002	–	IBC08	B2 B4

Portable tanks and bulk containers			EmS	Stowage and segregation	Properties and observations	UN No.
	Tank instructions	Provisions				
(12)	(13) 4.2.5 4.3	(14) 4.2.5	(15) 5.4.3.2 7.3	(16) 7.1 7.2	(17)	(18)
–	T4	TP1 TP8	F-E, S-D	Category B.	Miscibility with water depends upon the composition. Harmful by inhalation.	1306
–	T2	TP1	F-E, S-D	Category A.	See entry above.	1306
–	T4	TP1	F-E, S-D	Category B.	Colourless liquids. Flashpoint: 17°C to 23°C c.c. Explosive limits: 1.1% to 7%. Immiscible with water.	1307
–	T2	TP1	F-E, S-D	Category A.	See entry above. Flashpoint: 23°C to 30°C c.c.	1307
–	–	–	F-E, S-D	Category D.	Finely divided zirconium metal in a flammable liquid. Immiscible with water. Spillage is liable to self-ignition.	1308
–	–	–	F-E, S-D	Category B.	See entry above.	1308
–	–	–	F-E, S-D	Category B.	See entry above.	1308
–	T3	TP33	F-G, S-G	Category A. Keep as dry as reasonably practicable. "Away from" liquid halogenated hydrocarbons. "Separated from" class 5.1, acids, alkalis and iron oxide.	If uncoated, it possesses the property of evolving hydrogen gas when in contact with water, especially seawater; if treated with oil or wax, it does not at ordinary temperatures. Reacts readily with acids and caustic alkalis, evolving hydrogen, a flammable gas. Reacts readily with iron oxide, producing a thermite effect. May form explosive mixtures with oxidizing substances. In the event of breakage of receptacles, the scattered powder is readily ignited by sparks or open flame and may give rise to an explosive atmosphere.	1309
–	T1	TP33	F-G, S-G	Category A. Keep as dry as reasonably practicable. "Away from" liquid halogenated hydrocarbons. "Separated from" class 5.1, acids, alkalis and iron oxide.	See entry above.	1309
–	–	–	F-B, S-J	Category D. "Away from" class 3 and heavy metals and their salts.	Desensitized explosive. Substance in pure form consists of yellow crystals. Explosive and sensitive to friction in the dry state. May form extremely sensitive compounds with heavy metals or their salts. Harmful if swallowed or by skin contact.	1310
–	T1	TP33	F-A, S-I	Category A.	White, translucent lumps. Camphor-like odour. Insoluble in water. Harmful by ingestion.	1312
–	T1	TP33	F-A, S-I	Category A.	Yellowish-white, amorphous powder or lumps. Insoluble in water. Liable to spontaneous heating. Irritating to skin and mucous membranes.	1313
–	T1	TP33	F-A, S-I	Category A.	Yellowish-white, amorphous powder or lumps. Insoluble in water. Liable to spontaneous heating. Irritating to skin and mucous membranes.	1314
–	T1	TP33	F-A, S-I	Category A.	Dark brownish-black solid. Insoluble in water. Readily combustible; may ignite spontaneously if contaminated with vegetable fibres (such as cotton). Irritating to skin and mucous membranes.	1318
–	–	–	F-B, S-J	Category E. "Away from" class 3 and heavy metals and their salts.	Desensitized explosive. Substance when pure consists of yellow crystals. Slightly soluble in water. May form extremely sensitive compounds with heavy metals or their salts. Toxic if swallowed, by skin contact or by inhalation.	1320
–	–	–	F-B, S-J	Category E. "Away from" class 3 and heavy metals and their salts.	Desensitized explosives. Explosive and sensitive to friction in the dry state. May form extremely sensitive compounds with heavy metals or their salts. Toxic if swallowed, by skin contact or by inhalation.	1321
–	–	–	F-B, S-J	Category E. "Away from" class 3 and heavy metals and their salts.	Desensitized explosive. Explosive when dry. May form extremely sensitive compounds with heavy metals or their salts. Harmful if swallowed or by skin contact.	1322
–	T3	TP33	F-G, S-G	Category A.	Alloy derived from cerium or mischmetal, with the addition of 10% to 65% iron. Emits sparks when struck.	1323

UN No.	Proper Shipping Name (PSN)	Class or division	Subsidiary risk(s)	Packing group	Special provisions	Limited and excepted quantity provisions		Packing		IBC	
						Limited quantities	Excepted quantities	Instructions	Provisions	Instructions	Provisions
(1)	(2) 3.1.2	(3) 2.0	(4) 2.0	(5) 2.0.1.3	(6) 3.3	(7a) 3.4	(7b) 3.5	(8) 4.1.4	(9) 4.1.4	(10) 4.1.4	(11) 4.1.4
1324	FILMS, NITROCELLULOSE BASE gelatin coated, except scrap	4.1	–	III	–	5 kg	E1	P002	PP15	–	–
1325	FLAMMABLE SOLID, ORGANIC, N.O.S.	4.1	–	II	274 915	1 kg	E2	P002	–	IBC08	B2 B4
1325	FLAMMABLE SOLID, ORGANIC, N.O.S.	4.1	–	III	223 274 915	5 kg	E1	P002	–	IBC08	B3
1326	HAFNIUM POWDER, WETTED with not less than 25% water (a visible excess of water must be present) (a) mechanically produced, particle size less than 53 microns; (b) chemically produced, particle size less than 840 microns	4.1	–	II	916	1 kg	E2	P410	PP31 PP40	IBC06	B2
1327	HAY, STRAW or BHUSA	4.1	–	–	29 281 954	3 kg	E0	P003	PP19	IBC08	B6
1328	HEXAMETHYLENETETRAMINE	4.1	–	III	–	5 kg	E1	P002	–	IBC08	B3
1330	MANGANESE RESINATE	4.1	–	III	–	5 kg	E1	P002	–	IBC06	–
1331	MATCHES, "STRIKE ANYWHERE"	4.1	–	III	293	5 kg	E1	P407	PP27	–	–
1332	METALDEHYDE	4.1	–	III	–	5 kg	E1	P002 LP02	–	IBC08	B3
1333	CERIUM slabs, ingots or rods	4.1	–	II	–	1 kg	E2	P002	–	IBC08	B2 B4
△ 1334	NAPHTHALENE, CRUDE or NAPHTHALENE, REFINED	4.1	–	III	948 967	5 kg	E1	P002 LP02	–	IBC08	B3
1336	NITROGUANIDINE (PICRITE), WETTED with not less than 20% water, by mass	4.1	–	I	28	0	E0	P406	PP31	–	–

Portable tanks and bulk containers			EmS	Stowage and segregation	Properties and observations	UN No.
	Tank instructions	Provisions				
(12)	(13) 4.2.5 4.3	(14) 4.2.5	(15) 5.4.3.2 7.3	(16) 7.1 7.2	(17)	(18)
–	–	–	F-G, S-I	Category D. "Away from" class 3.	Ignites readily. When involved in a fire, evolves toxic fumes; in closed compartments, these fumes may form an explosive mixture with air.	1324
–	T3	TP33	F-A, S-G	Category B.	–	1325
–	T1	TP33	F-A, S-G	Category B.	–	1325
–	T3	TP33	F-A, S-J	Category E. "Separated from" class 5.1.	Insoluble in water. Liable to spontaneous combustion when dry. Forms explosive mixtures with oxidizing substances.	1326
–	–	–	F-A, S-I	Category A. "Away from" animal or vegetable oils. Unless carried in closed cargo transport units, bales shall be properly covered by tarpaulins or the like. Cargo spaces shall be clean, dry and free from oil or grease. Ventilator cowls leading into the cargo space shall have spark-preventing screens. All other openings, entrances and hatches leading to the cargo space shall be securely closed. During temporary interruption of loading, when the hatch remains uncovered, a fire-watch shall be kept. During loading or discharge, smoking in the vicinity shall be prohibited and fire-fighting appliances kept ready for immediate operation.	Ignites readily. Liable to spontaneous combustion when wet, damp or contaminated with oil. Refuse for shipment when loose, damp, wet or contaminated with oil.	1327
–	T1	TP33	F-A, S-G	Category A.	White, crystalline powder. Soluble in water.	1328
–	T1	TP33	F-A, S-I	Category A.	Very dark brown solid. Insoluble in water. Liable to spontaneous heating. Irritating to skin, eyes and mucous membranes.	1330
–	–	–	F-A, S-I	Category B.	Ignite by friction; prepared surface is not required.	1331
–	T1	TP33	F-A, S-G	Category A.	White crystals, powder or tablets. Insoluble in water. Harmful if swallowed or by dust inhalation.	1332
–	–	–	F-G, S-P	Category A. "Separated from" classes 3 and 5.1.	Contains 94–99% rare earth metals. In contact with water or moist air, evolves hydrogen, a flammable gas. Emits sparks when scratched or struck.	1333
–	T1 BK2 BK3	TP33	F-A, S-G	Category A. When transported in BK3 bulk container, see 7.6.2.12 and 7.7.3.9.	Crystalline flakes or powder with a persistent odour. Evolves flammable vapours at, or below, its melting point.	1334 △
–	–	–	F-B, S-J	Category E. "Away from" class 3 and heavy metals and their salts.	Desensitized explosive. White solid. When involved in a fire, evolves toxic fumes; in closed compartments, these fumes may form an explosive mixture with air. May form extremely sensitive compounds with heavy metals or their salts.	1336

UN No.	Proper Shipping Name (PSN)	Class or division	Subsidiary risk(s)	Packing group	Special provisions	Limited and excepted quantity provisions		Packing		IBC	
						Limited quantities	Excepted quantities	Instructions	Provisions	Instructions	Provisions
(1)	(2) 3.1.2	(3) 2.0	(4) 2.0	(5) 2.0.1.3	(6) 3.3	(7a) 3.4	(7b) 3.5	(8) 4.1.4	(9) 4.1.4	(10) 4.1.4	(11) 4.1.4
1337	NITROSTARCH, WETTED with not less than 20% water, by mass	4.1	–	I	28	0	E0	P406	PP31	–	–
1338	PHOSPHORUS, AMORPHOUS	4.1	–	III	–	5 kg	E1	P410	–	IBC08	B3
1339	PHOSPHORUS HEPTASULPHIDE free from yellow or white phosphorus	4.1	–	II	–	1 kg	E2	P410	PP31	IBC04	–
1340	PHOSPHORUS PENTASULPHIDE free from yellow or white phosphorus	4.3	4.1	II	–	500 g	E2	P410	PP31 PP40	IBC04	–
1341	PHOSPHORUS SESQUISULPHIDE free from yellow or white phosphorus	4.1	–	II	–	1 kg	E2	P410	PP31	IBC04	–
1343	PHOSPHORUS TRISULPHIDE free from yellow or white phosphorus	4.1	–	II	–	1 kg	E2	P410	PP31	IBC04	–
1344	TRINITROPHENOL (PICRIC ACID), WETTED with not less than 30% water, by mass	4.1	–	I	28	0	E0	P406	PP26 PP31	–	–
1345	RUBBER SCRAP powdered or granulated, not exceeding 840 microns and rubber content exceeding 45% or RUBBER SHODDY powdered or granulated, not exceeding 840 microns and rubber content exceeding 45%	4.1	–	II	223 917	1 kg	E2	P002	–	IBC08	B2 B4
1346	SILICON POWDER, AMORPHOUS	4.1	–	III	32	5 kg	E1	P002 LP02	–	IBC08	B3
△ 1347	SILVER PICRATE, WETTED with not less than 30% water, by mass	4.1	–	I	28 900	0	E0	P406	PP25 PP26 PP31	–	–
1348	SODIUM DINITRO-*o*-CRESOLATE, WETTED with not less than 15% water, by mass	4.1	6.1 P	I	28	0	E0	P406	PP26 PP31	–	–
1349	SODIUM PICRAMATE, WETTED with not less than 20% water, by mass	4.1	–	I	28	0	E0	P406	PP26 PP31	–	–

Portable tanks and bulk containers			EmS	Stowage and segregation	Properties and observations	UN No.
	Tank instructions	Provisions				
(12)	(13) 4.2.5 4.3	(14) 4.2.5	(15) 5.4.3.2 7.3	(16) 7.1 7.2	(17)	(18)
–	–	–	F-B, S-J	Category D. "Away from" class 3 and heavy metals and their salts.	Desensitized explosive. Orange powder. Explosive and sensitive to friction in the dry state. When involved in a fire, evolves toxic fumes; in closed compartments these fumes may form an explosive mixture with air. May form extremely sensitive compounds with heavy metals or their salts.	1337
–	T1	TP33	F-A, S-G	Category A. "Separated from" class 5.1.	Reddish-brown powder. Insoluble in water. Ignites readily by friction. When involved in a fire, evolves irritating fumes. Forms explosive mixtures with oxidizing substances. Harmful if swallowed or by dust inhalation.	1338
–	T3	TP33	F-G, S-G	Category B. "Separated from" class 5.1	Yellow solid. Ignites readily by friction. Develops heat in contact with moist air, evolving toxic and flammable gases. Forms explosive mixtures with oxidizing substances. Harmful if swallowed or by dust inhalation.	1339
–	T3	TP33	F-G, S-N	Category D.	Yellow solid. Ignites readily by friction. Develops heat in contact with moist air, evolving toxic and flammable gases. Forms explosive mixtures with oxidizing substances. Harmful if swallowed or by dust inhalation.	1340
–	T3	TP33	F-A, S-G	Category B. "Separated from" class 5.1.	Yellow solid. Ignites readily by friction. Develops heat in contact with moist air, evolving toxic and flammable gases. Forms explosive mixtures with oxidizing substances. Harmful if swallowed or by dust inhalation.	1341
–	T3	TP33	F-G, S-G	Category B. "Separated from" class 5.1.	Yellow solid. Ignites readily by friction. Develops heat in contact with moist air, evolving toxic and flammable gases. Forms explosive mixtures with oxidizing substances. Harmful if swallowed or by dust inhalation.	1343
–	–	–	F-B, S-J	Category E. "Away from" class 3 and heavy metals and their salts.	Desensitized explosive. Substance in pure form consists of yellow crystals. Soluble in water. Explosive and sensitive to friction in the dry state. May form extremely sensitive compounds with heavy metals or their salts. Harmful if swallowed or by skin contact.	1344
–	T3	TP33	F-A, S-I	Category A.	Liable to spontaneous heating.	1345
–	T1	TP33	F-A, S-G	Category A. "Separated from" class 5.1.	Dark brown, non-metallic powder. Burns in air, when ignited; readily flammable when mixed with oxidizing substances.	1346
–	–	–	F-B, S-J	Category D. "Away from" class 3 and heavy metals and their salts.	Desensitized explosive. Yellow crystals. Soluble in water. Explosive and sensitive to friction in the dry state. Harmful if swallowed or by skin contact. May form extremely sensitive compounds with heavy metals or their salts. Transport of SILVER PICRATE dry or wetted with less than 30% water, by mass is prohibited.	1347 △
–	–	–	F-B, S-J	Category E. "Away from" class 3 and heavy metals and their salts.	Desensitized explosive. Substance in pure form consists of yellow powder. Explosive and sensitive to friction in the dry state. May form extremely sensitive compounds with heavy metals or their salts. When involved in a fire, evolves toxic fumes; in closed compartments, these fumes may form an explosive mixture with air. Toxic if swallowed, by skin contact or by inhalation.	1348
–	–	–	F-B, S-J	Category E. "Away from" class 3 and heavy metals and their salts.	Desensitized explosive. Substance in pure form consists of yellow powder. Explosive and sensitive to friction in the dry state. May form extremely sensitive compounds with heavy metals or their salts. When involved in a fire, evolves toxic fumes; in closed compartments, these fumes may form an explosive mixture with air. Harmful if swallowed or by skin contact.	1349

UN No.	Proper Shipping Name (PSN)	Class or division	Subsidiary risk(s)	Packing group	Special provisions	Limited and excepted quantity provisions		Packing		IBC	
						Limited quantities	Excepted quantities	Instructions	Provisions	Instructions	Provisions
(1)	(2) 3.1.2	(3) 2.0	(4) 2.0	(5) 2.0.1.3	(6) 3.3	(7a) 3.4	(7b) 3.5	(8) 4.1.4	(9) 4.1.4	(10) 4.1.4	(11) 4.1.4
△ 1350	SULPHUR	4.1	–	III	242 967	5 kg	E1	P002 LP02	–	IBC08	B3
△ 1352	TITANIUM POWDER, WETTED with not less than 25% water (a visible excess of water must be present) (a) mechanically produced, particle size less than 53 microns; (b) chemically produced, particle size less than 840 microns	4.1	–	II	28 916	1 kg	E2	P410	PP31 PP40	IBC06	B2
1353	FIBRES or FABRICS IMPREGNATED WITH WEAKLY NITRATED NITROCELLULOSE, N.O.S.	4.1	–	III	–	5 kg	E1	P410	–	IBC08	B3
1354	TRINITROBENZENE, WETTED with not less than 30% water, by mass	4.1	–	I	28	0	E0	P406	PP31	–	–
1355	TRINITROBENZOIC ACID, WETTED with not less than 30% water, by mass	4.1	–	I	28	0	E0	P406	PP31	–	–
1356	TRINITROTOLUENE (TNT), WETTED with not less than 30% water, by mass	4.1	–	I	28	0	E0	P406	PP31	–	–
1357	UREA NITRATE, WETTED with not less than 20% water, by mass	4.1	–	I	28 227 919	0	E0	P406	PP31	–	–
1358	ZIRCONIUM POWDER, WETTED with not less than 25% water (a visible excess of water must be present) (a) mechanically produced, particle size less than 53 microns; (b) chemically produced, particle size less than 840 microns	4.1	–	II	916	1 kg	E2	P410	PP31 PP40	IBC06	B2
1360	CALCIUM PHOSPHIDE	4.3	6.1	I	–	0	E0	P403	PP31	–	–

Portable tanks and bulk containers			EmS	Stowage and segregation	Properties and observations	UN No.
	Tank instructions	Provisions				
(12)	(13) 4.2.5 4.3	(14) 4.2.5	(15) 5.4.3.2 7.3	(16) 7.1 7.2	(17)	(18)
–	T1 BK2 BK3	TP33	F-A, S-G	Category A. Protected from sources of heat. "Separated from" class 5.1. When transported in BK3 bulk container, see 7.6.2.12 and 7.7.3.9.	When involved in a fire, evolves toxic, very irritating and suffocating gas. The dust forms an explosive mixture with air which may be ignited by static electricity. Forms explosive mixtures with oxidizing substances. Corrosive to steel, in particular in the presence of moisture. The provisions of this Code should not apply to sulphur when it is formed to a specific shape (such as prills, granules, pellets, pastilles or flakes).	**1350** △
–	T3	TP33	F-A, S-J	Category E. "Separated from" class 5.1.	Grey powder. Forms explosive mixtures with oxidizing substances.	**1352** △
–	–	–	F-A, S-I	Category D.	Toe board used in the manufacture of boots and shoes. When involved in a fire, evolves toxic fumes; in closed compartments, these fumes may form an explosive mixture with air.	**1353**
–	–	–	F-B, S-J	Category E. "Away from" class 3 and heavy metals and their salts.	Desensitized explosive. Substance in pure form consists of yellow crystals. When involved in a fire, evolves toxic fumes; in closed compartments these fumes may form an explosive mixture with air. Explosive and sensitive to friction in the dry state. May form extremely sensitive compounds with heavy metals or their salts. Harmful if swallowed or by skin contact.	1354
–	–	–	F-B, S-J	Category E. "Away from" class 3 and heavy metals and their salts.	Desensitized explosive. Substance in pure form consists of yellow crystals. Soluble in water. When involved in a fire, evolves toxic fumes; in closed compartments these fumes may form an explosive mixture with air. Explosive and sensitive to friction in the dry state. Harmful if swallowed or by skin contact. May form extremely sensitive compounds with heavy metals or their salts.	**1355**
–	–	–	F-B, S-J	Category E. "Away from" class 3 and heavy metals and their salts.	Desensitized explosive. Substance in pure form consists of yellow crystals. When involved in a fire, evolves toxic fumes; in closed compartments these fumes may form an explosive mixture with air. Explosive and sensitive to friction in the dry state. May form extremely sensitive compounds with heavy metals or their salts. Harmful if swallowed or by skin contact.	1356
–	–	–	F-B, S-J	Category E. "Away from" class 3 and heavy metals and their salts.	Desensitized explosive. Substance in pure form consists of white crystals. Soluble in water. When involved in a fire, evolves toxic fumes; in closed compartments these fumes may form an explosive mixture with air. Explosive and sensitive to friction in the dry state. May form extremely sensitive compounds with heavy metals or their salts.	**1357**
–	T3	TP33	F-G, S-J	Category E. "Separated from" class 5.1.	Grey powder. Insoluble in water. Liable to spontaneous combustion when dry. Forms explosive mixtures with oxidizing substances.	1358
–	–	–	F-G, S-N	Category E. Under deck in a mechanically ventilated space. Clear of living quarters. "Separated from" acids.	Red to brown crystals. Reacts with acids or decomposes slowly in contact with water or damp air, evolving phosphine, a spontaneously flammable and highly toxic gas. Reacts violently with oxidizing substances. Toxic if swallowed, by skin contact or by inhalation.	**1360**

UN No.	Proper Shipping Name (PSN)	Class or division	Subsidiary risk(s)	Packing group	Special provisions	Limited and excepted quantity provisions		Packing		IBC	
						Limited quantities	Excepted quantities	Instructions	Provisions	Instructions	Provisions
(1)	(2) 3.1.2	(3) 2.0	(4) 2.0	(5) 2.0.1.3	(6) 3.3	(7a) 3.4	(7b) 3.5	(8) 4.1.4	(9) 4.1.4	(10) 4.1.4	(11) 4.1.4
1361	CARBON animal or vegetable origin	4.2	–	II	925	0	E2	P002	PP12	IBC06	–
1361	CARBON animal or vegetable origin	4.2	–	III	223 925	0	E1	P002 LP02	PP12	IBC08	B3
1362	CARBON, ACTIVATED	4.2	–	III	223 925	0	E1	P002	PP11 PP31	IBC08	B3
△ 1363	COPRA	4.2	–	III	29 926	0	E1	P003 LP02	PP20	IBC08	B3 B6
1364	COTTON WASTE, OILY	4.2	–	III	29	0	E1	P003 LP02	PP19	IBC08	B3 B6
1365	COTTON, WET	4.2	–	III	29	0	E1	P003	PP19	IBC08	B3 B6
△ 1369	*p*-NITROSODIMETHYLANILINE	4.2	–	II	927	0	E2	P410	–	IBC06	B2
1372	FIBRES ANIMAL or FIBRES VEGETABLE burnt, wet or damp	4.2	–	III	117	0	E1	P410	–	–	–
1373	FIBRES or FABRICS, ANIMAL or VEGETABLE or SYNTHETIC, N.O.S. with oil	4.2	–	III	–	0	E1	P410	PP31	IBC08	B3
△ 1374	FISHMEAL, UNSTABILIZED or FISHSCRAP, UNSTABILIZED High hazard. Unrestricted moisture content, Unrestricted fat content in excess of 12%, by mass; unrestricted fat content in excess of 15%, by mass, in the case of anti-oxidant treated fishmeal or fishscrap	4.2	–	II	300 928	0	E2	P410	PP31 PP40	IBC08	B2 B4
△ 1374	FISHMEAL, UNSTABILIZED or FISHSCRAP, UNSTABILIZED Not anti-oxidant treated. Moisture content: more than 5% but not more than 12%, by mass. Fat content: not more than 12%, by mass	4.2	–	III	29 300 907 928	0	E1	P410	PP31	IBC08	B2 B3

Portable tanks and bulk containers			EmS	Stowage and segregation	Properties and observations	UN No.	
	Tank instructions	Provisions					
(12)	(13) 4.2.5 4.3	(14) 4.2.5	(15) 5.4.3.2 7.3	(16) 7.1 7.2	(17)	(18)	
–	T3	TP33	F-A, S-J	Category A. Keep as cool as possible.	Black powder or granules. Liable to heat slowly and ignite spontaneously in air. The material as offered for shipment should have been sufficiently heat-treated and should be cooled down to ambient temperature before packing.	1361	
–	T1	TP33	F-A, S-J	Category A. Keep as cool as possible.	See entry above.	1361	
–	T1	TP33	F-A, S-J	Category A. Keep as cool as possible.	Black powder or granules. Liable to heat slowly and ignite spontaneously in air. The material as offered for shipment should have been sufficiently heat-treated and should be cooled down to ambient temperature before packing.	1362	
–	BK2	–	F-A, S-J	Category A. Keep as dry as reasonably practicable. Protected from sources of heat. Provide a good through ventilation for bagged cargo. Double strip stowage is recommended. The illustration in 7.6.2.7.2.3 shows how this can be achieved. During the voyage regular temperature readings shall be taken at varying depths in the hold and recorded. If the temperature of the cargo exceeds the ambient temperature and continues to increase, ventilation shall be closed down.	Dried kernels of coconuts, with a penetrating rancid odour which may taint other cargoes.	1363	△
–	–	–	F-A, S-J	Category A. "Separated from" animal or vegetable oil.	Fibres of vegetable origin.	1364	
–	–	–	F-A, S-J	Category A.	Readily combustible, liable to ignite spontaneously according to moisture content.	1365	
–	T3	TP33	F-A, S-J	Category D. Segregation from foodstuffs as in 7.3.4.2, 7.6.3.1.2 or 7.7.3.6.	Dark green, crystalline solid, insoluble in water. Ignites spontaneously in air when dry. Harmful if swallowed.	1369	△
–	–	–	F-A, S-J	Category A.	Liable to ignite spontaneously according to moisture content.	1372	
–	T1	TP33	F-A, S-J	Category A.	Liable to ignite spontaneously according to the oil content.	1373	
–	T3	TP33	F-A, S-J	Category B. "Separated by a complete compartment or hold from" class 1 except from division 1.4. For special stowage provisions, see 7.4.1.3 and 7.6.2.7.2. Protected from sources of heat.	Brown to greenish-brown product derived from oily fish. Strong odour which may affect other cargo. Liable to heat and ignite spontaneously.	1374	△
–	T1	TP33	F-A, S-J	Category A. For special stowage provisions, see 7.4.1.3 and 7.6.2.7.2. Protected from sources of heat.	See entry above.	1374	△

3

DGL

3

DGL

UN No.	Proper Shipping Name (PSN)	Class or division	Subsidiary risk(s)	Packing group	Special provisions	Limited and excepted quantity provisions		Packing		IBC	
						Limited quantities	Excepted quantities	Instructions	Provisions	Instructions	Provisions
(1)	(2) 3.1.2	(3) 2.0	(4) 2.0	(5) 2.0.1.3	(6) 3.3	(7a) 3.4	(7b) 3.5	(8) 4.1.4	(9) 4.1.4	(10) 4.1.4	(11) 4.1.4
1376	IRON OXIDE, SPENT or IRON SPONGE, SPENT obtained from coal gas purification	4.2	–	III	223	0	E1	P002 LP02	–	IBC08	B3
1378	METAL CATALYST, WETTED with a visible excess of liquid	4.2	–	II	274	0	E2	P410	PP31 PP39 PP40	IBC01	–
1379	PAPER, UNSATURATED OIL TREATED incompletely dried (including carbon paper)	4.2	–	III	–	0	E1	P410	PP31	IBC08	B3
1380	PENTABORANE	4.2	6.1	I	–	0	E0	P601	–	–	–
1381	PHOSPHORUS, WHITE or YELLOW, DRY or UNDER WATER or IN SOLUTION	4.2	6.1 P	I	–	0	E0	P405	PP31	–	–
1382	POTASSIUM SULPHIDE, ANHYDROUS or POTASSIUM SULPHIDE with less than 30% water of crystallization	4.2	–	II	–	0	E2	P410	PP31 PP40	IBC06	B2
1383	PYROPHORIC METAL, N.O.S. or PYROPHORIC ALLOY, N.O.S.	4.2	–	I	274	0	E0	P404	PP31	–	–
1384	SODIUM DITHIONITE (SODIUM HYDROSULPHITE)	4.2	–	II	–	0	E2	P410	PP31	IBC06	B2
1385	SODIUM SULPHIDE, ANHYDROUS or SODIUM SULPHIDE with less than 30% water of crystallization	4.2	–	II	–	0	E2	P410	PP31	IBC06	B2
△ 1386	SEED CAKE, containing vegetable oil (a) mechanically expelled seeds, containing more than 10% oil or more than 20% oil and moisture combined	4.2	–	III	29 929	0	E1	P003 LP02	PP20	IBC08	B3 B6
△ 1386	SEED CAKE, containing vegetable oil (b) solvent extractions and expelled seeds, containing not more than 10% of oil and when the amount of moisture is higher than 10%, not more than 20% of oil and moisture combined	4.2	–	III	29 929	0	E1	P003 LP02	PP20	IBC08	B3 B6
1387	WOOL WASTE, WET	4.2	–	III	117	0	E1	P410	–	–	–

Portable tanks and bulk containers			EmS	Stowage and segregation	Properties and observations	UN No.
	Tank instructions	Provisions				
(12)	(13) 4.2.5 4.3	(14) 4.2.5	(15) 5.4.3.2 7.3	(16) 7.1 7.2	(17)	(18)
–	T1 BK2	TP33	F-G, S-P	Category E.	Obtained from coal gas purification. Strong odour which may taint other cargo. Liable to heat and ignite spontaneously. May evolve hydrogen sulphide, sulphur dioxide and hydrogen cyanide, which are toxic gases. This substance should have been cooled and weathered for not less than eight weeks before shipment, unless packed in a metal drum.	1376
–	T3	TP33	F-H, S-M	Category C.	Liable to ignite spontaneously if dry.	1378
–	–	–	F-A, S-J	Category A.	Liable to ignite spontaneously. The provisions of this Code should not apply to manufactured articles properly aged.	1379
–	–	–	F-G, S-L	Category D.	Colourless liquid. Boiling point range: 48°C to 63°C. Ignites spontaneously in air. Decomposes in contact with water, evolving hydrogen, a flammable gas. Toxic if swallowed, by skin contact or by inhalation.	1380
–	T9	TP3 TP31	F-A, S-J	Category E.	Ignites spontaneously in air. Melting point: 44°C. Toxic if swallowed, by skin contact or by inhalation. Receptacles are usually filled with substance in the liquid state which subsequently solidifies. A sufficient ullage should be allowed.	1381
–	T3	TP33	F-A, S-J	Category A. "Separated from" acids.	Black solid, absorbs moisture to become crystalline. Liable to ignite spontaneously. In contact with acids, evolves hydrogen sulphide, a toxic and flammable gas. Reacts violently with acids.	1382
–	T21	TP7 TP33	F-G, S-M	Category D.	Liable to ignite spontaneously in air. If shaken, may produce sparks. In contact with water, evolves hydrogen, a flammable gas.	1383
–	T3	TP33	F-A, S-J	Category E. Keep as dry as reasonably practicable.	White or grey crystalline powder. Liable to heat and ignite spontaneously in air and to evolve sulphur dioxide, an irritating gas.	1384
–	T3	TP33	F-A, S-J	Category A. "Separated from" acids.	Black solid, absorbs moisture to become crystalline. Liable to ignite spontaneously. In contact with acids, evolves hydrogen sulphide, a toxic and flammable gas. Reacts violently with acids.	1385
–	BK2	–	F-A, S-J	Category E. Keep dry. Protected from sources of heat. For special stowage provisions, see 7.6.2.7.2.	Residue remaining after oil has been expelled mechanically from oil-bearing seeds. Used mainly as animal feed or fertilizer. The most common seed cakes include those derived from coconut (copra), cottonseed, groundnut (peanut), linseed, maize (hominy chop), niger seed, palm kernel, rape seed, rice bran, soya bean and sunflower seed and they may be shipped in the form of cake, flakes, pellets, meal, etc. May self-heat slowly and, if wet or containing an excessive proportion of unoxidized oil, ignite spontaneously. Before shipment, this cargo should be properly aged. The duration of ageing varies with the oil content. Smoking and the use of naked lights should be prohibited during loading and unloading and on entry to the cargo space(s) at any time.	1386 △
–	BK2	–	F-A, S-J	Category A. Keep dry. Protected from sources of heat. For special stowage provisions, see 7.6.2.7.2.	Residue remaining after oil has been extracted by a solvent process or expelled mechanically from oil-bearing seeds. Used mainly as animal feed or fertilizer. The most common seed cakes include those derived from coconut (copra), cottonseed, groundnut (peanut), linseed, maize (hominy chop), niger seed, palm kernel, rape seed, rice bran, soya bean and sunflower seed and they may be shipped in the form of cake, flakes, pellets, meal, etc. May self-heat slowly and, if wet or containing an excessive proportion of unoxidized oil, ignite spontaneously. The seed cake should be substantially free from flammable solvent. Before shipment, this cargo should be properly aged. The duration of ageing varies with the oil content. Smoking and the use of naked lights should be prohibited during loading and unloading and on entry to the cargo space(s) at any other time.	1386 △
–	–	–	F-A, S-J	Category A.	Liable to ignite spontaneously in air according to moisture content.	1387

UN No.	Proper Shipping Name (PSN)	Class or division	Subsidiary risk(s)	Packing group	Special provisions	Limited and excepted quantity provisions		Packing		IBC	
						Limited quantities	Excepted quantities	Instructions	Provisions	Instructions	Provisions
(1)	(2) 3.1.2	(3) 2.0	(4) 2.0	(5) 2.0.1.3	(6) 3.3	(7a) 3.4	(7b) 3.5	(8) 4.1.4	(9) 4.1.4	(10) 4.1.4	(11) 4.1.4
1389	ALKALI METAL AMALGAM, LIQUID	4.3	–	I	182	0	E0	P402	PP31	–	–
1390	ALKALI METAL AMIDE	4.3	–	II	182	500 g	E2	P410	PP31 PP40	IBC07	B2
1391	ALKALI METAL DISPERSION or ALKALINE EARTH METAL DISPERSION	4.3	–	I	182 183	0	E0	P402	PP31	–	–
1392	ALKALINE EARTH METAL AMALGAM, LIQUID	4.3	–	I	183	0	E0	P402	PP31	–	–
1393	ALKALINE EARTH METAL ALLOY, N.O.S.	4.3	–	II	183	500 g	E2	P410	PP31 PP40	IBC07	B2
1394	ALUMINIUM CARBIDE	4.3	–	II	–	500 g	E2	P410	PP31 PP40	IBC07	B2
1395	ALUMINIUM FERROSILICON POWDER	4.3	6.1	II	932	500 g	E2	P410	–	IBC05	B2
1396	ALUMINIUM POWDER, UNCOATED	4.3	–	II	–	500 g	E2	P410	PP40	IBC07	B2
1396	ALUMINIUM POWDER, UNCOATED	4.3	–	III	223	1 kg	E1	P410	–	IBC08	B4
1397	ALUMINIUM PHOSPHIDE	4.3	6.1	I	–	0	E0	P403	PP31	–	–
1398	ALUMINIUM SILICON POWDER, UNCOATED	4.3	–	III	37 223 932	1 kg	E1	P410	–	IBC08	B4
1400	BARIUM	4.3	–	II	–	500 g	E2	P410	PP31 PP40	IBC07	B2
1401	CALCIUM	4.3	–	II	–	500 g	E2	P410	PP31 PP40	IBC07	B2
△ 1402	CALCIUM CARBIDE	4.3	–	I	951	0	E0	P403	PP31	IBC04	B1
1402	CALCIUM CARBIDE	4.3	–	II	951	500 g	E2	P410	PP40	IBC07	B2

Portable tanks and bulk containers			EmS	Stowage and segregation	Properties and observations	UN No.
	Tank instructions	Provisions				
(12)	(13) 4.2.5 4.3	(14) 4.2.5	(15) 5.4.3.2 7.3	(16) 7.1 7.2	(17)	(18)
–	–	–	F-G, S-N	Category D. "Separated from" acids.	Si very liquid, consisting of metal alloyed with mercury. Reacts with moisture, water or acids, evolving hydrogen, a flammable gas. When heated, evolves toxic vapours.	1389
–	T3	TP33	F-G, S-O	Category E. Clear of living quarters. "Separated from" acids.	Small crystals. Decomposes in contact with water or acids, evolving ammonia vapour and producing highly caustic alkaline solutions.	1390
–	–	–	F-G, S-N	Category D. "Separated from" acids.	Finely divided alkali metal or alkaline earth metal, suspended in a liquid. Reacts violently with moisture, water or acids, evolving hydrogen, which may be ignited by the heat of the reaction.	1391
–	–	–	F-G, S-N	Category D. "Separated from" acids.	Consists of metal alloyed with mercury. Contains 2% to 10% alkaline earth metals and may contain up to 98% mercury. Reacts with moisture, water or acids, evolving hydrogen, a flammable gas. When heated, evolves toxic vapours.	1392
–	T3	TP33	F-G, S-N	Category E. "Separated from" acids.	When containing a substantial proportion of alkaline earth metals, readily decomposed by water and reacts violently with acids, evolving hydrogen, which may be ignited by the heat of the reaction.	1393
–	T3	TP33	F-G, S-N	Category A. "Separated from" acids.	Yellow crystals or powder. In contact with water, rapidly evolves methane, a flammable gas. Reacts violently with acids.	1394
–	T3 BK2	TP33	F-G, S-N	Category A. Only to be loaded under dry weather conditions. Under deck in a mechanically ventilated space. Clear of living quarters. "Away from" liquid halogenated hydrocarbons. "Separated from" acids and alkalis.	In contact with water, caustic alkalis or acids, evolves hydrogen, a flammable gas. Impurities may, under similar circumstances, produce phosphine and arsine, which are highly toxic gases.	1395
–	T3	TP33	F-G, S-O	Category A. "Away from" liquid halogenated hydrocarbons. "Separated from" acids and alkalis.	In contact with water, caustic alkalis or acids, evolves hydrogen, a flammable gas. When finely divided aluminium dust is scattered, it is easily ignited by naked lights, causing explosion. May explode when in contact with oxidizing substances. Reacts with liquid halogenated hydrocarbons.	1396
–	T1	TP33	F-G, S-O	Category A. "Away from" liquid halogenated hydrocarbons. "Separated from" acids and alkalis.	See entry above.	1396
–	–	–	F-G, S-N	Category E. Under deck in a mechanically ventilated space. Clear of living quarters. "Separated from" acids.	Crystals or powder. Reacts with acids or decomposes slowly in contact with water or damp air, evolving phosphine, a spontaneously flammable and highly toxic gas. Reacts violently with oxidizing substances. Toxic if swallowed, by skin contact or by inhalation.	1397
–	T1 BK2	TP33	F-G, S-N	Category A. Only to be loaded under dry weather conditions. Under deck in a mechanically ventilated space. Clear of living quarters. "Away from" liquid halogenated hydrocarbons. "Separated from" acids and alkalis.	In contact with water, caustic alkalis or acids, generates heat and evolves hydrogen, a flammable gas. May also evolve silanes, which are toxic and may ignite spontaneously.	1398
–	T3	TP33	F-G, S-O	Category E. "Separated from" acids.	Readily decomposes in water and reacts violently with acids, evolving hydrogen, which may be ignited by the heat of the reaction. Harmful if swallowed or by dust inhalation.	1400
–	T3	TP33	F-G, S-O	Category E. "Separated from" acids.	Readily decomposes in water and reacts violently with acids, evolving hydrogen, which may be ignited by the heat of the reaction.	1401
–	BK2	–	F-G, S-N	Category B. "Separated from" acids.	Solid. In contact with water, rapidly evolves acetylene, a highly flammable gas, which may be ignited by the heat of the reaction. Acetylene forms highly explosive compounds with salts of some heavy metals. Reacts violently with acids.	1402 △
–	T3 BK2	TP33	F-G, S-N	Category B. "Separated from" acids.	See entry above.	1402

UN No. (1)	Proper Shipping Name (PSN) (2) 3.1.2	Class or division (3) 2.0	Subsidiary risk(s) (4) 2.0	Packing group (5) 2.0.1.3	Special provisions (6) 3.3	Limited and excepted quantity provisions: Limited quantities (7a) 3.4	Limited and excepted quantity provisions: Excepted quantities (7b) 3.5	Packing: Instructions (8) 4.1.4	Packing: Provisions (9) 4.1.4	IBC: Instructions (10) 4.1.4	IBC: Provisions (11) 4.1.4
1403	CALCIUM CYANAMIDE with more than 0.1% calcium carbide	4.3	–	III	38 934	1 kg	E1	P410	–	IBC08	B4
1404	CALCIUM HYDRIDE	4.3	–	I	–	0	E0	P403	PP31	–	–
1405	CALCIUM SILICIDE	4.3	–	II	932	500 g	E2	P410	PP31	IBC07	B2
1405	CALCIUM SILICIDE	4.3	–	III	223 932	1 kg	E1	P410	PP31 PP40	IBC08	B4
1407	CAESIUM	4.3	–	I	–	0	E0	P403	PP31	IBC04	B1
1408	FERROSILICON with 30% or more but less than 90% silicon	4.3	6.1	III	39 223 932	1 kg	E1	P003	PP20	IBC08	B4 B6
1409	METAL HYDRIDES, WATER-REACTIVE, N.O.S.	4.3	–	I	274	0	E0	P403	PP31	–	–
1409	METAL HYDRIDES, WATER-REACTIVE, N.O.S.	4.3	–	II	274	500 g	E2	P410	PP40	IBC04	–
1410	LITHIUM ALUMINIUM HYDRIDE	4.3	–	I	–	0	E0	P403	PP31	–	–
1411	LITHIUM ALUMINIUM HYDRIDE, ETHEREAL	4.3	3	I	–	0	E0	P402	–	–	–
1413	LITHIUM BOROHYDRIDE	4.3	–	I	–	0	E0	P403	PP31	–	–
1414	LITHIUM HYDRIDE	4.3	–	I	–	0	E0	P403	PP31	–	–
1415	LITHIUM	4.3	–	I	–	0	E0	P403	PP31	IBC04	B1
1417	LITHIUM SILICON	4.3	–	II	–	500 g	E2	P410	PP31 PP40	IBC07	B2
1418	MAGNESIUM POWDER or MAGNESIUM ALLOYS POWDER	4.3	4.2	I	–	0	E0	P403	PP31	–	–

Portable tanks and bulk containers			EmS	Stowage and segregation	Properties and observations	UN No.
	Tank instructions	Provisions				
(12)	(13) 4.2.5 4.3	(14) 4.2.5	(15) 5.4.3.2 7.3	(16) 7.1 7.2	(17)	(18)
–	T1	TP33	F-G, S-N	Category A. "Separated from" acids.	Powder or granules. Contains calcium carbide as an impurity. In contact with water, evolves ammonia and acetylene, which is a highly flammable gas. Reacts vigorously with acids.	**1403**
–	–	–	F-G, S-O	Category E. "Separated from" acids.	Solid. In contact with water, acids or moisture, evolves hydrogen, which may be ignited by the heat of the reaction.	**1404**
–	T3	TP33	F-G, S-N	Category B. Only to be loaded under dry weather conditions. Under deck in a mechanically ventilated space. "Separated from" acids.	In contact with water, evolves hydrogen, a flammable gas. If calcium carbide is present as an impurity, acetylene will also be evolved. In contact with acids, evolves silane, a spontaneously flammable gas.	**1405**
–	T1	TP33	F-G, S-N	Category B. Only to be loaded under dry weather conditions. Under deck in a mechanically ventilated space. "Separated from" acids.	See entry above.	**1405**
–	–	–	F-G, S-N	Category D. "Separated from" acids.	White, ductile, soft metal. Reacts violently with moisture, water or acids, evolving hydrogen, which may be ignited by the heat of the reaction. Highly reactive, sometimes with explosive effect.	**1407**
–	T1 BK2	TP33	F-G, S-N	Category A. Only to be loaded under dry weather conditions. Keep as dry as reasonably practicable. Under deck in a mechanically ventilated cargo space. Clear of living quarters. "Separated from" acids and alkalis.	In contact with moisture, water, alkalis or acids, may evolve hydrogen, a flammable gas, which may form explosive mixtures with air, and also arsine and phosphine, which are highly toxic gases. These gases are evolved in proportions which, under mechanically ventilated conditions, make the poison hazard by far predominant over the explosion hazard. The rate of gas evolution is greatest from freshly broken surfaces, so is liable to increase whenever the cargo is disturbed, such as during loading. Toxic if swallowed, by skin contact or by vapour inhalation.	**1408**
–	–	–	F-G, S-L	Category D. "Separated from" acids.	Solids. React with water, moisture or acids, evolving hydrogen, which may be ignited by the heat of the reaction.	**1409**
–	T3	TP33	F-G, S-L	Category D. "Separated from" acids.	See entry above.	**1409**
–	–	–	F-G, S-M	Category E. "Separated from" acids.	White powder. In contact with water, acids or moisture, evolves hydrogen, which may be ignited by the heat of the reaction.	**1410**
–	–	–	F-G, S-M	Category D. Clear of living quarters.	Clear, colourless solution of lithium aluminium hydride in ether. Reacts readily with water, evolving hydrogen, a flammable gas. Evaporates readily to leave a residue which is easily ignited by a spark or friction.	**1411**
–	–	–	F-G, S-O	Category E. "Separated from" acids.	Crystalline, hygroscopic solid. In contact with water, acids or moisture, evolves hydrogen, which may be ignited by the heat of the reaction.	**1413**
–	–	–	F-G, S-N	Category E. "Separated from" acids.	Solid. In contact with water, acids or moisture, evolves hydrogen, which may be ignited by the heat of the reaction.	**1414**
–	–	–	<u>F-G</u>, S-N	Category E. "Separated from" acids.	White, ductile, soft metal. Floats on water. Readily decomposes in water and reacts violently with acids, evolving hydrogen, which may be ignited by the heat of the reaction. For fire-fighting purposes, dry lithium chloride powder, dry sodium chloride or graphite powder should be carried on board when this substance is transported.	**1415**
–	T3	TP33	F-G, S-N	Category A. Only to be loaded under dry weather conditions. Under deck in a mechanically ventilated space.	Shiny lumps, crystals or powder, with sharp irritating odour. Reacts readily with water, evolving hydrogen and silane, flammable gases. Enough heat may be generated to ignite the gas mixture in air.	**1417**
–	–	–	<u>F-G</u>, S-O	Category A. "Away from" liquid halogenated hydrocarbons. "Separated from" acids.	In contact with moisture, water or acids, evolves hydrogen, a flammable gas. Magnesium dust is easily ignited, causing explosion. May explode when in contact with oxidizing substances. For fire-fighting purposes, dry lithium chloride powder, dry sodium chloride or graphite powder should be carried on board when this substance is transported. Reacts with liquid halogenated hydrocarbons.	**1418**

UN No.	Proper Shipping Name (PSN)	Class or division	Subsidiary risk(s)	Packing group	Special provisions	Limited and excepted quantity provisions		Packing		IBC	
						Limited quantities	Excepted quantities	Instructions	Provisions	Instructions	Provisions
(1)	(2) 3.1.2	(3) 2.0	(4) 2.0	(5) 2.0.1.3	(6) 3.3	(7a) 3.4	(7b) 3.5	(8) 4.1.4	(9) 4.1.4	(10) 4.1.4	(11) 4.1.4
1418	MAGNESIUM POWDER or MAGNESIUM ALLOYS POWDER	4.3	4.2	II	–	0	E2	P410	PP40	IBC05	B2
1418	MAGNESIUM POWDER or MAGNESIUM ALLOYS POWDER	4.3	4.2	III	223	0	E1	P410	–	IBC08	B4
1419	MAGNESIUM ALUMINIUM PHOSPHIDE	4.3	6.1	I	–	0	E0	P403	PP31	–	–
1420	POTASSIUM METAL ALLOYS, LIQUID	4.3	–	I	–	0	E0	P402	PP31	–	–
1421	ALKALI METAL ALLOY, LIQUID, N.O.S.	4.3	–	I	182	0	E0	P402	PP31	–	–
1422	POTASSIUM SODIUM ALLOYS, LIQUID	4.3	–	I	–	0	E0	P402	PP31	–	–
1423	RUBIDIUM	4.3	–	I	–	0	E0	P403	PP31	IBC04	B1
1426	SODIUM BOROHYDRIDE	4.3	–	I	–	0	E0	P403	PP31	–	–
1427	SODIUM HYDRIDE	4.3	–	I	–	0	E0	P403	PP31	–	–
1428	SODIUM	4.3	–	I	–	0	E0	P403	PP31	IBC04	B1
1431	SODIUM METHYLATE	4.2	8	II	–	0	E2	P410	PP31	IBC05	B2
1432	SODIUM PHOSPHIDE	4.3	6.1	I	–	0	E0	P403	PP31	–	–
1433	STANNIC PHOSPHIDE	4.3	6.1	I	–	0	E0	P403	PP31	–	–
1435	ZINC ASHES	4.3	–	III	223 935	1 kg	E1	P002	–	IBC08	B4
1436	ZINC POWDER or ZINC DUST	4.3	4.2	I	–	0	E0	P403	–	–	–
1436	ZINC POWDER or ZINC DUST	4.3	4.2	II	–	0	E2	P410	PP40	IBC07	B2

Portable tanks and bulk containers			EmS	Stowage and segregation	Properties and observations	UN No.
	Tank instructions	Provisions				
(12)	(13) 4.2.5 4.3	(14) 4.2.5	(15) 5.4.3.2 7.3	(16) 7.1 7.2	(17)	(18)
–	T3	TP33	F-G, S-O	Category A. "Away from" liquid halogenated hydrocarbons. "Separated from" acids.	In contact with moisture, water or acids, evolves hydrogen, a flammable gas. Magnesium dust is easily ignited, causing explosion. May explode when in contact with oxidizing substances. For fire-fighting purposes, dry lithium chloride powder, dry sodium chloride or graphite powder should be carried on board when this substance is transported. Reacts with liquid halogenated hydrocarbons.	1418
–	T1	TP33	F-G, S-O	Category A. "Away from" liquid halogenated hydrocarbons. "Separated from" acids.	See entry above.	1418
–	–	–	F-G, S-N	Category E. Under deck in a mechanically ventilated space. Clear of living quarters. "Separated from" acids.	Solid. Reacts with acids or decomposes slowly in contact with water or damp air, evolving phosphine, a spontaneously flammable and highly toxic gas. Reacts violently with oxidizing substances. Toxic if swallowed, by skin contact or by inhalation.	1419
–	–	–	F-G, S-L	Category D. "Separated from" acids.	Soft, silvery metal liquid. Floats on water. Reacts violently with moisture, water or acids, evolving hydrogen, which may be ignited by the heat of the reaction. Highly reactive, sometimes with explosive effect.	1420
–	–	–	F-G, S-L	Category D. "Separated from" acids.	Flows like mercury at ordinary temperatures. Not volatile. Reacts violently with moisture, water or acids, evolving hydrogen, a flammable gas, and developing considerable heat, which may ignite the gas.	1421
–	T9	TP3 TP7 TP31	F-G, S-L	Category D. "Separated from" acids.	Soft, silvery metal liquid. Floats on water. Reacts violently with moisture, water or acids, evolving hydrogen, which may be ignited by the heat of the reaction. Highly reactive, sometimes with explosive effect.	1422
–	–	–	F-G, S-N	Category D. "Separated from" acids.	Silvery-white, ductile, soft metal. Melting point: 39°C. Floats on water. Reacts violently with moisture, water or acids, evolving hydrogen, which may be ignited by the heat of the reaction. Highly reactive, sometimes with explosive effect.	1423
–	–	–	F-G, S-O	Category E. "Separated from" acids.	Crystalline powder. In contact with water, acids or moisture, evolves hydrogen, which may be ignited by the heat of the reaction.	1426
–	–	–	F-G, S-O	Category E. "Separated from" acids.	White powder. In contact with water, acids or moisture, evolves hydrogen, which may be ignited by the heat of the reaction.	1427
–	T9	TP7 TP33	F-G, S-N	Category D. "Separated from" acids.	White, ductile, soft metal. Floats on water. Reacts violently with moisture, water or acids, evolving hydrogen, which may be ignited by the heat of the reaction. Highly reactive, sometimes with explosive effect.	1428
–	T3	TP33	F-A, S-L	Category B.	White, amorphous, free-flowing, hygroscopic powder. Decomposed by water to form methanol, a flammable liquid, which may be ignited by the heat of the reaction. Causes burns to skin, eyes and mucous membranes.	1431
–	–	–	F-G, S-N	Category E. Under deck in a mechanically ventilated space. Clear of living quarters. "Separated from" acids.	Solid. Reacts with acids or decomposes slowly in contact with water or damp air, evolving phosphine, a spontaneously flammable and highly toxic gas. Reacts violently with oxidizing substances. Toxic if swallowed, by skin contact or by inhalation.	1432
–	–	–	F-G, S-N	Category E. Under deck in a mechanically ventilated space. Clear of living quarters. "Separated from" acids.	Silver-white solid. Reacts with acids or decomposes slowly in contact with water or damp air, evolving phosphine, a spontaneously flammable and highly toxic gas. Reacts violently with oxidizing substances. Toxic if swallowed, by skin contact or by inhalation.	1433
–	T1 BK2	TP33	F-G, S-O	Category A.	In contact with moisture or water, liable to evolve dangerous gases, including hydrogen, a flammable gas.	1435
–	–	–	F-G, S-O	Category A. "Separated from" acids and alkalis.	In contact with water, alkalis or acids, evolves hydrogen, a flammable gas. Zinc dust is easily ignited, causing explosion. May explode when in contact with oxidizing substances.	1436
–	T3	TP33	F-G, S-O	Category A. "Separated from" acids and alkalis.	See entry above.	1436

UN No.	Proper Shipping Name (PSN)	Class or division	Subsidiary risk(s)	Packing group	Special provisions	Limited and excepted quantity provisions		Packing		IBC	
						Limited quantities	Excepted quantities	Instructions	Provisions	Instructions	Provisions
(1)	(2) 3.1.2	(3) 2.0	(4) 2.0	(5) 2.0.1.3	(6) 3.3	(7a) 3.4	(7b) 3.5	(8) 4.1.4	(9) 4.1.4	(10) 4.1.4	(11) 4.1.4
1436	**ZINC POWDER or ZINC DUST**	4.3	4.2	III	223	0	E1	P410	–	IBC08	B4
1437	**ZIRCONIUM HYDRIDE**	4.1	–	II	–	1 kg	E2	P410	PP31 PP40	IBC04	–
1438	**ALUMINIUM NITRATE**	5.1	–	III	–	5 kg	E1	P002 LP02	–	IBC08	B3
1439	**AMMONIUM DICHROMATE**	5.1	–	II	–	1 kg	E2	P002	–	IBC08	B2 B4
1442	**AMMONIUM PERCHLORATE**	5.1	–	II	152	1 kg	E2	P002	–	IBC06	B2
1444	**AMMONIUM PERSULPHATE**	5.1	–	III	–	5 kg	E1	P002 LP02	–	IBC08	B3
1445	**BARIUM CHLORATE, SOLID**	5.1	6.1	II	–	1 kg	E2	P002	–	IBC06	B2
1446	**BARIUM NITRATE**	5.1	6.1	II	–	1 kg	E2	P002	–	IBC08	B2 B4
1447	**BARIUM PERCHLORATE, SOLID**	5.1	6.1	II	–	1 kg	E2	P002	–	IBC06	B2
1448	**BARIUM PERMANGANATE**	5.1	6.1	II	–	1 kg	E2	P002	–	IBC06	B2
1449	**BARIUM PEROXIDE**	5.1	6.1	II	–	1 kg	E2	P002	–	IBC06	B2
△ 1450	**BROMATES, INORGANIC, N.O.S.**	5.1	–	II	274 350	1 kg	E2	P002	–	IBC08	B2 B4
1451	**CAESIUM NITRATE**	5.1	–	III	–	5 kg	E1	P002 LP02	–	IBC08	B3

Portable tanks and bulk containers			EmS	Stowage and segregation	Properties and observations	UN No.
	Tank instructions	Provisions				
(12)	(13) 4.2.5 4.3	(14) 4.2.5	(15) 5.4.3.2 7.3	(16) 7.1 7.2	(17)	(18)
–	T1	TP33	F-G, S-O	Category A. "Separated from" acids and alkalis.	In contact with water, alkalis or acids, evolves hydrogen, a flammable gas. Zinc dust is easily ignited, causing explosion. May explode when in contact with oxidizing substances.	1436
–	T3	TP33	F-A, S-G	Category E.	Black coloured powder.	1437
–	T1 BK2	TP33	F-A, S-Q	Category A.	Colourless or white crystals. Deliquescent. Soluble in water. Slightly corrosive. Mixtures with combustible material are readily ignited and may burn fiercely. Harmful if swallowed.	1438
–	T3	TP33	F-H, S-Q	Category A. "Separated from" strong acids.	Orange needles. Soluble in water. Mixtures with combustible material are readily ignited and may burn fiercely. May ignite spontaneously in contact with strong acids. Harmful if swallowed.	1439
–	T3	TP33	F-H, S-Q	Category E. "Separated from" cyanides and hydrogen peroxide.	White crystals or powder. Soluble in water. When heated, decomposes readily, even with explosion, evolving toxic fumes. Forms highly explosive mixtures with combustible material or powdered metals. These mixtures are sensitive to friction and are liable to ignite.	1442
–	T1	TP33	F-A, S-Q	Category A.	White crystals or powder. Soluble in water. Mixtures with combustible material are sensitive to friction and are liable to ignite.	1444
–	T3	TP33	F-H, S-Q	Category A. "Separated from" ammonium compounds and cyanides.	Colourless crystals or powder. Reacts vigorously with sulphuric acid. Reacts fiercely with cyanides when heated or by friction. May form explosive mixtures with combustible material, powdered metals or ammonium compounds. These mixtures are sensitive to friction and are liable to ignite. When involved in a fire, may cause an explosion. Toxic if swallowed, by skin contact or by dust inhalation.	1445
–	T3 BK2	TP33	F-A, S-Q	Category A.	White crystals. Mixtures with combustible material are readily ignited and may burn fiercely. Toxic if swallowed, by skin contact or by dust inhalation.	1446
–	T3	TP33	F-H, S-Q	Category A. "Separated from" ammonium compounds and cyanides.	White crystals or powder, soluble in water. Reacts vigorously with sulphuric acid. Reacts fiercely with cyanides when heated or by friction. May form explosive mixtures with combustible material, powdered metals or ammonium compounds. These mixtures are sensitive to friction and are liable to ignite. When involved in a fire, may cause an explosion. Toxic if swallowed, by skin contact or by dust inhalation.	1447
–	T3	TP33	F-H, S-Q	Category D. "Separated from" ammonium compounds, cyanides and peroxides.	Brownish-violet crystals. Soluble in water. Reacts vigorously with sulphuric acid and hydrogen peroxide. Reacts fiercely with cyanides when heated or by friction. May form explosive mixtures with combustible material, powdered metals or ammonium compounds. These mixtures are sensitive to friction and are liable to ignite. When involved in a fire, may cause an explosion. Toxic if swallowed, by skin contact or by dust inhalation.	1448
–	T3	TP33	F-G, S-Q	Category A. Keep as dry as reasonably practicable. "Separated from" permanganates, acids and class 4.1.	White powder. Particularly if wetted with small quantities of water, a mixture with combustible material may ignite following impact or friction. When involved in a fire, or in contact with water or acids, decomposes, evolving oxygen. Toxic if swallowed, by skin contact or by dust inhalation.	1449
–	T3	TP33	F-H, S-Q	Category A. "Separated from" ammonium compounds and cyanides.	Solids. React vigorously with sulphuric acid. React fiercely with cyanides when heated or by friction, and may form explosive mixtures with combustible material, powdered metals or ammonium compounds. These mixtures are sensitive to friction and are liable to ignite. When involved in a fire, may cause an explosion. Transport of ammonium bromate and mixtures of a bromate with an ammonium salt is **prohibited**.	1450 △
–	T1	TP33	F-A, S-Q	Category A.	White powder. Mixtures with combustible material are readily ignited and may burn fiercely. Harmful if swallowed.	1451

UN No.	Proper Shipping Name (PSN)	Class or division	Subsidiary risk(s)	Packing group	Special provisions	Limited and excepted quantity provisions		Packing		IBC	
						Limited quantities	Excepted quantities	Instructions	Provisions	Instructions	Provisions
(1)	(2) 3.1.2	(3) 2.0	(4) 2.0	(5) 2.0.1.3	(6) 3.3	(7a) 3.4	(7b) 3.5	(8) 4.1.4	(9) 4.1.4	(10) 4.1.4	(11) 4.1.4
1452	CALCIUM CHLORATE	5.1	–	II	–	1 kg	E2	P002	–	IBC08	B2 B4
1453	CALCIUM CHLORITE	5.1	–	II	–	1 kg	E2	P002	–	IBC08	B2 B4
△ 1454	CALCIUM NITRATE	5.1	–	III	208 967	5 kg	E1	P002 LP02	–	IBC08	B3
1455	CALCIUM PERCHLORATE	5.1	–	II	–	1 kg	E2	P002	–	IBC06	B2
1456	CALCIUM PERMANGANATE	5.1	–	II	–	1 kg	E2	P002	–	IBC06	B2
1457	CALCIUM PEROXIDE	5.1	–	II	–	1 kg	E2	P002	–	IBC06	B2
1458	CHLORATE AND BORATE MIXTURE	5.1	–	II	–	1 kg	E2	P002	–	IBC08	B2 B4
1458	CHLORATE AND BORATE MIXTURE	5.1	–	III	223	5 kg	E1	P002 LP02	–	IBC08	B3
1459	CHLORATE AND MAGNESIUM CHLORIDE MIXTURE, SOLID	5.1	–	II	–	1 kg	E2	P002	–	IBC08	B2 B4
1459	CHLORATE AND MAGNESIUM CHLORIDE MIXTURE, SOLID	5.1	–	III	223	5 kg	E1	P002 LP02	–	IBC08	B3
△ 1461	CHLORATES, INORGANIC, N.O.S.	5.1	–	II	274 351	1 kg	E2	P002	–	IBC06	B2

Portable tanks and bulk containers			EmS	Stowage and segregation	Properties and observations	UN No.
	Tank instructions	Provisions				
(12)	(13) 4.2.5 4.3	(14) 4.2.5	(15) 5.4.3.2 7.3	(16) 7.1 7.2	(17)	(18)
–	T3	TP33	F-H, S-Q	Category A. "Separated from" ammonium compounds and cyanides.	White to yellowish deliquescent crystals. Soluble in water. Reacts vigorously with sulphuric acid. Reacts fiercely with cyanides when heated or by friction. May form explosive mixtures with combustible material, powdered metals or ammonium compounds. These mixtures are sensitive to friction and are liable to ignite. When involved in a fire, may cause an explosion.	**1452**
–	T3	TP33	F-H, S-Q	Category A. "Separated from" ammonium compounds and cyanides.	White deliquescent crystals. Soluble in water. Sensitive to heat. Reacts vigorously with sulphuric acid. Reacts fiercely with cyanides when heated or by friction. May form explosive mixtures with combustible material, powdered metals or ammonium compounds. These mixtures are sensitive to friction and are liable to ignite. When involved in a fire, may cause an explosion.	**1453**
–	T1 BK2 BK3	TP33	F-A, S-Q	Category A. When transported in BK3 bulk container, see 7.6.2.12 and 7.7.3.9.	White deliquescent solid, soluble in water. Mixtures with combustible material are readily ignited and may burn fiercely. Harmful if swallowed.	**1454** △
–	T3	TP33	F-H, S-Q	Category A. "Separated from" ammonium compounds and cyanides.	White crystals or powder. Reacts vigorously with sulphuric acid. Reacts fiercely with cyanides when heated or by friction. May form explosive mixtures with combustible material, powdered metals or ammonium compounds. These mixtures are sensitive to friction and are liable to ignite. When involved in a fire, may cause an explosion.	**1455**
–	T3	TP33	F-H, S-Q	Category D. "Separated from" ammonium compounds, cyanides and peroxides.	Violet deliquescent crystals. Soluble in water. Occurs in hydrated form. Reacts vigorously with sulphuric acid and hydrogen peroxide. Reacts fiercely with cyanides when heated or by friction. May form explosive mixtures with combustible material, powdered metals or ammonium compounds. These mixtures are sensitive to friction and are liable to ignite. When involved in a fire, may cause an explosion.	**1456**
–	T3	TP33	F-G, S-Q	Category A. Keep as dry as reasonably practicable. "Separated from" permanganates, acids and class 4.1.	White or yellowish powder. Particularly if wetted with small quantities of water, a mixture with combustible material may ignite following impact or friction. When involved in a fire, or on contact with water or acids, decomposes evolving oxygen.	**1457**
–	T3	TP33	F-H, S-Q	Category A. "Separated from" ammonium compounds and cyanides.	Solid. Reacts vigorously with sulphuric acid. Reacts fiercely with cyanides when heated or by friction. May form explosive mixtures with combustible material, powdered metals or ammonium compounds. These mixtures are sensitive to friction and are liable to ignite. When involved in a fire, may cause an explosion.	**1458**
–	T1	TP33	F-H, S-Q	Category A. "Separated from" ammonium compounds and cyanides.	See entry above.	**1458**
–	T3	TP33	F-H, S-Q	Category A. "Separated from" ammonium compounds and cyanides.	Deliquescent solid. Reacts vigorously with sulphuric acid. Reacts fiercely with cyanides when heated or by friction. May form explosive mixtures with combustible material, powdered metals or ammonium compounds. These mixtures are sensitive to friction and are liable to ignite. When involved in a fire, may cause an explosion.	**1459**
–	T1	TP33	F-H, S-Q	Category A. "Separated from" ammonium compounds and cyanides.	See entry above.	**1459**
–	T3	TP33	F-H, S-Q	Category A. "Separated from" ammonium compounds and cyanides.	Solids. React vigorously with sulphuric acid. React fiercely with cyanides when heated or by friction. May form explosive mixtures with combustible material, powdered metals or ammonium compounds. These mixtures are sensitive to friction and are liable to ignite. When involved in a fire, may cause an explosion. Transport of ammonium chlorate and mixtures of a chlorate with an ammonium salt is prohibited.	**1461** △

UN No.	Proper Shipping Name (PSN)	Class or division	Subsidiary risk(s)	Packing group	Special provisions	Limited and excepted quantity provisions		Packing		IBC	
						Limited quantities	Excepted quantities	Instructions	Provisions	Instructions	Provisions
(1)	(2) 3.1.2	(3) 2.0	(4) 2.0	(5) 2.0.1.3	(6) 3.3	(7a) 3.4	(7b) 3.5	(8) 4.1.4	(9) 4.1.4	(10) 4.1.4	(11) 4.1.4
△ 1462	CHLORITES, INORGANIC, N.O.S.	5.1	–	II	274 352	1 kg	E2	P002	–	IBC06	B2
1463	CHROMIUM TRIOXIDE, ANHYDROUS	5.1	6.1 8	II	–	1 kg	E2	P002	PP31	IBC08	B2 B4
1465	DIDYMIUM NITRATE	5.1	–	III	–	5 kg	E1	P002 LP02	–	IBC08	B3
1466	FERRIC NITRATE	5.1	–	III	–	5 kg	E1	P002 LP02	–	IBC08	B3
1467	GUANIDINE NITRATE	5.1	–	III	–	5 kg	E1	P002 LP02	–	IBC08	B3
1469	LEAD NITRATE	5.1	6.1 P	II	–	1 kg	E2	P002	–	IBC08	B2 B4
1470	LEAD PERCHLORATE, SOLID	5.1	6.1 P	II	–	1 kg	E2	P002	–	IBC06	B2
△ 1471	LITHIUM HYPOCHLORITE, DRY or LITHIUM HYPOCHLORITE MIXTURE	5.1	–	II	–	1 kg	E2	P002	–	IBC08	B2 B4
△ 1471	LITHIUM HYPOCHLORITE, DRY or LITHIUM HYPOCHLORITE MIXTURE	5.1	–	III	223	5 kg	E1	P002 LP02	–	IBC08	B3

Portable tanks and bulk containers			EmS	Stowage and segregation	Properties and observations	UN No.
	Tank instructions	Provisions				
(12)	(13) 4.2.5 4.3	(14) 4.2.5	(15) 5.4.3.2 7.3	(16) 7.1 7.2	(17)	(18)
–	T3	TP33	F-H, S-Q	Category A. "Separated from" ammonium compounds and cyanides.	Solids. React vigorously with sulphuric acid. React fiercely with cyanides when heated or by friction. May form explosive mixtures with combustible material, powdered metals or ammonium compounds. These mixtures are sensitive to friction and are liable to ignite. When involved in a fire, may cause an explosion. Transport of ammonium chlorite and mixtures of a chlorite with an ammonium salt is prohibited.	1462 △
–	T3	TP33	F-A, S-Q	Category A. Segregation as for class 5.1 but "separated from" classes 4.1 and 7.	Dark purplish-red deliquescent crystals. Soluble in water. Mixtures with combustible material may ignite spontaneously and may even explode. In the presence of moisture, corrosive to most metals. Causes burns to skin, eyes and mucous membranes.	1463
–	T1	TP33	F-A, S-Q	Category A.	Hygroscopic solid. Mixture of neodymium nitrate and praseodymium nitrate. Mixtures with combustible material are readily ignited and may burn fiercely. Harmful if swallowed.	1465
–	T1	TP33	F-A, S-Q	Category A.	Violet deliquescent crystals. Soluble in water. Melting point: 47°C. Mixtures with combustible material are readily ignited and may burn fiercely. Solutions in water are slightly corrosive to most metals. Harmful if swallowed.	1466
–	T1	TP33	F-A, S-Q	Category A. "Separated from" chlorates.	White granules. Soluble in water. Mixtures with combustible material are sensitive to friction and are liable to ignite. NITROGUANIDINE is a different substance.	1467
–	T3 BK2	TP33	F-A, S-Q	Category A.	White crystals. Soluble in water. Mixtures with combustible material are readily ignited and may burn fiercely. Toxic if swallowed, by skin contact or by dust inhalation.	1469
–	T3	TP33	F-H, S-Q	Category A. "Separated from" ammonium compounds and cyanides.	White crystals or powder. Soluble in water. Reacts vigorously with sulphuric acid. Reacts fiercely with cyanides when heated or by friction. May form explosive mixtures with combustible material, powdered metals or ammonium compounds. These mixtures are sensitive to friction and are liable to ignite. When involved in a fire, may cause an explosion. Toxic if swallowed, by skin contact or by dust inhalation.	1470
–	T3	TP33	F-H, S-Q	Category A. Ventilation may be required. The possible need to open hatches in case of fire to provide maximum ventilation and to apply water in an emergency, and the consequent risk to the stability of the ship through flooding of the cargo spaces, shall be considered before loading. "Separated from" ammonium compounds, acids, cyanides, hydrogen peroxide and liquid organic substances. Protected from sources of heat.	White powder with pungent odour. Soluble in water. Critical ambient temperature of decomposition may be as low as 60°C. May cause fire in contact with organic material or ammonium compounds. Reacts with acids, evolving chlorine, an irritating, corrosive and toxic gas. In the presence of moisture, corrosive to most metals. Dust irritates mucous membranes.	1471 △
–	T1	TP33	F-H, S-Q	Category A. Ventilation may be required. The possible need to open hatches in case of fire to provide maximum ventilation and to apply water in an emergency, and the consequent risk to the stability of the ship through flooding of the cargo spaces, shall be considered before loading. "Separated from" ammonium compounds, acids, cyanides, hydrogen peroxide and liquid organic substances. Protected from sources of heat.	See entry above.	1471 △

UN No.	Proper Shipping Name (PSN)	Class or division	Subsidiary risk(s)	Packing group	Special provisions	Limited and excepted quantity provisions		Packing		IBC	
						Limited quantities	Excepted quantities	Instructions	Provisions	Instructions	Provisions
(1)	(2) 3.1.2	(3) 2.0	(4) 2.0	(5) 2.0.1.3	(6) 3.3	(7a) 3.4	(7b) 3.5	(8) 4.1.4	(9) 4.1.4	(10) 4.1.4	(11) 4.1.4
1472	LITHIUM PEROXIDE	5.1	–	II	–	1 kg	E2	P002	–	IBC06	B2
1473	MAGNESIUM BROMATE	5.1	–	II	–	1 kg	E2	P002	–	IBC08	B2 B4
△ 1474	MAGNESIUM NITRATE	5.1	–	III	332 967	5 kg	E1	P002 LP02	–	IBC08	B3
1475	MAGNESIUM PERCHLORATE	5.1	–	II	–	1 kg	E2	P002	–	IBC06	B2
1476	MAGNESIUM PEROXIDE	5.1	–	II	–	1 kg	E2	P002	–	IBC06	B2
1477	NITRATES, INORGANIC, N.O.S.	5.1	–	II	–	1 kg	E2	P002	–	IBC08	B2 B4
1477	NITRATES, INORGANIC, N.O.S.	5.1	–	III	223	5 kg	E1	P002 LP02	–	IBC08	B3
1479	OXIDIZING SOLID, N.O.S.	5.1	–	I	274 900	0	E0	P503	–	IBC05	B1
1479	OXIDIZING SOLID, N.O.S.	5.1	–	II	274 900	1 kg	E2	P002	–	IBC08	B2 B4
1479	OXIDIZING SOLID, N.O.S.	5.1	–	III	223 274 900	5 kg	E1	P002 LP02	–	IBC08	B3
1481	PERCHLORATES, INORGANIC, N.O.S.	5.1	–	II	–	1 kg	E2	P002	–	IBC06	B2
1481	PERCHLORATES, INORGANIC, N.O.S.	5.1	–	III	223	5 kg	E1	P002 LP02	–	IBC08	B3
△ 1482	PERMANGANATES, INORGANIC, N.O.S.	5.1	–	II	274 353	1 kg	E2	P002	–	IBC06	B2

Portable tanks and bulk containers			EmS	Stowage and segregation	Properties and observations	UN No.
	Tank instructions	Provisions				
(12)	(13) 4.2.5 4.3	(14) 4.2.5	(15) 5.4.3.2 7.3	(16) 7.1 7.2	(17)	(18)
–	T3	TP33	F-G, S-Q	Category A. Keep as dry as reasonably practicable. "Separated from" permanganates, acids and class 4.1.	White powder. Soluble in water. Solution in water is an alkaline corrosive liquid. Particularly if wetted with small quantities of water, a mixture with combustible material may ignite following impact or friction. When involved in a fire, or in contact with water or acids, decomposes, evolving oxygen.	1472
–	T3	TP33	F-H, S-Q	Category A. "Separated from" ammonium compounds and cyanides.	White deliquescent crystals or crystalline powder. Soluble in water. Reacts vigorously with sulphuric acid. Reacts fiercely with cyanides when heated or by friction. May form explosive mixtures with combustible material, powdered metals or ammonium compounds. These mixtures are sensitive to friction and are liable to ignite. When involved in a fire, may cause an explosion.	1473
–	T1 BK2 BK3	TP33	F-A, S-Q	Category A. When transported in BK3 bulk container, see 7.6.2.12 and 7.7.3.9.	White deliquescent crystals, soluble in water. Mixtures with combustible material are readily ignited and may burn fiercely. Harmful if swallowed.	1474 △
–	T3	TP33	F-H, S-Q	Category A. "Separated from" ammonium compounds and cyanides.	White crystals or powder. Reacts vigorously with sulphuric acid. Reacts fiercely with cyanides when heated or by friction. May form explosive mixtures with combustible material, powdered metals or ammonium compounds. These mixtures are sensitive to friction and are liable to ignite. When involved in a fire, may cause an explosion.	1475
–	T3	TP33	F-G, S-Q	Category A. Keep as dry as reasonably practicable. "Separated from" permanganates, acids and class 4.1.	White powder. Particularly if wetted with small quantities of water, a mixture with combustible material may ignite following impact or friction. When involved in a fire, or in contact with water or acids, decomposes, evolving oxygen. Harmful if swallowed.	1476
–	T3	TP33	F-A, S-Q	Category A. "Separated from" ammonium compounds and cyanides.	Solids. Solid mixtures with combustible material are readily ignited and may burn fiercely. Harmful if swallowed.	1477
–	T1	TP33	F-A, S-Q	Category A. "Separated from" ammonium compounds and cyanides.	See entry above.	1477
–	–	–	F-A, S-Q	Category D. "Separated from" powdered metals, ammonium compounds, cyanides and peroxides.	–	1479
–	T3	TP33	F-A, S-Q	Category B. "Separated from" powdered metals, ammonium compounds, cyanides and peroxides.	–	1479
–	T1	TP33	F-A, S-Q	Category B. "Separated from" powdered metals, ammonium compounds, cyanides and peroxides.	–	1479
–	T3	TP33	F-H, S-Q	Category A. "Separated from" ammonium compounds and cyanides.	Solids. React vigorously with sulphuric acid. React fiercely with cyanides when heated or by friction. May form explosive mixtures with combustible material, powdered metals or ammonium compounds. These mixtures are sensitive to friction and are liable to ignite. When involved in a fire, may cause an explosion.	1481
–	T1	TP33	F-H, S-Q	Category A. "Separated from" ammonium compounds and cyanides.	See entry above.	1481
–	T3	TP33	F-H, S-Q	Category D. "Separated from" ammonium compounds, cyanides and peroxides.	Solids. React vigorously with sulphuric acid. React fiercely with cyanides when heated or by friction. May form explosive mixtures with combustible material, powdered metals or ammonium compounds. These mixtures are sensitive to friction and are liable to ignite. When involved in a fire, may cause an explosion. Transport of ammonium permanganate and mixtures of a permanganate with an ammonium salt is **prohibited**.	1482 △

UN No.	Proper Shipping Name (PSN)	Class or division	Subsidiary risk(s)	Packing group	Special provisions	Limited and excepted quantity provisions		Packing		IBC	
						Limited quantities	Excepted quantities	Instructions	Provisions	Instructions	Provisions
(1)	(2) 3.1.2	(3) 2.0	(4) 2.0	(5) 2.0.1.3	(6) 3.3	(7a) 3.4	(7b) 3.5	(8) 4.1.4	(9) 4.1.4	(10) 4.1.4	(11) 4.1.4
1482	PERMANGANATES, INORGANIC, N.O.S.	5.1	–	III	223 274 353	5 kg	E1	P002	–	IBC08	B3
1483	PEROXIDES, INORGANIC, N.O.S.	5.1	–	II	–	1 kg	E2	P002	–	IBC06	B2
1483	PEROXIDES, INORGANIC, N.O.S.	5.1	–	III	223	5 kg	E1	P002 LP02	–	IBC08	B3
1484	POTASSIUM BROMATE	5.1	–	II	–	1 kg	E2	P002	–	IBC08	B2 B4
1485	POTASSIUM CHLORATE	5.1	–	II	–	1 kg	E2	P002	–	IBC08	B2 B4
△ 1486	POTASSIUM NITRATE	5.1	–	III	964 967	5 kg	E1	P002 LP02	–	IBC08	B3
1487	POTASSIUM NITRATE AND SODIUM NITRITE MIXTURE	5.1	–	II	–	1 kg	E2	P002	–	IBC08	B2 B4
1488	POTASSIUM NITRITE	5.1	–	II	–	1 kg	E2	P002	–	IBC08	B2 B4
1489	POTASSIUM PERCHLORATE	5.1	–	II	–	1 kg	E2	P002	–	IBC06	B2
1490	POTASSIUM PERMANGANATE	5.1	–	II	–	1 kg	E2	P002	–	IBC08	B2 B4
1491	POTASSIUM PEROXIDE	5.1	–	I	–	0	E0	P503	–	IBC06	B1

Portable tanks and bulk containers			EmS	Stowage and segregation	Properties and observations	UN No.
	Tank instructions	Provisions				
(12)	(13) 4.2.5 4.3	(14) 4.2.5	(15) 5.4.3.2 7.3	(16) 7.1 7.2	(17)	(18)
–	T1	TP33	F-H, S-Q	Category D. "Separated from" ammonium compounds, cyanides and peroxides.	Solids. React vigorously with sulphuric acid. React fiercely with cyanides when heated or by friction. May form explosive mixtures with combustible material, powdered metals or ammonium compounds. These mixtures are sensitive to friction and are liable to ignite. When involved in a fire, may cause an explosion. Transport of ammonium permanganate and mixtures of a permanganate with an ammonium salt is **prohibited.**	**1482**
–	T3	TP33	F-G, S-Q	Category A. Keep as dry as reasonably practicable. "Separated from" permanganates, acids and class 4.1.	Particularly if wetted with small quantities of water, a mixture with combustible material may ignite following impact or friction. When involved in a fire, or in contact with water or acids, decomposes, evolving oxygen.	**1483**
–	T1	TP33	F-G, S-Q	Category A. Keep as dry as reasonably practicable. "Separated from" permanganates, acids and class 4.1.	See entry above.	**1483**
–	T3	TP33	F-H, S-Q	Category A. "Separated from" ammonium compounds and cyanides.	White crystals or powder. Soluble in water. Reacts vigorously with sulphuric acid. Reacts fiercely with cyanides when heated or by friction. May form explosive mixtures with combustible materials, powdered metals or ammonium compounds. These mixtures are sensitive to friction and are liable to ignite. When involved in a fire, may cause an explosion.	**1484**
–	T3 BK2	TP33	F-H, S-Q	Category A. "Separated from" ammonium compounds and cyanides.	White crystals or powder. Soluble in water. Reacts vigorously with sulphuric acid. Reacts fiercely with cyanides when heated or by friction. May form explosive mixtures with combustible material, powdered metals or ammonium compounds. These mixtures are sensitive to friction and are liable to ignite. When involved in a fire, may cause an explosion.	**1485**
–	T1 BK2 BK3	TP33	F-A, S-Q	Category A. When transported in BK3 bulk container, see 7.6.2.12 and 7.7.3.9.	White crystals or powder. Soluble in water. Mixtures with combustible material are readily ignited and may burn fiercely. Harmful if swallowed.	**1486** △
–	T3	TP33	F-A, S-Q	Category A. "Separated from" ammonium compounds and cyanides.	Deliquescent solid. Soluble in water. May cause fire in contact with organic material such as wood, cotton or straw. Mixtures with ammonium compounds or cyanides may explode. Harmful if swallowed. May be shipped in the form of fused solid block or lumps.	**1487**
–	T3	TP33	F-A, S-Q	Category A. "Separated from" ammonium compounds and cyanides.	White or slightly yellowish deliquescent crystals or sticks. Soluble in water. Mixtures with combustible material are readily ignited and may burn fiercely. Mixtures with ammonium compounds or cyanides may explode. Harmful if swallowed.	**1488**
–	T3	TP33	F-H, S-Q	Category A. "Separated from" ammonium compounds and cyanides.	White crystals or powder, soluble in water. Reacts vigorously with sulphuric acid. Reacts fiercely with cyanides when heated or by friction. May form explosive mixtures with combustible material, powdered metals or ammonium compounds. These mixtures are sensitive to friction and are liable to ignite. When involved in a fire, may cause an explosion.	**1489**
–	T3	TP33	F-H, S-Q	Category D. "Separated from" ammonium compounds, cyanides and peroxides.	Dark purple crystals or powder. Soluble in water. Reacts vigorously with sulphuric acid and hydrogen peroxide. Reacts fiercely with cyanides when heated or by friction. May form explosive mixtures with combustible material, powdered metals or ammonium compounds. These mixtures are sensitive to friction and are liable to ignite. When involved in a fire, may cause an explosion.	**1490**
–	–	–	F-G, S-Q	Category B. Keep as dry as reasonably practicable. "Separated from" permanganates, acids and class 4.1.	Yellow powder. Particularly if wetted with small quantities of water, a mixture with combustible material may ignite, following impact or friction. When involved in a fire, or in contact with water or acids, decomposes, evolving oxygen. Highly irritating to skin, eyes and mucous membranes.	**1491**

3 DGL

UN No.	Proper Shipping Name (PSN)	Class or division	Subsidiary risk(s)	Packing group	Special provisions	Limited and excepted quantity provisions		Packing		IBC	
						Limited quantities	Excepted quantities	Instructions	Provisions	Instructions	Provisions
(1)	(2) 3.1.2	(3) 2.0	(4) 2.0	(5) 2.0.1.3	(6) 3.3	(7a) 3.4	(7b) 3.5	(8) 4.1.4	(9) 4.1.4	(10) 4.1.4	(11) 4.1.4
1492	POTASSIUM PERSULPHATE	5.1	–	III	–	5 kg	E1	P002 LP02	–	IBC08	B3
1493	SILVER NITRATE	5.1	–	II	–	1 kg	E2	P002	–	IBC08	B2 B4
1494	SODIUM BROMATE	5.1	–	II	–	1 kg	E2	P002	–	IBC08	B2 B4
1495	SODIUM CHLORATE	5.1	–	II	–	1 kg	E2	P002	–	IBC08	B2 B4
1496	SODIUM CHLORITE	5.1	–	II	–	1 kg	E2	P002	–	IBC08	B2 B4
△ 1498	SODIUM NITRATE	5.1	–	III	964 967	5 kg	E1	P002 LP02	–	IBC08	B3
△ 1499	SODIUM NITRATE AND POTASSIUM NITRATE MIXTURE	5.1	–	III	964 967	5 kg	E1	P002 LP02	–	IBC08	B3
1500	SODIUM NITRITE	5.1	6.1	III	–	5 kg	E1	P002	–	IBC08	B3
1502	SODIUM PERCHLORATE	5.1	–	II	–	1 kg	E2	P002	–	IBC06	B2
1503	SODIUM PERMANGANATE	5.1	–	II	–	1 kg	E2	P002	–	IBC06	B2
1504	SODIUM PEROXIDE	5.1	–	I	–	0	E0	P503	–	IBC05	B1
1505	SODIUM PERSULPHATE	5.1	–	III	–	5 kg	E1	P002 LP02	–	IBC08	B3

Portable tanks and bulk containers			EmS	Stowage and segregation	Properties and observations	UN No.
	Tank instructions	Provisions				
(12)	(13) 4.2.5 4.3	(14) 4.2.5	(15) 5.4.3.2 7.3	(16) 7.1 7.2	(17)	(18)
–	T1	TP33	F-A, S-Q	Category A. "Separated from" ammonium compounds other than AMMONIUM PERSULPHATE (UN 1444), and cyanides.	White crystals or powder. Soluble in water. Mixtures with combustible material are sensitive to friction and are liable to ignite. Reacts fiercely with cyanides when heated or by friction. May form explosive mixture with powdered metals or ammonium compounds.	1492
–	T3	TP33	F-A, S-Q	Category A.	Colourless crystals. Soluble in water. Mixtures with combustible material are readily ignited and may burn fiercely. Harmful if swallowed. Irritating to skin and mucous membranes.	1493
–	T3	TP33	F-H, S-Q	Category A. "Separated from" ammonium compounds and cyanides.	White deliquescent crystals. Soluble in water. Reacts vigorously with sulphuric acid. Reacts fiercely with cyanides when heated or by friction. May form explosive mixtures with combustible material, powdered metals or ammonium compounds. These mixtures are sensitive to friction and are liable to ignite. When involved in a fire, may cause an explosion.	1494
–	T3 BK2	TP33	F-H, S-Q	Category A. "Separated from" ammonium compounds and cyanides.	Colourless deliquescent crystals. Soluble in water. Reacts vigorously with sulphuric acid. Reacts fiercely with cyanides when heated or by friction. May form explosive mixtures with combustible material, powdered metals or ammonium compounds. These mixtures are sensitive to friction and are liable to ignite. When involved in a fire, may cause an explosion.	1495
–	T3	TP33	F-H, S-Q	Category A. "Separated from" ammonium compounds and cyanides.	Colourless deliquescent solid. Soluble in water. Reacts vigorously with sulphuric acid. Reacts fiercely with cyanides when heated or by friction. May form explosive mixtures with combustible material, powdered metals or ammonium compounds. These mixtures are sensitive to friction and are liable to ignite. When involved in a fire, may cause an explosion.	1496
–	T1 BK2 BK3	TP33	F-A, S-Q	Category A. When transported in BK3 bulk container, see 7.6.2.12 and 7.7.3.9.	Colourless deliquescent solid. Soluble in water. Mixtures with combustible material are readily ignited and may burn fiercely. Harmful if swallowed. This substance in the impure form is known as Chile Saltpetre.	1498 △
–	T1 BK2 BK3	TP33	F-A, S-Q	Category A. When transported in BK3 bulk container, see 7.6.2.12 and 7.7.3.9.	Colourless, hygroscopic solid. Soluble in water. Mixtures with combustible material are readily ignited and may burn fiercely. Harmful if swallowed. Mixture prepared as a fertilizer.	1499 △
–	T1	TP33	F-A, S-Q	Category A. "Separated from" ammonium compounds and cyanides.	Colourless deliquescent solid. Soluble in water. Mixtures with combustible material are readily ignited and may burn fiercely. Mixtures with ammonium compounds or cyanides may explode. Decomposes if heated, giving off toxic nitrous fumes and gases supporting combustion. Harmful if swallowed or by dust inhalation.	1500
–	T3	TP33	F-H, S-Q	Category A. "Separated from" ammonium compounds and cyanides.	Colourless crystals or powder, soluble in water. Reacts vigorously with sulphuric acid. Reacts fiercely with cyanides when heated or by friction. May form explosive mixtures with combustible material, powdered metals or ammonium compounds. These mixtures are sensitive to friction and are liable to ignite. When involved in a fire, may cause an explosion.	1502
–	T3	TP33	F-H, S-Q	Category D. "Separated from" ammonium compounds, cyanides and peroxides.	Red crystals or powder. Soluble in water. Reacts vigorously with sulphuric acid and hydrogen peroxide. Reacts fiercely with cyanides when heated or by friction. May form explosive mixtures with combustible material, powdered metals or ammonium compounds. These mixtures are sensitive to friction and are liable to ignite. When involved in a fire, may cause an explosion.	1503
–	–	–	F-G, S-Q	Category B. Keep as dry as reasonably practicable. "Separated from" permanganates, acids and class 4.1.	Pale yellow coarse powder or granules. Particularly if wetted with small quantities of water, a mixture with combustible material may ignite, following impact or friction. When involved in a fire, or in contact with water or acids, decomposes, evolving oxygen. Highly irritating to skin, eyes and mucous membranes.	1504
–	T1	TP33	F-A, S-Q	Category A. "Separated from" ammonium compounds, other than AMMONIUM PERSULPHATE (UN 1444), and cyanides.	Colourless crystals or powder. Soluble in water. Mixtures with combustible material are sensitive to friction and are liable to ignite. Reacts fiercely with cyanides when heated or by friction. May form explosive mixture with powdered metals or ammonium compounds.	1505

UN No.	Proper Shipping Name (PSN)	Class or division	Subsidiary risk(s)	Packing group	Special provisions	Limited and excepted quantity provisions		Packing		IBC	
						Limited quantities	Excepted quantities	Instructions	Provisions	Instructions	Provisions
(1)	(2) 3.1.2	(3) 2.0	(4) 2.0	(5) 2.0.1.3	(6) 3.3	(7a) 3.4	(7b) 3.5	(8) 4.1.4	(9) 4.1.4	(10) 4.1.4	(11) 4.1.4
1506	STRONTIUM CHLORATE	5.1	–	II	–	1 kg	E2	P002	–	IBC08	B2 B4
1507	STRONTIUM NITRATE	5.1	–	III	–	5 kg	E1	P002 LP02	–	IBC08	B3
1508	STRONTIUM PERCHLORATE	5.1	–	II	–	1 kg	E2	P002	–	IBC06	B2
1509	STRONTIUM PEROXIDE	5.1	–	II	–	1 kg	E2	P002	–	IBC06	B2
1510	TETRANITROMETHANE	6.1	5.1	I	354	0	E0	P602	–	–	–
1511	UREA HYDROGEN PEROXIDE	5.1	8	III	–	5 kg	E1	P002	–	IBC08	B3
△ 1512	ZINC AMMONIUM NITRITE	5.1	–	–	900	–	–	–	–	–	–
1513	ZINC CHLORATE	5.1	–	II	–	1 kg	E2	P002	–	IBC08	B2 B4
1514	ZINC NITRATE	5.1	–	II	–	1 kg	E2	P002	–	IBC08	B2 B4
1515	ZINC PERMANGANATE	5.1	–	II	–	1 kg	E2	P002	–	IBC06	B2
1516	ZINC PEROXIDE	5.1	–	II	–	1 kg	E2	P002	–	IBC06	B2
1517	ZIRCONIUM PICRAMATE, WETTED with not less than 20% water, by mass	4.1	–	I	28	0	E0	P406	PP26 PP31	–	–
△ 1541	ACETONE CYANOHYDRIN, STABILIZED	6.1	– P	I	354	0	E0	P602	–	–	–
1544	ALKALOIDS, SOLID, N.O.S. or ALKALOIDS SALTS, SOLID, N.O.S.	6.1	–	I	43 274	0	E5	P002	–	IBC07	B1

Portable tanks and bulk containers			EmS	Stowage and segregation	Properties and observations	UN No.
	Tank instructions	Provisions				
(12)	(13) 4.2.5 4.3	(14) 4.2.5	(15) 5.4.3.2 7.3	(16) 7.1 7.2	(17)	(18)
–	T3	TP33	F-H, S-Q	Category A. "Separated from" ammonium compounds and cyanides.	Colourless deliquescent solid, soluble in water. Reacts vigorously with sulphuric acid. Reacts fiercely with cyanides when heated or by friction. May form explosive mixtures with combustible material, powdered metals or ammonium compounds. These mixtures are sensitive to friction and are liable to ignite. When involved in a fire, may cause an explosion.	**1506**
–	T1	TP33	F-A S-Q	Category A.	Colourless solid. Soluble in water. Mixtures with combustible material are readily ignited and may burn fiercely. Harmful if swallowed.	**1507**
–	T3	TP33	F-H, S-Q	Category A. "Separated from" ammonium compounds and cyanides.	Colourless crystals or powder, soluble in water. Reacts vigorously with sulphuric acid. Reacts fiercely with cyanides when heated or by friction. May form explosive mixtures with combustible material, powdered metals or ammonium compounds. These mixtures are sensitive to friction and are liable to ignite. When involved in a fire, may cause an explosion.	**1508**
–	T3	TP33	F-G, S-Q	Category A. Keep as dry as reasonably practicable. "Separated from" permanganates, acids and class 4.1.	Colourless powder. Particularly if wetted with small quantities of water, a mixture with combustible materials may ignite following impact or friction. When involved in a fire, or in contact with water or acids, decomposes, evolving oxygen.	**1509**
–	–	–	F-H, S-Q	Category D. Clear of living quarters. "Separated from" class 4.1.	Colourless liquid with a pungent odour. Freezing point: 12.5°C. Insoluble in water. Mixtures with combustible material are readily ignited, burn fiercely and may also explode by friction or shock. Highly toxic if swallowed, by skin contact or by inhalation.	1510
–	T1	TP33	F-A, S-Q	Category A. Keep as dry as reasonably practicable.	White crystals or powder. Soluble in water. Mixtures with combustible material are sensitive to friction and are liable to ignite. Irritating to skin, eyes and mucous membranes.	**1511**
–	–	–	–	–	Transport is **prohibited**.	**1512** △
–	T3	TP33	F-H, S-Q	Category A. "Separated from" ammonium compounds and cyanides.	Colourless or yellowish crystals. Soluble in water. Reacts vigorously with sulphuric acid. Reacts fiercely with cyanides when heated or by friction. May form explosive mixtures with combustible material, powdered metals or ammonium compounds. These mixtures are sensitive to friction and are liable to ignite. When involved in a fire, may cause an explosion.	**1513**
–	T3	TP33	F-H, S-Q	Category A.	Colourless solid. Soluble in water. Melting point: 36°C. Mixtures with combustible material are readily ignited and may burn fiercely. Solutions in water are slightly corrosive. Harmful if swallowed.	1514
–	T3	TP33	F-H, S-Q	Category D. "Separated from" ammonium compounds, cyanides and peroxides.	Violet-brown or black crystals or powder. Soluble in water. Reacts vigorously with sulphuric acid and hydrogen peroxide. Reacts fiercely with cyanides when heated or by friction. May form explosive mixtures with combustible material, powdered metals or ammonium compounds. These mixtures are sensitive to friction and are liable to ignite. When involved in a fire, may cause an explosion.	**1515**
–	T3	TP33	F-G, S-Q	Category A. Keep as dry as reasonably practicable. "Separated from" permanganates, acids and class 4.1.	White powder. Particularly if wetted with small quantities of water, a mixture with combustible material may ignite following impact or friction. When involved in a fire, or in contact with water or acids, decomposes, evolving oxygen.	1516
–	–	–	F-B, S-J	Category D. "Away from" class 3 and heavy metals and their salts.	Desensitized explosive. Highly explosive in the dry state or if insufficiently wetted. May react violently in contact with heavy metals or their salts.	**1517**
–	T20	TP2 TP13 TP37	F-A, S-A	Category D. Protected from sources of heat. Clear of living quarters. "Separated from" acids and alkalis.	Colourless to amber liquid evolving toxic vapour. Miscible with water. Unstable in contact with acids and alkalis, evolving hydrogen cyanide, a highly toxic and flammable gas. Highly toxic if swallowed, by skin contact or by inhalation.	1541 △
–	T6	TP33	F-A, S-A	Category A.	A wide range of toxic solids, generally of vegetable origin. Toxic if swallowed, by skin contact or by inhalation.	**1544**

UN No.	Proper Shipping Name (PSN)	Class or division	Subsidiary risk(s)	Packing group	Special provisions	Limited and excepted quantity provisions		Packing		IBC	
						Limited quantities	Excepted quantities	Instructions	Provisions	Instructions	Provisions
(1)	(2) 3.1.2	(3) 2.0	(4) 2.0	(5) 2.0.1.3	(6) 3.3	(7a) 3.4	(7b) 3.5	(8) 4.1.4	(9) 4.1.4	(10) 4.1.4	(11) 4.1.4
1544	ALKALOIDS, SOLID, N.O.S. or ALKALOIDS SALTS, SOLID, N.O.S.	6.1	–	II	43 274	500 g	E4	P002	–	IBC08	B2 B4
1544	ALKALOIDS, SOLID, N.O.S. or ALKALOIDS SALTS, SOLID, N.O.S.	6.1	–	III	43 223 274	5 kg	E1	P002 LP02	–	IBC08	B3
1545	ALLYL ISOTHIOCYANATE, STABILIZED	6.1	3	II	–	100 mL	E4	P001	–	IBC02	–
1546	AMMONIUM ARSENATE	6.1	–	II	–	500 g	E4	P002	–	IBC08	B2 B4
1547	ANILINE	6.1	–	II	279	100 mL	E4	P001	–	IBC02	–
1548	ANILINE HYDROCHLORIDE	6.1	–	III	–	5 kg	E1	P002 LP02	–	IBC08	B3
1549	ANTIMONY COMPOUND, INORGANIC, SOLID, N.O.S.	6.1	–	III	45 274	5 kg	E1	P002 LP02	–	IBC08	B3
1550	ANTIMONY LACTATE	6.1	–	III	–	5 kg	E1	P002 LP02	–	IBC08	B3
1551	ANTIMONY POTASSIUM TARTRATE	6.1	–	III	–	5 kg	E1	P002 LP02	–	IBC08	B3
1553	ARSENIC ACID, LIQUID	6.1	–	I	–	0	E5	P001	PP31	–	–
1554	ARSENIC ACID, SOLID	6.1	–	II	–	500 g	E4	P002	–	IBC08	B2 B4
1555	ARSENIC BROMIDE	6.1	–	II	–	500 g	E4	P002	–	IBC08	B2 B4
1556	ARSENIC COMPOUND, LIQUID, N.O.S. inorganic, including: Arsenates, n.o.s., Arsenites, n.o.s., and Arsenic sulphides, n.o.s.	6.1	–	I	43 274	0	E5	P001	–	–	–
1556	ARSENIC COMPOUND, LIQUID, N.O.S. inorganic, including: Arsenates, n.o.s., Arsenites, n.o.s., and Arsenic sulphides, n.o.s.	6.1	–	II	43 274	100 mL	E4	P001	–	IBC02	–
1556	ARSENIC COMPOUND, LIQUID, N.O.S. inorganic, including: Arsenates, n.o.s., Arsenites, n.o.s., and Arsenic sulphides, n.o.s.	6.1	–	III	43 223 274	5 L	E1	P001 LP01	–	IBC03	–
1557	ARSENIC COMPOUND, SOLID, N.O.S. inorganic, including: Arsenates, n.o.s.; Arsenites, n.o.s.; and Arsenic sulphides, n.o.s.	6.1	–	I	43 274	0	E5	P002	–	IBC07	B1

Portable tanks and bulk containers			EmS	Stowage and segregation	Properties and observations	UN No.
	Tank instructions	Provisions				
(12)	(13) 4.2.5 4.3	(14) 4.2.5	(15) 5.4.3.2 7.3	(16) 7.1 7.2	(17)	(18)
–	T3	TP33	F-A, S-A	Category A.	A wide range of toxic solids, generally of vegetable origin. Toxic if swallowed, by skin contact or by inhalation.	1544
–	T1	TP33	F-A, S-A	Category A.	See entry above.	1544
–	T7	TP2	F-E, S-D	Category D. Clear of living quarters.	Colourless liquid evolving toxic vapour which is irritating and causes tears. Flashpoint: 46°C c.c. Toxic if swallowed, by skin contact or by inhalation.	1545
–	T3	TP33	F-A, S-A	Category A. "Separated from" alkalis.	White powder or crystals. Soluble in water. Reacts with alkalis, evolving ammonia gas. Toxic if swallowed, by skin contact or by dust inhalation.	1546
–	T7	TP2	F-A, S-A	Category A. Clear of living quarters. "Separated from" acids.	Colourless, oily, volatile liquid. Reacts with acids. Toxic if swallowed, by skin contact or by inhalation.	1547
–	T1	TP33	F-A, S-A	Category A.	White, crystalline solid. Soluble in water. Decomposes to aniline in contact with alkalis. Toxic if swallowed, by skin contact or by inhalation.	1548
–	T1	TP33	F-A, S-A	Category A.	A wide range of toxic solids. Toxic if swallowed, by skin contact or by inhalation.	1549
–	T1	TP33	F-A, S-A	Category A.	White powder or crystals. Toxic if swallowed, by skin contact or by dust inhalation.	1550
–	T1	TP33	F-A, S-A	Category A.	Colourless crystals or white powder. Toxic if swallowed, by skin contact or by dust inhalation.	1551
–	T20	TP2 TP7 TP13	F-A, S-A	Category B. "Away from" powdered metals.	White, deliquescent crystals which readily become liquid. Melting point: approximately 35°C. Miscible with water. In contact with metals, may evolve arsine, an extremely toxic gas. Highly toxic if swallowed, by skin contact or by inhalation.	1553
–	T3	TP33	F-A, S-A	Category A.	White crystals with a relatively high melting point. Soluble in water. Toxic if swallowed, by skin contact or by dust inhalation.	1554
–	T3	TP33	F-A, S-A	Category A. Keep as cool as reasonably practicable. Clear of living quarters.	White, deliquescent crystals. Melting point: approximately 33°C. Decomposed by water, evolving hydrogen bromide, an irritating and corrosive gas, apparent as white fumes. Toxic if swallowed, by skin contact or by dust inhalation.	1555
–	T14	TP2 TP13 TP27	F-A, S-A	Category B. Clear of living quarters. For arsenic sulphides, "separated from" acids.	A wide variety of toxic liquids. In contact with acids, arsenic sulphide evolves hydrogen sulphide, a toxic and flammable gas. Toxic if swallowed, by skin contact or by inhalation.	1556
–	T11	TP2 TP13 TP27	F-A, S-A	Category B. Clear of living quarters. For arsenic sulphides, "separated from" acids.	See entry above.	1556
–	T7	TP2 TP28	F-A, S-A	Category B. Clear of living quarters. For arsenic sulphides, "separated from" acids.	See entry above.	1556
–	T6	TP33	F-A, S-A	Category A. For arsenic sulphides, "separated from" acids.	A wide variety of toxic solids. In contact with acids, arsenic sulphide evolves hydrogen sulphide, a toxic and flammable gas. Toxic if swallowed, by skin contact or by dust inhalation.	1557

UN No. (1)	Proper Shipping Name (PSN) (2) 3.1.2	Class or division (3) 2.0	Subsidiary risk(s) (4) 2.0	Packing group (5) 2.0.1.3	Special provisions (6) 3.3	Limited and excepted quantity provisions: Limited quantities (7a) 3.4	Excepted quantities (7b) 3.5	Packing: Instructions (8) 4.1.4	Packing: Provisions (9) 4.1.4	IBC: Instructions (10) 4.1.4	IBC: Provisions (11) 4.1.4
1557	ARSENIC COMPOUND, SOLID, N.O.S. inorganic, including: Arsenates, n.o.s.; Arsenites, n.o.s.; and Arsenic sulphides, n.o.s.	6.1	–	II	43 274	500 g	E4	P002	–	IBC08	B2 B4
1557	ARSENIC COMPOUND, SOLID, N.O.S. inorganic, including: Arsenates, n.o.s.; Arsenites, n.o.s.; and Arsenic sulphides, n.o.s.	6.1	–	III	43 223 274	5 kg	E1	P002 LP02	–	IBC08	B3
1558	ARSENIC	6.1	–	II	–	500 g	E4	P002	–	IBC08	B2 B4
1559	ARSENIC PENTOXIDE	6.1	–	II	–	500 g	E4	P002	–	IBC08	B2 B4
1560	ARSENIC TRICHLORIDE	6.1	–	I	–	0	E5	P602	–	–	–
1561	ARSENIC TRIOXIDE	6.1	–	II	–	500 g	E4	P002	–	IBC08	B2 B4
1562	ARSENICAL DUST	6.1	–	II	–	500 g	E4	P002	–	IBC08	B2 B4
1564	BARIUM COMPOUND, N.O.S.	6.1	–	II	177 274	500 g	E4	P002	–	IBC08	B2 B4
1564	BARIUM COMPOUND, N.O.S.	6.1	–	III	177 223 274	5 kg	E1	P002 LP02	–	IBC08	B3
1565	BARIUM CYANIDE	6.1	– P	I	–	0	E5	P002	PP31	IBC07	B1
1566	BERYLLIUM COMPOUND, N.O.S.	6.1	–	II	274	500 g	E4	P002	–	IBC08	B2 B4
1566	BERYLLIUM COMPOUND, N.O.S.	6.1	–	III	223 274	5 kg	E1	P002 LP02	–	IBC08	B3
1567	BERYLLIUM POWDER	6.1	4.1	II	–	500 g	E4	P002	–	IBC08	B2 B4
1569	BROMOACETONE	6.1	3 P	II	–	0	E4	P602	–	–	–
1570	BRUCINE	6.1	–	I	43	0	E5	P002	–	IBC07	B1
1571	BARIUM AZIDE, WETTED with not less than 50% water, by mass	4.1	6.1	I	28	0	E0	P406	PP31	–	–
1572	CACODYLIC ACID	6.1	–	II	–	500 g	E4	P002	–	IBC08	B2 B4
1573	CALCIUM ARSENATE	6.1	– P	II	–	500 g	E4	P002	–	IBC08	B2 B4
1574	CALCIUM ARSENATE AND CALCIUM ARSENITE MIXTURE, SOLID	6.1	– P	II	–	500 g	E4	P002	–	IBC08	B2 B4

Portable tanks and bulk containers			EmS	Stowage and segregation	Properties and observations	UN No.
	Tank instructions	Provisions				
(12)	(13) 4.2.5 4.3	(14) 4.2.5	(15) 5.4.3.2 7.3	(16) 7.1 7.2	(17)	(18)
–	T3	TP33	F-A, S-A	Category A. For arsenic sulphides, "separated from" acids.	A wide variety of toxic solids. In contact with acids, arsenic sulphide evolves hydrogen sulphide, a toxic and flammable gas. Toxic if swallowed, by skin contact or by dust inhalation.	1557
–	T1	TP33	F-A, S-A	Category A. For arsenic sulphides, "separated from" acids.	See entry above.	1557
–	T3	TP33	F-A, S-A	Category A.	Silvery, brittle, crystalline solid with the appearance of a metal. Toxic if swallowed, by skin contact or by dust inhalation.	1558
–	T3	TP33	F-A, S-A	Category A.	White, deliquescent powder. Soluble in water. Toxic if swallowed, by skin contact or by dust inhalation.	1559
–	T14	TP2 TP13	F-A, S-A	Category B. Clear of living quarters.	Colourless, oily liquid. Fumes in moist air, evolving hydrogen chloride, an irritating and corrosive gas, apparent as white fumes. Reacts with water. Highly toxic if swallowed, by skin contact or by inhalation.	1560
–	T3	TP33	F-A, S-A	Category A.	White powder. Slightly soluble in water. Toxic if swallowed, by skin contact or by dust inhalation.	1561
–	T3	TP33	F-A, S-A	Category A.	Fine powder. Toxic if swallowed, by skin contact or by dust inhalation.	1562
–	T3	TP33	F-A, S-A	Category A.	White powder, lumps or crystals. Toxic if swallowed, by skin contact or by inhalation.	1564
–	T1	TP33	F-A, S-A	Category A.	See entry above.	1564
–	T6	TP33	F-A, S-A	Category A. Clear of living quarters. "Separated from" acids.	White crystals or powder. Soluble in water. Reacts with acids or acid fumes, evolving hydrogen cyanide, a highly toxic and flammable gas. Highly toxic if swallowed, by skin contact or by dust inhalation.	1565
–	T3	TP33	F-A, S-A	Category A.	A wide range of toxic solids. Toxic if swallowed, by skin contact or by dust inhalation.	1566
–	T1	TP33	F-A, S-A	Category A.	See entry above.	1566
–	T3	TP33	F-G, S-G	Category A.	White, metallic powder. Toxic if swallowed, by skin contact or by dust inhalation.	1567
–	T20	TP2 TP13	F-E, S-D	Category D. Clear of living quarters.	When pure, colourless liquid evolving irritating vapour ("Tear Gas"). Flashpoint: approximately 45°C c.c. Toxic if swallowed, by skin contact or by inhalation.	1569
–	T6	TP33	F-A, S-A	Category A.	White crystals or powder. Highly toxic if swallowed, by skin contact or by dust inhalation.	1570
–	–	–	F-B, S-J	Category D. "Away from" class 3 and heavy metals and their salts.	Desensitized explosive. White crystals or powder. Explosive and sensitive to friction in the dry state. Toxic if swallowed, by skin contact or by dust inhalation. May form extremely sensitive compounds with heavy metals or their salts.	1571
–	T3	TP33	F-A, S-A	Category E. "Separated from" acids.	Colourless crystals or white powder with an offensive odour. Soluble in water. May react with acids, evolving dimethylarsine, an extremely toxic gas. Toxic if swallowed, by skin contact or by dust inhalation.	1572
–	T3	TP33	F-A, S-A	Category A.	White powder. Slightly soluble in water. Toxic if swallowed, by skin contact or by dust inhalation.	1573
–	T3	TP33	F-A, S-A	Category A.	White powder. Toxic if swallowed, by skin contact or by dust inhalation.	1574

UN No.	Proper Shipping Name (PSN)	Class or division	Subsidiary risk(s)	Packing group	Special provisions	Limited and excepted quantity provisions		Packing		IBC	
						Limited quantities	Excepted quantities	Instructions	Provisions	Instructions	Provisions
(1)	(2) 3.1.2	(3) 2.0	(4) 2.0	(5) 2.0.1.3	(6) 3.3	(7a) 3.4	(7b) 3.5	(8) 4.1.4	(9) 4.1.4	(10) 4.1.4	(11) 4.1.4
1575	CALCIUM CYANIDE	6.1	– P	I	–	0	E5	P002	PP31	IBC07	B1
1577	CHLORODINITROBENZENES, LIQUID	6.1	– P	II	279	100 mL	E4	P001	–	IBC02	–
1578	CHLORONITROBENZENES, SOLID	6.1	–	II	279	500 g	E4	P002	–	IBC08	B2 B4
1579	4-CHLORO-*o*-TOLUIDINE HYDROCHLORIDE, SOLID	6.1	–	III	–	5 kg	E1	P002 LP02	–	IBC08	B3
1580	CHLOROPICRIN	6.1	– P	I	354	0	E0	P601	–	–	–
△ 1581	CHLOROPICRIN AND METHYL BROMIDE MIXTURE with more than 2% chloropicrin	2.3	–	–	–	0	E0	P200	–	–	–
△ 1582	CHLOROPICRIN AND METHYL CHLORIDE MIXTURE	2.3	–	–	–	0	E0	P200	–	–	–
1583	CHLOROPICRIN MIXTURE, N.O.S.	6.1	–	I	43 274 315	0	E5	P602	–	–	–
1583	CHLOROPICRIN MIXTURE, N.O.S.	6.1	–	II	43 274	100 mL	E4	P001	–	IBC02	–
1583	CHLOROPICRIN MIXTURE, N.O.S.	6.1	–	III	43 223 274	5 L	E1	P001 LP01	–	IBC03	–
1585	COPPER ACETOARSENITE	6.1	– P	II	–	500 g	E4	P002	–	IBC08	B2 B4
1586	COPPER ARSENITE	6.1	– P	II	–	500 g	E4	P002	–	IBC08	B2 B4
1587	COPPER CYANIDE	6.1	– P	II	–	500 g	E4	P002	–	IBC08	B2 B4
1588	CYANIDES, INORGANIC, SOLID, N.O.S.	6.1	– P	I	47 274	0	E5	P002	–	IBC07	B1
1588	CYANIDES, INORGANIC, SOLID, N.O.S.	6.1	– P	II	47 274	500 g	E4	P002	–	IBC08	B2 B4
1588	CYANIDES, INORGANIC, SOLID, N.O.S.	6.1	– P	III	47 223 274	5 kg	E1	P002 LP02	–	IBC08	B3
1589	CYANOGEN CHLORIDE, STABILIZED	2.3	8 P	–	–	0	E0	P200	–	–	–
1590	DICHLOROANILINES, LIQUID	6.1	– P	II	279	100 mL	E4	P001	–	IBC02	–

Portable tanks and bulk containers			EmS	Stowage and segregation	Properties and observations	UN No.
	Tank instructions	Provisions				
(12)	(13) 4.2.5 4.3	(14) 4.2.5	(15) 5.4.3.2 7.3	(16) 7.1 7.2	(17)	(18)
–	T6	TP33	F-A, S-A	Category A. Clear of living quarters. "Separated from" acids.	White crystals or powder. Decomposes slowly in water to form a weak hydrogen cyanide solution. Reacts with acids or acid fumes, evolving hydrogen cyanide, a highly toxic and flammable gas. Highly toxic if swallowed, by skin contact or by dust inhalation.	**1575**
–	T7	TP2	F-A, S-A	Category A. "Separated from" class 3.	Colourless liquids. May explode if involved in a fire. Toxic if swallowed, by skin contact or by inhalation.	1577
–	T3	TP33	F-A, S-A	Category A.	Yellow crystals. Melting point: approximately 30°C to 80°C. Toxic if swallowed, by skin contact or by dust inhalation.	**1578**
–	T1	TP33	F-A, S-A	Category A.	Dry solid or paste. Toxic if swallowed, by skin contact or by dust inhalation.	1579
–	T22	TP2 TP13 TP37	F-A, S-A	Category D. Clear of living quarters.	Colourless, oily liquid. Highly toxic if swallowed, by skin contact or by inhalation.	**1580**
–	T50	–	F-C, S-U	Category D. Protected from sources of heat. Clear of living quarters.	Extremely volatile liquid evolving highly toxic vapours. Highly toxic by skin contact or by inhalation. Causes burns to skin and eyes; vapour irritating to mucous membranes.	1581 △
–	T50	–	F-C, S-U	Category D. Protected from sources of heat. Clear of living quarters.	Extremely volatile liquid evolving highly toxic vapours. Highly toxic by skin contact or by inhalation. Causes burns to skin and eyes; vapour irritating to mucous membranes.	**1582** △
–	–	–	F-A, S-A	Category C. Clear of living quarters.	A wide range of liquid mixtures. May evolve highly toxic vapour. Toxic if swallowed, by skin contact or by inhalation.	1583
–	–	–	F-A, S-A	Category C. Clear of living quarters.	See entry above.	**1583**
–	–	–	F-A, S-A	Category C. Clear of living quarters.	See entry above.	1583
–	T3	TP33	F-A, S-A	Category A.	Green powder. Insoluble in water. Toxic if swallowed, by skin contact or by dust inhalation.	**1585**
–	T3	TP33	F-A, S-A	Category A.	Yellowish-green powder. Insoluble in water. Toxic if swallowed, by skin contact or by dust inhalation.	1586
–	T3	TP33	F-A, S-A	Category A. "Separated from" acids.	Green powder. Slightly soluble in water. Reacts with acids or acid fumes, evolving hydrogen cyanide, a highly toxic and flammable gas. Toxic if swallowed, by skin contact or by dust inhalation.	**1587**
–	T6	TP33	F-A, S-A	Category A. "Separated from" acids.	Solids. May be soluble in water. On contact with water, may form a weak hydrogen cyanide solution. React with acids or acid fumes, evolving hydrogen cyanide, a highly toxic and flammable gas. Toxic if swallowed, by skin contact or by dust inhalation. The provisions of this Code shall not apply to complex ferricyanides and ferrocyanides.	1588
–	T3	TP33	F-A, S-A	Category A. "Separated from" acids.	See entry above.	**1588**
–	T1	TP33	F-A, S-A	Category A. "Separated from" acids.	See entry above.	1588
–	–	–	F-C, S-U	Category D. Clear of living quarters.	Liquefied, non-flammable, toxic and corrosive gas with an irritating odour. Produces severe tearing of the eyes. On contact with water, reacts violently to give off highly toxic and corrosive fumes. Much heavier than air (2.1). Boiling point: 13°C. Toxic by skin contact or by inhalation. Highly irritating to skin, eyes and mucous membranes.	**1589**
–	T7	TP2	F-A, S-A	Category A. Clear of living quarters.	Colourless liquid with a penetrating odour. Liquid mixtures of various isomers of dichloroanilines, some of which in the pure state may be solid, with a melting point varying from 24°C to 72°C. Toxic if swallowed, by skin contact or by inhalation.	1590

UN No.	Proper Shipping Name (PSN)	Class or division	Subsidiary risk(s)	Packing group	Special provisions	Limited and excepted quantity provisions		Packing		IBC	
						Limited quantities	Excepted quantities	Instructions	Provisions	Instructions	Provisions
(1)	(2) 3.1.2	(3) 2.0	(4) 2.0	(5) 2.0.1.3	(6) 3.3	(7a) 3.4	(7b) 3.5	(8) 4.1.4	(9) 4.1.4	(10) 4.1.4	(11) 4.1.4
1591	*o*-DICHLOROBENZENE	6.1	–	III	279	5 L	E1	P001 LP01	–	IBC03	–
1593	DICHLOROMETHANE	6.1	–	III	–	5 L	E1	P001 LP01	–	IBC03	B8
1594	DIETHYL SULPHATE	6.1	–	II	–	100 mL	E4	P001	–	IBC02	–
1595	DIMETHYL SULPHATE	6.1	8	I	354	0	E0	P602	–	–	–
1596	DINITROANILINES	6.1	–	II	–	500 g	E4	P002	–	IBC08	B2 B4
1597	DINITROBENZENES, LIQUID	6.1	–	II	–	100 mL	E4	P001	–	IBC03	–
1597	DINITROBENZENES, LIQUID	6.1	–	III	223	5 L	E1	P001 LP01	–	IBC03	–
1598	DINITRO-*o*-CRESOL	6.1	– P	II	43	500 g	E4	P002	–	IBC08	B2 B4
1599	DINITROPHENOL SOLUTION	6.1	– P	II	–	100 mL	E4	P001	–	IBC02	–
1599	DINITROPHENOL SOLUTION	6.1	– P	III	223	5 L	E1	P001 LP01	–	IBC03	–
1600	DINITROTOLUENES, MOLTEN	6.1	–	II	–	0	E0	–	–	–	–
1601	DISINFECTANT, SOLID, TOXIC, N.O.S.	6.1	–	I	274	0	E5	P002	–	IBC07	B1
1601	DISINFECTANT, SOLID, TOXIC, N.O.S.	6.1	–	II	274	500 g	E4	P002	–	IBC08	B2 B4
1601	DISINFECTANT, SOLID, TOXIC, N.O.S.	6.1	–	III	223 274	5 kg	E1	P002 LP02	–	IBC08	B3
1602	DYE, LIQUID, TOXIC, N.O.S. or DYE INTERMEDIATE, LIQUID, TOXIC, N.O.S.	6.1	–	I	274	0	E5	P001	–	–	–
1602	DYE, LIQUID, TOXIC, N.O.S. or DYE INTERMEDIATE, LIQUID, TOXIC, N.O.S.	6.1	–	II	274	100 mL	E4	P001	–	IBC02	–
1602	DYE, LIQUID, TOXIC, N.O.S. or DYE INTERMEDIATE, LIQUID, TOXIC, N.O.S.	6.1	–	III	223 274	5 L	E1	P001 LP01	–	IBC03	–
1603	ETHYL BROMOACETATE	6.1	3	II	–	100 mL	E4	P001	–	IBC02	–
1604	ETHYLENEDIAMINE	8	3	II	–	1 L	E2	P001	–	IBC02	–
1605	ETHYLENE DIBROMIDE	6.1	–	I	354	0	E0	P602	–	–	–
1606	FERRIC ARSENATE	6.1	– P	II	–	500 g	E4	P002	–	IBC08	B2 B4

Portable tanks and bulk containers			EmS	Stowage and segregation	Properties and observations	UN No.
	Tank instructions	Provisions				
(12)	(13) 4.2.5 4.3	(14) 4.2.5	(15) 5.4.3.2 7.3	(16) 7.1 7.2	(17)	(18)
–	T4	TP1	F-A, S-A	Category A.	Volatile liquid. Melting point: approximately –17°C. Toxic if swallowed, by skin contact or by inhalation.	1591
–	T7	TP2	F-A, S-A	Category A.	Colourless, volatile liquid with heavy vapours. Boiling point: 40°C. When involved in a fire, evolves extremely toxic fumes (phosgene). Toxic if swallowed, by skin contact or by inhalation.	1593
–	T7	TP2	F-A, S-A	Category C.	Colourless, oily liquid. Readily hydrolysed by moisture to sulphuric acid, which is a corrosive liquid. Toxic if swallowed, by skin contact or by inhalation.	1594
–	T20	TP2 TP13 TP35	F-A, S-B	Category D. Clear of living quarters.	Colourless, volatile liquid evolving toxic vapours. In the presence of moisture, corrosive to most metals. Highly toxic if swallowed, by skin contact or by inhalation. Causes burns to skin, eyes and mucous membranes.	1595
–	T3	TP33	F-A, S-A	Category A. "Separated from" class 3.	Yellow crystals in pure form. Insoluble in water. May explode if involved in a fire. Toxic if swallowed, by skin contact or by inhalation.	1596
–	T7	TP2	F-A, S-A	Category A. "Separated from" class 3.	Yellow solutions. May explode if involved in a fire. Toxic if swallowed, by skin contact or by inhalation.	1597
–	T7	TP2	F-A, S-A	Category A. "Separated from" class 3.	See entry above.	1597
–	T3	TP33	F-A, S-A	Category A.	Yellow crystals or crystallized mass. Slightly soluble in water. Toxic if swallowed, by skin contact or by dust inhalation.	1598
–	T7	TP2	F-A, S-A	Category A. "Away from" heavy metals and their salts.	Substance when pure consists of yellow crystals. Slightly soluble in water. May form extremely sensitive compounds with heavy metals or their salts. Toxic if swallowed, by skin contact or by inhalation.	1599
–	T4	TP1	F-A, S-A	Category A. "Away from" heavy metals and their salts.	See entry above.	1599
–	T7	TP3	F-A, S-A	Category C.	Molten liquid. This entry covers the 2,3-, 2,4-, 2,5-, 2,6-, 3,4- and 3,5-isomers having melting points between 52°C and 93°C. Toxic if swallowed, by skin contact or by inhalation.	1600
–	T6	TP33	F-A, S-A	Category A. Clear of living quarters.	A wide range of toxic solids. Toxic if swallowed, by skin contact or by inhalation.	1601
–	T3	TP33	F-A, S-A	Category A. Clear of living quarters.	See entry above.	1601
–	T1	TP33	F-A, S-A	Category A. Clear of living quarters.	See entry above.	1601
–	–	–	F-A, S-A	Category A.	A wide range of toxic liquids. Toxic if swallowed, by skin contact or by inhalation.	1602
–	–	–	F-A, S-A	Category A.	See entry above.	1602
–	–	–	F-A, S-A	Category A.	See entry above.	1602
–	T7	TP2	F-E, S-D	Category D. Clear of living quarters.	Colourless, flammable liquid evolving irritating vapour ("Tear Gas"). Flashpoint: 58°C c c. Toxic if swallowed, by skin contact or by inhalation.	1603
–	T7	TP2	F-E, S-C	Category A. Clear of living quarters. "Separated from" acids.	Volatile, colourless, hygroscopic flammable liquid with an ammonia-like odour. Flashpoint: 34°C c.c. Miscible with water. Causes burns to skin, eyes and mucous membranes. Reacts violently with acids.	1604
–	T20	TP2 TP13 TP37	F-A, S-A	Category D. Clear of living quarters.	Colourless, volatile liquid. Highly toxic if swallowed, by skin contact or by inhalation.	1605
–	T3	TP33	F-A, S-A	Category A.	Green crystals or powder. Insoluble in water. Toxic if swallowed, by skin contact or by dust inhalation.	1606

UN No.	Proper Shipping Name (PSN)	Class or division	Subsidiary risk(s)	Packing group	Special provisions	Limited and excepted quantity provisions		Packing		IBC	
						Limited quantities	Excepted quantities	Instructions	Provisions	Instructions	Provisions
(1)	(2) 3.1.2	(3) 2.0	(4) 2.0	(5) 2.0.1.3	(6) 3.3	(7a) 3.4	(7b) 3.5	(8) 4.1.4	(9) 4.1.4	(10) 4.1.4	(11) 4.1.4
1607	FERRIC ARSENITE	6.1	– P	II	–	500 g	E4	P002	–	IBC08	B2 B4
1608	FERROUS ARSENATE	6.1	– P	II	–	500 g	E4	P002	–	IBC08	B2 B4
1611	HEXAETHYL TETRAPHOSPHATE	6.1	– P	II	–	100 mL	E4	P001	–	IBC02	–
1612	HEXAETHYL TETRAPHOSPHATE AND COMPRESSED GAS MIXTURE	2.3	–	–	–	0	E0	P200	–	–	–
△ 1613	HYDROCYANIC ACID, AQUEOUS SOLUTION (HYDROGEN CYANIDE, AQUEOUS SOLUTION) with not more than 20% hydrogen cyanide	6.1	– P	I	900	0	E5	P601	–	–	–
△ 1614	HYDROGEN CYANIDE, STABILIZED containing less than 3% water and absorbed in a porous inert material	6.1	– P	I	–	0	E5	P099	–	–	–
1616	LEAD ACETATE	6.1	– P	III	–	5 kg	E1	P002 LP02	–	IBC08	B3
1617	LEAD ARSENATES	6.1	– P	II	–	500 g	E4	P002	–	IBC08	B2 B4
1618	LEAD ARSENITES	6.1	– P	II	–	500 g	E4	P002	–	IBC08	B2 B4
1620	LEAD CYANIDE	6.1	– P	II	–	500 g	E4	P002	–	IBC08	B2 B4
1621	LONDON PURPLE	6.1	– P	II	43	500 g	E4	P002	–	IBC08	B2 B4
1622	MAGNESIUM ARSENATE	6.1	– P	II	–	500 g	E4	P002	–	IBC08	B2 B4
1623	MERCURIC ARSENATE	6.1	– P	II	–	500 g	E4	P002	–	IBC08	B2 B4
1624	MERCURIC CHLORIDE	6.1	– P	II	–	500 g	E4	P002	–	IBC08	B2 B4
1625	MERCURIC NITRATE	6.1	– P	II	–	500 g	E4	P002	–	IBC08	B2 B4
1626	MERCURIC POTASSIUM CYANIDE	6.1	– P	I	–	0	E5	P002	PP31	IBC07	B1
1627	MERCUROUS NITRATE	6.1	– P	II	–	500 g	E4	P002	–	IBC08	B2 B4
1629	MERCURY ACETATE	6.1	– P	II	–	500 g	E4	P002	–	IBC08	B2 B4
1630	MERCURY AMMONIUM CHLORIDE	6.1	– P	II	–	500 g	E4	P002	–	IBC08	B2 B4
1631	MERCURY BENZOATE	6.1	– P	II	–	500 g	E4	P002	–	IBC08	B2 B4
1634	MERCURY BROMIDES	6.1	– P	II	–	500 g	E4	P002	–	IBC08	B2 B4
1636	MERCURY CYANIDE	6.1	– P	II	–	500 g	E4	P002	–	IBC08	B2 B4

Portable tanks and bulk containers			EmS	Stowage and segregation	Properties and observations	UN No.
	Tank instructions	Provisions				
(12)	(13) 4.2.5 4.3	(14) 4 2.5	(15) 5.4.3.2 7.3	(16) 7.1 7.2	(17)	(18)
–	T3	TP33	F-A, S-A	Category A.	Brown or yellow powder. Insoluble in water. Toxic if swallowed, by skin contact or by dust inhalation.	1607
–	T3	TP33	F-A, S-A	Category A.	Green powder. Insoluble in water. Toxic if swallowed, by skin contact or by dust inhalation.	1608
–	T7	TP2	F-A, S-A	Category E. Clear of living quarters.	Yellow liquid. Miscible with water. Toxic if swallowed, by skin contact or by inhalation.	1611
–	–	–	F-C, S-U	Category D. Clear of living quarters.	Toxic if swallowed, by skin contact or by inhalation.	1612
–	T14	TP2 TP13	F-A, S-A	Category D. Clear of living quarters.	Colourless liquid evolving extremely toxic vapour with a bitter almond odour. Miscible with water. Highly toxic if swallowed, by skin contact or by inhalation. Transport of HYDROCYANIC ACID, AQUEOUS SOLUTION with more than 20% hydrogen cyanide and of HYDROGEN CYANIDE, AQUEOUS SOLUTION with more than 20% hydrogen cyanide is **prohibited**.	1613 △
–	–	–	F-A, <u>S-U</u>	Category D. Protected from sources of heat. Clear of living quarters.	Very volatile, colourless liquid, evolving extremely toxic flammable vapours, absorbed in a porous inert material. Miscible with water. Highly toxic if swallowed, by skin contact or by inhalation.	1614 △
–	T1	TP33	F-A, S-A	Category A.	White crystals, or brown or grey lumps. Soluble in water. Toxic if swallowed, by skin contact or by inhalation.	1616
–	T3	TP33	F-A, S-A	Category A.	White crystals or powder. Insoluble in water. Toxic if swallowed, by skin contact or by dust inhalation.	1617
–	T3	TP33	F-A, S-A	Category A.	White powder. Insoluble in water. Toxic if swallowed, by skin contact or by dust inhalation.	1618
–	T3	TP33	F-A, S-A	Category A. "Separated from" acids.	White powder. Slightly soluble in water. Reacts with acids or acid fumes, evolving hydrogen cyanide, a highly toxic and flammable gas. Toxic if swallowed, by skin contact or by dust inhalation.	1620
–	T3	TP33	F-A, S-A	Category A.	Mixture of arsenic trioxide, lime and ferric oxide, used as an insecticide. Insoluble in water. Toxic if swallowed, by skin contact or by dust inhalation.	1621
–	T3	TP33	F-A, S-A	Category A.	White crystals or powder. Insoluble in water. Toxic if swallowed, by skin contact or by dust inhalation.	1622
–	T3	TP33	F-A, S-A	Category A.	Yellow crystals or powder. Insoluble in water. Toxic if swallowed, by skin contact or by dust inhalation.	1623
–	T3	TP33	F-A, S-A	Category A.	White crystals or powder. Soluble in water. Toxic if swallowed, by skin contact or by dust inhalation.	1624
–	T3	TP33	F-A, S-A	Category A.	White, deliquescent crystals or powder. Soluble in water. Toxic if swallowed, by skin contact or by dust inhalation.	1625
–	T6	TP33	F-A, S-A	Category A. "Separated from" acids.	Colourless crystals. Soluble in water. Reacts with acids, evolving hydrogen cyanide, a highly toxic and flammable gas. Highly toxic if swallowed, by skin contact or by dust inhalation.	1626
–	T3	TP33	F-A, S-A	Category A.	Crystals or powder. Toxic if swallowed, by skin contact or by dust inhalation.	1627
–	T3	TP33	F-A, S-A	Category A.	White crystals or powder. Toxic if swallowed, by skin contact or by dust inhalation.	1629
–	T3	TP33	F-A, S-A	Category A.	White crystals or powder. Insoluble in water. Toxic if swallowed, by skin contact or by dust inhalation.	1630
–	T3	TP33	F-A, S-A	Category A.	White crystals. Toxic if swallowed, by skin contact or by dust inhalation.	1631
–	T3	TP33	F-A, S-A	Category A.	White crystals or powder. Toxic if swallowed, by skin contact or by dust inhalation.	1634
–	T3	TP33	F-A, S-A	Category A. "Separated from" acids.	White crystals or powder. Soluble in water. Reacts with acids or acid fumes, evolving hydrogen cyanide, a highly toxic and flammable gas. Toxic if swallowed, by skin contact or by dust inhalation.	1636

UN No.	Proper Shipping Name (PSN)	Class or division	Subsidiary risk(s)	Packing group	Special provisions	Limited and excepted quantity provisions		Packing		IBC	
						Limited quantities	Excepted quantities	Instructions	Provisions	Instructions	Provisions
(1)	(2) 3.1.2	(3) 2.0	(4) 2.0	(5) 2.0.1.3	(6) 3.3	(7a) 3.4	(7b) 3.5	(8) 4.1.4	(9) 4.1.4	(10) 4.1.4	(11) 4.1.4
1637	MERCURY GLUCONATE	6.1	– P	II	–	500 g	E4	P002	–	IBC08	B2 B4
1638	MERCURY IODIDE	6.1	– P	II	–	500 g	E4	P002	–	IBC08	B2 B4
1639	MERCURY NUCLEATE	6.1	– P	II	–	500 g	E4	P002	–	IBC08	B2 B4
1640	MERCURY OLEATE	6.1	– P	II	–	500 g	E4	P002	–	IBC08	B2 B4
1641	MERCURY OXIDE	6.1	– P	II	–	500 g	E4	P002	–	IBC08	B2 B4
△ 1642	MERCURY OXYCYANIDE, DESENSITIZED	6.1	– P	II	900	500 g	E4	P002	–	IBC08	B2 B4
1643	MERCURY POTASSIUM IODIDE	6.1	– P	II	–	500 g	E4	P002	–	IBC08	B2 B4
1644	MERCURY SALICYLATE	6.1	– P	II	–	500 g	E4	P002	–	IBC08	B2 B4
1645	MERCURY SULPHATE	6.1	– P	II	–	500 g	E4	P002	–	IBC08	B2 B4
1646	MERCURY THIOCYANATE	6.1	– P	II	–	500 g	E4	P002	–	IBC08	B2 B4
1647	METHYL BROMIDE AND ETHYLENE DIBROMIDE MIXTURE, LIQUID	6.1	– P	I	354	0	E0	P602	–	–	–
1648	ACETONITRILE	3	–	II	–	1 L	E2	P001	–	IBC02	–
△ 1649	MOTOR FUEL ANTI-KNOCK MIXTURE	6.1	– P	I	–	0	E5	P602	–	–	–
1650	*beta*-NAPHTHYLAMINE, SOLID	6.1	–	II	–	500 g	E4	P002	–	IBC08	B2 B4
1651	NAPHTHYLTHIOUREA	6.1	–	II	43	500 g	E4	P002	–	IBC08	B2 B4
1652	NAPHTHYLUREA	6.1	–	II	–	500 g	E4	P002	–	IBC08	B2 B4
1653	NICKEL CYANIDE	6.1	– P	II	–	500 g	E4	P002	–	IBC08	B2 B4
1654	NICOTINE	6.1	–	II	–	100 mL	E4	P001	–	IBC02	–
1655	NICOTINE COMPOUND, SOLID, N.O.S. or NICOTINE PREPARATION, SOLID, N.O.S.	6.1	–	I	43 274	0	E5	P002	–	IBC07	B1
1655	NICOTINE COMPOUND, SOLID, N.O.S. or NICOTINE PREPARATION, SOLID, N.O.S.	6.1	–	II	43 274	500 g	E4	P002	–	IBC08	B2 B4

Portable tanks and bulk containers			EmS	Stowage and segregation	Properties and observations	UN No.
	Tank instructions	Provisions				
(12)	(13) 4.2.5 4.3	(14) 4.2.5	(15) 5.4.3.2 7.3	(16) 7.1 7.2	(17)	(18)
–	T3	TP33	F-A, S-A	Category A.	Solid. Soluble in water. Toxic if swallowed, by skin contact or by dust inhalation.	1637
–	T3	TP33	F-A, S-A	Category A.	Red crystals or powder. Insoluble in water Toxic if swallowed, by skin contact or by dust inhalation.	1638
–	T3	TP33	F-A, S-A	Category A.	Brown powder containing about 20% mercury. Toxic if swallowed, by skin contact or by dust inhalation.	1639
–	T3	TP33	F-A, S-A	Category A.	Yellow oily paste. Insoluble in water. Toxic if swallowed, by skin contact or by inhalation.	1640
–	T3	TP33	F-A, S-A	Category A.	Orange powder. Insoluble in water. Toxic if swallowed, by skin contact or by dust inhalation.	1641
–	T3	TP33	F-A, S-A	Category A. "Separated from" acids and class 3.	White crystals or powder. Reacts with acids or acid fumes, evolving hydrogen cyanide, a highly toxic and flammable gas. May explode if involved in a fire. Toxic if swallowed, by skin contact or by dust inhalation. Should be sufficiently phlegmatized (mercury oxycyanide–mercury cyanide mixtures containing not less than 65% by mass of mercury cyanide can be regarded as adequately phlegmatized). Transport of MERCURY OXYCYANIDE pure is **prohibited.**	1642 △
–	T3	TP33	F-A, S-A	Category A.	Yellow, deliquescent crystals or powder. Soluble in water. Toxic if swallowed, by skin contact or by dust inhalation.	1643
–	T3	TP33	F-A, S-A	Category A.	White powder. Insoluble in water. Toxic if swallowed, by skin contact or by dust inhalation.	1644
–	T3	TP33	F-A, S-A	Category A.	White crystals or powder. Decomposes in water, forming sulphuric acid. Toxic if swallowed, by skin contact or by dust inhalation.	1645
–	T3	TP33	F-A, S-A	Category A.	White powder. Insoluble in water. Toxic if swallowed, by skin contact or by dust inhalation.	1646
–	T20	TP2 TP13	F-A, S-A	Category D. Clear of living quarters.	Solutions of methyl bromide gas, evolving toxic vapour. Methyl bromide has a boiling point of approximately 4°C. Highly toxic if swallowed, by skin contact or by inhalation.	1647
–	T7	TP2	F-E, S-D	Category B. Clear of living quarters.	Colourless, volatile liquid. Flashpoint: 2°C c.c. Explosive limits: 3% to 16%. Miscible with water. When involved in a fire, evolves toxic cyanide fumes. Harmful if swallowed, by skin contact or by inhalation.	1648
–	T14	TP2 TP13	F-A, S-A	Category D. Protected from sources of heat. Clear of living quarters.	Volatile liquids evolving toxic vapour. Mixture of tetraethyllead or tetramethyllead with ethylene dibromide and ethylene dichloride. Insoluble in water. Highly toxic if swallowed, by skin contact or by inhalation.	1649 △
–	T3	TP33	F-A, S-A	Category A.	White crystals. Toxic if swallowed, by skin contact or by inhalation.	1650
–	T3	TP33	F-A, S-A	Category A.	White crystals or powder. Toxic if swallowed, by skin contact or by dust inhalation.	1651
–	T3	TP33	F-A, S-A	Category A.	Crystals or powder. Toxic if swallowed, by skin contact or by dust inhalation.	1652
–	T3	TP33	F-A, S-A	Category A. "Separated from" acids.	Green crystals or powder. Insoluble in water. Reacts with acids or acid fumes, evolving hydrogen cyanide, a highly toxic and flammable gas. Toxic if swallowed, by skin contact or by dust inhalation.	1653
–	–	–	F-A, S-A	Category A.	Thick colourless oil, turning brown on exposure to air. Miscible with water. Toxic if swallowed, by skin contact or by inhalation.	1654
–	T6	TP33	F-A, S-A	Category B.	A wide variety of toxic solids. Toxic if swallowed, by skin contact or by dust inhalation	1655
–	T3	TP33	F-A, S-A	Category A.	See entry above.	1655

3

DGL

UN No.	Proper Shipping Name (PSN)	Class or division	Subsidiary risk(s)	Packing group	Special provisions	Limited and excepted quantity provisions: Limited quantities	Limited and excepted quantity provisions: Excepted quantities	Packing: Instructions	Packing: Provisions	IBC: Instructions	IBC: Provisions
(1)	(2) 3.1.2	(3) 2.0	(4) 2.0	(5) 2.0.1.3	(6) 3.3	(7a) 3.4	(7b) 3.5	(8) 4.1.4	(9) 4.1.4	(10) 4.1.4	(11) 4.1.4
1655	NICOTINE COMPOUND, SOLID, N.O.S. or NICOTINE PREPARATION, SOLID, N.O.S.	6.1	–	III	43 223 274	5 kg	E1	P002 LP02	–	IBC08	B3
1656	NICOTINE HYDROCHLORIDE, LIQUID or SOLUTION	6.1	–	II	43	100 mL	E4	P001	–	IBC02	–
1656	NICOTINE HYDROCHLORIDE, LIQUID or SOLUTION	6.1	–	III	43 223	5 L	E1	P001 LP01	–	IBC03	–
1657	NICOTINE SALICYLATE	6.1	–	II	–	500 g	E4	P002	–	IBC08	B2 B4
1658	NICOTINE SULPHATE SOLUTION	6.1	–	II	–	100 mL	E4	P001	–	IBC02	–
1658	NICOTINE SULPHATE SOLUTION	6.1	–	III	223	5 L	E1	P001 LP01	–	IBC03	–
1659	NICOTINE TARTRATE	6.1	–	II	–	500 g	E4	P002	–	IBC08	B2 B4
1660	NITRIC OXIDE, COMPRESSED	2.3	5.1/8	–	–	0	E0	P200	–	–	–
1661	NITROANILINES (*o*-, *m*-, *p*-)	6.1	–	II	279	500 g	E4	P002	–	IBC08	B2 B4
1662	NITROBENZENE	6.1	–	II	279	100 mL	E4	P001	–	IBC02	–
1663	NITROPHENOLS (*o*-, *m*-, *p*-)	6.1	–	III	279	5 kg	E1	P002 LP02	–	IBC08	B3
1664	NITROTOLUENES, LIQUID	6.1	–	II	–	100 mL	E4	P001	–	IBC02	–
1665	NITROXYLENES, LIQUID	6.1	–	II	–	100 mL	E4	P001	–	IBC02	–
1669	PENTACHLOROETHANE	6.1	– P	II	–	100 mL	E4	P001	–	IBC02	–
1670	PERCHLOROMETHYL MERCAPTAN	6.1	– P	I	354	0	E0	P602	–	–	–
1671	PHENOL, SOLID	6.1	–	II	279	500 g	E4	P002	–	IBC08	B2 B4
1672	PHENYLCARBYLAMINE CHLORIDE	6.1	–	I	–	0	E5	P602	–	–	–
1673	PHENYLENEDIAMINES (*o*-, *m*-, *p*-)	6.1	–	III	279	5 kg	E1	P002 LP02	–	IBC08	B3
1674	PHENYLMERCURIC ACETATE	6.1	– P	II	43	500 g	E4	P002	–	IBC08	B2 B4
1677	POTASSIUM ARSENATE	6.1	–	II	–	500 g	E4	P002	–	IBC08	B2 B4
1678	POTASSIUM ARSENITE	6.1	–	II	–	500 g	E4	P002	–	IBC08	B2 B4

Portable tanks and bulk containers			EmS	Stowage and segregation	Properties and observations	UN No.
	Tank instructions	Provisions				
(12)	(13) 4.2.5 4.3	(14) 4.2.5	(15) 5.4.3.2 7.3	(16) 7.1 7.2	(17)	(18)
–	T1	TP33	F-A, S-A	Category A.	A wide variety of toxic solids. Toxic if swallowed, by skin contact or by dust inhalation.	1655
–	–	–	F-A, S-A	Category A.	Miscible with water. Toxic if swallowed, by skin contact or by inhalation.	1656
–	–	–	F-A, S-A	Category A.	See entry above.	1656
–	T3	TP33	F-A, S-A	Category A.	White crystals. Soluble in water. Toxic if swallowed, by skin contact or by dust inhalation.	1657
–	T7	TP2	F-A, S-A	Category A.	Miscible with water. Toxic if swallowed, by skin contact or by inhalation.	1658
–	T7	TP2	F-A, S-A	Category A.	See entry above.	1658
–	T3	TP33	F-A, S-A	Category A.	White crystals. Soluble in water. Toxic if swallowed, by skin contact or by dust inhalation.	1659
–	–	–	F-C, S-W	Category D. Clear of living quarters. Segregation as for class 5.1 but "separated from" class 7.	Non-flammable, toxic and corrosive gas. Strong oxidizing agent. On contact with air, gives off brown fumes which are toxic by inhalation, with delayed effect similar to phosgene. Heavier than air (1.04). Highly irritating to skin, eyes and mucous membranes.	1660
–	T3	TP33	F-A, S-A	Category A.	Yellow crystals. Toxic if swallowed, by skin contact or by dust inhalation. *ortho*-NITROANILINES may be carried in the molten state.	1661
–	T7	TP2	F-A, S-A	Category A. Clear of living quarters.	Oily liquid, evolving toxic vapour. Melting point: approximately 6°C. Toxic if swallowed, by skin contact or by inhalation.	1662
–	T1	TP33	F-A, S-A	Category A.	Yellow crystals. Some isomers may have a melting point as low as 44°C. Toxic if swallowed, by skin contact or by dust inhalation. May be carried in the molten state.	1663
–	T7	TP2	F-A, S-A	Category A.	Yellow liquids. Melting points: *ortho*-NITROTOLUENE: –4°C, *meta*-NITROTOLUENE: 15°C. Toxic if swallowed, by skin contact or by inhalation.	1664
–	T7	TP2	F-A, S-A	Category A.	Yellow liquids. Melting points: 2-NITRO-3-XYLENE: 14°C to 16°C, 3-NITRO-2-XYLENE: 7°C to 9°C, 4-NITRO-3-XYLENE: 2°C. Immiscible with water. Toxic if swallowed, by skin contact or by inhalation.	1665
–	T7	TP2	F-A, S-A	Category A. Clear of living quarters.	Colourless liquid. Toxic if swallowed, by skin contact or by inhalation.	1669
–	T20	TP2 TP13 TP37	F-A, S-A	Category D. Clear of living quarters.	Yellow, oily, volatile liquid evolving irritating vapour ("Tear Gas"). Slowly decomposes in contact with water, producing hydrochloric acid. Reacts with iron or steel, evolving carbon tetrachloride. Corrosive to most metals. Highly toxic if swallowed, by skin contact or by inhalation.	1670
–	T3	TP33	F-A, S-A	Category A.	Colourless or white crystals or crystallized mass. Melting point: 43°C (pure product). Soluble in water. Toxic if swallowed, by skin contact or by vapour inhalation. Rapidly absorbed through the skin.	1671
–	T14	TP2 TP13	F-A, S-A	Category D. Clear of living quarters.	Pale yellow, oily liquid with an irritating unpleasant odour. Highly toxic if swallowed, by skin contact or by inhalation.	1672
–	T1	TP33	F-A, S-A	Category A.	White crystals or powder. Toxic if swallowed, by skin contact or by dust inhalation. May be carried in the molten state.	1673
–	T3	TP33	F-A, S-A	Category A.	Toxic if swallowed, by skin contact or by dust inhalation.	1674
–	T3	TP33	F-A, S-A	Category A.	Colourless crystals or white powder. Soluble in water. Toxic if swallowed, by skin contact or by dust inhalation.	1677
–	T3	TP33	F-A, S-A	Category A.	White powder. Soluble in water. Toxic if swallowed, by skin contact or by dust inhalation.	1678

UN No. (1)	Proper Shipping Name (PSN) (2) 3.1.2	Class or division (3) 2.0	Subsidiary risk(s) (4) 2.0	Packing group (5) 2.0.1.3	Special provisions (6) 3.3	Limited and excepted quantity provisions: Limited quantities (7a) 3.4	Excepted quantities (7b) 3.5	Packing: Instructions (8) 4.1.4	Provisions (9) 4.1.4	IBC: Instructions (10) 4.1.4	Provisions (11) 4.1.4
1679	POTASSIUM CUPROCYANIDE	6.1	– P	II	–	500 g	E4	P002	–	IBC08	B2 B4
1680	POTASSIUM CYANIDE, SOLID	6.1	– P	I	–	0	E5	P002	PP31	IBC07	B1
1683	SILVER ARSENITE	6.1	– P	II	–	500 g	E4	P002	–	IBC08	B2 B4
1684	SILVER CYANIDE	6.1	– P	II	–	500 g	E4	P002	–	IBC08	B2 B4
1685	SODIUM ARSENATE	6.1	–	II	–	500 g	E4	P002	–	IBC08	B2 B4
1686	SODIUM ARSENITE, AQUEOUS SOLUTION	6.1	–	II	43	100 mL	E4	P001	–	IBC02	–
1686	SODIUM ARSENITE, AQUEOUS SOLUTION	6.1	–	III	43 223	5 L	E1	P001 LP01	–	IBC03	–
1687	SODIUM AZIDE	6.1	–	II	–	500 g	E4	P002	–	IBC08	B2 B4
1688	SODIUM CACODYLATE	6.1	–	II	–	500 g	E4	P002	–	IBC08	B2 B4
1689	SODIUM CYANIDE, SOLID	6.1	– P	I	–	0	E5	P002	PP31	IBC07	B1
1690	SODIUM FLUORIDE, SOLID	6.1	–	III	–	5 kg	E1	P002 LP02	–	IBC08	B3
1691	STRONTIUM ARSENITE	6.1	–	II	–	500 g	E4	P002	–	IBC08	B2 B4
1692	STRYCHNINE or STRYCHNINE SALTS	6.1	– P	I	43	0	E5	P002	–	IBC07	B1
1693	TEAR GAS SUBSTANCE, LIQUID, N.O.S.	6.1	–	I	274	0	E5	P001	PP31	–	–
1693	TEAR GAS SUBSTANCE, LIQUID, N.O.S.	6.1	–	II	274	0	E4	P001	PP31	IBC02	–
1694	BROMOBENZYL CYANIDES, LIQUID	6.1	–	I	138	0	E5	P001	PP31	–	–
1695	CHLOROACETONE, STABILIZED	6.1	3/8 P	I	354	0	E0	P602	–	–	–
1697	CHLOROACETOPHENONE, SOLID	6.1	–	II	–	0	E4	P002	–	IBC08	B2 B4
1698	DIPHENYLAMINE CHLOROARSINE	6.1	– P	I	–	0	E5	P002	PP31	–	–

Portable tanks and bulk containers			EmS	Stowage and segregation	Properties and observations	UN No.
	Tank instructions	Provisions				
(12)	(13) 4.2.5 4.3	(14) 4.2.5	(15) 5.4.3.2 7.3	(16) 7.1 7.2	(17)	(18)
–	T3	TP33	F-A, S-A	Category A. "Separated from" acids.	White crystals or powder. Soluble in water. Reacts with acids or acid fumes, evolving hydrogen cyanide, a highly toxic and flammable gas. Toxic if swallowed, by skin contact or by dust inhalation.	**1679**
–	T6	TP33	F-A, S-A	Category B. "Separated from" acids.	White, deliquescent crystals or lumps. Soluble in water. Reacts with acids or acid fumes, evolving hydrogen cyanide, a highly toxic and flammable gas. Highly toxic if swallowed, by skin contact or by dust inhalation.	**1680**
–	T3	TP33	F-A, S-A	Category A.	Yellow powder. Insoluble in water. Toxic if swallowed, by skin contact or by dust inhalation.	**1683**
–	T3	TP33	F-A, S-A	Category A. Clear of living quarters. "Separated from" acids.	White powder. Insoluble in water. Reacts with acids or acid fumes, evolving hydrogen cyanide, a highly toxic and flammable gas. Toxic if swallowed, by skin contact or by dust inhalation.	**1684**
–	T3	TP33	F-A, S-A	Category A.	Colourless crystals. Soluble in water. Toxic if swallowed, by skin contact or by dust inhalation.	**1685**
–	T7	TP2	F-A, S-A	Category A.	Colourless. Toxic if swallowed, by skin contact or by inhalation.	**1686**
–	T4	TP2	F-A, S-A	Category A.	See entry above.	**1686**
–	–	–	F-A, S-A	Category A. "Away from" heavy metals and their salts. "Separated from" class 3 and from acids.	Colourless crystals. May react vigorously with acids to form hydrazoic acid, which is an explosive. May form extremely sensitive compounds with heavy metals or their salts. May explode if involved in a fire. Toxic if swallowed, by skin contact or by dust inhalation.	**1687**
–	T3	TP33	F-A, S-A	Category A. "Separated from" acids.	White, deliquescent solid with a foul odour. Reacts with acids, evolving dimethylarsine, an extremely toxic gas. Soluble in water. Toxic if swallowed, by skin contact or by dust inhalation.	**1688**
–	T6	TP33	F-A, S-A	Category B. "Separated from" acids.	White, deliquescent crystals or lumps. Soluble in water. Reacts with acids or acid fumes, evolving hydrogen cyanide, a highly toxic and flammable gas. Highly toxic if swallowed, by skin contact or by dust inhalation.	**1689**
–	T1	TP33	F-A, S-A	Category A. "Separated from" acids.	White crystals or powder. React with acids, evolving hydrogen fluoride, a toxic, irritating and corrosive gas, apparent as white fumes. Toxic if swallowed, by skin contact or by inhalation.	**1690**
–	T3	TP33	F-A, S-A	Category A.	White powder. Soluble in water. Toxic if swallowed, by skin contact or by dust inhalation.	**1691**
–	T6	TP33	F-A, S-A	Category A.	White crystals or powder. Strychnine is slightly soluble; the salts are soluble in water. Highly toxic if swallowed, by skin contact or by dust inhalation.	**1692**
–	–	–	F-A, S-A	Category D. Clear of living quarters.	"Tear gas substance" is a generic term for substances which, in minute quantities dispersed in air, cause extreme eye irritation and profuse tears. Toxic if swallowed, by skin contact or by inhalation.	**1693**
–	–	–	F-A, S-A	Category D. Clear of living quarters.	See entry above.	**1693**
–	T14	TP2 TP13	F-A, S-A	Category D. Keep as cool as reasonably practicable. Clear of living quarters. "Separated from" acids.	Volatile liquids evolving irritating vapour ("Tear Gas"). Melting points: *ortho*-BROMOBENZYL CYANIDE 1°C. Highly toxic if swallowed, by skin contact or by inhalation.	**1694**
–	T20	TP2 TP13 TP35	F-E, S-C	Category D. Clear of living quarters. Segregation as for class 3 but "away from" class 4.1.	Flammable, corrosive, colourless liquid, evolving irritating vapour ("Tear Gas"). Miscible with water. Flashpoint: 25°C c.c. Highly toxic if swallowed, by skin contact or by inhalation.	**1695**
–	T3	TP33	F-A, S-A	Category D. Keep as cool as reasonably practicable. Clear of living quarters.	White crystals evolving irritating vapour ("Tear Gas"). Melting point may be as low as 20°C. Toxic if swallowed, by skin contact or by inhalation.	**1697**
–	T6	TP33	F-A, S-A	Category D. Clear of living quarters.	Volatile, yellow crystals evolving irritating vapour ("Tear Gas"). Highly toxic if swallowed, by skin contact or by inhalation.	**1698**

UN No.	Proper Shipping Name (PSN)	Class or division	Subsidiary risk(s)	Packing group	Special provisions	Limited and excepted quantity provisions		Packing		IBC	
						Limited quantities	Excepted quantities	Instructions	Provisions	Instructions	Provisions
(1)	(2) 3.1.2	(3) 2.0	(4) 2.0	(5) 2.0.1.3	(6) 3.3	(7a) 3.4	(7b) 3.5	(8) 4.1.4	(9) 4.1.4	(10) 4.1.4	(11) 4.1.4
1699	DIPHENYLCHLOROARSINE, LIQUID	6.1	– P	I	–	0	E5	P001	PP31	–	–
1700	TEAR GAS CANDLES	6.1	4.1	II	–	0	E0	P600	–	–	–
1701	XYLYL BROMIDE, LIQUID	6.1	–	II	–	0	E4	P001	PP31	IBC02	–
1702	1,1,2,2-TETRACHLOROETHANE	6.1	– P	II	–	100 mL	E4	P001	–	IBC02	–
1704	TETRAETHYL DITHIOPYROPHOSPHATE	6.1	– P	II	43	100 mL	E4	P001	–	IBC02	–
1707	THALLIUM COMPOUND, N.O.S.	6.1	– P	II	43 274	500 g	E4	P002	–	IBC08	B2 B4
1708	TOLUIDINES, LIQUID	6.1	–	II	279	100 mL	E4	P001	–	IBC02	–
1709	2,4-TOLUYLENEDIAMINE, SOLID	6.1	–	III	–	5 kg	E1	P002 LP02	–	IBC08	B3
1710	TRICHLOROETHYLENE	6.1	–	III	–	5 L	E1	P001 LP01	–	IBC03	–
1711	XYLIDINES, LIQUID	6.1	–	II	–	100 mL	E4	P001	–	IBC02	–
1712	ZINC ARSENATE or ZINC ARSENITE or ZINC ARSENATE, ZINC ARSENITE MIXTURE	6.1	–	II	–	500 g	E4	P002	–	IBC08	B2 B4
1713	ZINC CYANIDE	6.1	– P	I	–	0	E5	P002	–	IBC07	B1
1714	ZINC PHOSPHIDE	4.3	6.1	I	–	0	E0	P403	PP31	–	–
1715	ACETIC ANHYDRIDE	8	3	II	–	1 L	E2	P001	–	IBC02	–
1716	ACETYL BROMIDE	8	–	II	–	1 L	E2	P001	–	IBC02	B20
1717	ACETYL CHLORIDE	3	8	II	–	1 L	E2	P001	–	IBC02	B20
1718	BUTYL ACID PHOSPHATE	8	–	III	–	5 L	E1	P001 LP01	–	IBC03	–
1719	CAUSTIC ALKALI LIQUID, N.O.S.	8	–	II	274	1 L	E2	P001	–	IBC02	–
1719	CAUSTIC ALKALI LIQUID, N.O.S.	8	–	III	223 274	5 L	E1	P001	–	IBC03	–

Portable tanks and bulk containers			EmS	Stowage and segregation	Properties and observations	UN No.
	Tank instructions	Provisions				
(12)	(13) 4.2.5 4.3	(14) 4.2.5	(15) 5.4.3.2 7.3	(16) 7.1 7.2	(17)	(18)
–	–	–	F-A, S-A	Category D. Clear of living quarters.	When pure, colourless liquid. The commercial product may be a dark brown liquid. Volatile liquid evolving an irritating vapour ("Tear Gas"). Highly toxic if swallowed, by skin contact or by inhalation.	1699
–	–	–	F-A, S-G	Category D. Clear of living quarters.	Devices containing tear-producing substances which, in minute quantities dispersed in air, cause extreme eye irritation and profuse tears.	1700
–	T7	TP2 TP13	F-A, S-A	Category D. Clear of living quarters.	Colourless liquid, evolving irritating vapour ("Tear Gas"). Toxic if swallowed, by skin contact or by inhalation.	1701
–	T7	TP2	F-A, S-A	Category A. Clear of living quarters.	Colourless liquid with a chloroform-like odour. Toxic if swallowed, by skin contact or by inhalation.	1702
–	T7	TP2	F-A, S-A	Category D. Clear of living quarters.	Colourless liquid. In the presence of moisture, corrosive to most metals. Toxic if swallowed, by skin contact or by inhalation.	1704
–	T3	TP33	F-A, S-A	Category A.	White crystals or powder. Toxic if swallowed, by skin contact or by dust inhalation.	1707
–	T7	TP2	F-A, S-A	Category A.	Colourless liquids. Toxic if swallowed, by skin contact or by inhalation.	1708
–	T1	TP33	F-A, S-A	Category A.	White crystals or powder. Toxic if swallowed, by skin contact or by dust inhalation.	1709
–	T4	TP1	F-A, S-A	Category A. Clear of living quarters.	Colourless liquid with a chloroform-like odour. When involved in a fire, evolves extremely toxic fumes (phosgene). Toxic if swallowed, by skin contact or by inhalation.	1710
–	T7	TP2	F-A, S-A	Category A.	Toxic if swallowed, by skin contact or by inhalation.	1711
–	T3	TP33	F-A, S-A	Category A.	Crystalline solid. Insoluble in water. Toxic if swallowed, by skin contact or by dust inhalation.	1712
–	T6	TP33	F-A, S-A	Category A. "Separated from" acids.	White crystals or powder. Insoluble in water. Reacts with acids or acid fumes, evolving hydrogen cyanide, a highly toxic and flammable gas. Highly toxic if swallowed, by skin contact or by dust inhalation.	1713
–	–	–	F-G, S-N	Category E. If under deck, in a mechanically ventilated space. Clear of living quarters. "Separated from" acids.	Grey crystals or powder. Reacts with acids or decomposes slowly in contact with water or damp air, evolving phosphine, a spontaneously flammable and highly toxic gas. Reacts violently with oxidizing substances.	1714
–	T7	TP2	F-E, S-C	Category A. Clear of living quarters.	Colourless, flammable liquid with an irritating odour. Flashpoint: 54°C c.c. Immiscible with water. In the presence of moisture, corrosive to most metals. Vapour irritates mucous membranes.	1715
–	T8	TP2	F-A, S-B	Category C. Clear of living quarters.	Colourless liquid. Reacts violently with water, evolving hydrogen bromide, an irritating and corrosive gas apparent as white fumes. In the presence of moisture, highly corrosive to most metals. Vapour irritates mucous membranes.	1716
–	T8	TP2	F-E, S-C	Category B. Clear of living quarters.	Colourless liquid. Flashpoint: 5°C c.c. Boiling point: 51°C. Reacts violently with water, evolving hydrogen chloride, an irritating and corrosive gas apparent as white fumes. In the presence of moisture, highly corrosive to most metals. Causes burns to skin, eyes and mucous membranes.	1717
–	T4	TP1	F-A, S-B	Category A.	Yellow liquid. Insoluble in water. Mildly corrosive to most metals.	1718
–	T11	TP2 TP27	F-A, S-B	Category A. "Separated from" acids. "Away from" ammonium salts.	Corrosive to aluminium, zinc and tin. Reacts violently with acids. Reacts with ammonium salts, evolving ammonia gas. Causes burns to skin, eyes and mucous membranes.	1719
–	T7	TP1 TP28	F-A, S-B	Category A. "Separated from" acids. "Away from" ammonium salts.	See entry above.	1719

UN No.	Proper Shipping Name (PSN)	Class or division	Subsidiary risk(s)	Packing group	Special provisions	Limited and excepted quantity provisions		Packing		IBC	
						Limited quantities	Excepted quantities	Instructions	Provisions	Instructions	Provisions
(1)	(2) 3.1.2	(3) 2.0	(4) 2.0	(5) 2.0.1.3	(6) 3.3	(7a) 3.4	(7b) 3.5	(8) 4.1.4	(9) 4.1.4	(10) 4.1.4	(11) 4.1.4
1722	ALLYL CHLOROFORMATE	6.1	3/8	I	–	0	E5	P001	–	–	–
1723	ALLYL IODIDE	3	8	II	–	1 L	E2	P001	–	IBC02	–
△ 1724	ALLYLTRICHLOROSILANE, STABILIZED	8	3	II	–	0	E0	P010	–	–	–
1725	ALUMINIUM BROMIDE, ANHYDROUS	8	–	II	937	1 kg	E2	P002	–	IBC08	B2 B4
1726	ALUMINIUM CHLORIDE, ANHYDROUS	8	–	II	937	1 kg	E2	P002	–	IBC08	B2 B4
△ 1727	AMMONIUM HYDROGENDIFLUORIDE, SOLID	8	–	II	–	1 kg	E2	P002	–	IBC08	B2 B4
△ 1728	AMYLTRICHLOROSILANE	8	–	II	–	0	E0	P010	–	–	–
1729	ANISOYL CHLORIDE	8	–	II	–	1 kg	E2	P002	–	IBC08	B2 B4
1730	ANTIMONY PENTACHLORIDE, LIQUID	8	–	II	–	1 L	E2	P001	–	IBC02	–
1731	ANTIMONY PENTACHLORIDE SOLUTION	8	–	II	–	1 L	E2	P001	–	IBC02	–
1731	ANTIMONY PENTACHLORIDE SOLUTION	8	–	III	223	5 L	E1	P001 LP01	–	IBC03	–
1732	ANTIMONY PENTAFLUORIDE	8	6.1	II	–	1 L	E2	P001	–	IBC02	–

Portable tanks and bulk containers (12)	Tank instructions (13) 4.2.5 4.3	Provisions (14) 4.2.5	EmS (15) 5.4.3.2 7.3	Stowage and segregation (16) 7.1 7.2	Properties and observations (17)	UN No. (18)
–	T14	TP2 TP13	F-E, S-C	Category D. Clear of living quarters. Segregation as for class 3 but "away from" class 4.1.	Colourless, flammable liquid, extremely irritating odour, causes tears. Flashpoint: 31°C c.c. When involved in a fire, evolves toxic gases. In the presence of moisture, corrosive to most metals. Highly toxic if swallowed, by skin contact or by inhalation. Causes burns to skin, eyes and mucous membranes.	1722
–	T7	TP2 TP13	F-E, S-C	Category B. Clear of living quarters.	Yellow liquid with an irritating odour. Flashpoint: 5°C c.c. Immiscible with water. In the presence of moisture, corrosive to most metals. Causes burns to skin, eyes and mucous membranes.	1723
–	T10	TP2 TP7 TP13	F-E, S-C	Category C. Clear of living quarters.	Colourless, flammable liquid with a pungent odour. Flashpoint: 35°C c.c. Reacts violently with water, evolving hydrogen chloride, an irritating and corrosive gas, apparent as white fumes. When involved in a fire, evolves toxic gases. In the presence of moisture, highly corrosive to most metals. Causes burns to skin, eyes and mucous membranes.	1724 △
–	T3	TP33	F-A, S-B	Category A. Clear of living quarters.	White to yellowish hygroscopic crystals. Forms corrosive vapours in moist air. Reacts violently with water, evolving heat and hydrogen bromide, an irritating and corrosive gas apparent as white fumes. In the presence of moisture, highly corrosive to most metals. Highly irritating to skin, eyes and mucous membranes. The solid hydrated form of this substance is not subject to the provisions of this Code.	1725
–	T3	TP33	F-A, S-B	Category A. Clear of living quarters.	White to yellowish hygroscopic crystals. Forms corrosive vapours in moist air. Reacts violently with water, evolving heat and hydrogen chloride, an irritating and corrosive gas apparent as white fumes. In the presence of moisture, highly corrosive to most metals. Highly irritating to skin, eyes and mucous membranes. The solid hydrated form of this substance is not subject to the provisions of this Code.	1726
–	T3	TP33	F-A, S-B	Category A. Protected from sources of heat. Clear of living quarters. "Separated from" acids.	White deliquescent crystals. Decomposed by heat or acids, evolving hydrogen fluoride, a toxic, extremely irritating and corrosive gas, apparent as white fumes. In the presence of moisture, highly corrosive to glass, other siliceous materials and most metals. Causes burns to skin and mucous membranes.	1727 △
–	T10	TP2 TP7 TP13	F-A, S-B	Category C. Clear of living quarters.	Colourless liquid with a pungent odour. Reacts violently with water, evolving hydrogen chloride, an irritating and corrosive gas apparent as white fumes. When involved in a fire, evolves toxic gases. In the presence of moisture, highly corrosive to most metals. Vapour irritates mucous membranes.	1728 △
–	T3	TP33	F-A, S-B	Category C. Clear of living quarters.	Crystalline powder. Melting point: 22°C. Reacts violently with water, evolving hydrogen chloride, an irritating and corrosive gas apparent as white fumes. In the presence of moisture, highly corrosive to most metals. Vapour irritates mucous membranes.	1729
–	T7	TP2	F-A, S-B	Category C. Clear of living quarters.	Yellow, oily liquid with an offensive odour. May solidify by absorption of moisture. Reacts violently with water, evolving hydrogen chloride, an irritating and corrosive gas apparent as white fumes. In the presence of moisture, highly corrosive to most metals. Causes burns to skin, eyes and mucous membranes.	1730
–	T7	TP2	F-A, S-B	Category C. Clear of living quarters.	Yellow liquid with an offensive odour. Corrosive to most metals. Causes burns to skin, eyes and mucous membranes.	1731
–	T4	TP1	F-A, S-B	Category C. Clear of living quarters.	See entry above.	1731
–	T7	TP2	F-A, S-B	Category D. Clear of living quarters. Segregation as for class 5.1 but "away from" classes 4.1, 5.1 and 7.	Colourless liquid with a pungent odour. When anhydrous, mildly corrosive to glass, other siliceous materials and most metals. Reacts violently with water, evolving hydrogen fluoride, an irritating gas, highly corrosive to glass and other siliceous materials and most metals. Powerful oxidant, may cause fire in contact with readily flammable organic substances. Toxic if swallowed, by skin contact or by inhalation. Causes burns to skin and mucous membranes.	1732

UN No.	Proper Shipping Name (PSN)	Class or division	Subsidiary risk(s)	Packing group	Special provisions	Limited and excepted quantity provisions		Packing		IBC	
						Limited quantities	Excepted quantities	Instructions	Provisions	Instructions	Provisions
(1)	(2) 3.1.2	(3) 2.0	(4) 2.0	(5) 2.0.1.3	(6) 3.3	(7a) 3.4	(7b) 3.5	(8) 4.1.4	(9) 4.1.4	(10) 4.1.4	(11) 4.1.4
1733	ANTIMONY TRICHLORIDE	8	–	II	–	1 kg	E2	P002	–	IBC08	B2 B4
1736	BENZOYL CHLORIDE	8	–	II	–	1 L	E2	P001	–	IBC02	B20
1737	BENZYL BROMIDE	6.1	8	II	–	0	E4	P001	–	IBC02	B20
1738	BENZYL CHLORIDE	6.1	8	II	–	0	E4	P001	–	IBC02	B20
1739	BENZYL CHLOROFORMATE	8	– P	I	–	0	E0	P001	–	–	–
△ 1740	HYDROGENDIFLUORIDES, SOLID, N.O.S.	8	–	II	–	1 kg	E2	P002	–	IBC08	B2 B4
△ 1740	HYDROGENDIFLUORIDES, SOLID, N.O.S.	8	–	III	223	5 kg	E1	P002 LP02	–	IBC08	B3
△ 1741	BORON TRICHLORIDE	2.3	8	–	–	0	E0	P200	–	–	–
1742	BORON TRIFLUORIDE ACETIC ACID COMPLEX, LIQUID	8	–	II	–	1 L	E2	P001	–	IBC02	B20
1743	BORON TRIFLUORIDE PROPIONIC ACID COMPLEX, LIQUID	8	–	II	–	500 mL	E2	P001	–	IBC02	B20
1744	BROMINE or BROMINE SOLUTION	8	6.1	I	–	0	E0	P804	–	–	–
△ 1745	BROMINE PENTAFLUORIDE	5.1	6.1/8	I	–	0	E0	P200	–	–	–

3

DGL

Portable tanks and bulk containers			EmS	Stowage and segregation	Properties and observations	UN No.
	Tank instructions	Provisions				
(12)	(13) 4.2.5 4.3	(14) 4.2.5	(15) 5.4.3.2 7.3	(16) 7.1 7.2	(17)	(18)
–	T3	TP33	F-A, S-B	Category C. Clear of living quarters.	Reacts slowly with water, evolving hydrogen chloride, an irritating and corrosive gas. In the presence of moisture, corrosive to most metals.	**1733**
–	T8	TP2 TP13	F-A, S-B	Category C. Clear of living quarters.	Colourless liquid, very irritating odour, causes tears. Reacts violently with water, evolving hydrogen chloride, an irritating and corrosive gas apparent as white fumes. In the presence of moisture, highly corrosive to most metals. Vapour irritates mucous membranes.	**1736**
–	T8	TP2 TP13	F-A, S-B	Category D. Keep as dry as reasonably practicable. Clear of living quarters.	Colourless liquid with a pungent odour, causes tears. In the presence of moisture, corrosive to most metals. Toxic if swallowed, by skin contact or by inhalation. Causes burns to skin, eyes and mucous membranes.	**1737**
–	T8	TP2 TP13	F-A, S-B	Category D. Keep as dry as reasonably practicable. Clear of living quarters.	Colourless liquid with a pungent odour. Causes tears. Immiscible with water, but hydrolyses slowly in contact with it. In the presence of moisture, corrosive to most metals. Toxic if swallowed, by skin contact or by vapour inhalation. Causes burns to skin, eyes and mucous membranes.	**1738**
–	T10	TP2 TP13	F-A, S-B	Category D. Clear of living quarters.	Colourless liquid with an irritating odour. Reacts with water. When involved in a fire, evolves toxic gases. In the presence of moisture, highly corrosive to most metals. Causes severe burns to skin, eyes and mucous membranes.	**1739**
–	T3	TP33	F-A, S-B	Category A. Protected from sources of heat. Clear of living quarters. "Separated from" acids.	Crystalline solids. Decomposed by heat or acid, evolving hydrogen fluoride, an extremely irritating and corrosive gas. In the presence of moisture, corrosive to glass, other siliceous materials and most metals. Cause burns to skin, eyes and mucous membranes.	**1740** △
–	T1	TP33	F-A, S-B	Category A. Protected from sources of heat. Clear of living quarters. "Separated from" acids.	See entry above.	**1740** △
–	–	–	F-C, S-U	Category D. Protected from sources of heat. Clear of living quarters.	Non-flammable, toxic and corrosive gas. Forms dense white corrosive fumes in moist air. Reacts violently with water, evolving hydrogen chloride, an irritating and corrosive gas apparent as white fumes. In the presence of moisture, highly corrosive to most metals. Much heavier than air (2.35). Highly irritating to skin, eyes and mucous membranes.	**1741** △
–	T8	TP2	F-A, S-B	Category A.	Highly corrosive to most metals. Causes burns to skin, eyes and mucous membranes.	**1742**
–	T8	TP2	F-A, S-B	Category A.	Highly corrosive to most metals. Causes burns to skin, eyes and mucous membranes.	**1743**
–	T22	TP2 TP10 TP13	F-A, S-B	Category D. Keep as cool as reasonably practicable. Clear of living quarters. Segregation as for class 5.1, but "separated from" classes 4.1, 5.1 and 7.	Very dark brown, heavy liquid with an extremely irritating odour. Density: 3.1 (pure product). Boiling point: 59°C. Powerful oxidant; may cause fire in contact with organic materials such as wood, cotton or straw. Highly corrosive to most metals. Solutions have the same properties to a lesser degree, depending on concentration. Toxic if swallowed, by skin contact or by inhalation. Causes burns to skin, eyes and mucous membranes.	**1744**
–	T22	TP2 TP13	F-A, S-B	Category D. Protected from sources of heat. Clear of living quarters. Segregation as for class 5.1 but "separated from" classes 4.1 and 7.	Colourless, heavy liquid with an extremely irritating odour. Boiling point: 40°C. Powerful oxidant; may cause fire in contact with organic material such as wood, cotton or straw. Reacts violently with water, evolving hydrogen fluoride, a toxic, extremely corrosive gas apparent as white fumes. In contact with acids or acid fumes, evolves highly toxic fumes of bromine, fluorine and their compounds. Highly corrosive to most metals. Toxic if swallowed, by skin contact or by inhalation. Causes burns to skin, eyes and mucous membranes.	**1745** △

	UN No.	Proper Shipping Name (PSN)	Class or division	Subsidiary risk(s)	Packing group	Special provisions	Limited and excepted quantity provisions		Packing		IBC	
							Limited quantities	Excepted quantities	Instructions	Provisions	Instructions	Provisions
	(1)	(2) 3.1.2	(3) 2.0	(4) 2.0	(5) 2.0.1.3	(6) 3.3	(7a) 3.4	(7b) 3.5	(8) 4.1.4	(9) 4.1.4	(10) 4.1.4	(11) 4.1.4
△	1746	BROMINE TRIFLUORIDE	5.1	6.1/8	I	–	0	E0	P200	–	–	–
△	1747	BUTYLTRICHLOROSILANE	8	3	II	–	0	E0	P010	–	–	–
△	1748	CALCIUM HYPOCHLORITE, DRY or CALCIUM HYPOCHLORITE MIXTURE, DRY with more than 39% available chlorine (8.8% available oxygen)	5.1	–	II	314	1 kg	E2	P002	PP85	–	–
△	1748	CALCIUM HYPOCHLORITE, DRY or CALCIUM HYPOCHLORITE MIXTURE, DRY with more than 39% available chlorine (8.8% available oxygen)	5.1	–	III	316	5 kg	E1	P002	PP85	–	–
	1749	CHLORINE TRIFLUORIDE	2.3	5.1/8	–	–	0	E0	P200	–	–	–
	1750	CHLOROACETIC ACID SOLUTION	6.1	8	II	–	100 mL	E4	P001	–	IBC02	–
	1751	CHLOROACETIC ACID, SOLID	6.1	8	II	–	500 g	E4	P002	–	IBC08	B2 B4
	1752	CHLOROACETYL CHLORIDE	6.1	8	I	354	0	E0	P602	–	–	–

Portable tanks and bulk containers			EmS	Stowage and segregation	Properties and observations	UN No.
	Tank instructions	Provisions				
(12)	(13) 4.2.5 4.3	(14) 4.2.5	(15) 5.4.3.2 7.3	(16) 7.1 7.2	(17)	(18)
–	T22	TP2 TP13	F-A, S-B	Category D. Protected from sources of heat. Clear of living quarters. Segregation as for class 5.1 but "separated from" classes 4.1 and 7.	Colourless, heavy liquid with an extremely irritating odour. Powerful oxidant; may cause fire in contact with organic material such as wood, cotton or straw. Reacts violently with water, evolving hydrogen fluoride, a toxic, extremely corrosive gas apparent as white fumes. In contact with acids or acid fumes, evolves highly toxic fumes of bromine, fluorine and their compounds. Highly corrosive to most metals. Toxic if swallowed, by skin contact or by inhalation. Causes burns to skin, eyes and mucous membranes.	1746 △
–	T10	TP2 TP7 TP13	F-E, S-C	Category C. Clear of living quarters.	Colourless, flammable liquid with a pungent odour. Flashpoint: 52°C c.c. Reacts violently with water, evolving hydrogen chloride, an irritating and corrosive gas apparent as white fumes. When involved in a fire, evolves toxic gases. In the presence of moisture, highly corrosive to most metals. Vapour irritates mucous membranes.	1747 △
–	–	–	F-H, S-Q	Category D. Protected from sources of heat. Cargo transport units shall be shaded from direct sunlight. Packages in cargo transport units shall be stowed so as to allow for adequate air circulation throughout the cargo. "Separated from" ammonium compounds, acids, cyanides, hydrogen peroxides and liquid organic substances.	White or yellowish solid (powder, granules or tablets) with chlorine-like odour. Soluble in water. May cause fire in contact with organic material or ammonium compounds. Substances are liable to exothermic decomposition at elevated temperatures. This condition may lead to fire or explosion. Decomposition can be initiated by heat or by impurities (e.g., powdered metals (iron, manganese, cobalt, magnesium) and their compounds). Liable to heat slowly. Reacts with acids, evolving chlorine, an irritating, corrosive and toxic gas. In the presence of moisture, corrosive to most metals. Dust irritates mucous membranes.	1748 △
–	–	–	F-H, S-Q	Category D. Protected from sources of heat. Cargo transport units shall be shaded from direct sunlight. Packages in cargo transport units shall be stowed so as to allow for adequate air circulation throughout the cargo. "Separated from" ammonium compounds, acids, cyanides, hydrogen peroxides and liquid organic substances.	See entry above.	1748 △
–	–	–	F-C, S-W	Category D. Clear of living quarters. Segregation as for class 5.1 but "separated from" class 7.	Non-flammable, toxic and corrosive gas. Forms dense, white, corrosive fumes in moist air. Reacts violently with water, evolving hydrogen fluoride, an irritating and corrosive gas apparent as white fumes. Corrosive to glass and to most metals. Powerful oxidizing agent which may cause fires with combustible materials. Much heavier than air. Highly irritating to skin, eyes and mucous membranes.	1749
–	T7	TP2	F-A, S-B	Category C. Clear of living quarters.	Colourless liquid. Corrosive to most metals. Toxic if swallowed, by skin contact or by inhalation. Causes burns to skin, eyes and mucous membranes.	1750
–	T3	TP33	F-A, S-B	Category C. Clear of living quarters.	Colourless, very deliquescent crystals. Melting point may be as low as 50°C. In the presence of moisture, corrosive to most metals. Toxic if swallowed, by skin contact or by dust inhalation. Causes burns to skin, eyes and mucous membranes.	1751
–	T20	TP2 TP13 TP35	F-A, S-B	Category D. Clear of living quarters.	Colourless liquid, with extremely irritating odour, causing tears. Reacts violently with water, evolving hydrogen chloride, an irritating and corrosive gas apparent as white fumes. In the presence of moisture, highly corrosive to most metals. Highly toxic if swallowed, by skin contact or by inhalation. Causes burns to skin, eyes and mucous membranes.	1752

UN No.	Proper Shipping Name (PSN)	Class or division	Subsidiary risk(s)	Packing group	Special provisions	Limited and excepted quantity provisions		Packing		IBC	
						Limited quantities	Excepted quantities	Instructions	Provisions	Instructions	Provisions
(1)	(2) 3.1.2	(3) 2.0	(4) 2.0	(5) 2.0.1.3	(6) 3.3	(7a) 3.4	(7b) 3.5	(8) 4.1.4	(9) 4.1.4	(10) 4.1.4	(11) 4.1.4
1753	CHLOROPHENYL-TRICHLOROSILANE	8	– P	II	–	0	E2	P010	–	–	–
1754	CHLOROSULPHONIC ACID (with or without sulphur trioxide)	8	–	I	–	0	E0	P001	–	–	–
1755	CHROMIC ACID SOLUTION	8	–	II	–	1 L	E2	P001	–	IBC02	B20
1755	CHROMIC ACID SOLUTION	8	–	III	223	5 L	E1	P001 LP01	–	IBC03	–
1756	CHROMIC FLUORIDE, SOLID	8	–	II	–	1 kg	E2	P002	–	IBC08	B2 B4
1757	CHROMIC FLUORIDE SOLUTION	8	–	II	–	1 L	E2	P001	–	IBC02	–
1757	CHROMIC FLUORIDE SOLUTION	8	–	III	223	5 L	E1	P001 LP01	–	IBC03	–
1758	CHROMIUM OXYCHLORIDE	8	–	I	–	0	E0	P001	–	–	–
1759	CORROSIVE SOLID, N.O.S.	8	–	I	274	0	E0	P002	–	IBC07	B1
1759	CORROSIVE SOLID, N.O.S.	8	–	II	274	1 kg	E2	P002	–	IBC08	B2 B4
1759	CORROSIVE SOLID, N.O.S.	8	–	III	223 274	5 kg	E1	P002 LP02	–	IBC08	B3
1760	CORROSIVE LIQUID, N.O.S.	8	–	I	274	0	E0	P001	–	–	–
1760	CORROSIVE LIQUID, N.O.S.	8	–	II	274	1 L	E2	P001	–	IBC02	–
1760	CORROSIVE LIQUID, N.O.S.	8	–	III	223 274	5 L	E1	P001 LP01	–	IBC03	–
1761	CUPRIETHYLENEDIAMINE SOLUTION	8	6.1 P	II	–	1 L	E2	P001	–	IBC02	–
1761	CUPRIETHYLENEDIAMINE SOLUTION	8	6.1 P	III	223	5 L	E1	P001	–	IBC03	–
△ 1762	CYCLOHEXENYLTRICHLORO-SILANE	8	–	II	–	0	E0	P010	–	–	–

3 DGL

Portable tanks and bulk containers			EmS	Stowage and segregation	Properties and observations	UN No.
	Tank instructions	Provisions				
(12)	(13) 4.2.5 4.3	(14) 4.2.5	(15) 5.4.3.2 7.3	(16) 7.1 7.2	(17)	(18)
–	T10	TP2 TP7	F-A, S-B	Category C. Clear of living quarters.	Colourless liquid with a pungent odour. Reacts violently with water, evolving hydrogen chloride, an irritating and corrosive gas apparent as white fumes. When involved in a fire, evolves toxic gases. In the presence of moisture, highly corrosive to most metals. Irritating to skin, eyes and mucous membranes.	1753
–	T20	TP2	F-A, S-B	Category C. Clear of living quarters.	Colourless liquid with a pungent odour. Reacts violently with water, evolving hydrogen chloride, an irritating and corrosive gas apparent as white fumes. In the presence of moisture, highly corrosive to most metals. Causes severe burns to skin, eyes and mucous membranes.	1754
–	T8	TP2	F-A, S-B	Category C. Clear of living quarters. Segregation as for class 5.1 but "away from" classes 4.1, 5.1 and 7.	Orange liquid. Powerful oxidant. May cause fire in contact with organic materials such as wood, cotton or straw. Highly corrosive to most metals. Causes burns to skin, eyes and mucous membranes.	1755
–	T4	TP1	F-A, S-B	Category C. Clear of living quarters. Segregation as for class 5.1 but "away from" classes 4.1, 5.1 and 7.	See entry above.	1755
–	T3	TP33	F-A, S-B	Category A. "Separated from" acids.	Green or violet crystals. Slightly soluble in water. Reacts with strong acids, evolving hydrogen fluoride, an extremely irritating and corrosive gas. Mildly corrosive to most metals. Causes burns to skin, eyes and mucous membranes.	1756
–	T7	TP2	F-A, S-B	Category A.	Green liquid. Reacts with strong acids, evolving hydrogen fluoride, an extremely irritating and corrosive gas. Mildly corrosive to most metals. Causes burns to skin, eyes and mucous membranes.	1757
–	T4	TP1	F-A, S-B	Category A.	See entry above.	1757
–	T10	TP2	F-A, S-B	Category C. Clear of living quarters. Segregation as for class 5.1, but "separated from" classes 4.1, 5.1 and 7.	Dark red liquid. Reacts violently with water, evolving hydrogen chloride and chlorine, both highly irritating and corrosive gases apparent as white fumes. Oxidant; may cause fire in contact with organic materials such as wood, cotton or straw. Highly corrosive to most metals. In the presence of moisture, highly corrosive to most metals. Causes severe burns to skin, eyes and mucous membranes.	1758
–	T6	TP33	F-A, S-B	Category B.	Causes burns to skin, eyes and mucous membranes.	1759
–	T3	TP33	F-A, S-B	Category A.	See entry above.	1759
–	T1	TP33	F-A, S-B	Category A.	See entry above.	1759
–	T14	TP2 TP27	F-A, S-B	Category B. Clear of living quarters.	Causes burns to skin, eyes and mucous membranes.	1760
–	T11	TP2 TP27	F-A, S-B	Category B. Clear of living quarters.	See entry above.	1760
–	T7	TP1 TP28	F-A, S-B	Category A. Clear of living quarters.	See entry above.	1760
–	T7	TP2	F-A, S-B	Category A.	Dark purple liquid with an ammonia-like odour. Corrosive to copper, aluminium, zinc and tin. Toxic if swallowed, by skin contact or by inhalation. Causes burns to skin, eyes and mucous membranes.	1761
–	T7	TP1 TP28	F-A, S-B	Category A.	See entry above.	1761
–	T10	TP2 TP7 TP13	F-A, S-B	Category C. Clear of living quarters.	Colourless liquid with a pungent odour. Reacts violently with water, evolving hydrogen chloride, an irritating and corrosive gas apparent as white fumes. When involved in a fire, evolves toxic gases. In the presence of moisture, highly corrosive to most metals. Causes burns to skin, eyes and mucous membranes.	1762 △

UN No. (1)	Proper Shipping Name (PSN) (2) 3.1.2	Class or division (3) 2.0	Subsidiary risk(s) (4) 2.0	Packing group (5) 2.0.1.3	Special provisions (6) 3.3	Limited and excepted quantity provisions: Limited quantities (7a) 3.4	Limited and excepted quantity provisions: Excepted quantities (7b) 3.5	Packing: Instructions (8) 4.1.4	Packing: Provisions (9) 4.1.4	IBC: Instructions (10) 4.1.4	IBC: Provisions (11) 4.1.4
△ 1763	CYCLOHEXYLTRICHLORO-SILANE	8	–	II	–	0	E0	P010	–	–	–
1764	DICHLOROACETIC ACID	8	–	II	–	1 L	E2	P001	–	IBC02	B20
1765	DICHLOROACETYL CHLORIDE	8	–	II	–	1 L	E2	P001	–	IBC02	–
△ 1766	DICHLOROPHENYL-TRICHLOROSILANE	8	– P	II	–	0	E0	P010	–	–	–
△ 1767	DIETHYLDICHLOROSILANE	8	3	II	–	0	E0	P010	–	–	–
1768	DIFLUOROPHOSPHORIC ACID, ANHYDROUS	8	–	II	–	1 L	E2	P001	–	IBC02	B20
△ 1769	DIPHENYLDICHLOROSILANE	8	–	II	–	0	E0	P010	–	–	–
1770	DIPHENYLMETHYL BROMIDE	8	–	II	–	1 kg	E2	P002	–	IBC08	B2 B4
△ 1771	DODECYLTRICHLOROSILANE	8	–	II	–	0	E0	P010	–	–	–
1773	FERRIC CHLORIDE, ANHYDROUS	8	–	III	–	5 kg	E1	P002 LP02	–	IBC08	B3
1774	FIRE EXTINGUISHER CHARGES corrosive liquid	8	–	II	–	1 L	E0	P001	PP4	–	–
1775	FLUOROBORIC ACID	8	–	II	–	1 L	E2	P001	–	IBC02	–
1776	FLUOROPHOSPHORIC ACID, ANHYDROUS	8	–	II	–	1 L	E2	P001	–	IBC02	B20
1777	FLUOROSULPHONIC ACID	8	–	I	–	0	E0	P001	–	–	–
1778	FLUOROSILICIC ACID	8	–	II	–	1 L	E2	P001	–	IBC02	B20

Portable tanks and bulk containers			EmS	Stowage and segregation	Properties and observations	UN No.	
	Tank instructions	Provisions					
(12)	(13) 4.2.5 4.3	(14) 4.2.5	(15) 5.4.3.2 7.3	(16) 7.1 7.2	(17)	(18)	
–	T10	TP2 TP7 TP13	F-A, S-B	Category C. Clear of living quarters.	Colourless liquid with a pungent odour. Reacts violently with water, evolving hydrogen chloride, an irritating and corrosive gas apparent as white fumes. When involved in a fire, evolves toxic gases. In the presence of moisture, highly corrosive to most metals. Vapour irritates mucous membranes.	**1763**	△
–	T8	TP2	F-A, S-B	Category A.	Colourless liquid. Melting point: −4°C. Highly corrosive to most metals. Causes burns to skin, eyes and mucous membranes.	**1764**	
–	T7	TP2	F-A, S-B	Category D. Clear of living quarters.	Colourless liquid with an extremely irritating odour, causing tears. Reacts violently with water, evolving hydrogen chloride, an irritating and corrosive gas apparent as white fumes. In the presence of moisture, highly corrosive to most metals. Causes burns to skin, eyes and mucous membranes.	**1765**	
–	T10	TP2 TP7 TP13	F-A, S-B	Category C. Clear of living quarters.	Colourless liquid with a pungent odour. Reacts violently with water, evolving hydrogen chloride, an irritating and corrosive gas apparent as white fumes. When involved in a fire, evolves toxic gases. In the presence of moisture, highly corrosive to most metals. Irritating to skin, eyes and mucous membranes.	**1766**	△
–	T10	TP2 TP7 TP13	F-E, S-C	Category C. Clear of living quarters.	Colourless, flammable liquid with a pungent odour. Flashpoint: 25°C c.c. Reacts violently with water, evolving hydrogen chloride, an irritating and corrosive gas apparent as white fumes. When involved in a fire, evolves toxic gases. In the presence of moisture, highly corrosive to most metals. Vapour irritates mucous membranes.	**1767**	△
–	T8	TP2	F-A, S-B	Category A. Clear of living quarters.	Colourless liquid. In the presence of moisture, highly corrosive to glass and other siliceous materials. Harmful if swallowed.	**1768**	
–	T10	TP2 TP7 TP13	F-A, S-B	Category C. Clear of living quarters.	Colourless liquid with a pungent odour. Reacts violently with water, evolving hydrogen chloride, an irritating and corrosive gas apparent as white fumes. When involved in a fire, evolves toxic gases. In the presence of moisture, highly corrosive to most metals. Vapour irritates mucous membranes.	**1769**	△
–	T3	TP33	F-A, S-B	Category D. Clear of living quarters.	Solid with an irritating odour. Causes tears. Melting point: 45°C. In the presence of moisture, corrosive to most metals. Vapour irritates mucous membranes.	**1770**	
–	T10	TP2 TP7 TP13	F-A, S-B	Category C. Clear of living quarters.	Colourless liquid with a pungent odour. Reacts violently with water, evolving hydrogen chloride, an irritating and corrosive gas apparent as white fumes. When involved in a fire, evolves toxic gases. In the presence of moisture, highly corrosive to most metals. Causes burns to skin, eyes and mucous membranes.	**1771**	△
–	T1	TP33	F-A, S-B	Category A.	Brown solid. In the presence of moisture, highly corrosive to most metals. The provisions of this Code should not apply to the solid hydrated form.	**1773**	
–	–	–	F-A, S-B	Category A.	Usually, diluted sulphuric acid in small glass receptacles.	**1774**	
–	T7	TP2	F-A, S-B	Category A.	Colourless, clear liquid. Corrosive to most metals. May cause severe burns to skin, eyes and mucous membranes if containing free hydrofluoric acid.	**1775**	
–	T8	TP2	F-A, S-B	Category A.	Colourless liquid. In the presence of moisture, highly corrosive to glass, other siliceous materials and most metals. Causes burns to skin, eyes and mucous membranes.	**1776**	
–	T10	TP2	F-A, S-B	Category D. Clear of living quarters.	Colourless liquid with a pungent odour. Reacts violently with water, evolving hydrogen fluoride, an extremely irritating and corrosive gas apparent as white fumes. In the presence of moisture, highly corrosive to glass, other siliceous materials and most metals. Causes severe burns to skin, eyes and mucous membranes.	**1777**	
–	T8	TP2	F-A, S-B	Category A.	Colourless liquid. Highly corrosive to most metals. May cause severe burns to skin, eyes and mucous membranes if containing free hydrofluoric acid.	**1778**	

UN No.	Proper Shipping Name (PSN)	Class or division	Subsidiary risk(s)	Packing group	Special provisions	Limited and excepted quantity provisions		Packing		IBC	
						Limited quantities	Excepted quantities	Instructions	Provisions	Instructions	Provisions
(1)	(2) 3.1.2	(3) 2.0	(4) 2.0	(5) 2.0.1.3	(6) 3.3	(7a) 3.4	(7b) 3.5	(8) 4.1.4	(9) 4.1.4	(10) 4.1.4	(11) 4.1.4
1779	FORMIC ACID with more than 85% acid, by mass	8	3	II	–	1 L	E2	P001	–	IBC02	–
1780	FUMARYL CHLORIDE	8	–	II	–	1 L	E2	P001	–	IBC02	–
△ 1781	HEXADECYLTRICHLORO-SILANE	8	–	II	–	0	E0	P010	–	–	–
1782	HEXAFLUOROPHOSPHORIC ACID	8	–	II	–	1 L	E2	P001	–	IBC02	B20
1783	HEXAMETHYLENEDIAMINE SOLUTION	8	–	II	–	1 L	E2	P001	–	IBC02	–
1783	HEXAMETHYLENEDIAMINE SOLUTION	8	–	III	223	5 L	E1	P001 LP01	–	IBC03	–
△ 1784	HEXYLTRICHLOROSILANE	8	–	II	–	0	E0	P010	–	–	–
1786	HYDROFLUORIC ACID AND SULPHURIC ACID MIXTURE	8	6.1	I	–	0	E0	P001	–	–	–
1787	HYDRIODIC ACID	8	–	II	–	1 L	E2	P001	–	IBC02	–
1787	HYDRIODIC ACID	8	–	III	223	5 L	E1	P001 LP01	–	IBC03	–
1788	HYDROBROMIC ACID	8	–	II	–	1 L	E2	P001	–	IBC02	–
1788	HYDROBROMIC ACID	8	–	III	223	5 L	E1	P001 LP01	–	IBC03	–
1789	HYDROCHLORIC ACID	8	–	II	–	1 L	E2	P001	–	IBC02	B20
1789	HYDROCHLORIC ACID	8	–	III	223	5 L	E1	P001 LP01	–	IBC03	–
1790	HYDROFLUORIC ACID solution, with more than 60% hydrogen fluoride	8	6.1	I	–	0	E0	P802	PP79 PP81	–	–
1790	HYDROFLUORIC ACID solution, with not more than 60% hydrogen fluoride	8	6.1	II	–	1 L	E2	P001	PP81	IBC02	B20
1791	HYPOCHLORITE SOLUTION	8	–	II	–	1 L	E2	P001	PP10	IBC02	B5

Portable tanks and bulk containers			EmS	Stowage and segregation	Properties and observations	UN No.
	Tank instructions	Provisions				
(12)	(13) 4.2.5 4.3	(14) 4.2.5	(15) 5.4.3.2 7.3	(16) 7.1 7.2	(17)	(18)
–	T7	TP2	F-E, S-C	Category A. Clear of living quarters.	Colourless flammable liquid with a pungent odour. Pure FORMIC ACID: flashpoint 42°C c.c. Corrosive to most metals. Causes burns to skin, eyes and mucous membranes.	1779
–	T7	TP2	F-A, S-B	Category C. Clear of living quarters.	Yellow liquid. Reacts violently with water, evolving hydrogen chloride, an irritating and corrosive gas apparent as white fumes. In the presence of moisture, highly corrosive to most metals. Causes burns to skin, eyes and mucous membranes.	1780
–	T10	TP2 TP7 TP13	F-A, S-B	Category C. Clear of living quarters.	Colourless liquid with a pungent odour. Reacts violently with water, evolving hydrogen chloride, an irritating and corrosive gas apparent as white fumes. When involved in a fire, evolves toxic gases. In the presence of moisture, highly corrosive to most metals. Vapour irritates mucous membranes.	1781 △
–	T8	TP2	F-A, S-B	Category A.	Colourless liquid. In the presence of moisture, highly corrosive to glass, other siliceous materials and most metals. Causes burns to skin, eyes and mucous membranes. Harmful if swallowed.	1782
–	T7	TP2	F-A, S-B	Category A.	Colourless liquid. Causes burns to skin, eyes and mucous membranes.	1783
–	T4	TP1	F-A, S-B	Category A.	See entry above.	1783
–	T10	TP2 TP7 TP13	F-A, S-B	Category C. Clear of living quarters.	Colourless liquid with a pungent odour. Reacts violently with water, evolving hydrogen chloride, an irritating and corrosive gas apparent as white fumes. When involved in a fire, evolves toxic gases. In the presence of moisture, highly corrosive to most metals. Causes burns to skin, eyes and mucous membranes.	1784 △
–	T10	TP2 TP13	F-A, S-B	Category D. Clear of living quarters.	Colourless syrupy liquid with a pungent odour. Mixture consists of between 70% and 80% by mass of acids and contains not less than 25% by mass of hydrofluoric acid. Reacts violently with water, developing heat. Highly corrosive to glass, other siliceous materials and most metals. Toxic if swallowed, by skin contact or by inhalation. Causes severe burns to skin and mucous membranes.	1786
–	T7	TP2	F-A, S-B	Category C.	Colourless liquid. An aqueous solution of the gas hydrogen iodide. Highly corrosive to most metals. Causes burns to skin, eyes and mucous membranes.	1787
–	T4	TP1	F-A, S-B	Category C.	See entry above.	1787
–	T7	TP2	F-A, S-B	Category C.	Colourless liquid. An aqueous solution of the gas hydrogen bromide. Highly corrosive to most metals. Causes burns to skin, eyes and mucous membranes.	1788
–	T4	TP1	F-A, S-B	Category C.	See entry above.	1788
–	T8	TP2	F-A, S-B	Category C.	Colourless liquid. An aqueous solution of the gas hydrogen chloride. Highly corrosive to most metals. Causes burns to skin, eyes and mucous membranes.	1789
–	T4	TP1	F-A, S-B	Category C.	See entry above.	1789
–	T10	TP2 TP13	F-A, S-B	Category D. Keep as cool as reasonably practicable. Clear of living quarters.	Colourless liquid with an irritating odour. Highly corrosive to glass, other siliceous materials and most metals. Toxic if swallowed, by skin contact or by inhalation. Both the liquid and its fumes cause severe burns to skin, eyes and mucous membranes.	1790
–	T8	TP2	F-A, S-B	Category D. Keep as cool as reasonably practicable. Clear of living quarters.	See entry above.	1790
–	T7	TP2 TP24	F-A, S-B	Category B. "Away from" acids.	Liquid with chlorine odour. In contact with acids, evolves very irritating and corrosive gases. Mildly corrosive to most metals. Causes burns to skin, eyes and mucous membranes.	1791

UN No.	Proper Shipping Name (PSN)	Class or division	Subsidiary risk(s)	Packing group	Special provisions	Limited and excepted quantity provisions		Packing		IBC	
						Limited quantities	Excepted quantities	Instructions	Provisions	Instructions	Provisions
(1)	(2) 3.1.2	(3) 2.0	(4) 2.0	(5) 2.0.1.3	(6) 3.3	(7a) 3.4	(7b) 3.5	(8) 4.1.4	(9) 4.1.4	(10) 4.1.4	(11) 4.1.4
1791	HYPOCHLORITE SOLUTION	8	–	III	223	5 L	E1	P001 LP01	–	IBC03	–
△ 1792	IODINE MONOCHLORIDE, SOLID	8	–	II	–	1 kg	E2	P002	–	IBC08	B2 B4
1793	ISOPROPYL ACID PHOSPHATE	8	–	III	–	5 L	E1	P001 LP01	–	IBC02	–
1794	LEAD SULPHATE with more than 3% free acid	8	–	II	–	1 kg	E2	P002	–	IBC08	B2 B4
1796	NITRATING ACID MIXTURE with more than 50% nitric acid	8	5.1	I	–	0	E0	P001	–	–	–
1796	NITRATING ACID MIXTURE with not more than 50% nitric acid	8	–	II	–	1 L	E2	P001	–	IBC02	B20
1798	NITROHYDROCHLORIC ACID	8	–	I	–	0	E0	P802	–	–	–
△ 1799	NONYLTRICHLOROSILANE	8	–	II	–	0	E0	P010	–	–	–
△ 1800	OCTADECYLTRICHLORO-SILANE	8	–	II	–	0	E0	P010	–	–	–
△ 1801	OCTYLTRICHLOROSILANE	8	–	II	–	0	E0	P010	–	–	–
1802	PERCHLORIC ACID with not more than 50% acid, by mass	8	5.1	II	–	1 L	E2	P001	–	IBC02	–
1803	PHENOLSULPHONIC ACID, LIQUID	8	–	II	–	1 L	E2	P001	–	IBC02	–
△ 1804	PHENYLTRICHLOROSILANE	8	–	II	–	0	E0	P010	–	–	–
1805	PHOSPHORIC ACID SOLUTION	8	–	III	223	5 L	E1	P001 LP01	–	IBC03	–
1806	PHOSPHORUS PENTACHLORIDE	8	–	II	–	1 kg	E2	P002	–	IBC08	B2 B4
1807	PHOSPHORUS PENTOXIDE	8	–	II	–	1 kg	E2	P002	–	IBC08	B2 B4

Portable tanks and bulk containers			EmS	Stowage and segregation	Properties and observations	UN No.	
	Tank instructions	Provisions					
(12)	(13) 4.2.5 4.3	(14) 4.2.5	(15) 5.4.3.2 7.3	(16) 7.1 7.2	(17)	(18)	
–	T4	TP2 TP24	F-A, S-B	Category B. "Away from" acids.	Liquid with chlorine odour. In contact with acids, evolves very irritating and corrosive gases. Mildly corrosive to most metals. Causes burns to skin, eyes and mucous membranes.	1791	
–	T7	TP2	F-A, S-B	Category D. Clear of living quarters. Segregation as for class 5.1, but "separated from" classes 4.1, 5.1 and 7.	Red, brown or black crystals. Reacts violently with water, evolving irritating and corrosive gases apparent as white fumes. Powerful oxidant; may cause fire in contact with organic materials such as wood, cotton or straw. In the presence of moisture, highly corrosive to most metals. Vapour irritates mucous membranes.	1792	△
–	T4	TP1	F-A, S-B	Category A.	Oily liquid. Mildly corrosive to most metals	1793	
–	T3	TP33	F-A, S-B	Category A.	May be dry solid or slurry. Corrosive to most metals. Harmful if swallowed.	1794	
–	T10	TP2 TP13	F-A, S-Q	Category D. Clear of living quarters. "Separated from" class 4.1.	Mixture of concentrated nitric and sulphuric acids. Oxidant; may cause fire in contact with organic materials such as wood, cotton or straw, developing highly toxic gas (brown fumes). Highly corrosive to most metals. Causes severe burns to skin, eyes and mucous membranes.	1796	
–	T8	TP2 TP13	F-A, S-B	Category D. Clear of living quarters.	See entry above.	1796	
–	T10	TP2 TP13	F-A, S-B	Category D. Clear of living quarters. Segregation as for class 5.1, but "separated from" classes 4.1, 5.1 and 7.	Yellow liquid; a mixture of nitric acid and hydrochloric acid, usually in the proportion of 1:3. Powerful oxidant; may cause fire in contact with organic materials such as wood, cotton or straw, evolving suffocating and highly toxic gases. Highly corrosive to all metals. Causes severe burns to skin, eyes and mucous membranes.	1798	
–	T10	TP2 TP7 TP13	F-A, S-B	Category C. Clear of living quarters.	Colourless liquid with a pungent odour. Reacts violently with water, evolving hydrogen chloride, an irritating and corrosive gas apparent as white fumes. When involved in a fire, evolves toxic gases. In the presence of moisture, highly corrosive to most metals. Causes burns to skin, eyes and mucous membranes.	1799	△
–	T10	TP2 TP7 TP13	F-A, S-B	Category C. Clear of living quarters.	Colourless liquid with a pungent odour. Reacts violently with water, evolving hydrogen chloride, an irritating and corrosive gas apparent as white fumes. When involved in a fire, evolves toxic gases. In the presence of moisture, highly corrosive to most metals. Vapour irritates mucous membranes.	1800	△
–	T10	TP2 TP7 TP13	F-A, S-B	Category C. Clear of living quarters.	Colourless liquid with a pungent odour. Reacts violently with water, evolving hydrogen chloride, an irritating and corrosive gas apparent as white fumes. When involved in a fire, evolves toxic gases. In the presence of moisture, highly corrosive to most metals. Vapour irritates mucous membranes.	1801	△
–	T7	TP2	F-H, S-Q	Category C. "Separated from" class 4.1.	Colourless liquid. Oxidant. Highly corrosive to most metals.	1802	
–	T7	TP2	F-A, S-B	Category C. For metal drums, category B.	Yellow, oily liquid. Corrosive to most metals.	1803	
–	T10	TP2 TP7 TP13	F-A, S-B	Category C. Clear of living quarters.	Colourless liquid with a pungent odour. Reacts violently with water, evolving hydrogen chloride, an irritating and corrosive gas apparent as white fumes. When involved in a fire, evolves toxic gases. In the presence of moisture, highly corrosive to most metals. Vapour irritates mucous membranes.	1804	△
–	T4	TP1	F-A, S-B	Category A.	Miscible in water. Mildly corrosive to most metals.	1805	
–	T3	TP33	F-A, S-B	Category C. Clear of living quarters. Segregation as for class 5.1 but "away from" classes 4.1, 5.1 and 7.	Colourless, crystalline powder. Reacts violently with water, evolving hydrogen chloride, an irritating and corrosive gas apparent as white fumes. Powerful oxidant; may cause fire in contact with organic materials such as wood, cotton or straw. In the presence of moisture, highly corrosive to most metals.	1806	
–	T3	TP33	F-A, S-B	Category A.	Crystalline powder, very deliquescent. Reacts violently with water and organic materials such as wood, cotton or straw, generating heat. In the presence of moisture, mildly corrosive to most metals.	1807	

UN No. (1)	Proper Shipping Name (PSN) (2) 3.1.2	Class or division (3) 2.0	Subsidiary risk(s) (4) 2.0	Packing group (5) 2.0.1.3	Special provisions (6) 3.3	Limited and excepted quantity provisions: Limited quantities (7a) 3.4	Limited and excepted quantity provisions: Excepted quantities (7b) 3.5	Packing: Instructions (8) 4.1.4	Packing: Provisions (9) 4.1.4	IBC: Instructions (10) 4.1.4	IBC: Provisions (11) 4.1.4
1808	PHOSPHORUS TRIBROMIDE	8	–	II	–	1 L	E2	P001	–	IBC02	–
1809	PHOSPHORUS TRICHLORIDE	6.1	8	I	354	0	E0	P602	–	–	–
1810	PHOSPHORUS OXYCHLORIDE	6.1	8	I	354	0	E0	P602	–	–	–
△ 1811	POTASSIUM HYDROGEN DIFLUORIDE, SOLID	8	6.1	II	–	1 kg	E2	P002	–	IBC08	B2 B4
1812	POTASSIUM FLUORIDE, SOLID	6.1	–	III	–	5 kg	E1	P002 LP02	–	IBC08	B3
1813	POTASSIUM HYDROXIDE, SOLID	8	–	II	–	1 kg	E2	P002	–	IBC08	B2 B4
1814	POTASSIUM HYDROXIDE SOLUTION	8	–	II	–	1 L	E2	P001	–	IBC02	–
1814	POTASSIUM HYDROXIDE SOLUTION	8	–	III	223	5 L	E1	P001 LP01	–	IBC03	–
1815	PROPIONYL CHLORIDE	3	8	II	–	1 L	E2	P001	–	IBC02	–
△ 1816	PROPYLTRICHLOROSILANE	8	3	II	–	0	E0	P010	–	–	–
1817	PYROSULPHURYL CHLORIDE	8	–	II	–	1 L	E2	P001	–	IBC02	–
△ 1818	SILICON TETRACHLORIDE	8	–	II	–	0	E0	P010	–	–	–
1819	SODIUM ALUMINATE SOLUTION	8	–	II	–	1 L	E2	P001	–	IBC02	–
1819	SODIUM ALUMINATE SOLUTION	8	–	III	223	5 L	E1	P001 LP01	–	IBC03	–

Portable tanks and bulk containers			EmS	Stowage and segregation	Properties and observations	UN No.
	Tank instructions	Provisions				
(12)	(13) 4.2.5 4.3	(14) 4.2.5	(15) 5.4.3.2 7.3	(16) 7.1 7.2	(17)	(18)
–	T7	TP2	F-A, S-B	Category C. Clear of living quarters	Colourless liquid with a pungent odour. Reacts violently with water, evolving hydrogen bromide, an irritating and corrosive gas apparent as white fumes. In the presence of moisture, highly corrosive to most metals. Causes burns to skin, eyes and mucous membranes.	1808
–	T20	TP2 TP13 TP35	F-A, S-B	Category D. Clear of living quarters.	Colourless liquid with a pungent odour. Reacts violently with water, evolving hydrogen chloride, an irritating and corrosive gas apparent as white fumes. In the presence of moisture, highly corrosive to most metals. Highly toxic if swallowed, by skin contact or by inhalation. Causes burns to skin, eyes and mucous membranes.	1809
–	T20	TP2 TP13 TP37	F-A, S-B	Category D. Clear of living quarters.	Colourless liquid with a pungent odour. Reacts violently with water, evolving hydrogen chloride, an irritating and corrosive gas apparent as white fumes. In the presence of moisture, highly corrosive to most metals. Causes burns to skin, eyes and mucous membranes. Highly toxic if swallowed, by skin contact or by inhalation.	1810
–	T3	TP33	F-A, S-B	Category A. Protected from sources of heat. Clear of living quarters. "Separated from" acids.	White crystalline solid. Decomposed by heat or acids, evolving hydrogen fluoride, a toxic, extremely irritating and corrosive gas apparent as white fumes. In the presence of moisture, highly corrosive to glass, other siliceous materials and most metals. Toxic if swallowed, by skin contact or by inhalation. Causes burns to skin, eyes and mucous membranes.	1811 △
–	T1	TP33	F-A, S-A	Category A. "Separated from" acids.	White, deliquescent crystals or powder. Decomposed by acids, evolving hydrogen fluoride, an irritating and corrosive gas. Toxic if swallowed, by skin contact or by inhalation.	1812
–	T3	TP33	F-A, S-B	Category A. "Separated from" acids.	White pellets, flakes, lumps or solid blocks deliquescent. Reacts with ammonium salts, evolving ammonia gas. In the presence of moisture, corrosive to aluminium, zinc and tin. Causes burns to skin, eyes and mucous membranes. Reacts violently with acids.	1813
–	T7	TP2	F-A, S-B	Category A. "Separated from" acids.	Colourless liquid. Reacts with ammonium salts, evolving ammonia gas Corrosive to aluminium, zinc and tin. Causes burns to skin, eyes and mucous membranes. Reacts violently with acids.	1814
–	T4	TP1	F-A, S-B	Category A. "Separated from" acids.	See entry above.	1814
–	T7	TP1	F-E, S-C	Category B. Clear of living quarters.	Colourless liquid. Flashpoint: 12°C c.c. Reacts violently with water, evolving hydrogen chloride, an irritating and corrosive gas, apparent as white fumes. In the presence of moisture, highly corrosive to most metals. Causes burns to skin, eyes and mucous membranes.	1815
–	T10	TP2 TP7 TP13	F-E, S-C	Category C. Clear of living quarters.	Colourless, flammable liquid, with a pungent odour. Flashpoint: 38°C c.c. Reacts violently with water, evolving hydrogen chloride, an irritating and corrosive gas apparent as white fumes. When involved in a fire, evolves toxic gases. In the presence of moisture, highly corrosive to most metals. Vapour irritates mucous membranes.	1816 △
–	T8	TP2	F-A, S-B	Category C. Clear of living quarters.	Colourless liquid with a pungent odour. Reacts violently with water, evolving hydrogen chloride, an irritating and corrosive gas apparent as white fumes. In the presence of moisture, highly corrosive to most metals. Vapour irritates mucous membranes.	1817
–	T10	TP2 TP7 TP13	F-A, S-B	Category C. Clear of living quarters. See 7.2.6.3.2.	Colourless, extremely mobile liquid with a suffocating odour. Reacts violently with water, evolving hydrogen chloride, an irritating and corrosive gas apparent as white fumes. In the presence of moisture, highly corrosive to most metals. Vapour irritates mucous membranes.	1818 △
–	T7	TP2	F-A, S-B	Category A. "Separated from" acids.	Colourless liquid. Reacts with ammonium salts, evolving ammonia gas. Corrosive to aluminium, zinc and tin. Causes burns to skin, eyes and mucous membranes. Reacts violently with acids.	1819
–	T4	TP1	F-A, S-B	Category A. "Separated from" acids.	See entry above.	1819

UN No.	Proper Shipping Name (PSN)	Class or division	Subsidiary risk(s)	Packing group	Special provisions	Limited and excepted quantity provisions		Packing		IBC	
						Limited quantities	Excepted quantities	Instructions	Provisions	Instructions	Provisions
(1)	(2) 3.1.2	(3) 2.0	(4) 2.0	(5) 2.0.1.3	(6) 3.3	(7a) 3.4	(7b) 3.5	(8) 4.1.4	(9) 4.1.4	(10) 4.1.4	(11) 4.1.4
1823	SODIUM HYDROXIDE, SOLID	8	–	II	–	1 kg	E2	P002	–	IBC08	B2 B4
1824	SODIUM HYDROXIDE SOLUTION	8	–	II	–	1 L	E2	P001	–	IBC02	–
1824	SODIUM HYDROXIDE SOLUTION	8	–	III	223	5 L	E1	P001 LP01	–	IBC03	–
1825	SODIUM MONOXIDE	8	–	II	–	1 kg	E2	P002	–	IBC08	B2 B4
1826	NITRATING ACID MIXTURE, SPENT with more than 50% nitric acid	8	5.1	I	113	0	E0	P001	–	–	–
1826	NITRATING ACID MIXTURE, SPENT with not more than 50% nitric acid	8	–	II	113	1 L	E2	P001	–	IBC02	B20
1827	STANNIC CHLORIDE, ANHYDROUS	8	–	II	–	1 L	E2	P001	–	IBC02	–
1828	SULPHUR CHLORIDES	8	–	I	–	0	E0	P602	–	–	–
1829	SULPHUR TRIOXIDE, STABILIZED	8	–	I	–	0	E0	P001	–	–	–
1830	SULPHURIC ACID with more than 51% acid	8	–	II	–	1 L	E2	P001	–	IBC02	B20
1831	SULPHURIC ACID, FUMING	8	6.1	I	–	0	E0	P602	–	–	–
1832	SULPHURIC ACID, SPENT	8	–	II	113	1 L	E2	P001	–	IBC02	B20
1833	SULPHUROUS ACID	8	–	II	–	1 L	E2	P001	–	IBC02	–
1834	SULPHURYL CHLORIDE	6.1	8	I	354	0	E0	P602	–	–	–
1835	TETRAMETHYLAMMONIUM HYDROXIDE SOLUTION	8	–	II	–	1 L	E2	P001	–	IBC02	–
1835	TETRAMETHYLAMMONIUM HYDROXIDE SOLUTION	8	–	III	223	5 L	E1	P001 LP01	–	IBC03	–

Portable tanks and bulk containers			EmS	Stowage and segregation	Properties and observations	UN No.
	Tank instructions	Provisions				
(12)	(13) 4.2.5 4.3	(14) 4.2.5	(15) 5.4.3.2 7.3	(16) 7.1 7.2	(17)	(18)
–	T3	TP33	F-A, S-B	Category A. "Separated from" acids.	White pellets, flakes, lumps or solid blocks, deliquescent. Reacts with ammonium salts, evolving ammonia gas. In the presence of moisture, corrosive to aluminium, zinc and tin. Causes burns to skin, eyes and mucous membranes. Reacts violently with acids.	1823
–	T7	TP2	F-A, S-B	Category A. "Separated from" acids.	Colourless liquid. Corrosive to aluminium, zinc and tin. Reacts with ammonium salts, evolving ammonia gas. Causes burns to skin, eyes and mucous membranes. Reacts violently with acids.	1824
–	T4	TP1	F-A, S-B	Category A. "Separated from" acids.	See entry above.	1824
–	T3	TP33	F-A, S-B	Category A. "Separated from" acids.	Deliquescent crystalline solid. Reacts violently with water and acids, generating heat. Reacts with ammonium salts, evolving ammonia gas. In the presence of moisture, corrosive to aluminium, zinc and tin. Causes burns to skin, eyes and mucous membranes.	1825
–	T10	TP2 TP13	F-A, S-Q	Category D. Clear of living quarters. "Separated from" class 4.1.	Usually a mixture of acids which has been used for nitration processes. Highly corrosive to most metals. Causes severe burns to skin, eyes and mucous membranes. Prohibited for shipment unless the mixture is (1) chemically stable; and (2) certified as containing no explosive impurities.	1826
–	T8	TP2	F-A, S-Q	Category D. Clear of living quarters.	See entry above.	1826
–	T7	TP2	F-A, S-B	Category C.	Colourless liquid. In the presence of water, corrosive to most metals. Vapour irritates mucous membranes.	1827
–	T20	TP2	F-A, S-B	Category C. Clear of living quarters.	Red liquids with a suffocating odour. React violently with water, evolving hydrogen chloride and sulphur dioxide, irritating and corrosive gases. In the presence of moisture, highly corrosive to most metals. Causes severe burns to skin, eyes and mucous membranes.	1828
–	T20	TP4 TP13 TP25 TP26	F-A, S-B	Category C. Clear of living quarters.	Very deliquescent solid. Melting point may be as low as 17°C. Reacts violently with water, generating heat. May cause fire in contact with organic materials such as wood, cotton or straw. In the presence of moisture, highly corrosive to most metals. Causes severe burns to skin, eyes and mucous membranes.	1829
–	T8	TP2	F-A, S-B	Category C. For steel drums, category B.	Colourless, oily liquid, mixture over 1.41 up to 1.84 relative density. In the presence of moisture, highly corrosive to most metals. Causes burns to skin, eyes and mucous membranes.	1830
–	T20	TP2 TP13	F-A, S-B	Category C. For steel drums, category B. Clear of living quarters.	Colourless, oily liquid, may be partly crystallized. Solution of varying quantities of sulphur trioxide in sulphuric acid. Reacts violently with water and organic material, generating heat. In the presence of moisture, highly corrosive to most metals. Toxic if swallowed, by skin contact or by inhalation. Causes severe burns to skin, eyes and mucous membranes.	1831
–	T8	TP2	F-A, S-B	Category C. For metal drums, category B.	Sulphuric acid, usually of high concentration, which has been used for chemical processes. Highly corrosive to most metals.	1832
–	T7	TP2	F-A, S-B	Category B. Clear of living quarters.	Solution of sulphur dioxide in water, with a suffocating odour. Corrosive to most metals. Vapour irritates mucous membranes.	1833
–	T20	TP2 TP13	F-A, S-B	Category D. Clear of living quarters.	Colourless liquid with a pungent odour. Boiling point: 69°C. Reacts violently with water, evolving hydrogen chloride, an irritating and corrosive gas apparent as white fumes. In the presence of moisture, highly corrosive to most metals. Causes severe burns to skin, eyes and mucous membranes. Highly toxic if swallowed, by skin contact or by inhalation.	1834
–	T7	TP2	F-A, S-B	Category A. "Separated from" acids.	Miscible with water. Reacts violently with acids.	1835
–	T7	TP2	F-A, S-B	Category A. "Separated from" acids.	See entry above.	1835

UN No.	Proper Shipping Name (PSN)	Class or division	Subsidiary risk(s)	Packing group	Special provisions	Limited and excepted quantity provisions		Packing		IBC	
						Limited quantities	Excepted quantities	Instructions	Provisions	Instructions	Provisions
(1)	(2) 3.1.2	(3) 2.0	(4) 2.0	(5) 2.0.1.3	(6) 3.3	(7a) 3.4	(7b) 3.5	(8) 4.1.4	(9) 4.1.4	(10) 4.1.4	(11) 4.1.4
1836	THIONYL CHLORIDE	8	–	I	–	0	E0	P802	–	–	–
1837	THIOPHOSPHORYL CHLORIDE	8	–	II	–	1 L	E2	P001	–	IBC02	–
1838	TITANIUM TETRACHLORIDE	6.1	8	I	354	0	E0	P602	–	–	–
1839	TRICHLOROACETIC ACID, SOLID	8	–	II	–	1 kg	E2	P002	–	IBC08	B2 B4
1840	ZINC CHLORIDE SOLUTION	8	–	III	223	5 L	E1	P001 LP01	–	IBC03	–
△ 1841	ACETALDEHYDE AMMONIA	9	–	III	–	5 kg	E1	P002 LP02	–	IBC08	B3 B6
1843	AMMONIUM DINITRO-*o*-CRESOLATE, SOLID	6.1	– P	II	–	500 g	E4	P002	–	IBC08	B2 B4
△ 1845	CARBON DIOXIDE, SOLID (DRY ICE)	9	–	–	–	0	E0	P003	PP18	–	–
1846	CARBON TETRACHLORIDE	6.1	– P	II	–	100 mL	E4	P001	–	IBC02	–
1847	POTASSIUM SULPHIDE, HYDRATED with not less than 30% water of crystallization	8	–	II	–	1 kg	E2	P002	–	IBC08	B2 B4
1848	PROPIONIC ACID with not less than 10% and less than 90% acid, by mass	8	–	III	–	5 L	E1	P001 LP01	–	IBC03	–
1849	SODIUM SULPHIDE, HYDRATED with not less than 30% water	8	–	II	–	1 kg	E2	P002	–	IBC08	B2 B4
1851	MEDICINE, LIQUID, TOXIC, N.O.S.	6.1	–	II	221	100 mL	E4	P001	–	–	–
1851	MEDICINE, LIQUID, TOXIC, N.O.S.	6.1	–	III	221 223	5 L	E1	P001 LP01	–	–	–
1854	BARIUM ALLOYS, PYROPHORIC	4.2	–	I	–	0	E0	P404	PP31	–	–
1855	CALCIUM, PYROPHORIC or CALCIUM ALLOYS, PYROPHORIC	4.2	–	I	–	0	E0	P404	PP31	–	–

Portable tanks and bulk containers			EmS	Stowage and segregation	Properties and observations	UN No.
	Tank instructions	Provisions				
(12)	(13) 4.2.5 4.3	(14) 4.2.5	(15) 5.4.3.2 7.3	(16) 7.1 7.2	(17)	(18)
–	T10	TP2 TP13	F-A, S-B	Category C. Clear of living quarters.	Yellow or red liquid. Boiling point: 79°C. Reacts violently with water, evolving hydrogen chloride and sulphur dioxide, irritating and corrosive gases. In the presence of moisture, highly corrosive to most metals. Causes severe burns to skin, eyes and mucous membranes.	**1836**
–	T7	TP2	F-A, S-B	Category C. Clear of living quarters.	Colourless liquid with a pungent odour. Reacts violently with water, evolving hydrogen chloride, an irritating and corrosive gas apparent as white fumes. In the presence of moisture, highly corrosive to most metals. Vapour irritates mucous membranes.	**1837**
–	T20	TP2 TP13 TP37	F-A, S-B	Category D. Clear of living quarters.	Colourless liquid. Reacts violently with water, evolving hydrogen chloride, an irritating and corrosive gas apparent as white fumes. In the presence of moisture, highly corrosive to most metals. Highly toxic if swallowed, by skin contact or by inhalation. Causes burns to skin, eyes and mucous membranes.	**1838**
–	T3	TP33	F-A, S-B	Category A.	Colourless, deliquescent crystals. Melting point of the pure substance: 58°C. In the presence of moisture, corrosive to most metals. Causes burns to skin, eyes and mucous membranes.	**1839**
–	T4	TP1	F-A, S-B	Category A.	Colourless liquid. Mildly corrosive to most metals. Causes burns to skin eyes and mucous membranes.	**1840**
–	T1	TP33	F-A, S-B	Category A. Segregation from foodstuffs as in 7.3.4.2.2, 7.6.3.1.2 or 7.7.3.7.	White crystalline solid. Soluble in water. When heated, decomposes into ammonia and acetaldehyde.	**1841** △
–	T3	TP33	F-A, S-A	Category B. "Away from" heavy metals and their salts. "Separated from" classes 3 and 4.1. "Separated longitudinally by an intervening complete compartment or hold from" class 1.	May support combustion and burn without oxygen. When involved in a fire, evolves toxic fumes. Forms extremely sensitive explosive compounds with lead, silver or other heavy metals and their compounds. Toxic if swallowed, by skin contact or by inhalation.	**1843**
–	–	–	F-C, S-V	Category C. Clear of living quarters.	Non-flammable gas in a white solid form. Slowly evolves vapours which are heavier than air (1.5). Inhalation of vapours may lead to unconsciousness. Can cause severe burns when in contact with the skin.	**1845** △
–	T7	TP2	F-A, S-A	Category A. Clear of living quarters.	Colourless, volatile liquid with a heavy anaesthetic vapour. Non-flammable; when involved in a fire, evolves extremely toxic fumes (phosgene). Toxic if swallowed, by skin contact or by inhalation.	**1846**
–	T3	TP33	F-A, S-B	Category A. "Separated from" acids.	Crystalline solid. Melting point: 60°C. Reacts violently with acids, evolving hydrogen sulphide, a toxic and flammable gas. Mildly corrosive to most metals. Causes burns to skin, eyes and mucous membranes.	**1847**
–	T4	TP1	F-A, S-B	Category A.	Colourless liquid with a pungent odour. Miscible with water. Corrosive to lead and most other metals. Burns skin. Vapours irritate mucous membranes.	**1848**
–	T3	TP33	F-A, S-B	Category A. "Separated from" acids.	Yellow-pink or white deliquescent crystals, flakes or lumps. Melting point: 50°C. Soluble in water. Reacts violently with acids, evolving hydrogen sulphide, a toxic and flammable gas. Mildly corrosive to most metals. Causes burns to skin, eyes and mucous membranes.	**1849**
–	–	–	F-A, S-A	Category C. Clear of living quarters.	Toxic if swallowed, by skin contact or by inhalation.	**1851**
–	–	–	F-A, S-A	Category C. Clear of living quarters.	See entry above.	**1851**
–	T21	TP7 TP33	F-G, S-M	Category D.	Liable to ignite spontaneously in air. If shaken, may produce sparks. In contact with water, evolve hydrogen, a flammable gas.	**1854**
–	–	–	F-G, S-M	Category D.	Liable to ignite spontaneously in air. If shaken, may produce sparks. In contact with water, evolve hydrogen, a flammable gas.	**1855**

UN No. (1)	Proper Shipping Name (PSN) (2) 3.1.2	Class or division (3) 2.0	Subsidiary risk(s) (4) 2.0	Packing group (5) 2.0.1.3	Special provisions (6) 3.3	Limited and excepted quantity provisions		Packing		IBC	
						Limited quantities (7a) 3.4	Excepted quantities (7b) 3.5	Instructions (8) 4.1.4	Provisions (9) 4.1.4	Instructions (10) 4.1.4	Provisions (11) 4.1.4
1856	RAGS, OILY	4.2	–	–	29 117	0	E0	P003	PP19	IBC08	B3 B6
1857	TEXTILE WASTE, WET	4.2	–	III	117	0	E1	P410	–	–	–
1858	HEXAFLUOROPROPYLENE (REFRIGERANT GAS R 1216)	2.2	–	–	–	120 mL	E1	P200	–	–	–
1859	SILICON TETRAFLUORIDE	2.3	8	–	–	0	E0	P200	–	–	–
1860	VINYL FLUORIDE, STABILIZED	2.1	–	–	–	0	E0	P200	–	–	–
1862	ETHYL CROTONATE	3	–	II	–	1 L	E2	P001	–	IBC02	–
△ 1863	FUEL, AVIATION, TURBINE ENGINE	3	–	I	363	500 mL	E3	P001	–	–	–
△ 1863	FUEL, AVIATION, TURBINE ENGINE	3	–	II	363	1 L	E2	P001	–	IBC02	–
△ 1863	FUEL, AVIATION, TURBINE ENGINE	3	–	III	223 363	5 L	E1	P001 LP01	–	IBC03	–
1865	*n*-PROPYL NITRATE	3	–	II	26	1 L	E2	P001	–	–	–
1866	RESIN SOLUTION flammable	3	–	I	–	500 mL	E3	P001	–	–	–
1866	RESIN SOLUTION flammable	3	–	II	–	5 L	E2	P001	PP1	IBC02	–
1866	RESIN SOLUTION flammable	3	–	III	223 955	5 L	E1	P001 LP01	PP1	IBC03	–
1868	DECABORANE	4.1	6.1	II	–	1 kg	E2	P002	PP31	IBC06	B2
1869	MAGNESIUM or MAGNESIUM ALLOYS with more than 50% magnesium in pellets, turnings or ribbons	4.1	–	III	59 920	5 kg	E1	P002 LP02	–	IBC08	B3
1870	POTASSIUM BOROHYDRIDE	4.3	–	I	–	0	E0	P403	PP31	–	–
1871	TITANIUM HYDRIDE	4.1	–	II	–	1 kg	E2	P410	PP31 PP40	IBC04	–
1872	LEAD DIOXIDE	5.1	–	III	–	5 kg	E1	P002 LP02	–	IBC08	B3
△ 1873	PERCHLORIC ACID with more than 50% but not more than 72% acid, by mass	5.1	8	I	900	0	E0	P502	PP28	–	–
1884	BARIUM OXIDE	6.1	–	III	–	5 kg	E1	P002 LP02	–	IBC08	B3

Portable tanks and bulk containers			EmS	Stowage and segregation	Properties and observations	UN No.
	Tank instructions	Provisions				
(12)	(13) 4.2.5 4.3	(14) 4.2.5	(15) 5.4.3.2 7.3	(16) 7.1 7.2	(17)	(18)
–	–	–	F-A, S-J	Category A.	Liable to ignite spontaneously in air according to oil content.	1856
–	–	–	F-A, S-J	Category A.	Liable to ignite spontaneously in air according to moisture content.	1857
–	T50	–	F-C, S-V	Category A.	Non-flammable gas. Much heavier than air (5.2).	1858
–	–	–	F-C, S-U	Category D. Clear of living quarters.	Non-flammable, toxic and corrosive gas with a pungent odour. Corrosive to metals. In moist air, produces hydrogen fluoride. Much heavier than air (3.6). Highly irritating to skin, eyes and mucous membranes.	1859
–	–	–	F-D, S-U	Category E. Clear of living quarters.	Flammable gas. Explosive limits: 2.9% to 29%. Heavier than air (1.6).	1860
–	T4	TP2	F-E, S-D	Category B.	Colourless liquid with a pungent odour. Flashpoint: 2°C c.c. Immiscible with water.	1862
–	T11	TP1 TP8 TP28	F-E, S-E	Category E.	Boiling range: –14°C upwards. Immiscible with water.	1863 △
–	T4	TP1 TP8	F-E, S-E	Category B.	Immiscible with water.	1863 △
–	T2	TP1	F-E, S-E	Category A.	See entry above.	1863 △
–	–	–	F-E, S-D	Category D. Segregation as for class 5.1 but "away from" classes 4.1, 5.1 and 7.	White to straw-coloured liquid with an ether-like odour. Flashpoint: 20°C c.c. Explosive limits: 2% to 100%. Immiscible with water. Oxidizing material. May explode on heating. Harmful if swallowed or by inhalation.	1865
–	T11	TP1 TP8 TP28	F-E, S-E	Category E.	Miscibility with water depends upon the composition.	1866
–	T4	TP1 TP8	F-E, S-E	Category B.	See entry above.	1866
–	T2	TP1	F-E, S-E	Category A.	See entry above.	1866
–	T3	TP33	F-A, S-G	Category A. "Separated from" class 5.1.	Colourless crystals. Slightly soluble in water. Vapours may form explosive mixture in air. Forms explosive and extremely sensitive mixtures with oxidizing substances. Toxic if swallowed, by skin contact or by dust inhalation.	1868
–	T1	TP33	F-G, S-G	Category A. "Away from" liquid halogenated hydrocarbons. "Separated from" class 5.1, acids, alkalis and iron oxide.	Silvery white metal. Burns with an intense white light and heat. In contact with water, especially seawater, may evolve hydrogen, a flammable gas. Reacts readily with acids and caustic alkali, evolving hydrogen. Reacts readily with iron oxide, producing a thermite effect. Forms explosive mixtures with oxidizing substances.	1869
–	–	–	F-G, S-O	Category E. "Separated from" acids.	White, crystalline powder. In contact with water, acids or moisture evolves hydrogen, which may be ignited by the heat of the reaction.	1870
–	T3	TP33	F-A, S-G	Category E.	Dark grey powder or crystals.	1871
–	T1	TP33	F-A, S-Q	Category A.	Brown powder or crystals. Insoluble in water. Harmful if swallowed.	1872
–	T10	TP1	F-A, S-Q	Category D. "Separated from" class 4.1.	Colourless liquid. Mixtures with combustible material may ignite spontaneously and, when involved in a fire, by shock or by friction, may cause an explosion. Highly corrosive to most metals. Causes burns to skin, eyes and mucous membranes. Transport of PERCHLORIC ACID with more than 72% acid, by mass, is prohibited.	1873 △
–	T1	TP33	F-A, S-A	Category A.	White solid. Evolves heat in contact with water. Toxic if swallowed, by skin contact or by dust inhalation.	1884

UN No.	Proper Shipping Name (PSN)	Class or division	Subsidiary risk(s)	Packing group	Special provisions	Limited and excepted quantity provisions		Packing		IBC	
						Limited quantities	Excepted quantities	Instructions	Provisions	Instructions	Provisions
(1)	(2) 3.1.2	(3) 2.0	(4) 2.0	(5) 2.0.1.3	(6) 3.3	(7a) 3.4	(7b) 3.5	(8) 4.1.4	(9) 4.1.4	(10) 4.1.4	(11) 4.1.4
1885	BENZIDINE	6.1	–	II	–	500 g	E4	P002	–	IBC08	B2 B4
1886	BENZYLIDENE CHLORIDE	6.1	–	II	–	100 mL	E4	P001	–	IBC02	–
1887	BROMOCHLOROMETHANE	6.1	–	III	–	5 L	E1	P001 LP01	–	IBC03	–
1888	CHLOROFORM	6.1	–	III	–	5 L	E1	P001 LP01	–	IBC03	–
1889	CYANOGEN BROMIDE	6.1	8 P	I	–	0	E5	P002	PP31	–	–
1891	ETHYL BROMIDE	6.1	–	II	–	100 mL	E4	P001	–	IBC02	B8
1892	ETHYLDICHLOROARSINE	6.1	– P	I	354	0	E0	P602	–	–	–
1894	PHENYLMERCURIC HYDROXIDE	6.1	– P	II	–	500 g	E4	P002	–	IBC08	B2 B4
1895	PHENYLMERCURIC NITRATE	6.1	– P	II	–	500 g	E4	P002	–	IBC08	B2 B4
1897	TETRACHLOROETHYLENE	6.1	– P	III	–	5 L	E1	P001 LP01	–	IBC03	–
1898	ACETYL IODIDE	8	–	II	–	1 L	E2	P001	–	IBC02	–
1902	DIISOOCTYL ACID PHOSPHATE	8	–	III	–	5 L	E1	P001 LP01	–	IBC03	–
1903	DISINFECTANT, LIQUID, CORROSIVE, N.O.S.	8	–	I	274	0	E0	P001	–	–	–
1903	DISINFECTANT, LIQUID, CORROSIVE, N.O.S.	8	–	II	274	1 L	E2	P001	–	IBC02	–
1903	DISINFECTANT, LIQUID, CORROSIVE, N.O.S.	8	–	III	223 274	5 L	E1	P001 LP01	–	IBC03	–
1905	SELENIC ACID	8	–	I	–	0	E0	P002	–	IBC07	B1
1906	SLUDGE ACID	8	–	II	–	1 L	E2	P001	–	IBC02	–
1907	SODA LIME with more than 4% sodium hydroxide	8	–	III	62	5 kg	E1	P002 LP02	–	IBC08	B3

Portable tanks and bulk containers			EmS	Stowage and segregation	Properties and observations	UN No.
	Tank instructions	Provisions				
(12)	(13) 4.2.5 4.3	(14) 4.2.5	(15) 5.4.3.2 7.3	(16) 7.1 7.2	(17)	(18)
–	T3	TP33	F-A, S-A	Category A.	White, crystalline solid. Toxic if swallowed, by skin contact or by inhalation.	1885
–	T7	TP2	F-A, S-A	Category D. Clear of living quarters.	Colourless liquid evolving vapour which is irritating to eyes and skin ("Tear Gas"). Toxic if swallowed, by skin contact or by inhalation.	1886
–	T4	TP1	F-A, S-A	Category A.	Clear, colourless, volatile liquid with a chloroform-like odour. Immiscible with water. When involved in a fire, evolves extremely toxic fumes (phosgene). Toxic if swallowed, by skin contact or by inhalation.	1887
–	T7	TP2	F-A, S-A	Category A. Clear of living quarters.	Colourless, volatile liquid. Boiling point: 61°C. Non-flammable. When involved in a fire, evolves extremely toxic fumes (phosgene). Toxic if swallowed, by skin contact or by inhalation. Anaesthetic.	1888
–	T6	TP33	F-A, S-B	Category D. Clear of living quarters. "Separated from" acids.	Colourless crystals evolving toxic vapour which is irritating and causes tears. Melting point: approximately 52°C. Boiling point: approximately 62°C. In contact with water evolves hydrogen bromide and hydrogen cyanide, which are highly toxic, flammable and corrosive gases. Highly toxic if swallowed, by skin contact or by inhalation. Causes burns to skin, eyes and mucous membranes.	1889
–	T7	TP2 TP13	F-A, S-A	Category B in a mechanically ventilated space. Clear of living quarters.	Colourless volatile liquid evolving irritating vapour with a narcotic effect. Boiling point: 38°C. Vapour can be ignited by an electric spark or similar sources of ignition. Toxic if swallowed, by skin contact or by inhalation.	1891
–	T20	TP2 TP13 TP37	F-A, S-A	Category D. Clear of living quarters.	Colourless liquid evolving irritating vapour ("Tear Gas"). Highly toxic if swallowed, by skin contact or by inhalation.	1892
–	T3	TP33	F-A, S-A	Category A.	White crystals or powder. Soluble in water. Toxic if swallowed, by skin contact or by dust inhalation.	1894
–	T3	TP33	F-A, S-A	Category A.	White crystals or powder. Toxic if swallowed, by skin contact or by inhalation.	1895
–	T4	TP1	F-A, S-A	Category A. Clear of living quarters.	Colourless liquid with an ethereal odour. When involved in a fire, evolves extremely toxic fumes (phosgene). Toxic if swallowed, by skin contact or by inhalation.	1897
–	T7	TP2 TP13	F-A, S-B	Category C. Clear of living quarters.	Colourless liquid. Reacts violently with water, evolving hydrogen iodide, an irritating and corrosive gas apparent as white fumes. In the presence of moisture, highly corrosive to most metals. Vapour irritates mucous membranes.	1898
–	T4	TP1	F-A, S-B	Category A.	Oily liquid. Mildly corrosive to most metals.	1902
–	–	–	F-A, S-B	Category B.	A wide variety of corrosive liquids. Cause burns to skin, eyes and mucous membranes.	1903
–	–	–	F-A, S-B	Category B.	See entry above.	1903
–	–	–	F-A, S-B	Category A.	See entry above.	1903
–	T6	TP33	F-A, S-B	Category A.	White, very deliquescent crystalline solid. Melting point: 50°C. Soluble in water. Reacts violently with organic materials such as wood, cotton or straw. In the presence of moisture, corrosive to most metals. Causes severe burns to skin, eyes and mucous membranes.	1905
–	T8	TP2 TP28	F-A, S-B	Category C. For metal drums, category B.	Waste or spent sulphuric acid, usually a by-product of refining petroleum oils or crude benzenes. Highly corrosive to most metals.	1906
–	T1	TP33	F-A, S-B	Category A. "Separated from" acids.	Deliquescent, granulated mixture of sodium hydroxide and calcium hydroxide. Reacts violently with acids. Reacts with ammonium salts, evolving ammonia gas. In the presence of moisture, corrosive to aluminium, zinc and tin. Causes burns to skin, eyes and mucous membranes.	1907

UN No.	Proper Shipping Name (PSN)	Class or division	Subsidiary risk(s)	Packing group	Special provisions	Limited and excepted quantity provisions		Packing		IBC	
						Limited quantities	Excepted quantities	Instructions	Provisions	Instructions	Provisions
(1)	(2) 3.1.2	(3) 2.0	(4) 2.0	(5) 2.0.1.3	(6) 3.3	(7a) 3.4	(7b) 3.5	(8) 4.1.4	(9) 4.1.4	(10) 4.1.4	(11) 4.1.4
1908	CHLORITE SOLUTION	8	–	II	–	1 L	E2	P001	–	IBC02	–
1908	CHLORITE SOLUTION	8	–	III	223	5 L	E1	P001 LP01	–	IBC03	–
1910	CALCIUM OXIDE	8	–	–	960	–	–	–	–	–	–
1911	DIBORANE	2.3	2.1	–	–	0	E0	P200	–	–	–
1912	METHYL CHLORIDE AND METHYLENE CHLORIDE MIXTURE	2.1	–	–	228	0	E0	P200	–	–	–
1913	NEON, REFRIGERATED LIQUID	2.2	–	–	–	120 mL	E1	P203	–	–	–
1914	BUTYL PROPIONATES	3	–	III	–	5 L	E1	P001 LP01	–	IBC03	–
1915	CYCLOHEXANONE	3	–	III	–	5 L	E1	P001 LP01	–	IBC03	–
1916	2,2′-DICHLORODIETHYL ETHER	6.1	3	II	–	100 mL	E4	P001	–	IBC02	–
1917	ETHYL ACRYLATE, STABILIZED	3	–	II	–	1 L	E2	P001	–	IBC02	–
1918	ISOPROPYLBENZENE	3	–	III	–	5 L	E1	P001 LP01	–	IBC03	–
1919	METHYL ACRYLATE, STABILIZED	3	–	II	–	1 L	E2	P001	–	IBC02	–
1920	NONANES	3	–	III	–	5 L	E1	P001 LP01	–	IBC03	–
1921	PROPYLENEIMINE, STABILIZED	3	6.1	I	–	0	E0	P001	–	–	–
1922	PYRROLIDINE	3	8	II	–	1 L	E2	P001	–	IBC02	–
1923	CALCIUM DITHIONITE (CALCIUM HYDROSULPHITE)	4.2	–	II	–	0	E2	P410	PP31	IBC06	B2
1928	METHYLMAGNESIUM BROMIDE IN ETHYL ETHER	4.3	3	I	–	0	E0	P402	–	–	–
1929	POTASSIUM DITHIONITE (POTASSIUM HYDROSULPHITE)	4.2	–	II	–	0	E2	P410	PP31	IBC06	B2
1931	ZINC DITHIONITE (ZINC HYDROSULPHITE)	9	–	III	–	5 kg	E1	P002 LP02	–	IBC08	B3

Portable tanks and bulk containers			EmS	Stowage and segregation	Properties and observations	UN No.
	Tank instructions	Provisions				
(12)	(13) 4.2.5 4.3	(14) 4.2.5	(15) 5.4.3.2 7.3	(16) 7.1 7.2	(17)	(18)
–	T7	TP2 TP24	F-A, S-B	Category B. "Away from" acids. Segregation as for class 5.1 but "away from" classes 4.1, 5.1 and 7.	Colourless liquid. In contact with acids, evolves very irritating and corrosive gases. Oxidizing solution. May cause fire in contact with organic materials such as wood, cotton or straw. Mildly corrosive to most metals. Causes burns to skin, eyes and mucous membranes.	1908
–	T4	TP2 TP24	F-A, S-B	Category B. "Away from" acids. Segregation as for class 5.1 but "away from" classes 4.1, 5.1 and 7.	See entry above.	1908
–	–	–	–	–	Not subject to the provisions of this Code but may be subject to provisions governing the transport of dangerous goods by other modes.	1910
–	–	–	F-D, S-U	Category D. Clear of living quarters. "Separated from" chlorine.	Liquefied, flammable, toxic, colourless gas with an unpleasant odour. Explosive limits: 0.9% to 98%. Lighter than air (0.95). May decompose above −18°C with the formation of hydrogen and boron hydrides. Autoignition temperature: 90°C. Toxic by inhalation; forms boric acid and water by hydrolysis within the lungs.	1911
–	T50	–	F-D, S-U	Category D. Clear of living quarters.	Solution of the flammable gas methyl chloride, UN No. 1063, in the liquid methylene chloride.	1912
–	T75	TP5	F-C, S-V	Category D.	Liquefied, inert gas. Lighter than air (0.7).	1913
–	T2	TP1	F-E, S-D	Category A.	Colourless liquids. Flashpoint: 32°C c.c. Immiscible with water.	1914
–	T2	TP1	F-E, S-D	Category A.	Colourless liquid. Flashpoint: 38°C to 44°C c.c. Explosive limits: 1.1% to 9.4%. Immiscible with water.	1915
–	T7	TP2	F-E, S-D	Category A.	Colourless flammable liquid. Flashpoint: 55°C c.c. Immiscible with water, but reacts with it, forming corrosive and toxic fumes. Toxic if swallowed, by skin contact or by inhalation.	1916
–	T4	TP1 TP13	F-E, S-D	Category B. Clear of living quarters.	Colourless liquid with a pungent odour. Flashpoint: 16°C c.c. Explosive limits: 1.3% to 14%. Immiscible with water. Irritating to skin, eyes and mucous membranes.	1917
–	T2	TP1	F-E, S-E	Category A.	Colourless liquid with a chloroform-like odour. Flashpoint: 31°C c.c. Explosive limits: 0.9% to 6.5%. Immiscible with water.	1918
–	T4	TP1 TP13	F-E, S-D	Category B.	Colourless, volatile liquid with a pungent odour. Flashpoint: −3°C c.c. Explosive limits: 1.2% to 25%. Immiscible with water. Harmful by inhalation. Irritating to skin, eyes and mucous membranes.	1919
–	T2	TP1	F-E, S-E	Category A.	Colourless liquids. Explosive limits: 0.8% to 2.9%. *normal*-NONANE: flashpoint 31°C c.c. Immiscible with water. Irritating to skin, eyes and mucous membranes.	1920
–	T14	TP2 TP13	F-E, S-D	Category B. Clear of living quarters.	Colourless liquid with an ammoniacal odour. Flashpoint: −4°C o.c. Miscible with water. Toxic if swallowed, by skin contact or by inhalation. Causes burns to skin and eyes.	1921
–	T7	TP1	F-E, S-C	Category B. Clear of living quarters. "Separated from" acids.	Colourless to pale yellow liquid with an ammoniacal odour. Reacts violently with acids. Flashpoint: 3°C c.c. Miscible with water. Harmful by inhalation. Causes burns to skin, eyes and mucous membranes.	1922
–	T3	TP33	F-A, S-J	Category E. Keep as dry as reasonably practicable.	Liable to heat and ignite spontaneously in air and to evolve sulphur dioxide, an irritating gas.	1923
–	–	–	F-G, S-L	Category D.	Colourless, yellowish liquid. Decomposes violently in contact with water. Spillage will ignite spontaneously.	1928
–	T3	TP33	F-A, S-J	Category E. Keep as dry as reasonably practicable.	Liable to heat and ignite spontaneously in air and to evolve sulphur dioxide, an irritating gas.	1929
–	T1	TP33	F-A, S-J	Category A. Keep as dry as reasonably practicable. "Away from" class 6.2 and acids.	White, amorphous solid material. Soluble in water. Liable to heat on contact with moisture and heating results in evolution of sulphur dioxide, an intensely irritating gas. Also evolves sulphur dioxide on contact with acids.	1931

3

DGL

UN No.	Proper Shipping Name (PSN)	Class or division	Subsidiary risk(s)	Packing group	Special provisions	Limited and excepted quantity provisions		Packing		IBC	
						Limited quantities	Excepted quantities	Instructions	Provisions	Instructions	Provisions
(1)	(2) 3.1.2	(3) 2.0	(4) 2.0	(5) 2.0.1.3	(6) 3.3	(7a) 3.4	(7b) 3.5	(8) 4.1.4	(9) 4.1.4	(10) 4.1.4	(11) 4.1.4
1932	ZIRCONIUM, SCRAP	4.2	–	III	223	0	E1	P002 LP02	PP31	IBC08	B3
1935	CYANIDE SOLUTION, N.O.S.	6.1	– P	I	274	0	E5	P001	–	–	–
1935	CYANIDE SOLUTION, N.O.S.	6.1	– P	II	274	100 mL	E4	P001	–	IBC02	–
1935	CYANIDE SOLUTION, N.O.S.	6.1	– P	III	223 274	5 L	E1	P001 LP01	–	IBC03	–
1938	BROMOACETIC ACID SOLUTION	8	–	II	–	1 L	E2	P001	–	IBC02	–
1938	BROMOACETIC ACID SOLUTION	8	–	III	223	5 L	E1	P001 LP01	–	IBC03	–
△ 1939	PHOSPHORUS OXYBROMIDE	8	–	II	–	1 kg	E2	P002	–	IBC08	B2 B4
1940	THIOGLYCOLIC ACID	8	–	II	–	1 L	E2	P001	–	IBC02	–
△ 1941	DIBROMODIFLUOROMETHANE	9	–	III	–	5 L	E1	P001 LP01	–	–	–
△ 1942	AMMONIUM NITRATE with not more than 0.2% total combustible material, including any organic substance, calculated as carbon to the exclusion of any other added substance	5.1	–	III	900 952 967	5 kg	E1	P002 LP02	–	IBC08	B3
1944	MATCHES, SAFETY (book, card or strike on box)	4.1	–	III	293 294	5 kg	E1	P407	–	–	–
1945	MATCHES, WAX 'VESTA'	4.1	–	III	294	5 kg	E1	P407	–	–	–

Portable tanks and bulk containers			EmS	Stowage and segregation	Properties and observations	UN No.
	Tank instructions	Provisions				
(12)	(13) 4.2.5 4.3	(14) 4.2.5	(15) 5.4.3.2 7.3	(16) 7.1 7.2	(17)	(18)
–	T1	TP33	F-G, S-L	Category D.	Particle size larger than 840 microns. Readily flammable; may ignite spontaneously in air. In contact with water, may evolve hydrogen, a flammable gas.	**1932**
–	T14	TP2 TP13 TP27	F-A, S-A	Category B. Clear of living quarters. "Separated from" acids.	Liquid evolving toxic vapour. Reacts with acids or acid fumes, evolving hydrogen cyanide, a highly toxic and flammable gas. Toxic if swallowed, by skin contact or by inhalation.	**1935**
–	T11	TP2 TP13 TP27	F-A, S-A	Category A. Clear of living quarters. "Separated from" acids.	See entry above.	**1935**
–	T7	TP2 TP13 TP28	F-A, S-A	Category A. Clear of living quarters. "Separated from" acids.	See entry above.	**1935**
–	T7	TP2	F-A, S-B	Category A. Clear of living quarters.	Corrosive to most metals. Harmful if swallowed. Causes burns to eyes and skin.	**1938**
–	T7	TP2	F-A, S-B	Category A. Clear of living quarters.	See entry above.	**1938**
–	T3	TP33	F-A, S-B	Category C. Keep as cool as reasonably practicable. Clear of living quarters.	Colourless crystals. Melting point: 56°C. Reacts violently with water, evolving hydrogen bromide, a toxic and corrosive gas apparent as white fumes. Reacts violently with organic materials (such as wood, cotton, straw), causing fire. Decomposes when heated, evolving toxic and corrosive gases. When involved in a fire, evolves toxic and corrosive gases. In the presence of moisture, highly corrosive to most metals. Causes burns to skin, eyes and mucous membranes.	**1939** △
–	T7	TP2	F-A, S-B	Category A.	Colourless liquid with a strong, very unpleasant odour. Corrosive to most metals. Harmful if swallowed.	**1940**
–	T11	TP2	F-A, S-A	Category A. Protected from sources of heat.	Colourless, heavy liquid. Boiling point: 24°C. Immiscible with water. When involved in a fire, may evolve toxic fumes. Toxic if swallowed, by skin contact or by inhalation. Irritating to skin, eyes and mucous membranes.	**1941** △
–	T1 BK2 BK3	TP33	F-H, S-Q	Category C. Category A only if the special stowage provisions of 7.4.1.4 and 7.6.2.8.4. are complied with. Protected from sources of heat. "Separated from" class 4.1, combustible material (particularly liquids), bromates, chlorates, chlorites, hypochlorites, nitrites, perchlorates, permanganates and powdered metals. When transported in BK3 bulk container, see 7.6.2.12 and 7.7.3.9.	Crystals, granules or prills. Soluble in water. Supporter of combustion. A major fire aboard a ship carrying this substance may involve a risk of explosion in the event of contamination (e.g., by fuel oil) or strong confinement. An adjacent detonation may also involve the risk of explosion. If heated strongly, decomposes, giving off toxic gases and gases which support combustion. Transport of AMMONIUM NITRATE liable to self-heating sufficient to initiate decomposition is **prohibited**.	**1942** △
–	–	–	F-A, S-I	Category A.	Intended to be ignited on a specially prepared surface.	**1944**
–	–	–	F-A, S-I	Category B.	Ignite by friction; a prepared surface may be required.	**1945**

UN No.	Proper Shipping Name (PSN)	Class or division	Subsidiary risk(s)	Packing group	Special provisions	Limited and excepted quantity provisions		Packing		IBC	
						Limited quantities	Excepted quantities	Instructions	Provisions	Instructions	Provisions
(1)	(2) 3.1.2	(3) 2.0	(4) 2.0	(5) 2.0.1.3	(6) 3.3	(7a) 3.4	(7b) 3.5	(8) 4.1.4	(9) 4.1.4	(10) 4.1.4	(11) 4.1.4
△ 1950	AEROSOLS	2	– See SP63	–	63 190 277 327 344 959	See SP277	E0	P207 LP02	PP87 L2	–	–
1951	ARGON, REFRIGERATED LIQUID	2.2	–	–	–	120 mL	E1	P203	–	–	–
1952	ETHYLENE OXIDE AND CARBON DIOXIDE MIXTURE with not more than 9% ethylene oxide	2.2	–	–	–	120 mL	E1	P200	–	–	–
1953	COMPRESSED GAS, TOXIC, FLAMMABLE, N.O.S.	2.3	2.1	–	274	0	E0	P200	–	–	–
1954	COMPRESSED GAS, FLAMMABLE, N.O.S.	2.1	–	–	274	0	E0	P200	–	–	–
1955	COMPRESSED GAS, TOXIC, N.O.S.	2.3	–	–	274	0	E0	P200	–	–	–
1956	COMPRESSED GAS, N.O.S.	2.2	–	–	274	120 mL	E1	P200	–	–	–
1957	DEUTERIUM, COMPRESSED	2.1	–	–	–	0	E0	P200	–	–	–
1958	1,2-DICHLORO-1,1,2,2-TETRAFLUOROETHANE (REFRIGERANT GAS R 114)	2.2	–	–	–	120 mL	E1	P200	–	–	–
1959	1,1-DIFLUOROETHYLENE (REFRIGERANT GAS R 1132a)	2.1	–	–	–	0	E0	P200	–	–	–
1961	ETHANE, REFRIGERATED LIQUID	2.1	–	–	–	0	E0	P203	–	–	–
1962	ETHYLENE	2.1	–	–	–	0	E0	P200	–	–	–
1963	HELIUM, REFRIGERATED LIQUID	2.2	–	–	–	120 mL	E1	P203	–	–	–
1964	HYDROCARBON GAS MIXTURE, COMPRESSED, N.O.S.	2.1	–	–	274	0	E0	P200	–	–	–
1965	HYDROCARBON GAS MIXTURE, LIQUEFIED, N.O.S.	2.1	–	–	274	0	E0	P200	–	–	–
1966	HYDROGEN, REFRIGERATED LIQUID	2.1	–	–	–	0	E0	P203	–	–	–
1967	INSECTICIDE GAS, TOXIC, N.O.S.	2.3	–	–	274	0	E0	P200	–	–	–
1968	INSECTICIDE GAS, N.O.S.	2.2	–	–	274	120 mL	E1	P200	–	–	–

Portable tanks and bulk containers			EmS	Stowage and segregation	Properties and observations	UN No.
	Tank instructions	Provisions				
(12)	(13) 4.2.5 4.3	(14) 4.2.5	(15) 5.4.3.2 7.3	(16) 7.1 7.2	(17)	(18)
–	–	–	F-D, S-U	Protected from sources of heat. For AEROSOLS with a maximum capacity of 1 litre: Category A. Segregation as for class 9 but "separated from" class 1 except division 1.4. For AEROSOLS with a capacity above 1 litre: Category B. Segregation as for the appropriate sub-division of class 2. For WASTE AEROSOLS: Category C. Clear of living quarters. Segregation as for the appropriate sub-division of class 2.	–	1950 △
–	T75	TP5	F-C, S-V	Category D.	Liquefied, inert gas Heavier than air (1.4).	1951
–	–	–	F-C, S-V	Category A.	Liquefied, non-flammable gas with an ether-like odour. Explosive limits: 31% to 52%. Heavier than air (1.5).	1952
–	–	–	F-D, S-U	Category D. Clear of living quarters.	–	1953
–	–	–	F-D, S-U	Category D. Clear of living quarters.	–	1954
–	–	–	F-C, S-U	Category D. Clear of living quarters.	–	1955
–	–	–	F-C, S-V	Category A.	–	1956
–	–	–	F-D, S-U	Category E. Clear of living quarters.	Flammable, odourless gas. Much lighter than air (0.14).	1957
–	T50	–	F-C, S-V	Category A.	Liquefied, non-flammable gas with a chloroform-like odour. Much heavier than air (5.9). Boiling point: 4°C.	1958
–	–	–	F-D, S-U	Category E. Clear of living quarters.	Flammable gas. Explosive limits: 2.3% to 25%. Much heavier than air (2.2).	1959
–	T75	TP5	F-D, S-U	Category D. Clear of living quarters.	Liquefied, flammable gas with a faint odour. Explosive limits: 3% to 16%. Slightly heavier than air (1.05).	1961
–	–	–	F-D, S-U	Category E. Clear of living quarters.	Flammable gas. Explosive limits: 3% to 34%. Slightly lighter than air (0.98).	1962
–	T75	TP5 TP34	F-C, S-V	Category D.	Liquefied, inert gas. Much lighter than air (0.14).	1963
–	–	–	F-D, S-U	Category E. Clear of living quarters.	Flammable hydrocarbon gas mixture obtained from natural gas or by distillation of mineral oils or coal, etc. May contain propane, cyclopropane, propylene, butane, butylene, etc., in varying proportions. Heavier than air.	1964
–	T50	–	F-D, S-U	Category E. Clear of living quarters.	Liquefied flammable hydrocarbon gas obtained from natural gas or by distillation of mineral oils or coal, etc. May contain propane, cyclopropane, propylene, butane, butylene, etc., in varying proportions. Heavier than air.	1965
–	T75	TP5 TP23 TP34	F-D, S-U	Category D. Clear of living quarters. "Separated from" chlorine.	Liquefied, flammable, odourless gas. Explosive limits: 4% to 75%. Much lighter than air (0.07).	1966
–	–	–	F-C, S-U	Category D. Clear of living quarters.	Toxic mixtures of insecticides with liquefied gases. These mixtures may be flammable.	1967
–	–	–	F-C, S-V	Category A.	Non-flammable and non-toxic mixtures of insecticides with liquefied gases.	1968

UN No. (1)	Proper Shipping Name (PSN) (2) 3.1.2	Class or division (3) 2.0	Subsidiary risk(s) (4) 2.0	Packing group (5) 2.0.1.3	Special provisions (6) 3.3	Limited and excepted quantity provisions: Limited quantities (7a) 3.4	Excepted quantities (7b) 3.5	Packing: Instructions (8) 4.1.4	Provisions (9) 4.1.4	IBC: Instructions (10) 4.1.4	Provisions (11) 4.1.4
1969	ISOBUTANE	2.1	–	–	–	0	E0	P200	–	–	–
1970	KRYPTON, REFRIGERATED LIQUID	2.2	–	–	–	120 mL	E1	P203	–	–	–
1971	METHANE, COMPRESSED or NATURAL GAS, COMPRESSED with high methane content	2.1	–	–	–	0	E0	P200	–	–	–
1972	METHANE, REFRIGERATED LIQUID or NATURAL GAS, REFRIGERATED LIQUID with high methane content	2.1	–	–	–	0	E0	P203	–	–	–
1973	CHLORODIFLUOROMETHANE AND CHLOROPENTAFLUORO-ETHANE MIXTURE with a fixed boiling point, with approximately 49% chlorodifluoromethane (REFRIGERANT GAS R 502)	2.2	–	–	–	120 mL	E1	P200	–	–	–
1974	CHLORODIFLUOROBROMO-METHANE (REFRIGERANT GAS R 12B1)	2.2	–	–	–	120 mL	E1	P200	–	–	–
1975	NITRIC OXIDE AND DINITROGEN TETROXIDE MIXTURE (NITRIC OXIDE AND NITROGEN DIOXIDE MIXTURE)	2.3	5.1/8	–	–	0	E0	P200	–	–	–
1976	OCTAFLUOROCYCLOBUTANE (REFRIGERANT GAS RC 318)	2.2	–	–	–	120 mL	E1	P200	–	–	–
△ 1977	NITROGEN, REFRIGERATED LIQUID	2.2	–	–	345 346	120 mL	E1	P203	–	–	–
1978	PROPANE	2.1	–	–	–	0	E0	P200	–	–	–
1982	TETRAFLUOROMETHANE (REFRIGERANT GAS R 14)	2.2	–	–	–	120 mL	E1	P200	–	–	–
1983	1-CHLORO-2,2,2-TRIFLUORO-ETHANE (REFRIGERANT GAS R 133a)	2.2	–	–	–	120 mL	E1	P200	–	–	–
1984	TRIFLUOROMETHANE (REFRIGERANT GAS R 23)	2.2	–	–	–	120 mL	E1	P200	–	–	–
1986	ALCOHOLS, FLAMMABLE, TOXIC, N.O.S.	3	6.1	I	274	0	E0	P001	–	–	–
1986	ALCOHOLS, FLAMMABLE, TOXIC, N.O.S.	3	6.1	II	274	1 L	E2	P001	–	IBC02	–
1986	ALCOHOLS, FLAMMABLE, TOXIC, N.O.S.	3	6.1	III	223 274	5 L	E1	P001	–	IBC03	–
1987	ALCOHOLS, N.O.S.	3	–	II	274	1 L	E2	P001	–	IBC02	–
1987	ALCOHOLS, N.O.S.	3	–	III	223 274	5 L	E1	P001 LP01	–	IBC03	–
1988	ALDEHYDES, FLAMMABLE, TOXIC, N.O.S.	3	6.1	I	274	0	E0	P001	–	–	–

Portable tanks and bulk containers			EmS	Stowage and segregation	Properties and observations	UN No.
	Tank instructions	Provisions				
(12)	(13) 4.2.5 4.3	(14) 4.2.5	(15) 5.4.3.2 7.3	(16) 7.1 7.2	(17)	(18)
–	T50	–	F-D, S-U	Category E. Clear of living quarters.	Flammable hydrocarbon. Heavier than air.	**1969**
–	T75	TP5	F-C, S-V	Category D.	Liquefied, inert gas. Much heavier than air (2.9).	1970
–	–	–	F-D, S-U	Category E. Clear of living quarters.	Flammable gas. Explosive limits: 5% to 16%. Lighter than air (methane 0.55).	**1971**
–	T75	TP5	F-D, S-U	Category D. Clear of living quarters.	Liquefied, flammable gas. Explosive limits: 5% to 16%. Lighter than air (methane 0.55).	1972
–	T50	–	F-C, S-V	Category A.	Liquefied, non-flammable gas. Much heavier than air (4.2.)	**1973**
–	T50	–	F-C, S-V	Category A.	Liquefied, non-flammable gas. Much heavier than air (5.7).	1974
–	–	–	F-C, S-W	Category D. Clear of living quarters. Segregation as for class 5.1 but "separated from" class 7.	Non-flammable, toxic and corrosive, brown gas mixtures of varying composition with a pungent odour. Strong oxidizing agent. Heavier than air. Highly irritating to skin, eyes and mucous membranes. Toxic by inhalation, with delayed effect similar to phosgene.	**1975**
–	T50	–	F-C, S-V	Category A.	Liquefied, non-flammable gas. Much heavier than air (7.0).	1976
–	T75	TP5	F-C, S-V	Category D.	Liquefied, non-flammable, odourless gas. Lighter than air (0.97). Arrangements for the containment of the liquid nitrogen and fittings in use should be appropriate to the potential danger to the structure of the freight container or ship from the effect of misuse or accidental spillage.	**1977** △
–	T50	–	F-D, S-U	Category E. Clear of living quarters.	Flammable hydrocarbon gas. Explosive limits: 2.3% to 9.5%. Heavier than air (1.56).	1978
–	–	–	F-C, S-V	Category A.	Non-flammable gas. Much heavier than air (3.1).	**1982**
–	T50	–	F-C, S-V	Category A.	Liquefied, non-flammable gas. Much heavier than air (4.1). Boiling point: 7°C.	1983
–	–	–	F-C, S-V	Category A.	Liquefied, non-flammable gas. Much heavier than air (2.4).	**1984**
–	T14	TP2 TP13 TP27	F-E, S-D	Category E. Clear of living quarters.	Toxic if swallowed, by skin contact or by inhalation.	1986
–	T11	TP2 TP27	F-E, S-D	Category B. Clear of living quarters.	See entry above.	**1986**
–	T7	TP1 TP28	F-E, S-D	Category A.	See entry above.	1986
–	T7	TP1 TP8 TP28	F-E, S-D	Category B.	–	**1987**
–	T4	TP1 TP29	F-E, S-D	Category A.	–	1987
–	T14	TP2 TP13 TP27	F-E, S-D	Category E. Clear of living quarters.	Toxic if swallowed, by skin contact or by inhalation.	**1988**

UN No.	Proper Shipping Name (PSN)	Class or division	Subsidiary risk(s)	Packing group	Special provisions	Limited and excepted quantity provisions		Packing		IBC	
						Limited quantities	Excepted quantities	Instructions	Provisions	Instructions	Provisions
(1)	(2) 3.1.2	(3) 2.0	(4) 2.0	(5) 2.0.1.3	(6) 3.3	(7a) 3.4	(7b) 3.5	(8) 4.1.4	(9) 4.1.4	(10) 4.1.4	(11) 4.1.4
1988	ALDEHYDES, FLAMMABLE, TOXIC, N.O.S.	3	6.1	II	274	1 L	E2	P001	–	IBC02	–
1988	ALDEHYDES, FLAMMABLE, TOXIC, N.O.S.	3	6.1	III	223 274	5 L	E1	P001	–	IBC03	–
1989	ALDEHYDES, N.O.S.	3	–	I	274	0	E3	P001	–	–	–
1989	ALDEHYDES, N.O.S.	3	–	II	274	1 L	E2	P001	–	IBC02	–
1989	ALDEHYDES, N.O.S.	3	–	III	223 274	5 L	E1	P001 LP01	–	IBC03	–
1990	BENZALDEHYDE	9	–	III	–	5 L	E1	P001 LP01	–	IBC03	–
1991	CHLOROPRENE, STABILIZED	3	6.1	I	–	0	E0	P001	–	–	–
1992	FLAMMABLE LIQUID, TOXIC, N.O.S.	3	6.1	I	274	0	E0	P001	–	–	–
1992	FLAMMABLE LIQUID, TOXIC, N.O.S.	3	6.1	II	274	1 L	E2	P001	–	IBC02	–
1992	FLAMMABLE LIQUID, TOXIC, N.O.S.	3	6.1	III	223 274	5 L	E1	P001	–	IBC03	–
1993	FLAMMABLE LIQUID, N.O.S.	3	–	I	274	0	E3	P001	–	–	–
1993	FLAMMABLE LIQUID, N.O.S.	3	–	II	274	1 L	E2	P001	–	IBC02	–
1993	FLAMMABLE LIQUID, N.O.S.	3	–	III	223 274 955	5 L	E1	P001 LP01	–	IBC03	–
1994	IRON PENTACARBONYL	6.1	3	I	354	0	E0	P601	–	–	–
1999	TARS, LIQUID, including road oils, and cutback bitumens	3	–	II	–	5 L	E2	P001	–	IBC02	–
1999	TARS, LIQUID, including road oils, and cutback bitumens	3	–	III	955	5 L	E1	P001 LP01	–	IBC03	–
2000	CELLULOID in block, rods, rolls, sheets, tubes, etc., except scrap	4.1	–	III	223	5 kg	E1	P002 LP02	PP7	–	–
2001	COBALT NAPHTHENATES, POWDER	4.1	–	III	–	5 kg	E1	P002 LP02	–	IBC08	B3
2002	CELLULOID, SCRAP	4.2	–	III	223	0	E1	P002 LP02	PP8	IBC08	B3
2004	MAGNESIUM DIAMIDE	4.2	–	II	–	0	E2	P410	PP31	IBC06	–
2006	PLASTICS, NITROCELLULOSE-BASED, SELF-HEATING, N.O.S.	4.2	–	III	274	0	E1	P002	–	–	–
2008	ZIRCONIUM POWDER, DRY	4.2	–	I	–	0	E0	P404	PP31	–	–
2008	ZIRCONIUM POWDER, DRY	4.2	–	II	–	0	E2	P410	PP31	IBC06	B2

Portable tanks and bulk containers			EmS	Stowage and segregation	Properties and observations	UN No.
	Tank instructions	Provisions				
(12)	(13) 4.2.5 4.3	(14) 4.2.5	(15) 5.4.3.2 7.3	(16) 7.1 7.2	(17)	(18)
–	T11	TP2 TP27	F-E, S-D	Category B. Clear of living quarters.	Toxic if swallowed, by skin contact or by inhalation.	1988
–	T7	TP1 TP28	F-E, S-D	Category A.	See entry above.	1988
–	T11	TP1 TP27	F-E, S-D	Category E.	–	1989
–	T7	TP1 TP8 TP28	F-E, S-D	Category B.	–	1989
–	T4	TP1 TP29	F-E, S-D	Category A.	–	1989
–	T2	TP1	F-A, S-A	Category A.	Colourless or yellowish volatile oil with a bitter almond odour. Slightly soluble in water. Irritating to skin, eyes and mucous membranes.	1990
–	T14	TP2 TP6 TP13	F-E, S-D	Category D. Clear of living quarters.	Colourless liquid. Flashpoint: −20°C c.c. Explosive limits: 2.5% to 12%. Slightly soluble in water. Toxic if swallowed, by skin contact or by inhalation.	1991
–	T14	TP2 TP13 TP27	F-E, S-D	Category E. Clear of living quarters.	Flammable toxic liquid which is not specified by name in this class or, on account of its characteristics, in some other class. Toxic if swallowed, by skin contact or by inhalation.	1992
–	T7	TP2 TP13	F-E, S-D	Category B. Clear of living quarters.	See entry above.	1992
–	T7	TP1 TP28	F-E, S-D	Category A.	See entry above.	1992
–	T11	TP1 TP27	F-E, S-E	Category E.	–	1993
–	T7	TP1 TP8 TP28	F-E, S-E	Category B.	–	1993
–	T4	TP1 TP29	F-E, S-E	Category A.	–	1993
–	T22	TP2 TP13	F-E, S-D	Category D. Clear of living quarters.	Yellow to dark red, volatile flammable liquid. Flashpoint: −15°C c.c. Explosive limits: 3.7% to 12.5%. May react with water or steam, evolving carbon monoxide, which is a toxic gas. Highly toxic if swallowed, by skin contact or by inhalation.	1994
–	T3	TP3 TP29	F-E, S-E	Category B.	Mobile liquids prepared by mixing asphalt with petroleum distillate. Pungent odour. Immiscible with water.	1999
–	T1	TP3	F-E, S-E	Category A.	See entry above.	1999
–	–	–	F-A, S-I	Category A.	Ignites readily. When involved in a fire, evolves toxic fumes; in enclosed cargo spaces, these fumes may form an explosive mixture with air.	2000
–	T1	TP33	F-A, S-I	Category A.	Brown, amorphous powder. Insoluble in water. Readily combustible.	2001
–	–	–	F-A, S-J	Category D.	Ignites readily. When involved in a fire, evolves toxic fumes; in enclosed cargo spaces, these fumes may form an explosive mixture with air.	2002
–	T3	TP33	F-G, S-M	Category C.	White powder. Ignites spontaneously in air. Reacts violently in contact with water.	2004
–	–	–	F-A, S-G	Category C.	–	2006
–	T21	TP7 TP33	F-G, S-M	Category D.	Amorphous powder. Liable to ignite spontaneously in air. Forms explosive mixtures with oxidizing substances.	2008
–	T3	TP33	F-G, S-M	Category D.	See entry above.	2008

UN No.	Proper Shipping Name (PSN)	Class or division	Subsidiary risk(s)	Packing group	Special provisions	Limited and excepted quantity provisions		Packing		IBC	
						Limited quantities	Excepted quantities	Instructions	Provisions	Instructions	Provisions
(1)	(2) 3.1.2	(3) 2.0	(4) 2.0	(5) 2.0.1.3	(6) 3.3	(7a) 3.4	(7b) 3.5	(8) 4.1.4	(9) 4.1.4	(10) 4.1.4	(11) 4.1.4
2008	ZIRCONIUM POWDER, DRY	4.2	–	III	223	0	E1	P002 LP02	PP31	IBC08	B3
2009	ZIRCONIUM, DRY finished sheets, strip or coiled wire	4.2	–	III	223	0	E1	P002 LP02	PP31	–	–
2010	MAGNESIUM HYDRIDE	4.3	–	I	–	0	E0	P403	PP31	–	–
2011	MAGNESIUM PHOSPHIDE	4.3	6.1	I	–	0	E0	P403	PP31	–	–
2012	POTASSIUM PHOSPHIDE	4.3	6.1	I	–	0	E0	P403	PP31	–	–
2013	STRONTIUM PHOSPHIDE	4.3	6.1	I	–	0	E0	P403	PP31	–	–
△ 2014	HYDROGEN PEROXIDE, AQUEOUS SOLUTION with not less than 20% but not more than 60% hydrogen peroxide (stabilized as necessary)	5.1	8	II	–	1 L	E2	P504	PP10	IBC02	B5
△ 2015	HYDROGEN PEROXIDE, STABILIZED or HYDROGEN PEROXIDE, AQUEOUS SOLUTION, STABILIZED with more than 60% hydrogen peroxide	5.1	8	I	–	0	E0	P501	–	–	–
2016	AMMUNITION, TOXIC, NON-EXPLOSIVE without burster or expelling charge, non-fuzed	6.1	–	II	–	0	E0	P600	–	–	–
2017	AMMUNITION, TEAR-PRODUCING, NON-EXPLOSIVE without burster or expelling charge, non-fuzed	6.1	8	II	–	0	E0	P600	–	–	–
2018	CHLOROANILINES, SOLID	6.1	–	II	–	500 g	E4	P002	–	IBC08	B2 B4
2019	CHLOROANILINES, LIQUID	6.1	–	II	–	100 mL	E4	P001	–	IBC02	–
2020	CHLOROPHENOLS, SOLID	6.1	–	III	205	5 kg	E1	P002 LP02	–	IBC08	B3
2021	CHLOROPHENOLS, LIQUID	6.1	–	III	–	5 L	E1	P001 LP01	–	IBC03	–
2022	CRESYLIC ACID	6.1	8	II	–	100 mL	E4	P001	–	IBC02	–
2023	EPICHLOROHYDRIN	6.1	3 P	II	279	100 mL	E4	P001	–	IBC02	–

Portable tanks and bulk containers			EmS	Stowage and segregation	Properties and observations	UN No.
	Tank instructions	Provisions				
(12)	(13) 4.2.5 4.3	(14) 4.2.5	(15) 5.4.3.2 7.3	(16) 7.1 7.2	(17)	(18)
–	T1	TP33	F-G, S-M	Category D.	Amorphous powder. Liable to ignite spontaneously in air. Forms explosive mixtures with oxidizing substances.	2008
–	–	–	F-G, S-M	Category D.	Hard, silvery metal, liable to ignite spontaneously in air.	2009
–	–	–	F-G, S-O	Category E. "Separated from" acids.	White crystals. In contact with water, acids or moisture, evolves hydrogen, which may be ignited by the heat of the reaction.	2010
–	–	–	F-G, S-N	Category E. Under deck in a mechanically ventilated space. Clear of living quarters. "Separated from" acids.	Solid. Reacts with acids or decomposes slowly in contact with water or damp air, evolving phosphine, a spontaneously flammable and highly toxic gas. Reacts violently with oxidizing substances. Toxic if swallowed, by skin contact or by inhalation.	2011
–	–	–	F-G, S-N	Category E. Under deck in a mechanically ventilated space. Clear of living quarters. "Separated from" acids.	Solid. Reacts with acids or decomposes slowly in contact with water or damp air, evolving phosphine, a spontaneously flammable and highly toxic gas. Reacts violently with oxidizing substances. Toxic if swallowed, by skin contact or by inhalation.	2012
–	–	–	F-G, S-N	Category E. Under deck in a mechanically ventilated space. Clear of living quarters. "Separated from" acids.	Solid. Reacts with acids or decomposes slowly in contact with water or damp air, evolving phosphine, a spontaneously flammable and highly toxic gas. Reacts violently with oxidizing substances. Toxic if swallowed, by skin contact or by inhalation.	2013
–	T7	TP2 TP6 TP24	F-H, S-Q	Category D. Protected from sources of heat. "Separated from" permanganates and class 4.1. See 7.2.6.3.2.	Colourless liquid. Slowly decomposes, evolving oxygen; the rate of decomposition increases in contact with metals, except aluminium. In contact with combustible material, may cause fire or explosion. Causes burns to skin, eyes and mucous membranes. Even though stabilized, these solutions may evolve oxygen.	2014 △
–	T9	TP2 TP6 TP24	F-H, S-Q	Category D. Protected from sources of heat. "Separated from" permanganates and class 4.1.	Colourless liquid. Slowly decomposes, evolving oxygen; the rate of decomposition increases in contact with metals, except aluminium. Decomposes vigorously in contact with permanganates. When involved in a fire, mixtures with combustible material may be explosive. Causes burns to skin, eyes and mucous membranes. Even though stabilized, these solutions may evolve oxygen.	2015 △
–	–	–	F-A, S-A	Category E. Keep as dry as reasonably practicable. Clear of living quarters.	Contents may evolve toxic fumes or vapour. Gases evolved are toxic by skin contact or by inhalation.	2016
–	–	–	F-A, S-B	Category E. Keep as dry as reasonably practicable. Clear of living quarters.	Contents may evolve irritant gas or vapour with lachrymatory effects.	2017
–	T3	TP33	F-A, S-A	Category A.	Crystalline solid. Melting point of pure *para*-chloroaniline: 70°C approximately. Toxic if swallowed, by skin contact or by dust inhalation.	2018
–	T7	TP2	F-A, S-A	Category A. "Separated from" acids.	Colourless liquid. May be a mixture of two of the isomers (e.g., *ortho*- and *meta*-) of chloroaniline. Reacts with acids. Toxic if swallowed, by skin contact or by inhalation.	2019
–	T1	TP33	F-A, S-A	Category A.	A wide range of toxic solids. Toxic if swallowed, by skin contact or by dust inhalation.	2020
–	T4	TP1	F-A, S-A	Category A.	A wide range of toxic liquids. Toxic if swallowed, by skin contact or by inhalation.	2021
–	T7	TP2 TP13	F-A, S-B	Category B.	Colourless to brownish-yellow liquid mixture with a phenolic odour. Miscible with water. Toxic if swallowed, by skin contact or by inhalation. Causes burns to skin, eyes and mucous membranes. "Cresylic acid" is a generic name for mixtures of cresols and higher alkylphenols, in varying proportions. It generally contains more than 95% phenolic compounds.	2022
–	T7	TP2 TP13	F-E, S-D	Category A. Clear of living quarters.	Colourless flammable liquid with a chloroform-like odour. Flashpoint: approximately 32°C c.c. Toxic if swallowed, by skin contact or by inhalation.	2023

UN No.	Proper Shipping Name (PSN)	Class or division	Subsidiary risk(s)	Packing group	Special provisions	Limited and excepted quantity provisions		Packing		IBC	
						Limited quantities	Excepted quantities	Instructions	Provisions	Instructions	Provisions
(1)	(2) 3.1.2	(3) 2.0	(4) 2.0	(5) 2.0.1.3	(6) 3.3	(7a) 3.4	(7b) 3.5	(8) 4.1.4	(9) 4.1.4	(10) 4.1.4	(11) 4.1.4
2024	MERCURY COMPOUND, LIQUID, N.O.S.	6.1	– P	I	43 66 274	0	E5	P001	–	–	–
2024	MERCURY COMPOUND, LIQUID, N.O.S.	6.1	– P	II	43 66 274	100 mL	E4	P001	–	IBC02	–
2024	MERCURY COMPOUND, LIQUID, N.O.S.	6.1	– P	III	43 66 223 274	5 L	E1	P001 LP01	–	IBC03	–
2025	MERCURY COMPOUND, SOLID, N.O.S.	6.1	– P	I	43 66 274	0	E5	P002	–	IBC07	B1
2025	MERCURY COMPOUND, SOLID, N.O.S.	6.1	– P	II	43 66 274	500 g	E4	P002	–	IBC08	B2 B4
2025	MERCURY COMPOUND, SOLID, N.O.S.	6.1	– P	III	43 66 223 274	5 kg	E1	P002 LP02	–	IBC08	B3
2026	PHENYLMERCURIC COMPOUND, N.O.S.	6.1	– P	I	43 274	0	E5	P002	–	IBC07	B1
2026	PHENYLMERCURIC COMPOUND, N.O.S.	6.1	– P	II	43 274	500 g	E4	P002	–	IBC08	B2 B4
2026	PHENYLMERCURIC COMPOUND, N.O.S.	6.1	– P	III	43 223 274	5 kg	E1	P002 LP02	–	IBC08	B3
2027	SODIUM ARSENITE, SOLID	6.1	–	II	43	500 g	E4	P002	–	IBC08	B2 B4
2028	BOMBS, SMOKE, NON-EXPLOSIVE with corrosive liquid, without initiating device	8	–	II	–	0	E0	P803	–	–	–
2029	HYDRAZINE, ANHYDROUS	8	3/6.1	I	–	0	E0	P001	–	–	–
2030	HYDRAZINE, AQUEOUS SOLUTION with more than 37% hydrazine, by mass	8	6.1	I	–	0	E0	P001	–	–	–
2030	HYDRAZINE, AQUEOUS SOLUTION with more than 37% hydrazine, by mass	8	6.1	II	–	1 L	E2	P001	–	IBC02	–
2030	HYDRAZINE, AQUEOUS SOLUTION with more than 37% hydrazine, by mass	8	6.1	III	–	5 L	E1	P001 LP01	–	IBC03	–
2031	NITRIC ACID other than red fuming, with more than 70% nitric acid	8	5.1	I	–	0	E0	P001	PP81	–	–
2031	NITRIC ACID other than red fuming, with at least 65% but with not more than 70% nitric acid.	8	5.1	II	–	1 L	E2	P001	PP81	IBC02	B15 B20

Portable tanks and bulk containers			EmS	Stowage and segregation	Properties and observations	UN No.
	Tank instructions	Provisions				
(12)	(13) 4.2.5 4.3	(14) 4.2.5	(15) 5.4.3.2 7.3	(16) 7.1 7.2	(17)	(18)
–	–	–	F-A, S-A	Category B. Clear of living quarters.	Toxic if swallowed, by skin contact or by inhalation.	2024
–	–	–	F-A, S-A	Category B. Clear of living quarters.	See entry above.	2024
–	–	–	F-A, S-A	Category B. Clear of living quarters.	See entry above.	2024
–	T6	TP33	F-A, S-A	Category A.	Toxic if swallowed, by skin contact or by dust inhalation.	2025
–	T3	TP33	F-A, S-A	Category A.	See entry above.	2025
–	T1	TP33	F-A, S-A	Category A.	See entry above.	2025
–	T6	TP33	F-A, S-A	Category A.	Usually white crystals or powder. Toxic if swallowed, by skin contact or by dust inhalation.	2026
–	T3	TP33	F-A, S-A	Category A.	See entry above.	2026
–	T1	TP33	F-A, S-A	Category A.	See entry above.	2026
–	T3	TP33	F-A, S-A	Category A.	Greyish-white powder. Soluble in water. Reacts with oxidizing substances, evolving heat. Toxic if swallowed, by skin contact or by dust inhalation.	2027
–	–	–	F-A, S-B	Category E. Clear of living quarters.	Corrosive content evolves dense smoke when in contact with air. Corrosive content may cause acid burns to skin.	2028
–	–	–	F-E, <u>S-C</u>	Category D. Clear of living quarters. Segregation as for class 3, but "away from" class 4.1. "Separated from" acids.	Colourless, flammable liquid with an ammoniacal odour. Reacts violently with acids. Flashpoint: 52°C c.c. Miscible with water. Highly reactive reducing agent. Ignites spontaneously when in contact with porous materials such as earth, wood or cloth. Toxic if swallowed, by skin contact or by inhalation. Causes severe burns to skin, eyes and mucous membranes.	2029
–	T10	TP2 TP13	F-A, S-B	Category D. Clear of living quarters. "Separated from" acids.	Colourless liquid. Powerful reducing agent, burns readily. Toxic if swallowed, by skin contact or by inhalation. Causes burns to skin, eyes and mucous membranes. Reacts violently with acids.	2030
–	T7	TP2 TP13	F-A, S-B	Category D. Clear of living quarters. "Separated from" acids.	See entry above.	2030
–	T4	TP1	F-A, S-B	Category D. Clear of living quarters. "Separated from" acids.	See entry above.	2030
	T10	TP2 TP13	F-A, S-Q	Category D. Segregation as for class 5.1, but "separated from" classes 4.1, 5.1 and 7.	Colourless liquid. Powerful oxidant; may cause fire in contact with organic materials such as wood, cotton or straw, evolving highly toxic gases (brown fumes). Highly corrosive to most metals. Causes severe burns to skin, eyes and mucous membranes.	2031
–	T8	TP2	F-A, S-Q	Category D. Segregation as for class 5.1, but "separated from" classes 4.1, 5.1 and 7	Colourless liquid. Oxidant; may cause fire in contact with organic materials such as wood, cotton or straw, evolving highly toxic gases (brown fumes). Highly corrosive to most metals. Causes severe burns to skin, eyes and mucous membranes.	2031

UN No.	Proper Shipping Name (PSN)	Class or division	Subsidiary risk(s)	Packing group	Special provisions	Limited and excepted quantity provisions		Packing		IBC	
						Limited quantities	Excepted quantities	Instructions	Provisions	Instructions	Provisions
(1)	(2) 3.1.2	(3) 2.0	(4) 2.0	(5) 2.0.1.3	(6) 3.3	(7a) 3.4	(7b) 3.5	(8) 4.1.4	(9) 4.1.4	(10) 4.1.4	(11) 4.1.4
2031	NITRIC ACID other than red fuming, with less than 65% nitric acid	8	–	II	–	1 L	E2	P001	PP81	IBC02	B15 B20
2032	NITRIC ACID, RED FUMING	8	5.1/6.1	I	–	0	E0	P602	–	–	–
2033	POTASSIUM MONOXIDE	8	–	II	–	1 kg	E2	P002	–	IBC08	B2 B4
2034	HYDROGEN AND METHANE MIXTURE, COMPRESSED	2.1	–	–	–	0	E0	P200	–	–	–
2035	1,1,1-TRIFLUOROETHANE (REFRIGERANT GAS R 143a)	2.1	–	–	–	0	E0	P200	–	–	–
2036	XENON	2.2	–	–	–	120 mL	E1	P200	–	–	–
2037	RECEPTACLES, SMALL, CONTAINING GAS (GAS CARTRIDGES) without a release device, non-refillable	2	–	–	191 277 303 344	see SP277	E0	P003	PP17	–	–
2038	DINITROTOLUENES, LIQUID	6.1	–	II	–	100 mL	E4	P001	–	IBC02	B20
2044	2,2-DIMETHYLPROPANE	2.1	–	–	–	0	E0	P200	–	–	–
2045	ISOBUTYL ALDEHYDE (ISOBUTYRALDEHYDE)	3	–	II	–	1 L	E2	P001	–	IBC02	–
2046	CYMENES	3	– P	III	–	5 L	E1	P001 LP01	–	IBC03	–
2047	DICHLOROPROPENES	3	–	II	–	1 L	E2	P001	–	IBC02	–
2047	DICHLOROPROPENES	3	–	III	223	5 L	E1	P001 LP01	–	IBC03	–
2048	DICYCLOPENTADIENE	3	–	III	–	5 L	E1	P001 LP01	–	IBC03	–
△ 2049	DIETHYLBENZENE	3	–	III	–	5 L	E1	P001 LP01	–	IBC03	–
2050	DIISOBUTYLENES, ISOMERIC COMPOUNDS	3	–	II	–	1 L	E2	P001	–	IBC02	–
2051	2-DIMETHYLAMINOETHANOL	8	3	II	–	1 L	E2	P001	–	IBC02	–
2052	DIPENTENE	3	– P	III	–	5 L	E1	P001 LP01	–	IBC03	–
2053	METHYL ISOBUTYL CARBINOL	3	–	III	–	5 L	E1	P001 LP01	–	IBC03	–

Portable tanks and bulk containers			EmS	Stowage and segregation	Properties and observations	UN No.
	Tank instructions	Provisions				
(12)	(13) 4.2.5 4.3	(14) 4.2.5	(15) 5.4.3.2 7.3	(16) 7.1 7.2	(17)	(18)
–	T8	TP2	F-A, S-B	Category D.	Colourless liquid. Oxidant; may cause fire in contact with organic materials such as wood, cotton or straw, evolving highly toxic gases (brown fumes). Highly corrosive to most metals. Causes severe burns to skin, eyes and mucous membranes.	2031
–	T20	TP2 TP13	F-A, S-Q	Category D. Clear of living quarters. Segregation as for class 5.1, but "separated from" classes 4.1, 5.1 and 7.	Brown liquid. Powerful oxidant; may cause fire in contact with organic materials such as wood, cotton or straw. Highly corrosive to most metals. Toxic if swallowed, by skin contact or by vapour inhalation. Causes severe burns to skin, eyes and mucous membranes.	2032
–	T3	TP33	F-A, S-B	Category A. "Away from" ammonium salts. "Separated from" acids.	Deliquescent crystalline solid. Reacts violently with water, generating heat. Reacts with ammonium salts, evolving ammonia gas. Reacts violently with acids. In the presence of moisture, corrosive to aluminium, zinc and tin. Causes burns to skin, eyes and mucous membranes.	2033
–	–	–	F-D, S-U	Category E. Clear of living quarters. "Separated from" chlorine.	Flammable, odourless gas mixtures. Much lighter than air.	2034
–	T50	–	F-D, S-U	Category B. Clear of living quarters.	Flammable gas with a slight odour. Much heavier than air (2.9).	2035
–	–	–	F-C, S-V	Category A.	Liquefied, inert gas. Much heavier than air (4.5).	2036
–	–	–	F-D, S-U	Category B. Clear of living quarters.	Normally contain mixtures of liquefied butane and propane in various proportions for use in camping stoves, etc.	2037
–	T7	TP2	F-A, S-A	Category A.	Immiscible with water. A commercial grade consisting of a mixture of the 2,4-, 3,4- and 3,5-isomers is an oily liquid. Toxic if swallowed, by skin contact or by inhalation.	2038
–	–	–	F-D, S-U	Category E. Clear of living quarters.	Flammable hydrocarbon gas. Explosive limits: 1.4% to 7.2%. Heavier than air (2.48).	2044
–	T4	TP1	F-E, S-D	Category E. Clear of living quarters.	Colourless liquid with a characteristic pungent odour. Flashpoint: −24°C c.c. Explosive limits: 1% to 12%. Immiscible with water.	2045
–	T2	TP1	F-E, S-D	Category A.	Colourless liquids with an aromatic odour. Immiscible with water. Explosive limits: 0.7% to 5.6%.	2046
–	T4	TP1	F-E, S-D	Category B.	Colourless or yellow liquids with a sweet odour. Explosive limits: 5% to 14%. Immiscible with water. Irritating to skin, eyes and mucous membranes.	2047
–	T2	TP1	F-E, S-D	Category A.	See entry above.	2047
–	T2	TP1	F-E, S-D	Category A.	The pure substance is a solid with a melting point of 34°C. Flashpoint: 26°C to 38°C o.c. Commercial products are liquids. Immiscible with water. Harmful if swallowed.	2048
–	T2	TP1	F-E, S-D	Category A.	Colourless liquids. Flashpoint: 49°C to 56°C c.c. Immiscible with water. The commercial product is a mixture of isomers.	2049 △
–	T4	TP1	F-E, S-D	Category B.	Colourless liquids. Flashpoint: −18°C to 21°C c.c. Explosive limits: 0.8% to 4.8%. Immiscible with water.	2050
–	T7	TP2	F-E, S-C	Category A.	Colourless, flammable liquid with a fishy odour. Flashpoint: 31°C o.c. Miscible with water. Causes burns to skin, eyes and mucous membranes.	2051
–	T2	TP1	F-E, S-E	Category A.	Colourless liquid with a lemon-like odour. Flashpoint: 43°C c.c. Explosive limits: 0.7% to 6.1%. Immiscible with water.	2052
–	T2	TP1	F-E, S-D	Category A.	Colourless liquid. Flashpoint: 41°C c.c. Explosive limits: 1% to 5.5%. Miscible with water. Harmful by inhalation.	2053

3

DGL

UN No.	Proper Shipping Name (PSN)	Class or division	Subsidiary risk(s)	Packing group	Special provisions	Limited and excepted quantity provisions		Packing		IBC	
						Limited quantities	Excepted quantities	Instructions	Provisions	Instructions	Provisions
(1)	(2) 3.1.2	(3) 2.0	(4) 2.0	(5) 2.0.1.3	(6) 3.3	(7a) 3.4	(7b) 3.5	(8) 4.1.4	(9) 4.1.4	(10) 4.1.4	(11) 4.1.4
2054	MORPHOLINE	8	3	I	–	0	E0	P001	–	–	–
2055	STYRENE MONOMER, STABILIZED	3	–	III	–	5 L	E1	P001	–	IBC03	–
2056	TETRAHYDROFURAN	3	–	II	–	1 L	E2	P001	–	IBC02	–
2057	TRIPROPYLENE	3	–	II	–	1 L	E2	P001	–	IBC02	–
2057	TRIPROPYLENE	3	–	III	223	5 L	E1	P001 LP01	–	IBC03	–
2058	VALERALDEHYDE	3	–	II	–	1 L	E2	P001	–	IBC02	–
2059	NITROCELLULOSE SOLUTION, FLAMMABLE with not more than 12.6% nitrogen, by dry mass, and not more than 55% nitrocellulose	3	–	I	198	0	E0	P001	–	–	–
2059	NITROCELLULOSE SOLUTION, FLAMMABLE with not more than 12.6% nitrogen, by dry mass, and not more than 55% nitrocellulose	3	–	II	198	1 L	E0	P001	–	IBC02	–
2059	NITROCELLULOSE SOLUTION, FLAMMABLE with not more than 12.6% nitrogen, by dry mass, and not more than 55% nitrocellulose	3	–	III	198 223	5 L	E0	P001 LP01	–	IBC03	–
△ 2067	AMMONIUM NITRATE BASED FERTILIZER	5.1	–	III	186 306 307 900 967	5 kg	E1	P002 LP02	–	IBC08	B3
△ 2071	AMMONIUM NITRATE BASED FERTILIZER	9	–	III	186 193	5 kg	E1	P002 LP02	–	IBC08	B3
2073	AMMONIA SOLUTION relative density less than 0.880 at 15°C in water, with more than 35% but not more than 50% ammonia	2.2	–	–	–	120 mL	E1	P200	–	–	–
2074	ACRYLAMIDE, SOLID	6.1	–	III	–	5 kg	E1	P002 LP02	–	IBC08	B3
2075	CHLORAL, ANHYDROUS, STABILIZED	6.1	–	II	–	100 mL	E4	P001	–	IBC02	–

Portable tanks and bulk containers			EmS	Stowage and segregation	Properties and observations	UN No.	
	Tank instructions	Provisions					
(12)	(13) 4.2.5 4.3	(14) 4.2.5	(15) 5.4.3.2 7.3	(16) 7.1 7.2	(17)	(18)	
–	T10	TP2	F-E, S-C	Category A.	Colourless liquid with a fishy odour. Flashpoint: 38°C o.c. Explosive limits: 2% to 11.2%. Miscible with water. Harmful by skin contact or by inhalation. Corrosive to skin, eyes and mucous membranes.	2054	
–	T2	TP1	F-E, S-D	Category A.	Colourless, oily liquid. Flashpoint: 32°C c.c. Explosive limits: 1.1% to 6.1%. Immiscible with water. Irritating to skin, eyes and mucous membranes.	2055	
–	T4	TP1	F-E, S-D	Category B.	Colourless liquid with an ethereal odour. Flashpoint: below −18°C c.c. Explosive limits: 1.5% to 12%. Miscible with water.	2056	
–	T4	TP1	F-E, S-D	Category B.	Colourless liquid. Immiscible with water.	2057	
–	T2	TP1	F-E, S-D	Category A.	See entry above.	2057	
–	T4	TP1	F-E, S-D	Category B.	Colourless liquid. Flashpoint: 12°C c.c. Partially miscible with water. Irritating to skin, eyes and mucous membranes.	2058	
–	T11	TP1 TP8 TP27	F-E, S-D	Category E.	When involved in a fire, evolves toxic nitrous fumes.	2059	
–	T4	TP1 TP8	F-E, S-D	Category B.	See entry above.	2059	
–	T2	TP1	F-E, S-D	Category A.	See entry above.	2059	
–	T1 BK2 BK3	TP33	F-H, S-Q	Category C. Category A only if the special stowage provisions of 7.4.1.4 and 7.6.2.8.4 are complied with. "Separated from" class 4.1, combustible material (particularly liquids), bromates, chlorates, chlorites, hypochlorites, nitrites, perchlorates, permanganates and powdered metals. Protected from sources of heat. When transported in BK3 bulk container, see 7.6.2.12 and 7.7.3.9.	Crystals, granules or prills. Wholly or partly soluble in water. Supporters of combustion. A major fire aboard a ship carrying these substances may involve a risk of explosion in the event of contamination (e.g., by fuel oil) or strong confinement. An adjacent detonation may also involve a risk of explosion. If heated strongly, decompose, giving off toxic gases and gases which support combustion. Transport of AMMONIUM NITRATE liable to self-heating sufficient to initiate decomposition is **prohibited**.	2067	△
–	BK2	–	F-H, S-Q	Category A. For special stowage provisions, see 7.4.1.4 and 7.6.2.11.1.1.	Usually granules. Wholly or partly soluble in water. These mixtures may be subject to self-sustaining decomposition if heated. The temperature in such a reaction can reach 500°C. Decomposition, once initiated, may spread throughout the remainder, producing gases which are toxic. None of these mixtures is subject to the explosion hazard. Transport of AMMONIUM NITRATE liable to self-heating sufficient to initiate decomposition is **prohibited**.	2071	△
–	–	–	F-C, S-U	Category E. Clear of living quarters. "Separated from" chlorine. "Separated from" acids.	Solution in water of non-flammable gas with a pungent odour. Reacts violently with acids. Extremely dangerous to the eyes.	2073	
–	T1	TP33	F-A, S-A	Category A. Keep as cool as reasonably practicable.	Crystals or powder. Soluble in water. May polymerise violently on melting. Toxic if swallowed, by skin contact or by inhalation.	2074	
–	T7	TP2	F-A, S-A	Category D. Clear of living quarters.	Colourless, mobile liquid, evolving toxic vapours which are considerably heavier than air. Toxic if swallowed, by skin contact or by inhalation.	2075	

3

DGL

UN No.	Proper Shipping Name (PSN)	Class or division	Subsidiary risk(s)	Packing group	Special provisions	Limited and excepted quantity provisions		Packing		IBC	
						Limited quantities	Excepted quantities	Instructions	Provisions	Instructions	Provisions
(1)	(2) 3.1.2	(3) 2.0	(4) 2.0	(5) 2.0.1.3	(6) 3.3	(7a) 3.4	(7b) 3.5	(8) 4.1.4	(9) 4.1.4	(10) 4.1.4	(11) 4.1.4
2076	CRESOLS, LIQUID	6.1	8	II	–	100 mL	E4	P001	–	IBC02	–
2077	*alpha*-NAPHTHYLAMINE	6.1	–	III	–	5 kg	E1	P002 LP02	–	IBC08	B3
△ 2078	TOLUENE DIISOCYANATE	6.1	–	II	279	100 mL	E4	P001	–	IBC02	–
2079	DIETHYLENETRIAMINE	8	–	II	–	1 L	E2	P001	–	IBC02	–
△ 2186	HYDROGEN CHLORIDE, REFRIGERATED LIQUID	2.3	8	–	900	–	–	–	–	–	–
2187	CARBON DIOXIDE, REFRIGERATED LIQUID	2.2	–	–	–	120 mL	E1	P203	–	–	–
2188	ARSINE	2.3	2.1	–	–	0	E0	P200	–	–	–
△ 2189	DICHLOROSILANE	2.3	2.1/8	–	–	0	E0	P200	–	–	–
2190	OXYGEN DIFLUORIDE, COMPRESSED	2.3	5.1/8	–	–	0	E0	P200	–	–	–
2191	SULPHURYL FLUORIDE	2.3	–	–	–	0	E0	P200	–	–	–
2192	GERMANE	2.3	2.1	–	–	0	E0	P200	–	–	–
2193	HEXAFLUOROETHANE (REFRIGERANT GAS R 116)	2.2	–	–	–	120 mL	E1	P200	–	–	–
2194	SELENIUM HEXAFLUORIDE	2.3	8	–	–	0	E0	P200	–	–	–
2195	TELLURIUM HEXAFLUORIDE	2.3	8	–	–	0	E0	P200	–	–	–
2196	TUNGSTEN HEXAFLUORIDE	2.3	8	–	–	0	E0	P200	–	–	–
2197	HYDROGEN IODIDE, ANHYDROUS	2.3	8	–	–	0	E0	P200	–	–	–

Portable tanks and bulk containers			EmS	Stowage and segregation	Properties and observations	UN No.
	Tank instructions	Provisions				
(12)	(13) 4.2.5 4.3	(14) 4.2.5	(15) 5.4.3.2 7.3	(16) 7.1 7.2	(17)	(18)
–	T7	TP2	F-A, S-B	Category B.	Colourless to light yellow liquids. Miscible with water. Melting point of *meta*-CRESOL: 12°C. Toxic if swallowed, by skin contact or by inhalation. Cause burns to skin, eyes and mucous membranes.	2076
–	T1	TP33	F-A, S-A	Category A.	White crystals. Toxic if swallowed, by skin contact or by inhalation.	2077
–	T7	TP2 TP13	F-A, S-A	Category C. Protected from sources of heat. Clear of living quarters.	Colourless to pale yellow liquid with a pungent odour. Immiscible with water but reacts with it to form carbon dioxide. Melting point: 20°C (pure product). Toxic if swallowed, by skin contact or by inhalation. Irritating to skin, eyes and mucous membranes.	2078 △
–	T7	TP2	F-A, S-B	Category A. Clear of living quarters. "Separated from" acids.	Yellow hygroscopic liquid with ammoniacal odour. Soluble in water. Strongly alkaline, corrosive. Can form explosive mixtures with nitric acid. Reacts with oxidizing substances. Corrosive to copper and its alloys. Reacts violently with acids. Liquid and vapour can cause severe damage to skin and eyes.	2079
–	–	–	–	–	Transport is **prohibited**.	2186 △
–	T75	TP5	F-C, S-V	Category D.	Non-flammable, liquefied gas, colourless and odourless. Heavier than air (1.5). Cannot remain in the liquid state above 31°C.	2187
–	–	–	F-D, S-U	Category D. Clear of living quarters.	Flammable, toxic, colourless gas with a garlic odour. Explosive limits: 3.9% to 77.8%. Much heavier than air (2.8).	2188
–	–	–	F-D, S-U	Category D. Clear of living quarters. Segregation as for class 2.1 but "away from" class 4.3. See 7.2.6.3.2.	Flammable, toxic and corrosive gas. Reacts with water, evolving hydrogen chloride. Highly irritating to skin, eyes and mucous membranes.	2189 △
–	–	–	F-C, S-W	Category D. Keep as dry as reasonably practicable. Clear of living quarters. Segregation as for class 5.1 but "separated from" class 7.	Non-flammable, toxic and corrosive, colourless gas with a foul odour. Strong oxidizing agent. Reacts slowly with water or moist air to produce poisonous and corrosive fumes. Corrosive to glass and to most metals. Heavier than air (1.9). Highly irritating to skin, eyes and mucous membranes.	2190
–	–	–	F-C, S-U	Category D. Clear of living quarters.	Non-flammable, toxic, colourless, odourless gas. Reacts with water or moist air to produce toxic and corrosive fumes. Much heavier than air (3.5). Irritating to skin, eyes and mucous membranes.	2191
–	–	–	F-D, S-U	Category D. Clear of living quarters.	Flammable, toxic, colourless gas with a pungent odour. Much heavier than air (2.6).	2192
–	–	–	F-C, S-V	Category A.	Non-flammable, colourless and odourless gas. Much heavier than air (4.8). Cannot remain in liquid state above 24.3°C.	2193
–	–	–	F-C, S-U	Category D. Clear of living quarters.	Colourless, toxic and corrosive gas. Corrosive to glass and to most metals. Heavier than air. Highly irritating to skin, eyes and mucous membranes.	2194
–	–	–	F-C, S-U	Category D. Clear of living quarters.	Non-flammable, toxic and corrosive colourless gas with an unpleasant odour. Decomposes in water, evolving highly toxic and corrosive fumes. Corrosive to glass and to most metals. Much heavier than air (7.2). Highly irritating to skin, eyes and mucous membranes.	2195
–	–	–	F-C, S-U	Category D. Clear of living quarters.	Non-flammable, toxic and corrosive, colourless gas, or yellow liquid. Decomposes in water or moist air, evolving highly toxic and corrosive fumes. Corrosive to glass and to most metals. Much heavier than air (10.3). Boiling point: 19.5°C. Highly irritating to skin, eyes and mucous membranes.	2196
–	–	–	F-C, S-U	Category D. Clear of living quarters.	Non-flammable, toxic and corrosive colourless gas with a pungent odour. Highly corrosive in the presence of water. Much heavier than air (4.4). Highly irritating to skin, eyes and mucous membranes.	2197

UN No.	Proper Shipping Name (PSN)	Class or division	Subsidiary risk(s)	Packing group	Special provisions	Limited and excepted quantity provisions		Packing		IBC	
						Limited quantities	Excepted quantities	Instructions	Provisions	Instructions	Provisions
(1)	(2) 3.1.2	(3) 2.0	(4) 2.0	(5) 2.0.1.3	(6) 3.3	(7a) 3.4	(7b) 3.5	(8) 4.1.4	(9) 4.1.4	(10) 4.1.4	(11) 4.1.4
2198	PHOSPHORUS PENTAFLUORIDE	2.3	8	–	–	0	E0	P200	–	–	–
2199	PHOSPHINE	2.3	2.1	–	–	0	E0	P200	–	–	–
2200	PROPADIENE, STABILIZED	2.1	–	–	–	0	E0	P200	–	–	–
2201	NITROUS OXIDE, REFRIGERATED LIQUID	2.2	5.1	–	–	0	E0	P203	–	–	–
2202	HYDROGEN SELENIDE, ANHYDROUS	2.3	2.1	–	–	0	E0	P200	–	–	–
2203	SILANE	2.1	–	–	–	0	E0	P200	–	–	–
2204	CARBONYL SULPHIDE	2.3	2.1	–	–	0	E0	P200	–	–	–
2205	ADIPONITRILE	6.1	–	III	–	5 L	E1	P001 LP01	–	IBC03	–
△ 2206	ISOCYANATES, TOXIC, N.O.S. or ISOCYANATE SOLUTION, TOXIC, N.O.S.	6.1	–	II	274	100 mL	E4	P001	–	IBC02	–
△ 2206	ISOCYANATES, TOXIC, N.O.S. or ISOCYANATE SOLUTION, TOXIC, N.O.S.	6.1	–	III	223 274	5 L	E1	P001 LP01	–	IBC03	–
△ 2208	CALCIUM HYPOCHLORITE MIXTURE, DRY with more than 10% but not more than 39% available chlorine	5.1	–	III	314	5 kg	E1	P002	PP85	–	–
2209	FORMALDEHYDE SOLUTION with not less than 25% formaldehyde	8	–	III	–	5 L	E1	P001 LP01	–	IBC03	–
△ 2210	MANEB or MANEB PREPARATION with not less than 60% maneb	4.2	4.3 P	III	273	0	E1	P002	–	IBC06	–

Portable tanks and bulk containers			EmS	Stowage and segregation	Properties and observations	UN No.	
	Tank instructions	Provisions					
(12)	(13) 4.2.5 4.3	(14) 4.2.5	(15) 5.4.3.2 7.3	(16) 7.1 7.2	(17)	(18)	
–	–	–	F-C, S-U	Category D. Clear of living quarters.	Non-flammable, toxic and corrosive gas with an irritating odour. Reacts with water or moist air to produce toxic and corrosive fumes. Corrosive to glass and to most metals. Much heavier than air (4.3). Highly irritating to skin, eyes and mucous membranes.	2198	
–	–	–	F-D, S-U	Category D. Clear of living quarters.	Flammable, toxic, colourless gas with a garlic odour. Ignites spontaneously in air. Heavier than air (1.2). Irritating to skin, eyes and mucous membranes.	2199	
–	–	–	F-D, S-U	Category B. Clear of living quarters.	Liquefied, flammable, colourless gas. Explosive limits: 1.7% to 12%. Heavier than air (1.4). Boiling point: −34°C. Irritating to skin, eyes and mucous membranes.	2200	
–	T75	TP5 TP22	F-C, S-W	Category D. Clear of living quarters.	Liquefied, non-flammable, colourless gas with a slightly sweet odour. Strong oxidizing agent. Heavier than air (1.5). Cannot remain in liquid state above 36.5°C.	2201	
–	–	–	F-D, S-U	Category D. Clear of living quarters.	Flammable, toxic, colourless gas with a disagreeable odour. Much heavier than air (2.8). Highly irritating to skin, eyes and mucous membranes.	2202	
–	–	–	F-D, S-U	Category E. Clear of living quarters. "Separated from" bromine and chlorine.	Flammable, colourless gas with a foul odour. Explosive limits: 1% to 100%. Ignites spontaneously in air. Strong reducing agent which reacts violently with oxidizing substances. Heavier than air (1.1).	2203	
–	–	–	F-D, S-U	Category D. Clear of living quarters.	Flammable, toxic, colourless gas with a foul odour. Much heavier than air (2.1).	2204	
–	T3	TP1	F-A, S-A	Category A.	Colourless, odourless oil. Decomposes above 93°C, evolving hydrogen cyanide, a highly toxic and flammable gas. Toxic if swallowed, by skin contact or by inhalation.	2205	
–	T11	TP2 TP13 TP27	F-A, S-A	Category E. Protected from sources of heat. Clear of living quarters.	Liquids with a pungent odour. Immiscible with water but react with it to form carbon dioxide. Toxic if swallowed, by skin contact or by inhalation. If under deck, with mechanical ventilation, six air changes per hour, except when carried in closed containers, when two air changes per hour are required. Irritating to skin, eyes and mucous membranes.	2206	△
–	T7	TP1 TP13 TP28	F-A, S-A	Category E. Protected from sources of heat. Clear of living quarters.	See entry above.	2206	△
–	–	–	F-H, S-Q	Category D. Protected from sources of heat. Cargo transport units shall be shaded from direct sunlight. Packages in cargo transport units shall be stowed so as to allow for adequate air circulation throughout the cargo. "Separated from" ammonium compounds, acids, cyanides, hydrogen peroxides and liquid organic substances.	White or yellowish solid (powder, granules or tablets) with chlorine-like odour. Soluble in water. May cause fire in contact with organic material or ammonium compounds. Substances are liable to exothermic decomposition at elevated temperatures. This condition may lead to fire or explosion. Decomposition can be initiated by heat or by impurities (e.g., powdered metals (iron, manganese, cobalt, magnesium) and their compounds). Liable to heat slowly. Reacts with acids, evolving chlorine, an irritating, corrosive and toxic gas. In the presence of moisture, corrosive to most metals. Dust irritates mucous membranes.	2208	△
–	T4	TP1	F-A, S-B	Category A.	Colourless, clear liquid, with a suffocating pungent odour. Usually stabilized with methyl alcohol. Miscible with water. Causes burns to skin, eyes and mucous membranes.	2209	
–	T1	TP33	F-G, S-L	Category A. Segregation from foodstuffs as in 7.3.4.2.2, 7.6.3.1.2 or 7.7.3.7.	Yellow powder, liable to heat and to ignite spontaneously in air. May evolve toxic, irritating or flammable fumes when wet, when involved in a fire or in contact with acids. Used as fungicide.	2210	△

	UN No.	Proper Shipping Name (PSN)	Class or division	Subsidiary risk(s)	Packing group	Special provisions	Limited and excepted quantity provisions		Packing		IBC	
							Limited quantities	Excepted quantities	Instructions	Provisions	Instructions	Provisions
	(1)	(2) 3.1.2	(3) 2.0	(4) 2.0	(5) 2.0.1.3	(6) 3.3	(7a) 3.4	(7b) 3.5	(8) 4.1.4	(9) 4.1.4	(10) 4.1.4	(11) 4.1.4
△	2211	POLYMERIC BEADS, EXPANDABLE evolving flammable vapour	9	–	III	207 965	5 kg	E1	P002	PP14	IBC08	B3 B6
△	2212	BLUE ASBESTOS (crocidolite) or BROWN ASBESTOS (amosite, mysorite)	9	–	II	168	1 kg	E2	P002	PP37	IBC08	B2 B4
△	2213	PARAFORMALDEHYDE	4.1	–	III	967	5 kg	E1	P002 LP02	PP12	IBC08	B3
	2214	PHTHALIC ANHYDRIDE with more than 0.05% of maleic anhydride	8	–	III	169 939	5 kg	E1	P002 LP02	–	IBC08	B3
△	2215	MALEIC ANHYDRIDE	8	–	III	–	5 kg	E1	P002	–	IBC08	B3
△	2215	MALEIC ANHYDRIDE, MOLTEN	8	–	III	–	0	E0	–	–	–	–
△	2216	FISHMEAL (FISHSCRAP), STABILIZED Anti-oxidant treated. Moisture content greater than 5% but not exceeding 12%, by mass. Fat content not more than 15%	9	–	III	29 117 300 308 907 928 945	0	E1	P900	–	IBC08	B3

Portable tanks and bulk containers			EmS	Stowage and segregation	Properties and observations	UN No.	
	Tank instructions	Provisions					
(12)	(13) 4.2.5 4.3	(14) 4.2.5	(15) 5.4.3.2 7.3	(16) 7.1 7.2	(17)	(18)	
–	T1 BK2	TP33	F-A, S-I	Category E. Protected from sources of heat. When stowed under-deck, mechanical ventilation shall be in accordance with SOLAS regulation II-2/19 (II-2/54) for flammable liquids with flashpoint below 23°C c.c. Segregation as for class 3 but "separated from" class 1 except division 1.4S.	A moulding material in bead or granular form consisting predominantly of polystyrene, poly(methyl methacrylate) or other polymeric material and containing 5% to 8% of a volatile hydrocarbon which is predominantly pentane. During storage, a small proportion of this pentane is released to the atmosphere; this proportion increases at elevated temperatures.	2211	△
–	T3	TP33	F-A, S-A	Category A. Clear of living quarters. Segregation from foodstuffs as in 7.3.4.2.2, 7.6.3.1.2 or 7.7.3.7.	Mineral fibres of varying length. Non-combustible. Inhalation of the dust of asbestos fibres is dangerous and therefore exposure should be avoided at all times. Always prevent the generation of asbestos dust. Crocidolite (blue asbestos) should be regarded as the most hazardous type of asbestos. A safe level of airborne concentration of asbestos fibres may be obtained through effective packing. Cargo spaces or freight containers that have contained any type of raw asbestos should be carefully cleaned before discharging any remaining cargo, loading other cargo or carrying out repair or maintenance work. Whenever possible, cleaning of cargo spaces should be carried out whilst the ship is in a port where proper facilities and equipment, including proper respiratory apparatus and protective clothing, is available. Parts of the body that may have been exposed should be immediately and thoroughly washed. All waste material should be collected in impermeable and sealed bags for safe disposal ashore. If cleaning cannot be carried out at the discharge port, arrangements should be made in advance for cleaning to be carried out at the next port where necessary facilities are available. If cleaning of cargo spaces must be carried out at sea, the safety procedures followed and standard of equipment used must be at least as effective as those which would be employed in a port. Until such cleaning is undertaken, the cargo spaces in which the asbestos has been carried should be closed and access to those spaces should be prohibited.	2212	△
–	T1 BK2 BK3	TP33	F-A, S-G	Category A. When transported in BK3 bulk container, see 7.6.2.12 and 7.7.3.9.	White powder with a pungent odour. Evolves formaldehyde, particularly when heated, which is irritating to eyes and mucous membranes.	2213	△
–	T1	TP33	F-A, S-B	Category A.	White powder or flakes and lumps containing a high proportion of dust. Melting point: 131°C. The vapour of the molten substance has a flashpoint of 152°C c.c. and forms a flammable atmosphere with explosive limits of 1.7% to 10.4%. Causes burns to skin, eyes and mucous membranes. May be carried in the molten state. The molten substance can cause severe skin burns.	2214	
–	T1	TP33	F-A, S-B	Category A. Segregation from foodstuffs as in 7.3.4.2.1, 7.6.3.1.2 or 7.7.3.6. "Separated from" odour-absorbing cargoes.	White powder, needles, flakes, pellets, rods, briquettes, lumps or fused mass. Melting point: about 53°C. Fumes and dust are irritating to skin, eyes and mucous membranes. Inhalation can cause respiratory trouble.	2215	△
–	T4	TP3	F-A, S-B	Category A. Segregation from foodstuffs as in 7.3.4.2.1, 7.6.3.1.2 or 7.7.3.6. "Separated from" odour-absorbing cargoes.	Melting point: about 53°C. The vapour of the molten substance has a flashpoint of 103°C c.c. and forms a flammable atmosphere with explosive limits of 1.4% to 7.1%. Fumes are irritating to skin, eyes and mucous membranes.	2215	△
–	T1 BK2	TP33	F-A, S-J	Category B. "Separated from" class 6.2. "Separated by a complete compartment or hold from" class 1 except division 1.4. For special stowage provisions, see 7.4.1.3 and 7.6.2.7.2.	Brown to greenish-brown product obtained through heating and drying of oily fish. Strong odour which may affect other cargo. Liable to heat spontaneously unless of low fat content or effectively anti-oxidant treated.	2216	△

UN No.	Proper Shipping Name (PSN)	Class or division	Subsidiary risk(s)	Packing group	Special provisions	Limited and excepted quantity provisions		Packing		IBC	
						Limited quantities	Excepted quantities	Instructions	Provisions	Instructions	Provisions
(1)	(2) 3.1.2	(3) 2.0	(4) 2.0	(5) 2.0.1.3	(6) 3.3	(7a) 3.4	(7b) 3.5	(8) 4.1.4	(9) 4.1.4	(10) 4.1.4	(11) 4.1.4
2217	SEED CAKE with not more than 1.5% oil and not more than 11% moisture	4.2	–	III	29 117 142	0	E1	P002 LP02	PP20	IBC08	B3 B6
△ 2218	ACRYLIC ACID, STABILIZED	8	3	II	–	1 L	E2	P001	–	IBC02	–
2219	ALLYL GLYCIDYL ETHER	3	–	III	–	5 L	E1	P001 LP01	–	IBC03	–
2222	ANISOLE	3	–	III	–	5 L	E1	P001 LP01	–	IBC03	–
2224	BENZONITRILE	6.1	–	II	–	100 mL	E4	P001	–	IBC02	–
2225	BENZENESULPHONYL CHLORIDE	8	–	III	–	5 L	E1	P001 LP01	–	IBC03	–
2226	BENZOTRICHLORIDE	8	–	II	–	1 L	E2	P001	–	IBC02	–
2227	*n*-BUTYL METHACRYLATE, STABILIZED	3	–	III	–	5 L	E1	P001 LP01	–	IBC03	–
2232	2-CHLOROETHANAL	6.1	–	I	354	0	E0	P602	–	–	–
2233	CHLOROANISIDINES	6.1	–	III	–	5 kg	E1	P002 LP02	–	IBC08	B3
2234	CHLOROBENZOTRIFLUORIDES	3	–	III	–	5 L	E1	P001 LP01	–	IBC03	–
2235	CHLOROBENZYL CHLORIDES, LIQUID	6.1	– P	III	–	5 L	E1	P001 LP01	–	IBC03	–
2236	3-CHLORO-4-METHYLPHENYL ISOCYANATE, LIQUID	6.1	–	II	–	100 mL	E4	P001	–	IBC02	–
2237	CHLORONITROANILINES	6.1	– P	III	–	5 kg	E1	P002 LP02	–	IBC08	B3

Portable tanks and bulk containers			EmS	Stowage and segregation	Properties and observations	UN No.
	Tank instructions	Provisions				
(12)	(13) 4.2.5 4.3	(14) 4.2.5	(15) 5.4.3.2 7.3	(16) 7.1 7.2	(17)	(18)
–	BK2	–	F-A, S-J	Category A. Keep dry. Surface ventilation is required to assist in removing any residual solvent vapour. The cargo shall be stowed "away from" pipes and bulkheads which are liable to become heated (e.g., engine-room bulkhead).	Residue remaining after oil has been extracted by a solvent process from oil-bearing seeds. Used mainly as an animal feed or fertilizer. The most common seed cakes include those derived from coconut (copra), cottonseed, groundnut (peanut), linseed, maize (hominy chop), niger seed, palm kernel, rape seed, rice bran, soya bean and sunflower seed and they may be shipped in the form of cake, flakes, pellets, meal, etc. May self-heat slowly if wet and ignite spontaneously. Before shipment, this cargo should be properly aged. The duration of ageing varies with the oil content. The seed cake should be substantially free from flammable solvent. Smoking and the use of naked lights should not be allowed during loading and unloading, and on entry to cargo spaces at any other time.	2217
–	T7	TP2	F-E S-C	Category C. Protected from sources of heat. Clear of living quarters.	Colourless, flammable liquid with an acrid odour. Melting point: 13°C. Flashpoint: 54°C o.c. Miscible with water. May polymerize violently, which may cause fire and explosion, unless properly stabilized. Harmful if swallowed or by inhalation. Corrosive to skin, eyes and mucous membranes.	2218 △
–	T2	TP1	F-E, S-D	Category A.	Colourless liquid. Flashpoint: 48°C c.c. Miscible with water. Harmful by inhalation. Irritating to skin, eyes and mucous membranes.	2219
–	T2	TP1	F-E, S-D	Category A.	Colourless to yellow liquid. Flashpoint: 41°C c.c. Explosive limits: 0.8% to 6.3%. Immiscible with water. Irritating to skin, eyes and mucous membranes.	2222
–	T7	TP2	F-A, S-A	Category A. Clear of living quarters. "Separated from" acids.	Colourless liquid with an odour similar to oil of bitter almonds. Reacts with acids, evolving hydrogen cyanide, a highly toxic and flammable gas. Toxic if swallowed, by skin contact or by inhalation.	2224
–	T4	TP1	F-A, S-B	Category A. Clear of living quarters.	Colourless to slightly yellow liquid with a pungent odour. Melting point: 12°C. Immiscible with water. Decomposes slowly in water. Harmful if swallowed or by skin contact. Highly irritating to skin, eyes and mucous membranes.	2225
–	T7	TP2	F-A, S-B	Category A. Clear of living quarters.	Colourless to slightly yellow or brown fuming liquid. Reacts with water, evolving hydrogen chloride, an irritating and corrosive gas apparent as white fumes. In the presence of moisture, corrosive to most metals. Harmful if swallowed, by skin contact or by inhalation. Burns skin and eyes. Vapour irritates eyes and mucous membranes.	2226
–	T2	TP1	F-E, S-D	Category A.	Colourless liquid. Flashpoint: 41°C c.c. Explosive limits: 2% to 8%. Immiscible with water. Irritating to skin, eyes and mucous membranes.	2227
–	T20	TP2 TP13 TP37	F-A, S-A	Category D. Clear of living quarters.	Clear colourless liquid with a pungent odour. Miscible with water. Highly toxic if swallowed, by skin contact or by inhalation.	2232
–	T1	TP33	F-A, S-A	Category A.	Crystalline solid. Melting point: 52°C. Soluble in water. Toxic if swallowed, by skin contact or by dust inhalation.	2233
–	T2	TP1	F-E, S-D	Category A. Clear of living quarters.	Colourless liquids with an aromatic odour. Flashpoint: 33°C to 59°C c.c. On contact with moisture, can evolve hydrogen fluoride, which is a toxic and corrosive gas. Harmful by inhalation.	2234
–	T4	TP1	F-A, S-A	Category A.	Colourless liquid. Immiscible with water. Toxic if swallowed, by skin contact or by inhalation.	2235
–	–	–	F-A, S-A	Category B. Clear of living quarters.	Colourless liquid with a pungent odour. Immiscible with water. Reacts with water, evolving carbon dioxide. Toxic if swallowed, by skin contact or by inhalation. Irritating to skin, eyes and mucous membranes.	2236
–	T1	TP33	F-A, S-A	Category A.	Yellow or orange crystalline powders or needles. Insoluble in water. Toxic if swallowed, by skin contact or by dust inhalation.	2237

3 DGL

UN No.	Proper Shipping Name (PSN)	Class or division	Subsidiary risk(s)	Packing group	Special provisions	Limited and excepted quantity provisions		Packing		IBC	
						Limited quantities	Excepted quantities	Instructions	Provisions	Instructions	Provisions
(1)	(2) 3.1.2	(3) 2.0	(4) 2.0	(5) 2.0.1.3	(6) 3.3	(7a) 3.4	(7b) 3.5	(8) 4.1.4	(9) 4.1.4	(10) 4.1.4	(11) 4.1.4
2238	CHLOROTOLUENES	3	–	III	–	5 L	E1	P001 LP01	–	IBC03	–
2239	CHLOROTOLUIDINES, SOLID	6.1	–	III	–	5 kg	E1	P002 LP02	–	IBC08	B3
2240	CHROMOSULPHURIC ACID	8	–	I	–	0	E0	P001	–	–	–
2241	CYCLOHEPTANE	3	–	II	–	1 L	E2	P001	–	IBC02	–
2242	CYCLOHEPTENE	3	–	II	–	1 L	E2	P001	–	IBC02	–
2243	CYCLOHEXYL ACETATE	3	–	III	–	5 L	E1	P001 LP01	–	IBC03	–
2244	CYCLOPENTANOL	3	–	III	–	5 L	E1	P001 LP01	–	IBC03	–
2245	CYCLOPENTANONE	3	–	III	–	5 L	E1	P001 LP01	–	IBC03	–
2246	CYCLOPENTENE	3	–	II	–	1 L	E2	P001	–	IBC02	B8
2247	*n*-DECANE	3	–	III	–	5 L	E1	P001 LP01	–	IBC03	–
2248	DI-*n*-BUTYLAMINE	8	3	II	–	1 L	E2	P001	–	IBC02	–
2249	DICHLORODIMETHYL ETHER, SYMMETRICAL	6.1	3	I	76	0	E5	P099	–	–	–
△ 2250	DICHLOROPHENYL ISOCYANATES	6.1	–	II	–	500 g	E4	P002	–	IBC08	B2 B4
2251	BICYCLO[2.2.1]HEPTA-2,5-DIENE, STABILIZED (2,5-NORBORNADIENE, STABILIZED)	3	–	II	–	1 L	E2	P001	–	IBC02	–
2252	1,2-DIMETHOXYETHANE	3	–	II	–	1 L	E2	P001	–	IBC02	–
2253	*N*,*N*-DIMETHYLANILINE	6.1	–	II	–	100 mL	E4	P001	–	IBC02	–
2254	MATCHES, FUSEE	4.1	–	III	293	5 kg	E1	P407	–	–	–
2256	CYCLOHEXENE	3	–	II	–	1 L	E2	P001	–	IBC02	–

Portable tanks and bulk containers			EmS	Stowage and segregation	Properties and observations	UN No.
	Tank instructions	Provisions				
(12)	(13) 4.2.5 4.3	(14) 4.2.5	(15) 5.4.3.2 7.3	(16) 7.1 7.2	(17)	(18)
–	T2	TP1	F-E, S-D	Category A.	Colourless to brown liquids. Flashpoint: 43°C to 47°C c.c. Immiscible with water. When involved in a fire, evolve toxic gases. Harmful by skin contact or by inhalation. Irritating to eyes and mucous membranes.	2238
–	T1	TP33	F-A, S-A	Category A.	Crystalline solids. Some isomers may melt at low temperature: melting range between 0°C and 24°C. Toxic if swallowed, by skin contact or by inhalation.	2239
–	T10	TP2 TP13	F-A, S-B	Category B. Clear of living quarters. Segregation as for class 5.1, but "separated from" classes 4.1, 5.1 and 7.	A liquid mixture of sulphuric acid and a chromium compound (e.g., chromium trioxide or sodium dichromate) and sometimes also water. Highly corrosive to most metals. Causes severe burns to skin, eyes and mucous membranes.	2240
–	T4	TP1	F-E, S-D	Category B. Clear of living quarters.	Oily liquid. Immiscible with water. Narcotic.	2241
–	T4	TP1	F-E, S-D	Category B.	Oily liquid. Immiscible with water.	2242
–	T2	TP1	F-E, S-D	Category A.	Colourless liquid. Flashpoint: 56°C c.c. Immiscible with water. Irritating to skin, eyes and mucous membranes.	2243
–	T2	TP1	F-E, S-D	Category A.	Colourless, oily liquid. Flashpoint: 51°C c.c. Immiscible with water.	2244
–	T2	TP1	F-E, S-D	Category A.	Colourless liquid. Flashpoint: 31°C c.c. Immiscible with water.	2245
–	T7	TP2	F-E, S-D	Category E.	Colourless liquid. Flashpoint: −30°C c.c. Boiling point: 44°C. Immiscible with water. Irritating to skin, eyes and mucous membranes. Narcotic.	2246
–	T2	TP1	F-E, S-E	Category A.	Colourless liquid. Flashpoint: 47°C c.c. Explosive limits: 0.6% to 5.5%. Immiscible with water.	2247
–	T7	TP2	F-E, S-C	Category A.	Colourless, flammable liquid with an amine odour. Flashpoint: 39°C c.c. Partially miscible with water. Decomposes when heated, evolving flammable and toxic gases. Liquid is corrosive to skin, eyes and mucous membranes. Vapour irritates mucous membranes.	2248
–	–	–	F-E, S-D	Category D. Clear of living quarters.	Colourless, volatile, flammable liquid. Flashpoint: 42°C c.c. Immiscible with water. Decomposed by heat and water. Highly toxic if swallowed, by skin contact or by inhalation. The transport of this substance is prohibited except with special authorization granted by the competent authorities.	2249
–	T3	TP33	F-A, S-A	Category B. Protected from sources of heat. Clear of living quarters.	Colourless to yellowish crystalline solid with an irritating odour. Insoluble in water. Reacts with water, evolving carbon dioxide. Toxic if swallowed, by skin contact or by inhalation. May be carried in the molten state. Irritating to skin, eyes and mucous membranes.	2250 △
–	T7	TP2	F-E, S-D	Category D.	Colourless, volatile liquid. Flashpoint: below −18°C c.c. Explosive limits: 1.7% to 6.3%. Immiscible with water.	2251
–	T4	TP1	F-E, S-D	Category B.	Colourless liquid with an ethereal odour. Flashpoint: 1°C c.c. Miscible with water.	2252
–	T7	TP2	F-A, S-A	Category A.	Yellowish to brownish oily liquid. Combustible. Toxic if swallowed, by skin contact or by inhalation.	2253
–	–	–	F-A, S-I	Category A.	Matches, the heads of which are prepared with a friction-sensitive igniter composition and a pyrotechnic composition which burns with little or no flame, but with intense heat, regardless of wind or other weather conditions.	2254
–	T4	TP1	F-E, S-D	Category E.	Colourless liquid with an aromatic odour. Immiscible with water. Slightly irritating to skin, eyes and mucous membranes.	2256

UN No.	Proper Shipping Name (PSN)	Class or division	Subsidiary risk(s)	Packing group	Special provisions	Limited and excepted quantity provisions		Packing		IBC	
						Limited quantities	Excepted quantities	Instructions	Provisions	Instructions	Provisions
(1)	(2) 3.1.2	(3) 2.0	(4) 2.0	(5) 2.0.1.3	(6) 3.3	(7a) 3.4	(7b) 3.5	(8) 4.1.4	(9) 4.1.4	(10) 4.1.4	(11) 4.1.4
2257	POTASSIUM	4.3	–	I	–	0	E0	P403	PP31	IBC04	B1
2258	1,2-PROPYLENEDIAMINE	8	3	II	–	1 L	E2	P001	–	IBC02	–
2259	TRIETHYLENETETRAMINE	8	–	II	–	1 L	E2	P001	–	IBC02	–
2260	TRIPROPYLAMINE	3	8	III	–	5 L	E1	P001	–	IBC03	–
2261	XYLENOLS, SOLID	6.1	–	II	–	500 g	E4	P002	–	IBC08	B2 B4
2262	DIMETHYLCARBAMOYL CHLORIDE	8	–	II	–	1 L	E2	P001	–	IBC02	–
2263	DIMETHYLCYCLOHEXANES	3	–	II	–	1 L	E2	P001	–	IBC02	–
2264	*N,N*-DIMETHYL-CYCLOHEXYLAMINE	8	3	II	–	1 L	E2	P001	–	IBC02	–
2265	*N,N*-DIMETHYLFORMAMIDE	3	–	III	–	5 L	E1	P001 LP01	–	IBC03	–
△ 2266	DIMETHYL-*N*-PROPYLAMINE	3	8	II	–	1 L	E2	P001	–	IBC02	–
△ 2267	DIMETHYL THIOPHOSPHORYL CHLORIDE	6.1	8	II	–	100 mL	E4	P001	–	IBC02	–
2269	3,3′-IMINODIPROPYLAMINE	8	–	III	–	5 L	E1	P001 LP01	–	IBC03	–
2270	ETHYLAMINE, AQUEOUS SOLUTION with not less than 50% but not more than 70% ethylamine	3	8	II	–	1 L	E2	P001	–	IBC02	–
2271	ETHYL AMYL KETONES	3	–	III	–	5 L	E1	P001 LP01	–	IBC03	–
2272	*N*-ETHYLANILINE	6.1	–	III	–	5 L	E1	P001 LP01	–	IBC03	–

Portable tanks and bulk containers			EmS	Stowage and segregation	Properties and observations	UN No.
	Tank instructions	Provisions				
(12)	(13) 4.2.5 4.3	(14) 4.2.5	(15) 5.4.3.2 7.3	(16) 7.1 7.2	(17)	(18)
–	T9	TP7 TP33	F-G, S-N	Category D. "Separated from" acids.	Soft, silvery metal, solid or liquid. Floats on water. Reacts violently with moisture, water or acids, evolving hydrogen, which may be ignited by the heat of the reaction. Highly reactive, sometimes with explosive effect.	2257
–	T7	TP2	F-E, S-C	Category A. Clear of living quarters.	Colourless, flammable liquid with an ammoniacal odour. Flashpoint range: 33°C to 48°C c.c. Miscible with water. When involved in a fire, evolves toxic gases. Harmful by inhalation. Causes burns to skin and eyes. Irritating to mucous membranes.	2258
–	T7	TP2	F-A, S-B	Category B. Clear of living quarters. "Separated from" acids.	Moderately viscous, yellow combustible liquid with an ammoniacal odour. Miscible with water. Strongly alkaline. Can form explosive mixtures with nitric acid. When involved in a fire, evolves toxic gases. Corrosive to copper and copper alloys. Reacts violently with acids. Liquid and vapours cause burns to skin, eyes and mucous membranes. Causes skin allergy.	2259
–	T4	TP1	F-E, S-C	Category A. Clear of living quarters.	Colourless liquid. Flashpoint: 35°C c.c. Partially miscible with water. When involved in a fire, evolves toxic gases. Harmful by inhalation. Causes burns to skin and eyes. Irritating to mucous membranes.	2260
–	T3	TP33	F-A, S-A	Category A.	Crystals or needles. Toxic if swallowed, by skin contact or by inhalation.	2261
–	T7	TP2	F-A, S-B	Category A. Clear of living quarters.	Colourless to yellow liquid with a pungent odour. Immiscible with water. Reacts with water, evolving toxic and corrosive fumes. Causes tears. Causes burns to skin, eyes and mucous membranes.	2262
–	T4	TP1	F-E, S-D	Category B.	Colourless liquids. Flashpoint: 5°C to 16°C c.c. Immiscible with water.	2263
–	T7	TP2	F-E, S-C	Category A. Clear of living quarters.	Colourless, flammable liquid. Flashpoint: 43°C c.c. Partially miscible with water. Causes burns to skin, eyes and mucous membranes.	2264
–	T2	TP2	F-E, S-D	Category A.	Colourless liquid. Flashpoint: 58°C c.c. Explosive limits: 2.2% to 16%. Miscible with water. May react violently with oxidizing materials.	2265
–	T7	TP2 TP13	F-E, S-C	Category B. Clear of living quarters.	Colourless liquid with a fishy odour. Flashpoint: −11°C c.c. Miscible with water. Harmful by inhalation. Causes burns to skin, eyes and mucous membranes.	2266 △
–	T7	TP2	F-A, S-B	Category B. Protected from sources of heat.	Colourless, combustible liquid with a pungent odour. Reacts slowly with water, evolving hydrogen chloride, a corrosive gas apparent as white fumes. May decompose above 60°C, evolving flammable gases. Toxic if swallowed, by skin contact or by inhalation. Causes burns to skin, eyes and mucous membranes.	2267 △
–	T4	TP2	F-A, S-B	Category A.	Colourless combustible liquid. Miscible with water. Harmful if swallowed or by inhalation. Corrosive to skin, eyes and mucous membranes.	2269
–	T7	TP1	F-E, S-C	Category B. Clear of living quarters. "Separated from" acids.	Aqueous solution of a flammable gas with an ammonia-like odour. Explosive limits: 3.5% to 14%. ETHYLAMINE SOLUTION, concentration 50%: flashpoint −11°C c.c.; boiling point 56°C. Pure ETHYLAMINE: boiling point 17°C. Miscible with water. Harmful by inhalation. Causes burns to skin, eyes and mucous membranes. Reacts violently with acids.	2270
–	T2	TP1	F-E, S-D	Category A.	Colourless liquids. Vapour is much heavier than air (4.4). ETHYL *normal*-AMYL KETONE: flashpoint 43°C c.c. ETHYL *secondary*-AMYL KETONE: flashpoint 57°C c.c. Immiscible with water. Dissolves some types of plastics. Irritating to skin, eyes and mucous membranes.	2271
–	T4	TP1	F-A, S-A	Category A. "Separated from" acids and class 5.1.	Colourless to yellowish oily liquid. Reacts with acids, evolving highly toxic fumes of aniline and oxides of nitrogen. Reacts violently with oxidizing substances. Toxic if swallowed, by skin contact or by inhalation.	2272

UN No.	Proper Shipping Name (PSN)	Class or division	Subsidiary risk(s)	Packing group	Special provisions	Limited and excepted quantity provisions		Packing		IBC	
						Limited quantities	Excepted quantities	Instructions	Provisions	Instructions	Provisions
(1)	(2) 3.1.2	(3) 2.0	(4) 2.0	(5) 2.0.1.3	(6) 3.3	(7a) 3.4	(7b) 3.5	(8) 4.1.4	(9) 4.1.4	(10) 4.1.4	(11) 4.1.4
2273	2-ETHYLANILINE	6.1	–	III	–	5 L	E1	P001 LP01	–	IBC03	–
2274	*N*-ETHYL-*N*-BENZYLANILINE	6.1	–	III	–	5 L	E1	P001 LP01	–	IBC03	–
2275	2-ETHYLBUTANOL	3	–	III	–	5 L	E1	P001 LP01	–	IBC03	–
2276	2-ETHYLHEXYLAMINE	3	8	III	–	5 L	E1	P001	–	IBC03	–
2277	ETHYL METHACRYLATE, STABILIZED	3	–	II	–	1 L	E2	P001	–	IBC02	–
2278	*n*-HEPTENE	3	–	II	–	1 L	E2	P001	–	IBC02	–
2279	HEXACHLOROBUTADIENE	6.1	– P	III	–	5 L	E1	P001 LP01	–	IBC03	–
2280	HEXAMETHYLENEDIAMINE, MOLTEN	8	–	III	–	0	E1	–	–	–	–
2280	HEXAMETHYLENEDIAMINE, SOLID	8	–	III	–	5 kg	E1	P002 LP02	–	IBC08	B3
2281	HEXAMETHYLENE DIISOCYANATE	6.1	–	II	–	100 mL	E4	P001	–	IBC02	–
2282	HEXANOLS	3	–	III	–	5 L	E1	P001 LP01	–	IBC03	–
2283	ISOBUTYL METHACRYLATE, STABILIZED	3	–	III	–	5 L	E1	P001 LP01	–	IBC03	–
2284	ISOBUTYRONITRILE	3	6.1	II	–	1 L	E2	P001	–	IBC02	–
△ 2285	ISOCYANATOBENZO-TRIFLUORIDES	6.1	3	II	–	100 mL	E4	P001	–	IBC02	–
2286	PENTAMETHYLHEPTANE	3	–	III	–	5 L	E1	P001 LP01	–	IBC03	–
2287	ISOHEPTENES	3	–	II	–	1 L	E2	P001	–	IBC02	–
2288	ISOHEXENES	3	–	II	–	1 L	E2	P001	–	IBC02	B8
2289	ISOPHORONEDIAMINE	8	–	III	–	5 L	E1	P001 LP01	–	IBC03	–
2290	ISOPHORONE DIISOCYANATE	6.1	–	III	–	5 L	E1	P001 LP01	–	IBC03	–
2291	LEAD COMPOUND, SOLUBLE, N.O.S.	6.1	– P	III	199 274	5 kg	E1	P002 LP02	–	IBC08	B3
2293	4-METHOXY-4-METHYL-PENTAN-2-ONE	3	–	III	–	5 L	E1	P001 LP01	–	IBC03	–

Portable tanks and bulk containers			EmS	Stowage and segregation	Properties and observations	UN No.
	Tank instructions	Provisions				
(12)	(13) 4.2.5 4.3	(14) 4.2.5	(15) 5.4.3.2 7.3	(16) 7.1 7.2	(17)	(18)
–	T4	TP1	F-A, S-A	Category A. "Separated from" acids and class 5.1.	Brown liquid. Immiscible with water. Reacts with acids, evolving highly toxic fumes of aniline and oxides of nitrogen. Reacts violently with oxidizing substances. Toxic if swallowed, by skin contact or by inhalation.	2273
–	T4	TP1	F-A, S-A	Category A.	Light yellow, oily liquid. Immiscible with water. Toxic if swallowed, by skin contact or by inhalation.	2274
–	T2	TP1	F-E, S-D	Category A.	Colourless liquid. Flashpoint: 57°C o.c. Immiscible with water.	2275
–	T4	TP1	F-E, S-C	Category A. Clear of living quarters.	Colourless liquid. Flashpoint: 50°C c.c. Miscible with water. Irritating to skin, eyes and mucous membranes.	2276
–	T4	TP1	F-E, S-D	Category B.	Colourless liquid with a pungent odour. Flashpoint: 20°C o.c. Explosive limits: 1.8% to .... Immiscible with water. Irritating to skin, eyes and mucous membranes.	2277
–	T4	TP1	F-E, S-D	Category B.	Colourless liquid. Flashpoint: −3°C c.c. Immiscible with water.	2278
–	T4	TP1	F-A, S-A	Category A.	Colourless liquid. Immiscible with water. Toxic if swallowed, by skin contact or by inhalation.	2279
–	T4	TP1	F-A, S-B	Category A. Keep as cool as reasonably practicable.	White crystals or shiny flakes with a specific odour. Melting point: 29°C. Soluble in water; solution in water is a strong alkali. Decomposes when heated, evolving flammable and toxic gases. Causes burns to skin, eyes and mucous membranes.	2280
–	T1	TP33	F-A, S-B	Category A. Keep as cool as reasonably practicable.	See entry above.	2280
–	T7	TP2 TP13	F-A, S-A	Category C. Keep as dry as reasonably practicable. Clear of living quarters.	Colourless to light yellow liquid with a pungent odour. Immiscible with water but reacts with it, evolving heat and carbon dioxide gas. When heated, evolves toxic nitrous fumes. Toxic if swallowed, by skin contact or by inhalation. Irritating to skin, eyes and mucous membranes.	2281
–	T2	TP1	F-E, S-D	Category A.	Colourless liquids. *normal*-HEXANOL: flashpoint 57°C c.c. Miscible with water	2282
–	T2	TP1	F-E, S-D	Category A.	Colourless liquid. Flashpoint: 49°C c.c. Immiscible with water. Irritating to skin, eyes and mucous membranes.	2283
–	T7	TP2 TP13	F-E, S-D	Category E. Clear of living quarters.	Colourless liquid. Flashpoint: 8°C c.c. Immiscible with water. Toxic by skin contact or by inhalation.	2284
–	T7	TP2	F-E, S-D	Category D. Protected from sources of heat. Clear of living quarters.	Colourless or yellowish liquids with a pungent odour. Flashpoint of *ortho*- and *meta*-isomers: 56°C. Immiscible with water, but reacts with it to form carbon dioxide gas. Toxic if swallowed, by skin contact or by inhalation. Irritating to skin, eyes and mucous membranes.	2285 △
–	T2	TP1	F-E, S-D	Category A.	Colourless liquid. Flashpoint: 43°C c.c. Immiscible with water.	2286
–	T4	TP1	F-E, S-D	Category B.	Colourless liquids. Immiscible with water.	2287
–	T11	TP1	F-E, S-D	Category E.	Colourless liquids. Boiling range: 54°C to 69°C. Immiscible with water.	2288
–	T4	TP1	F-A, S-B	Category A.	Colourless, slightly hygroscopic liquid with a slight amine odour. Combustible. Miscible with water. Harmful if swallowed. Irritating to skin, eyes and mucous membranes.	2289
–	T4	TP2	F-A, S-A	Category B. Clear of living quarters.	Colourless or yellowish liquid. Immiscible with water. When involved in a fire, evolves nitrous fumes. Toxic if swallowed, by skin contact or by inhalation. Irritating to skin, eyes and mucous membranes.	2290
–	T1	TP33	F-A, S-A	Category A.	Colourless crystals or powder. Soluble in water. Toxic if swallowed, by skin contact or by dust inhalation.	2291
–	T2	TP1	F-E, S-D	Category A.	Colourless liquid. Flashpoint: 49°C c.c. Immiscible with water.	2293

UN No.	Proper Shipping Name (PSN)	Class or division	Subsidiary risk(s)	Packing group	Special provisions	Limited and excepted quantity provisions		Packing		IBC	
						Limited quantities	Excepted quantities	Instructions	Provisions	Instructions	Provisions
(1)	(2) 3.1.2	(3) 2.0	(4) 2.0	(5) 2.0.1.3	(6) 3.3	(7a) 3.4	(7b) 3.5	(8) 4.1.4	(9) 4.1.4	(10) 4.1.4	(11) 4.1.4
2294	*N*-METHYLANILINE	6.1	–	III	–	5 L	E1	P001 LP01	–	IBC03	–
2295	METHYL CHLOROACETATE	6.1	3	I	–	0	E5	P001	–	–	–
2296	METHYLCYCLOHEXANE	3	–	II	–	1 L	E2	P001	–	IBC02	–
△ 2297	METHYLCYCLOHEXANONE	3	–	III	–	5 L	E1	P001 LP01	–	IBC03	–
2298	METHYLCYCLOPENTANE	3	–	II	–	1 L	E2	P001	–	IBC02	–
2299	METHYL DICHLOROACETATE	6.1	–	III	–	5 L	E1	P001 LP01	–	IBC03	–
2300	2-METHYL-5-ETHYLPYRIDINE	6.1	–	III	–	5 L	E1	P001 LP01	–	IBC03	–
2301	2-METHYLFURAN	3	–	II	–	1 L	E2	P001	–	IBC02	–
2302	5-METHYLHEXAN-2-ONE	3	–	III	–	5 L	E1	P001 LP01	–	IBC03	–
2303	ISOPROPENYLBENZENE	3	–	III	–	5 L	E1	P001 LP01	–	IBC03	–
2304	NAPHTHALENE, MOLTEN	4.1	–	III	–	0	E0	–	–	–	–
2305	NITROBENZENESULPHONIC ACID	8	–	II	–	1 kg	E2	P002	–	IBC08	B2 B4
2306	NITROBENZOTRIFLUORIDES, LIQUID	6.1	– P	II	–	100 mL	E4	P001	–	IBC02	–
2307	3-NITRO-4-CHLORO-BENZOTRIFLUORIDE	6.1	– P	II	–	100 mL	E4	P001	–	IBC02	–
2308	NITROSYLSULPHURIC ACID, LIQUID	8	–	II	–	1 L	E2	P001	–	IBC02	B20
2309	OCTADIENE	3	–	II	–	1 L	E2	P001	–	IBC02	–
2310	PENTANE-2,4-DIONE	3	6.1	III	–	5 L	E1	P001	–	IBC03	–
2311	PHENETIDINES	6.1	–	III	279	5 L	E1	P001 LP01	–	IBC03	–
2312	PHENOL, MOLTEN	6.1	–	II	–	0	E0	–	–	–	–

Portable tanks and bulk containers			EmS	Stowage and segregation	Properties and observations	UN No.
	Tank instructions	Provisions				
(12)	(13) 4.2.5 4.3	(14) 4.2.5	(15) 5.4.3.2 7.3	(16) 7.1 7.2	(17)	(18)
–	T4	TP1	F-A, S-A	Category A.	Colourless to brown combustible liquid. Toxic if swallowed, by skin contact or by inhalation.	2294
–	T14	TP2 TP13	F-E, S-D	Category D.	Colourless, flammable liquid with a pungent odour. Flashpoint: 47°C c.c. Vapour much heavier than air (vapour density relative to air: 3.8). Immiscible with water. Highly toxic if swallowed, by skin contact or by inhalation.	2295
–	T4	TP1	F-E, S-D	Category B.	Colourless liquid. Flashpoint: −4°C c.c. Explosive limits: 1.2% to 6.7%. Immiscible with water. Irritating to skin, eyes and mucous membranes.	2296
–	T2	TP1	F-E, S-D	Category A.	Colourless to pale yellow liquids with a sweet odour. 2-METHYLCYCLOHEXANONE: flashpoint 46°C c.c. 3-METHYLCYCLOHEXANONE: flashpoint 51°C c.c. 4-METHYLCYCLOHEXANONE: flashpoint 40°C c.c. Immiscible with water.	2297 △
–	T4	TP1	F-E, S-D	Category B.	Colourless liquid. Flashpoint: below −10°C c.c. Explosive limits: 1% to 8.4%. Immiscible with water. Irritating to skin, eyes and mucous membranes.	2298
–	T4	TP1	F-A, S-A	Category A.	Liquid. Toxic if swallowed, by skin contact or by inhalation.	2299
–	T4	TP1	F-A, S-A	Category A.	Colourless liquid with a pungent odour. Toxic if swallowed, by skin contact or by inhalation.	2300
–	T4	TP1	F-E, S-D	Category E.	Colourless liquid with a sweetish odour. Flashpoint: −30°C c.c. Immiscible with water. When involved in a fire, evolves toxic gases. Harmful if swallowed or by inhalation. Irritating to skin, eyes and mucous membranes.	2301
–	T2	TP1	F-E, S-D	Category A.	Colourless liquid. Flashpoint: 43°C c.c. Immiscible with water.	2302
–	T2	TP1	F-E, S-D	Category A.	Colourless liquid. Flashpoint: 38°C to 54°C c.c. Explosive limits: 0.7% to 6.6%. Immiscible with water. Irritating to skin, eyes and mucous membranes.	2303
–	T1	TP3	F-A, S-H	Category C.	Molten liquid with a persistent odour. Melting point: 80°C. Evolves flammable vapours. As the melting point of naphthalene approximates very closely its flashpoint, care should be taken to avoid all possible causes of ignition. Contact between water and molten naphthalene above 110°C must be avoided, as the addition of water will cause violent foaming or even an explosion.	2304
–	T3	TP33	F-A, S-B	Category A.	Crystals. Soluble in water. Causes burns to skin, eyes and mucous membranes.	2305
–	T7	TP2	F-A, S-A	Category A. Clear of living quarters.	Pale straw-coloured, oily liquids with an aromatic odour. Immiscible with water. Toxic if swallowed, by skin contact or by inhalation.	2306
–	T7	TP2	F-A, S-A	Category A. Clear of living quarters.	Yellowish, oily liquid. Immiscible with water. Toxic if swallowed, by skin contact or by inhalation.	2307
–	T8	TP2	F-A, S-B	Category D. Clear of living quarters. Segregation as for class 5.1, but "separated from" classes 4.1, 5.1 and 7.	Clear, straw-coloured, oily liquid. Oxidant which may cause fire with organic materials (such as wood, straw, etc.). When involved in a fire, evolves toxic gases. In the presence of moisture, highly corrosive to most metals. Causes burns to skin, eyes and mucous membranes.	2308
–	T4	TP1	F-E, S-D	Category B.	Colourless liquid. Flashpoint: 9°C to 15°C c.c. Immiscible with water.	2309
–	T4	TP1	F-E, S-D	Category A.	Colourless liquid. Flashpoint: 34°C c.c. Explosive limits: 1.7% to .... Miscible with water. Toxic if swallowed, by skin contact or by inhalation.	2310
–	T4	TP1	F-A, S-A	Category A.	Colourless to yellowish liquids. Immiscible with water. Toxic if swallowed, by skin contact or by inhalation.	2311
–	T7	TP3	F-A, S-A	Category B. Clear of living quarters.	Molten liquid with a distinctive strong odour. Melting point: 10°C to 43°C (pure product). Toxic if swallowed, by skin contact or by inhalation. Rapidly absorbed through the skin.	2312

UN No.	Proper Shipping Name (PSN)	Class or division	Subsidiary risk(s)	Packing group	Special provisions	Limited and excepted quantity provisions		Packing		IBC	
						Limited quantities	Excepted quantities	Instructions	Provisions	Instructions	Provisions
(1)	(2) 3.1.2	(3) 2.0	(4) 2.0	(5) 2.0.1.3	(6) 3.3	(7a) 3.4	(7b) 3.5	(8) 4.1.4	(9) 4.1.4	(10) 4.1.4	(11) 4.1.4
2313	PICOLINES	3	–	III	–	5 L	E1	P001 LP01	–	IBC03	–
△ 2315	POLYCHLORINATED BIPHENYLS, LIQUID	9	– P	II	305	1 L	E2	P906	–	IBC02	–
2316	SODIUM CUPROCYANIDE, SOLID	6.1	– P	I	–	0	E5	P002	–	IBC07	B1
2317	SODIUM CUPROCYANIDE SOLUTION	6.1	– P	I	–	0	E5	P001	–	–	–
2318	SODIUM HYDROSULPHIDE with less than 25% water of crystallization	4.2	–	II	–	0	E2	P410	PP31	IBC06	B2
2319	TERPENE HYDROCARBONS, N.O.S.	3	–	III	–	5 L	E1	P001 LP01	–	IBC03	–
2320	TETRAETHYLENEPENTAMINE	8	–	III	–	5 L	E1	P001 LP01	–	IBC03	–
2321	TRICHLOROBENZENES, LIQUID	6.1	– P	III	–	5 L	E1	P001 LP01	–	IBC03	–
△ 2322	TRICHLOROBUTENE	6.1	– P	II	–	100 mL	E4	P001	–	IBC02	–
2323	TRIETHYL PHOSPHITE	3	–	III	–	5 L	E1	P001 LP01	–	IBC03	–
2324	TRIISOBUTYLENE	3	–	III	–	5 L	E1	P001 LP01	–	IBC03	–
2325	1,3,5-TRIMETHYLBENZENE	3	–	III	–	5 L	E1	P001 LP01	–	IBC03	–
2326	TRIMETHYLCYCLOHEXYL-AMINE	8	–	III	–	5 L	E1	P001 LP01	–	IBC03	–
2327	TRIMETHYLHEXAMETHYLENE-DIAMINES	8	–	III	–	5 L	E1	P001 LP01	–	IBC03	–
2328	TRIMETHYLHEXAMETHYLENE DIISOCYANATE	6.1	–	III	–	5 L	E1	P001 LP01	–	IBC03	–
2329	TRIMETHYL PHOSPHITE	3	–	III	–	5 L	E1	P001 LP01	–	IBC03	–
2330	UNDECANE	3	–	III	–	5 L	E1	P001 LP01	–	IBC03	–
2331	ZINC CHLORIDE, ANHYDROUS	8	–	III	–	5 kg	E1	P002 LP02	–	IBC08	B3
2332	ACETALDEHYDE OXIME	3	–	III	–	5 L	E1	P001 LP01	–	IBC03	–

Portable tanks and bulk containers			EmS	Stowage and segregation	Properties and observations	UN No.	
	Tank instructions	Provisions					
(12)	(13) 4.2.5 4.3	(14) 4.2.5	(15) 5.4.3.2 7.3	(16) 7.1 7.2	(17)	(18)	
–	T4	TP1	F-E, S-D	Category A. Clear of living quarters.	Colourless to yellow liquids with a pungent or sweet odour. Explosive limits: 1.3% to 8.7%. Miscible with water. Harmful by inhalation. *alpha*-Picoline flashpoint: 28°C c.c. *beta*-Picoline flashpoint: 40°C c.c. *gamma*-Picoline flashpoint 40°C c.c. Irritating to skin, eyes and mucous membranes.	2313	
–	T4	TP1	F-A, S-A	Category A. Segregation from foodstuffs as in 7.3.4.2.1, 7.6.3.1.2 or 7.7.3.6.	Colourless liquid (pure product) with perceptible odours. Immiscible with water. Harmful by ingestion or by skin contact. If spilled, can be a persistent hazard to the environment. This entry also covers articles, such as transformers and condensers, containing free liquid polychlorinated biphenyls.	2315	△
–	T6	TP33	F-A, S-A	Category A. "Separated from" acids.	White powder. Soluble in water. Reacts with acids or acid fumes, evolving hydrogen cyanide, a highly toxic and flammable gas. Highly toxic if swallowed, by skin contact or by dust inhalation.	2316	
–	T14	TP2 TP13	F-A, S-A	Category B. Clear of living quarters. "Separated from" acids.	Colourless liquid. Miscible with water. Decomposed by acids, evolving hydrogen cyanide, a highly toxic and flammable gas. Highly toxic if swallowed, by skin contact or by inhalation.	2317	
–	T3	TP33	F-A, S-J	Category A. "Separated from" acids.	Colourless needles to lemon-coloured flakes. Soluble in water. Reacts violently with acids.	2318	
–	T4	TP1 TP29	F-E, S-D	Category A.	Colourless or yellowish liquids. Flashpoint: 32°C to 49°C c.c. Immiscible with water.	2319	
–	T4	TP1	F-A, S-B	Category A. "Separated from" acids.	Viscous liquid. Miscible with water. When involved in a fire, evolves toxic gases. Causes burns to skin, eyes and mucous membranes. Reacts violently with acids.	2320	
–	T4	TP1	F-A, S-A	Category A.	Colourless liquids. Immiscible with water. Toxic if swallowed, by skin contact or by inhalation.	2321	
–	T7	TP2	F-A, S-A	Category A. Protected from sources of heat. Clear of living quarters.	Colourless liquid. Immiscible with water. When heated, develops toxic and irritant gases such as phosgene and hydrogen chloride and may also explode. Toxic if swallowed, by skin contact or by inhalation.	2322	△
–	T2	TP1	F-E, S-D	Category A.	Colourless liquid. Flashpoint: 44°C c.c. Immiscible with water. Irritating to skin, eyes and mucous membranes.	2323	
–	T4	TP1	F-E, S-D	Category A.	Colourless liquid. Immiscible with water.	2324	
–	T2	TP1	F-E, S-D	Category A.	Colourless liquid. Flashpoint: 44°C c.c. Immiscible with water. Harmful by inhalation.	2325	
–	T4	TP1	F-A, S-B	Category A.	Colourless, slightly hygroscopic, combustible liquid with a slight amine odour. Immiscible with water. Causes burns to skin, eyes and mucous membranes.	2326	
–	T4	TP1	F-A, S-B	Category A.	Colourless, slightly hygroscopic, combustible liquids. Miscible with water. Irritating to skin, eyes and mucous membranes.	2327	
–	T4	TP2 TP13	F-A, S-A	Category B.	Colourless or yellowish liquid. Reacts with water, evolving carbon dioxide. Toxic if swallowed, by skin contact or by inhalation. Irritating to skin, eyes and mucous membranes.	2328	
–	T2	TP1	F-E, S-D	Category A.	Colourless liquid. Flashpoint: 23°C c.c. Immiscible with water. Irritating to skin, eyes and mucous membranes.	2329	
–	T2	TP1	F-E, S-E	Category A.	Colourless liquid. Flashpoint: 60°C c.c. Immiscible with water.	2330	
–	T1	TP33	F-A, S-B	Category A.	White, deliquescent crystals. Soluble in water. Dust causes burns to skin, eyes and mucous membranes.	2331	
–	T4	TP1	F-E, S-D	Category A.	Colourless liquid. Flashpoint: 40°C c.c. Explosive limits: 4.2% to 52%. Freezing point 12°C. Miscible with water. Irritating to skin, eyes and mucous membranes.	2332	

UN No.	Proper Shipping Name (PSN)	Class or division	Subsidiary risk(s)	Packing group	Special provisions	Limited and excepted quantity provisions		Packing		IBC	
						Limited quantities	Excepted quantities	Instructions	Provisions	Instructions	Provisions
(1)	(2) 3.1.2	(3) 2.0	(4) 2.0	(5) 2.0.1.3	(6) 3.3	(7a) 3.4	(7b) 3.5	(8) 4.1.4	(9) 4.1.4	(10) 4.1.4	(11) 4.1.4
2333	ALLYL ACETATE	3	6.1	II	–	1 L	E2	P001	–	IBC02	–
2334	ALLYLAMINE	6.1	3	I	354	0	E0	P602	–	–	–
2335	ALLYL ETHYL ETHER	3	6.1	II	–	1 L	E2	P001	–	IBC02	–
2336	ALLYL FORMATE	3	6.1	I	–	0	E0	P001	–	–	–
2337	PHENYL MERCAPTAN	6.1	3	I	354	0	E0	P602	–	–	–
2338	BENZOTRIFLUORIDE	3	–	II	–	1 L	E2	P001	–	IBC02	–
2339	2-BROMOBUTANE	3	–	II	–	1 L	E2	P001	–	IBC02	–
2340	2-BROMOETHYL ETHYL ETHER	3	–	II	–	1 L	E2	P001	–	IBC02	–
2341	1-BROMO-3-METHYLBUTANE	3	–	III	–	5 L	E1	P001 LP01	–	IBC03	–
2342	BROMOMETHYLPROPANES	3	–	II	–	1 L	E2	P001	–	IBC02	–
2343	2-BROMOPENTANE	3	–	II	–	1 L	E2	P001	–	IBC02	–
2344	BROMOPROPANES	3	–	II	–	1 L	E2	P001	–	IBC02	–
2344	BROMOPROPANES	3	–	III	223	5 L	E1	P001 LP01	–	IBC03	–
2345	3-BROMOPROPYNE	3	–	II	905	1 L	E2	P001	–	IBC02	–
2346	BUTANEDIONE	3	–	II	–	1 L	E2	P001	–	IBC02	–
△ 2347	BUTYL MERCAPTAN	3	–	II	–	1 L	E2	P001	–	IBC02	–
2348	BUTYL ACRYLATES, STABILIZED	3	–	III	–	5 L	E1	P001 LP01	–	IBC03	–
2350	BUTYL METHYL ETHER	3	–	II	–	1 L	E2	P001	–	IBC02	–

Portable tanks and bulk containers			EmS	Stowage and segregation	Properties and observations	UN No.
	Tank instructions	Provisions				
(12)	(13) 4.2.5 4.3	(14) 4.2.5	(15) 5.4.3.2 7.3	(16) 7.1 7.2	(17)	(18)
–	T7	TP1 TP13	F-E, S-D	Category E. Clear of living quarters.	Colourless liquid. Flashpoint: 7°C c.c. Partially miscible with water. Toxic if swallowed, by skin contact or by inhalation. Harmful if swallowed.	2333
–	T20	TP2 TP13 TP35	F-E, S-D	Category D. Clear of living quarters.	Colourless to light yellow volatile liquid with a pungent odour. Flashpoint: −29°C c.c. Explosive limits: 2.2% to 22%. Boiling range: 55°C to 58°C. Miscible with water. When involved in a fire, evolves highly toxic gases. Highly toxic if swallowed, by skin contact or by inhalation.	2334
–	T7	TP1 TP13	F-E, S-D	Category E. Clear of living quarters.	Colourless liquid. Flashpoint: −11°C c.c. Vapour heavier than air. Immiscible with water. Narcotic. Toxic if swallowed, by skin contact or by inhalation.	2335
–	T14	TP2 TP13	F-E, S-D	Category E. Clear of living quarters.	Colourless liquid. Immiscible with water. Highly toxic if swallowed, by skin contact or by inhalation.	2336
–	T20	TP2 TP13 TP35	F-E, S-D	Category D. Clear of living quarters. "Separated from" acids.	Colourless flammable liquid with a foul odour. Flashpoint: 50°C c.c. Immiscible with water. In contact with acids or when involved in a fire, evolves highly toxic sulphurous fumes. Highly toxic if swallowed, by skin contact or by inhalation.	2337
–	T4	TP1	F-E, S-D	Category B. Clear of living quarters.	Colourless liquid with an aromatic odour. Flashpoint: 12°C c.c. Explosive limits: 2.7% to .... Immiscible with water. On contact with moisture or air evolves hydrogen fluoride, which is a toxic and corrosive gas. Harmful by inhalation. Irritating to skin, eyes and mucous membranes.	2338
–	T4	TP1	F-E, S-D	Category B. Clear of living quarters.	Colourless liquid with a pleasant odour. Flashpoint: 21°C c.c. Immiscible with water. When involved in a fire, evolves toxic fumes. Narcotic.	2339
–	T4	TP1	F-E, S-D	Category B. Clear of living quarters.	Colourless liquid with an ethereal odour. Partially miscible with water. Harmful by inhalation.	2340
–	T2	TP1	F-E, S-D	Category A.	Colourless liquid. Flashpoint: 23°C to 32°C c.c. Immiscible with water.	2341
–	T4	TP1	F-E, S-D	Category B.	Colourless liquids. Immiscible with water. Harmful by inhalation.	2342
–	T4	TP1	F-E, S-D	Category B.	Colourless or yellow liquid with a strong odour. Flashpoint: 21°C c.c. Immiscible with water. Harmful by inhalation.	2343
–	T4	TP1	F-E, S-D	Category B. Clear of living quarters.	Colourless liquids. Immiscible with water. When involved in a fire, evolve toxic fumes. Harmful by inhalation.	2344
–	T2	TP1	F-E, S-D	Category A.	See entry above.	2344
–	T4	TP1	F-E, S-D	Category D. Clear of living quarters.	Colourless to light amber liquid with a sharp odour. Flashpoint: 10°C c.c. Explosive limits: 3% to .... Vapour much heavier than air (4.1). The pure product is shock-sensitive and decomposes with explosive violence, and the possibility of detonation, when heated under confinement. Can be ignited by impact. Immiscible with water. Harmful by inhalation. Irritating to skin, eyes and mucous membranes. Causes tears.	2345
–	T4	TP1	F-E, S-D	Category B.	Greenish-yellow liquid with a strong odour. Flashpoint: 6°C c.c. Miscible with water.	2346
–	T4	TP1	F-E, S-D	Category B. "Separated from" acids. Segregation from foodstuffs as in 7.3.4.2.1, 7.6.3.1.2 or 7.7.3.6. "Separated from" odour-absorbing cargoes.	Colourless liquids with a foul odour. *tertiary*-BUTYL MERCAPTAN: flashpoint −26°C c.c. *secondary*-BUTYL MERCAPTAN: flashpoint −23°C c.c. 1-BUTANETHIOL (*normal*-BUTYL MERCAPTAN): flashpoint 12°C c.c. ISOBUTYL MERCAPTAN: flashpoint −9°C c.c. Immiscible with water. On contact with acids, emit highly toxic fumes.	2347 △
–	T2	TP1	F-E, S-D	Category A.	Colourless liquid with an unpleasant odour. Flashpoint: 36°C to 41°C c.c. Explosive limits: 1.2% to 9.9%. Immiscible with water. Harmful by inhalation. Irritating to skin, eyes and mucous membranes.	2348
–	T4	TP1	F-E, S-D	Category B.	Colourless liquid. Immiscible with water.	2350

UN No.	Proper Shipping Name (PSN)	Class or division	Subsidiary risk(s)	Packing group	Special provisions	Limited and excepted quantity provisions		Packing		IBC	
						Limited quantities	Excepted quantities	Instructions	Provisions	Instructions	Provisions
(1)	(2) 3.1.2	(3) 2.0	(4) 2.0	(5) 2.0.1.3	(6) 3.3	(7a) 3.4	(7b) 3.5	(8) 4.1.4	(9) 4.1.4	(10) 4.1.4	(11) 4.1.4
2351	BUTYL NITRITES	3	–	II	–	1 L	E2	P001	–	IBC02	–
2351	BUTYL NITRITES	3	–	III	223	5 L	E1	P001 LP01	–	IBC03	–
2352	BUTYL VINYL ETHER, STABILIZED	3	–	II	–	1 L	E2	P001	–	IBC02	–
2353	BUTYRYL CHLORIDE	3	8	II	–	1 L	E2	P001	–	IBC02	B20
2354	CHLOROMETHYL ETHYL ETHER	3	6.1	II	–	1 L	E2	P001	–	IBC02	–
2356	2-CHLOROPROPANE	3	–	I	–	0	E3	P001	–	–	–
2357	CYCLOHEXYLAMINE	8	3	II	–	1 L	E2	P001	–	IBC02	–
2358	CYCLOOCTATETRAENE	3	–	II	–	1 L	E2	P001	–	IBC02	–
2359	DIALLYLAMINE	3	6.1/8	II	–	1 L	E2	P001	–	IBC99	–
2360	DIALLYL ETHER	3	6.1	II	–	1 L	E2	P001	–	IBC02	–
2361	DIISOBUTYLAMINE	3	8	III	–	5 L	E1	P001	–	IBC03	–
2362	1,1-DICHLOROETHANE	3	–	II	–	1 L	E2	P001	–	IBC02	–
△ 2363	ETHYL MERCAPTAN	3	– P	I	–	0	E3	P001	–	–	–
2364	*n*-PROPYLBENZENE	3	–	III	–	5 L	E1	P001 LP01	–	IBC03	–
2366	DIETHYL CARBONATE	3	–	III	–	5 L	E1	P001 LP01	–	IBC03	–
2367	*alpha*-METHYL-VALERALDEHYDE	3	–	II	–	1 L	E2	P001	–	IBC02	–
2368	*alpha*-PINENE	3	–	III	–	5 L	E1	P001 LP01	–	IBC03	–
2370	1-HEXENE	3	–	II	–	1 L	E2	P001	–	IBC02	–

Portable tanks and bulk containers			EmS	Stowage and segregation	Properties and observations	UN No.
	Tank instructions	Provisions				
(12)	(13) 4.2.5 4.3	(14) 4.2.5	(15) 5.4.3.2 7.3	(16) 7.1 7.2	(17)	(18)
–	T4	TP1	F-E, S-D	Category B. Clear of living quarters.	Yellowish, volatile, oily liquids. Partially miscible with water. Decompose on exposure to air, light, water or heat, evolving toxic nitrous fumes. Harmful by inhalation.	2351
–	T2	TP1	F-E, S-D	Category A. Clear of living quarters.	See entry above.	2351
–	T4	TP1	F-E, S-D	Category B. Clear of living quarters.	Colourless, volatile liquid with a sharp ethereal odour. Flashpoint: −9°C c.c. Immiscible with water. Harmful by inhalation. Irritating to skin, eyes and mucous membranes.	2352
–	T8	TP2 TP13	F-E, S-C	Category C. Clear of living quarters.	Colourless liquid with a pungent odour. Reacts with water, evolving hydrogen chloride, an irritating and corrosive gas apparent as white fumes. In the presence of moisture, highly corrosive to most metals. Causes burns to skin, eyes and mucous membranes.	2353
–	T7	TP1 TP13	F-E, S-D	Category E. Clear of living quarters.	Colourless liquid with a pungent odour. Partially miscible with water. Fumes in air, evolving hydrogen chloride, which is an irritating and corrosive gas. Toxic by inhalation. Strong lachrymator.	2354
–	T11	TP2 TP13	F-E, S-D	Category E.	Colourless liquid. Flashpoint: −32°C c.c. Explosive limits: 2.8% to 10.7%. Boiling point: 35°C. Immiscible with water. On contact with heat or flame, emits highly toxic phosgene gas. Can react vigorously with oxidizing materials.	2356
–	T7	TP2	F-E, S-C	Category A. Clear of living quarters.	Colourless or yellowish flammable liquid with a fishy odour. Flashpoint: 27°C c.c. Explosive limits: 0.5% to 21.7%. Miscible with water. Harmful by inhalation. Causes burns to skin and eyes. Irritating to mucous membranes.	2357
–	T4	TP1	F-E, S-D	Category B.	Colourless liquid. Freezing point: −4°C. Immiscible with water.	2358
–	T7	TP1	F-E, S-C	Category B. Clear of living quarters. Segregation as for class 3 but "away from" class 4.1.	Colourless, volatile liquid with a disagreeable odour. Flashpoint: 7°C c.c. Partially miscible with water. Toxic if swallowed, by skin contact or by inhalation. Causes burns to skin, eyes and mucous membranes.	2359
–	T7	TP1 TP13	F-E, S-D	Category E.	Colourless, volatile liquid with a perceptible odour. Flashpoint: −11°C c.c. Immiscible with water. Toxic if swallowed, by skin contact or by inhalation.	2360
–	T4	TP1	F-E, S-C	Category A.	Colourless liquid with a fishy odour. Flashpoint: 29°C c.c. Immiscible with water. Harmful by inhalation. Causes burns to skin and eyes. Irritating to mucous membranes.	2361
–	T4	TP1	F-E, S-D	Category B. Clear of living quarters.	Colourless liquid with an aromatic, ethereal odour. Flashpoint: −10°C c.c. Explosive limits: 5.6% to .... Immiscible with water. When involved in a fire, emits toxic fumes of phosgene. Harmful by inhalation.	2362
–	T11	TP2 TP13	F-E, S-D	Category E. Segregation from foodstuffs as in 7.3.4.2, 7.6.3.1.2 or 7.7.3.6. "Separated from" odour-absorbing cargoes.	Volatile liquid with a strong unpleasant odour. Flashpoint: −45°C c.c. Explosive limits: 2.8% to 18.2%. Boiling point: 35°C. Immiscible with water.	2363 △
–	T2	TP1	F-E, S-D	Category A.	Colourless liquid. Flashpoint: 39°C c.c. Explosive limits: 0.8% to 6%. Immiscible with water.	2364
–	T2	TP1	F-E, S-D	Category A.	Colourless liquid. Flashpoint: 25°C to 31°C c.c. Vapour much heavier than air (4.1). Immiscible with water. Irritating to skin, eyes and mucous membranes.	2366
–	T4	TP1	F-E, S-D	Category B.	Colourless liquid. Flashpoint: 13°C c.c. Immiscible with water. Irritating to skin, eyes and mucous membranes.	2367
–	T2	TP1	F-E, S-E	Category A.	Colourless liquid with an odour of turpentine. Flashpoint: 33°C c.c. Explosive limits: 0.8% to 6%. Immiscible with water. Harmful by inhalation. Irritating to skin, eyes and mucous membranes.	2368
–	T4	TP1	F-E, S-D	Category E.	Colourless liquid. Explosive limits: 1.2% to 6.9%. Immiscible with water.	2370

UN No.	Proper Shipping Name (PSN)	Class or division	Subsidiary risk(s)	Packing group	Special provisions	Limited and excepted quantity provisions		Packing		IBC	
						Limited quantities	Excepted quantities	Instructions	Provisions	Instructions	Provisions
(1)	(2) 3.1.2	(3) 2.0	(4) 2.0	(5) 2.0.1.3	(6) 3.3	(7a) 3.4	(7b) 3.5	(8) 4.1.4	(9) 4.1.4	(10) 4.1.4	(11) 4.1.4
2371	ISOPENTENES	3	–	I	–	0	E3	P001	–	–	–
△ 2372	1,2-DI(DIMETHYLAMINO) ETHANE	3	–	II	–	1 L	E2	P001	–	IBC02	–
2373	DIETHOXYMETHANE	3	–	II	–	1 L	E2	P001	–	IBC02	–
2374	3,3-DIETHOXYPROPENE	3	–	II	–	1 L	E2	P001	–	IBC02	–
2375	DIETHYL SULPHIDE	3	–	II	–	1 L	E2	P001	–	IBC02	–
2376	2,3-DIHYDROPYRAN	3	–	II	–	1 L	E2	P001	–	IBC02	–
2377	1,1-DIMETHOXYETHANE	3	–	II	–	1 L	E2	P001	–	IBC02	–
2378	2-DIMETHYLAMINO-ACETONITRILE	3	6.1	II	–	1 L	E2	P001	–	IBC02	–
2379	1,3-DIMETHYLBUTYLAMINE	3	8	II	–	1 L	E2	P001	–	IBC02	–
2380	DIMETHYLDIETHOXYSILANE	3	–	II	–	1 L	E2	P001	–	IBC02	–
△ 2381	DIMETHYL DISULPHIDE	3	6.1	II	–	1 L	E2	P001	–	IBC02	–
2382	DIMETHYLHYDRAZINE, SYMMETRICAL	6.1	3 P	I	354	0	E0	P602	–	–	–
2383	DIPROPYLAMINE	3	8	II	–	1 L	E2	P001	–	IBC02	–
2384	DI-*n*-PROPYL ETHER	3	–	II	–	1 L	E2	P001	–	IBC02	–
2385	ETHYL ISOBUTYRATE	3	–	II	–	1 L	E2	P001	–	IBC02	–
2386	1-ETHYLPIPERIDINE	3	8	II	–	1 L	E2	P001	–	IBC02	–
2387	FLUOROBENZENE	3	–	II	–	1 L	E2	P001	–	IBC02	–
2388	FLUOROTOLUENES	3	–	II	–	1 L	E2	P001	–	IBC02	–
2389	FURAN	3	–	I	–	0	E3	P001	–	–	–
2390	2-IODOBUTANE	3	–	II	–	1 L	E2	P001	–	IBC02	–
2391	IODOMETHYLPROPANES	3	–	II	–	1 L	E2	P001	–	IBC02	–

Portable tanks and bulk containers			EmS	Stowage and segregation	Properties and observations	UN No.
	Tank instructions	Provisions				
(12)	(13) 4.2.5 4.3	(14) 4.2.5	(15) 5.4.3.2 7.3	(16) 7.1 7.2	(17)	(18)
–	T11	TP2	F-E, S-D	Category E.	Colourless, volatile liquid with a disagreeable odour. Flashpoint: below −18°C c.c. Immiscible with water. Irritating to skin, eyes and mucous membranes.	2371
–	T4	TP1	F-E, S-D	Category B.	Colourless liquid. Flashpoint: 21°C c.c. Miscible with water. Irritability to skin, eyes and mucous membranes.	2372 △
–	T4	TP1	F-E, S-D	Category B.	Colourless liquid. Flashpoint: below −5°C c.c. Miscible with water.	2373
–	T4	TP1	F-E, S-D	Category B.	Colourless liquid. Flashpoint: 15°C c.c. Partially miscible with water. Harmful by inhalation.	2374
–	T7	TP1 TP13	F-E, S-D	Category E.	Colourless, volatile liquid with an odour of garlic. Flashpoint: −10°C c.c. Immiscible with water.	2375
–	T4	TP1	F-E, S-D	Category B.	Colourless, volatile liquid with an ethereal odour. Flashpoint: −16°C c.c. Miscible with water.	2376
–	T7	TP1	F-E, S-D	Category B.	Colourless liquid with a strong aromatic odour. Miscible with water.	2377
–	T7	TP1	F-E, S-D	Category A. Clear of living quarters. "Separated from" acids.	Colourless liquid. Flashpoint: 35°C c.c. Immiscible with water. On contact with water and acids, evolves toxic fumes. Toxic if swallowed, by skin contact or by inhalation.	2378
–	T7	TP1	F-E, S-C	Category B. "Separated from" acids.	Colourless liquid with an ammonia-like odour. Flashpoint: 9°C to 13°C c.c. Immiscible with water. Reacts violently with acids. Harmful by inhalation. Causes burns to skin and eyes. Irritating to mucous membranes.	2379
–	T4	TP1	F-E, S-D	Category B.	Colourless liquid. Flashpoint: 13°C c.c. Miscible with water. Irritating to skin, eyes and mucous membranes.	2380
–	T7	TP2 TP13 TP39	F-E, S-D	Category B. Clear of living quarters.	Yellow liquid with an unpleasant odour. Flashpoint: 15°C c.c. Immiscible with water. When involved in a fire, evolves toxic gases. Toxic if swallowed, by skin contact or by inhalation.	2381 △
–	T20	TP2 TP13 TP37	F-E, S-D	Category D. Clear of living quarters. "Separated from" class 5.1. "Separated from" acids.	Colourless, flammable, volatile liquid with an ammonia-like odour. Miscible with water. Reacts violently with acids. May react dangerously with oxidizing substances. Flashpoint: −17°C c.c. Highly toxic if swallowed, by skin contact or by inhalation.	2382
–	T7	TP1	F-E, S-C	Category B.	Colourless liquid with a fishy odour. Flashpoint: 7°C c.c. Immiscible with water. Harmful by inhalation. Causes burns to skin, eyes and mucous membranes.	2383
–	T4	TP1	F-E, S-D	Category B.	Colourless liquid. Flashpoint (pure product): −21°C c.c. Explosive limits: 1.7% to .... Immiscible with water.	2384
–	T4	TP1	F-E, S-D	Category B.	Colourless, volatile liquid with an aromatic odour. Flashpoint: 21°C c.c. Immiscible with water.	2385
–	T7	TP1	F-E, S-C	Category B. "Separated from" acids.	Colourless liquid. Flashpoint: 19°C c.c. Immiscible with water. Reacts violently with acids. Harmful by inhalation. Causes burns to skin, eyes and mucous membranes. May cause lung damage.	2386
–	T4	TP1	F-E, S-D	Category B.	Colourless liquid with a benzene odour. Flashpoint: −15°C c.c. Immiscible with water. Harmful by inhalation.	2387
–	T4	TP1	F-E, S-D	Category B.	Colourless liquids. *ortho*-FLUOROTOLUENE: flashpoint 9°C c.c. *meta*-FLUOROTOLUENE: flashpoint 12°C c.c. *para*-FLUOROTOLUENE: flashpoint 10°C c.c. Immiscible with water.	2388
–	T12	TP2 TP13	F-E, S-D	Category E. Clear of living quarters.	Colourless liquid with a strong odour. Flashpoint: below −18°C c.c. Explosive limits: 1.3% to 14.3%. Boiling point: 31°C. Immiscible with water. Harmful if swallowed, by skin contact or by inhalation.	2389
–	T4	TP1	F-E, S-D	Category B.	Colourless liquid. Flashpoint: 21°C c.c. Immiscible with water.	2390
–	T4	TP1	F-E, S-D	Category B.	Colourless liquids. Immiscible with water.	2391

UN No.	Proper Shipping Name (PSN)	Class or division	Subsidiary risk(s)	Packing group	Special provisions	Limited and excepted quantity provisions		Packing		IBC	
						Limited quantities	Excepted quantities	Instructions	Provisions	Instructions	Provisions
(1)	(2) 3.1.2	(3) 2.0	(4) 2.0	(5) 2.0.1.3	(6) 3.3	(7a) 3.4	(7b) 3.5	(8) 4.1.4	(9) 4.1.4	(10) 4.1.4	(11) 4.1.4
2392	IODOPROPANES	3	–	III	–	5 L	E1	P001 LP01	–	IBC03	–
2393	ISOBUTYL FORMATE	3	–	II	–	1 L	E2	P001	–	IBC02	–
2394	ISOBUTYL PROPIONATE	3	–	III	–	5 L	E1	P001 LP01	–	IBC03	–
2395	ISOBUTYRYL CHLORIDE	3	8	II	–	1 L	E2	P001	–	IBC02	–
2396	METHACRYLALDEHYDE, STABILIZED	3	6.1	II	–	1 L	E2	P001	–	IBC02	–
2397	3-METHYLBUTAN-2-ONE	3	–	II	–	1 L	E2	P001	–	IBC02	–
2398	METHYL *tert*-BUTYL ETHER	3	–	II	–	1 L	E2	P001	–	IBC02	–
2399	1-METHYLPIPERIDINE	3	8	II	–	1 L	E2	P001	–	IBC02	–
2400	METHYL ISOVALERATE	3	–	II	–	1 L	E2	P001	–	IBC02	–
2401	PIPERIDINE	8	3	I	–	0	E0	P001	–	–	–
△ 2402	PROPANETHIOLS	3	–	II	–	1 L	E2	P001	–	IBC02	–
2403	ISOPROPENYL ACETATE	3	–	II	–	1 L	E2	P001	–	IBC02	–
2404	PROPIONITRILE	3	6.1	II	–	1 L	E2	P001	–	IBC02	–
2405	ISOPROPYL BUTYRATE	3	–	III	–	5 L	E1	P001 LP01	–	IBC03	–
2406	ISOPROPYL ISOBUTYRATE	3	–	II	–	1 L	E2	P001	–	IBC02	–
2407	ISOPROPYL CHLOROFORMATE	6.1	3/8	I	354	0	E0	P602	–	–	–
2409	ISOPROPYL PROPIONATE	3	–	II	–	1 L	E2	P001	–	IBC02	–
2410	1,2,3,6-TETRAHYDROPYRIDINE	3	–	II	–	1 L	E2	P001	–	IBC02	–
2411	BUTYRONITRILE	3	6.1	II	–	1 L	E2	P001	–	IBC02	–
2412	TETRAHYDROTHIOPHENE	3	–	II	–	1 L	E2	P001	–	IBC02	–
2413	TETRAPROPYL ORTHOTITANATE	3	–	III	–	5 L	E1	P001 LP01	–	IBC03	–

Portable tanks and bulk containers			EmS	Stowage and segregation	Properties and observations	UN No.
	Tank instructions	Provisions				
(12)	(13) 4.2.5 4.3	(14) 4.2.5	(15) 5.4.3.2 7.3	(16) 7.1 7.2	(17)	(18)
–	T2	TP1	F-E, S-D	Category A.	Colourless liquids. 1-IODOPROPANE: flashpoint 34°C c.c. 2-IODOPROPANE: flashpoint approx. 25°C c.c. Immiscible with water.	2392
–	T4	TP1	F-E, S-D	Category B.	Colourless liquid. Flashpoint: 5°C c.c. Explosive limits: 1.7% to 8%. Irritating to skin, eyes and mucous membranes.	2393
–	T2	TP1	F-E, S-D	Category B.	Colourless liquid. Flashpoint: 31°C c.c. Immiscible with water.	2394
–	T7	TP2	F-E, S-C	Category C. Clear of living quarters.	Colourless liquid with a pungent odour. Reacts with water, evolving hydrogen chloride, an irritating and corrosive gas apparent as white fumes. In the presence of moisture, highly corrosive to most metals. Causes burns to skin, eyes and mucous membranes.	2395
–	T7	TP1 TP13	F-E, S-D	Category E. Clear of living quarters.	Colourless liquid. Flashpoint: 2°C c.c. Miscible with water. Toxic by inhalation. Irritating to skin, eyes and mucous membranes.	2396
–	T4	TP1	F-E, S-D	Category B.	Colourless liquid. Flashpoint: –3°C c.c. Explosive limits: 1.5% to 8%. Immiscible with water.	2397
–	T7	TP1	F-E, S-D	Category E.	Colourless liquid. Flashpoint: below –18°C c.c. Explosive limits: 1.7% to 8.4%. Boiling point: 55°C. Immiscible with water.	2398
–	T7	TP1	F-E, S-C	Category B. "Separated from" acids.	Colourless liquid. Flashpoint: 3°C c.c. Miscible with water. Reacts violently with acids. Harmful by inhalation. Causes burns to skin, eyes and mucous membranes.	2399
–	T4	TP1	F-E, S-D	Category B.	Colourless liquid. Immiscible with water.	2400
–	T10	TP2	F-E, S-C	Category D. "Separated from" acids.	Colourless liquid with a fish-like odour. Miscible with water. Reacts violently with acids. Solution in water is a strong alkali and is corrosive. When involved in fire, evolves toxic nitrous fumes.	2401
–	T4	TP1 TP13	F-E, S-D	Category E. Segregation from foodstuffs as in 7.3.4.2, 7.6.3.1.2 or 7.7.3.6. "Separated from" odour-absorbing cargoes.	Colourless or yellowish liquids with a strong unpleasant odour. Flashpoint: below –18°C c.c. Boiling range: 53°C to 67°C. Immiscible with water.	2402 △
–	T4	TP1	F-E, S-D	Category B.	Colourless liquid. Flashpoint: 10°C c.c. Immiscible with water.	2403
–	T7	TP1 TP13	F-E, S-D	Category E. Clear of living quarters.	Colourless, volatile liquid with an ether-like odour. Flashpoint: 2°C c.c. Explosive limits: 3.1% to .... Miscible with water. When involved in a fire, evolves highly toxic cyanide fumes. Toxic if swallowed, by skin contact or by inhalation.	2404
–	T2	TP1	F-E, S-D	Category A.	Colourless liquid. Flashpoint: 25°C c.c. Immiscible with water. Irritating to skin, eyes and mucous membranes.	2405
–	T4	TP1	F-E, S-D	Category B.	Colourless liquid. Flashpoint: 20°C c.c. Immiscible with water. Narcotic. Irritating to skin, eyes and mucous membranes.	2406
–	–	–	F-E, S-C	Category D. Clear of living quarters. Segregation as for class 3 but "away from" class 4.1.	Colourless flammable liquid. Flashpoint: 16°C c.c. Decomposed by water, evolving hydrogen chloride, an irritating and corrosive gas apparent as white fumes. In the presence of moisture, corrosive to most metals. Highly toxic if swallowed, by skin contact or by inhalation. Causes burns to skin, eyes and mucous membranes.	2407
–	T4	TP1	F-E, S-D	Category B.	Colourless liquid. Flashpoint: 21°C c.c. Immiscible with water.	2409
–	T4	TP1	F-E, S-D	Category B.	Colourless liquid. Flashpoint: 16°C c.c. Miscible with water. Harmful by inhalation.	2410
–	T7	TP1 TP13	F-E, S-D	Category E. Clear of living quarters.	Colourless liquid. Flashpoint: 21°C c.c. Explosive limits: 1.6% to .... Immiscible with water. Toxic if swallowed, by skin contact or by inhalation.	2411
–	T4	TP1	F-E, S-D	Category B.	Colourless liquid with a pleasant odour. Flashpoint: 13°C c.c. Immiscible with water.	2412
–	T4	TP1	F-E, S-D	Category A.	Colourless liquid. Flashpoint: 38°C c.c.	2413

UN No.	Proper Shipping Name (PSN)	Class or division	Subsidiary risk(s)	Packing group	Special provisions	Limited and excepted quantity provisions		Packing		IBC	
						Limited quantities	Excepted quantities	Instructions	Provisions	Instructions	Provisions
(1)	(2) 3.1.2	(3) 2.0	(4) 2.0	(5) 2.0.1.3	(6) 3.3	(7a) 3.4	(7b) 3.5	(8) 4.1.4	(9) 4.1.4	(10) 4.1.4	(11) 4.1.4
2414	THIOPHENE	3	–	II	–	1 L	E2	P001	–	IBC02	–
2416	TRIMETHYL BORATE	3	–	II	–	1 L	E2	P001	–	IBC02	–
2417	CARBONYL FLUORIDE	2.3	8	–	–	0	E0	P200	–	–	–
2418	SULPHUR TETRAFLUORIDE	2.3	8	–	–	0	E0	P200	–	–	–
2419	BROMOTRIFLUOROETHYLENE	2.1	–	–	–	0	E0	P200	–	–	–
2420	HEXAFLUOROACETONE	2.3	8	–	–	0	E0	P200	–	–	–
2421	NITROGEN TRIOXIDE	2.3	5.1/8	–	–	0	E0	P200	–	–	–
2422	OCTAFLUOROBUT-2-ENE (REFRIGERANT GAS R 1318)	2.2	–	–	–	120 mL	E1	P200	–	–	–
2424	OCTAFLUOROPROPANE (REFRIGERANT GAS R 218)	2.2	–	–	–	120 mL	E1	P200	–	–	–
2426	AMMONIUM NITRATE, LIQUID (hot concentrated solution)	5.1	–	–	252 942	0	E0	–	–	–	–
2427	POTASSIUM CHLORATE, AQUEOUS SOLUTION	5.1	–	II	–	1 L	E2	P504	–	IBC02	–
2427	POTASSIUM CHLORATE, AQUEOUS SOLUTION	5.1	–	III	223	5 L	E1	P504	–	IBC02	–

Portable tanks and bulk containers			EmS	Stowage and segregation	Properties and observations	UN No.
	Tank instructions	Provisions				
(12)	(13) 4.2.5 4.3	(14) 4.2.5	(15) 5.4.3.2 7.3	(16) 7.1 7.2	(17)	(18)
–	T4	TP1	F-E, S-D	Category B. Clear of living quarters.	Colourless liquid with an unpleasant odour. Flashpoint: −9°C c.c. Explosive limits: 1.5% to 12.5%. Immiscible with water. Irritating to skin, eyes and mucous membranes.	2414
–	T7	TP1	F-E, S-D	Category B.	Colourless liquid. Reacts with water, evolving flammable vapours.	2416
–	–	–	F-C, S-U	Category D. Clear of living quarters.	Non-flammable, toxic and corrosive colourless gas with a pungent odour. Corrosive to glass and to most metals. Corrosive in the presence of water. Much heavier than air (2.3). Highly irritating to skin, eyes and mucous membranes.	2417
–	–	–	F-C, S-U	Category D. Clear of living quarters. "Separated from" acids.	Non-flammable, toxic and corrosive, colourless gas with a pungent odour. Reacts with water, moist air or acids to produce toxic and corrosive fumes. Corrosive to glass and to most metals. Much heavier than air (3.7). Highly irritating to skin, eyes and mucous membranes.	2418
–	–	–	F-D, S-U	Category B. Clear of living quarters.	Liquefied, flammable, colourless gas. Much heavier than air (5.6). Boiling point: −3°C.	2419
–	–	–	F-C, S-U	Category D. Clear of living quarters.	Non-flammable, toxic and corrosive, colourless, hygroscopic gas with an unpleasant odour. Reacts vigorously with water, evolving heat. Corrosive to glass and to most metals. Fumes in moist air. Much heavier than air (5.7). Highly irritating to skin, eyes and mucous membranes.	2420
–	–	–	F-C, S-W	Category D. Clear of living quarters. Segregation as for class 5.1 but "separated from" class 7.	Liquefied, non-flammable, toxic and corrosive gas. At lower temperatures, present as a blue liquid. Strong oxidizing agent. Much heavier than air (2.6). Boiling point: 3.5°C. Highly irritating to skin, eyes and mucous membranes.	2421
–	–	–	F-C, S-V	Category A.	Liquefied, non-flammable, colourless gas. Much heavier than air (6.9). Boiling point: 1.2°C.	2422
–	T50	–	F-C, S-V	Category A.	Liquefied, non-flammable, colourless gas. Much heavier than air (6.6). Boiling point: −36°C.	2424
–	T7	TP1 TP16 TP17	F-H, S-Q	Category D. "Separated from" combustible material, bromates, chlorates, chlorites, hypochlorites, nitrites, perchlorates, permanganates and powdered metals.	Hot aqueous solution of not more than 93% ammonium nitrate with not more than 0.2% combustible material (including organic material calculated as carbon) and free from any other added matter, containing at least 7% water, while the maximum content of chloride ions should not exceed 0.02%. May cause fire and explosion in contact with combustible material (e.g., wood, straw, cotton, oil, sugar, etc.), strong acids, and other class 5.1 substances and burn fiercely. Maximum allowable transport temperature of the solution 140°C. This temperature should be indicated on the transport unit. The acidity (pH) of the cargo when diluted with ten parts of water to one part of cargo, by mass, should be between 5.0 and 7.0. The concentration and temperature of the solution at the time of loading, its percentage of combustible materials and of chlorides, and the contents of free acid should be certified.	2426
–	T4	TP1	F-H, S-Q	Category B. "Separated from" ammonium compounds, cyanides and sulphur.	Colourless liquid. When involved in a fire, may cause an explosion. Leakage and subsequent evaporation of the water may present increased dangers as follows: 1. in contact with combustible material (particularly with fibrous material such as jute, cotton or sisal) or sulphur, danger of spontaneous combustion; 2. in contact with ammonium compounds, powdered metals or oils, danger of explosion.	2427
–	T4	TP1	F-H, S-Q	Category B. "Separated from" ammonium compounds, cyanides and sulphur.	See entry above.	2427

UN No.	Proper Shipping Name (PSN)	Class or division	Subsidiary risk(s)	Packing group	Special provisions	Limited and excepted quantity provisions		Packing		IBC	
						Limited quantities	Excepted quantities	Instructions	Provisions	Instructions	Provisions
(1)	(2) 3.1.2	(3) 2.0	(4) 2.0	(5) 2.0.1.3	(6) 3.3	(7a) 3.4	(7b) 3.5	(8) 4.1.4	(9) 4.1.4	(10) 4.1.4	(11) 4.1.4
2428	SODIUM CHLORATE, AQUEOUS SOLUTION	5.1	–	II	–	1 L	E2	P504	–	IBC02	–
2428	SODIUM CHLORATE, AQUEOUS SOLUTION	5.1	–	III	223	5 L	E1	P504	–	IBC02	–
2429	CALCIUM CHLORATE, AQUEOUS SOLUTION	5.1	–	II	–	1 L	E2	P504	–	IBC02	–
2429	CALCIUM CHLORATE, AQUEOUS SOLUTION	5.1	–	III	223	5 L	E1	P504	–	IBC02	–
2430	ALKYLPHENOLS, SOLID, N.O.S. (including C_2–C_{12} homologues)	8	–	I	–	0	E0	P002	–	IBC07	B1
2430	ALKYLPHENOLS, SOLID, N.O.S. (including C_2–C_{12} homologues)	8	–	II	–	1 kg	E2	P002	–	IBC08	B2 B4
2430	ALKYLPHENOLS, SOLID, N.O.S. (including C_2–C_{12} homologues)	8	–	III	223	5 kg	E1	P002 LP02	–	IBC08	B3
△ 2431	ANISIDINES	6.1	–	III	–	5 L	E1	P001 LP01	–	IBC03	–
2432	*N,N*-DIETHYLANILINE	6.1	–	III	279	5 L	E1	P001 LP01	–	IBC03	–
2433	CHLORONITROTOLUENES, LIQUID	6.1	– P	III	–	5 L	E1	P001 LP01	–	IBC03	–
△ 2434	DIBENZYLDICHLOROSILANE	8	–	II	–	0	E0	P010	–	–	–
△ 2435	ETHYLPHENYL-DICHLOROSILANE	8	–	II	–	0	E0	P010	–	–	–
2436	THIOACETIC ACID	3	–	II	–	1 L	E2	P001	–	IBC02	–
△ 2437	METHYLPHENYL-DICHLOROSILANE	8	–	II	–	0	E0	P010	–	–	–

Portable tanks and bulk containers			EmS	Stowage and segregation	Properties and observations	UN No.
	Tank instructions	Provisions				
(12)	(13) 4.2.5 4.3	(14) 4.2.5	(15) 5.4.3.2 7.3	(16) 7.1 7.2	(17)	(18)
–	T4	TP1	F-H, S-Q	Category B. "Separated from" ammonium compounds, cyanides and sulphur.	Colourless liquid. When involved in a fire, may cause an explosion. Leakage and subsequent evaporation of the water may present increased dangers as follows: 1. in contact with combustible material (particularly with fibrous material such as jute, cotton or sisal) or sulphur, danger of spontaneous combustion; 2. in contact with ammonium compounds, powdered metals or oils, danger of explosion.	2428
–	T4	TP1	F-H, S-Q	Category B. "Separated from" ammonium compounds, cyanides and sulphur.	See entry above.	2428
–	T4	TP1	F-H, S-Q	Category B. "Separated from" ammonium compounds, cyanides and sulphur.	Colourless liquid. When involved in a fire, may cause an explosion. Leakage and subsequent evaporation of the water may present increased dangers as follows: 1. in contact with combustible material (particularly with fibrous material such as jute, cotton or sisal) or sulphur, danger of spontaneous combustion; 2. in contact with ammonium compounds, powdered metals or oils, danger of explosion.	2429
–	T4	TP1	F-H, S-Q	Category B. "Separated from" ammonium compounds, cyanides and sulphur.	See entry above.	2429
–	T6	TP33	F-A, S-B	Category B.	A wide range of colourless to pale straw-coloured solids with penetrating odours (sometimes camphor-like). Some have low melting points. Insoluble in water. Cause burns to skin, eyes and mucous membranes.	2430
–	T3	TP33	F-A, S-B	Category B.	See entry above.	2430
–	T1	TP33	F-A, S-B	Category A.	See entry above.	2430
–	T4	TP1	F-A, S-A	Category A.	Reddish or yellowish oily liquid. Immiscible with water. Toxic if swallowed, by skin contact or by inhalation.	2431 △
–	T4	TP1	F-A, S-A	Category A.	Colourless to yellow-brown oily liquid. Combustible. Toxic if swallowed, by skin contact or by inhalation.	2432
–	T4	TP1	F-A, S-A	Category A. Segregation as for class 5.1 but "away from" classes 4.1, 5.1 and 7.	Immiscible with water. Oxidizing substance which may explode or burn fiercely when in contact with organic materials. Toxic if swallowed, by skin contact or by inhalation.	2433
–	T10	TP2 TP7 TP13	F-A, S-B	Category C. Clear of living quarters.	Colourless liquid with a pungent odour. Reacts violently with water, evolving hydrogen chloride, a corrosive gas apparent as white fumes. When involved in a fire, evolves toxic gases. In the presence of moisture, highly corrosive to most metals. Vapour irritating to skin, eyes and mucous membranes.	2434 △
–	T10	TP2 TP7 TP13	F-A, S-B	Category C.	Colourless liquid with a pungent odour. Reacts with water, evolving hydrogen chloride, a corrosive gas apparent as white fumes. When involved in a fire, evolves toxic gases. In the presence of moisture, highly corrosive to most metals. Causes burns to skin, eyes and mucous membranes.	2435 △
–	T4	TP1	F-E, S-D	Category B.	Colourless or yellow liquid with a pungent odour. Miscible with water. Harmful by inhalation.	2436
–	T10	TP2 TP7 TP13	F-A, S-B	Category C. Clear of living quarters.	Colourless liquid. Reacts with water, evolving hydrogen chloride, an irritating and corrosive gas apparent as white fumes. When involved in a fire, evolves toxic gases. In the presence of moisture, highly corrosive to most metals. Causes burns to skin, eyes and mucous membranes.	2437 △

	UN No. (1)	Proper Shipping Name (PSN) (2) 3.1.2	Class or division (3) 2.0	Subsidiary risk(s) (4) 2.0	Packing group (5) 2.0.1.3	Special provisions (6) 3.3	Limited and excepted quantity provisions: Limited quantities (7a) 3.4	Limited and excepted quantity provisions: Excepted quantities (7b) 3.5	Packing: Instructions (8) 4.1.4	Packing: Provisions (9) 4.1.4	IBC: Instructions (10) 4.1.4	IBC: Provisions (11) 4.1.4
△	2438	TRIMETHYLACETYL CHLORIDE	6.1	3/8	I	–	0	E5	P001	–	–	–
△	2439	SODIUM HYDROGEN-DIFLUORIDE	8	–	II	–	1 kg	E2	P002	–	IBC08	B2 B4
	2440	STANNIC CHLORIDE PENTAHYDRATE	8	–	III	–	5 kg	E1	P002 LP02	–	IBC08	B3
	2441	TITANIUM TRICHLORIDE, PYROPHORIC or TITANIUM TRICHLORIDE MIXTURE, PYROPHORIC	4.2	8	I	–	0	E0	P404	–	–	–
	2442	TRICHLOROACETYL CHLORIDE	8	–	II	–	0	E2	P001	–	–	–
	2443	VANADIUM OXYTRICHLORIDE	8	–	II	–	1 L	E2	P001	–	IBC02	–
	2444	VANADIUM TETRACHLORIDE	8	–	I	–	0	E0	P802	–	–	–
	2446	NITROCRESOLS, SOLID	6.1	–	III	–	5 kg	E1	P002 LP02	–	IBC08	B3
	2447	PHOSPHORUS, WHITE, MOLTEN	4.2	6.1 P	I	–	0	E0	–	–	–	–
	2448	SULPHUR, MOLTEN	4.1	–	III	–	0	E0	–	–	IBC01	–
	2451	NITROGEN TRIFLUORIDE	2.2	5.1	–	–	0	E0	P200	–	–	–
	2452	ETHYLACETYLENE, STABILIZED	2.1	–	–	–	0	E0	P200	–	–	–
	2453	ETHYL FLUORIDE (REFRIGERANT GAS R 161)	2.1	–	–	–	0	E0	P200	–	–	–
	2454	METHYL FLUORIDE (REFRIGERANT GAS R 41)	2.1	–	–	–	0	E0	P200	–	–	–
△	2455	METHYL NITRITE	2.2	–	–	900	–	–	–	–	–	–

Portable tanks and bulk containers			EmS	Stowage and segregation	Properties and observations	UN No.	
	Tank instructions	Provisions					
(12)	(13) 4.2.5 4.3	(14) 4.2.5	(15) 5.4.3.2 7.3	(16) 7.1 7.2	(17)	(18)	
–	T14	TP2 TP13	F-E, S-C	Category D. Protected from sources of heat. Clear of living quarters. Segregation as for class 3 but "away from" class 4.1.	Flammable liquid. Flashpoint: 19°C c.c. Boiling point: 108°C. Reacts with water, evolving hydrogen chloride, a corrosive gas apparent as white fumes. In the presence of moisture, corrosive to most metals. Highly toxic if swallowed, by skin contact or by inhalation. Causes burns to skin, eyes and mucous membranes.	**2438**	△
–	T3	TP33	F-A, S-B	Category A. Keep as cool as reasonably practicable. Protected from sources of heat. Clear of living quarters. "Separated from" acids.	White, crystalline powder. Soluble in water. Decomposed by heat or acids, evolving hydrogen fluoride, a toxic extremely irritating and corrosive gas. In the presence of moisture, highly corrosive to glass, other siliceous materials and most metals. Causes burns to skin, eyes and mucous membranes.	**2439**	△
–	T1	TP33	F-A, S-B	Category A.	White, deliquescent solid. Melting point: about 60°C. Soluble in water. In the presence of water, corrosive to most metals. Irritating to skin, eyes and mucous membranes.	**2440**	
–	–	–	F-G, S-M	Category D. Clear of living quarters.	Finely divided, violet, crystalline solid. May ignite on exposure to air or moisture. In the presence of moisture, corrosive to most metals. Causes burns to skin, eyes and mucous membranes.	**2441**	
–	T7	TP2	F-A, S-B	Category D. Clear of living quarters.	Liquid with a pungent odour, which fumes in moist air. Reacts violently with water, evolving hydrogen chloride, a corrosive gas apparent as white fumes. When involved in a fire, evolves toxic gases. In the presence of moisture, corrosive to most metals. Liquid and vapours cause burns to skin, eyes and mucous membranes.	**2442**	
–	T7	TP2	F-A, S-B	Category C. Clear of living quarters.	Yellow liquid. Decomposition occurs on exposure to moist air, forming red fumes of vanadic acid and hydrogen chloride, a corrosive gas apparent as white fumes. Reacts with, or dissolves, many organic compounds. In the presence of moisture, corrosive to most metals. Causes burns to skin, eyes and mucous membranes.	**2443**	
–	T10	TP2	F-A, S-B	Category C. Clear of living quarters.	Reddish-brown liquid. Decomposes under the influence of light, evolving chlorine, a highly toxic and irritating gas. Reacts violently with water, evolving hydrogen chloride, a corrosive gas apparent as white fumes. In the presence of moisture, corrosive to most metals. Liquid and vapours cause burns to skin, eyes and mucous membranes.	**2444**	
–	T1	TP33	F-A, S-A	Category A.	Yellow crystals. Melting point: 32°C or above. Slightly soluble in water. Toxic if swallowed, by skin contact or by inhalation.	**2446**	
–	T21	TP3 TP7 TP26	F-A, S-M	Category D.	Molten liquid. Melting point: 44°C. Ignites spontaneously in air. Toxic if swallowed, by skin contact or by inhalation. Shipped molten above its melting point.	**2447**	
–	T1	TP3	F-A, S-H	Category C. "Separated from" class 5.1.	Melting point: 119°C. Molten sulphur may contain hydrogen sulphide, which is highly poisonous in low concentrations. When involved in a fire, evolves toxic, very irritating and suffocating gas. Forms explosive and extremely sensitive mixtures with oxidizing substances. Shipped molten above its melting point.	**2448**	
–	–	–	F-C, S-W	Category D. Clear of living quarters.	Non-flammable, non-toxic, colourless, odourless gas. Strong oxidizing agent; reacts violently with many substances, e.g., grease, oil, etc. Much heavier than air (2.4). May cause slight eye irritation.	**2451**	
–	–	–	F-D, S-U	Category B. Clear of living quarters.	Liquefied, flammable, colourless gas with an odour similar to acetylene. Heavier than air (1.9). Boiling point: 8°C. Irritating to skin, eyes and mucous membranes.	**2452**	
–	–	–	F-D, S-U	Category E. Clear of living quarters.	Liquefied, flammable, colourless gas. Explosive limits: 5% to 10%. Heavier than air (1.7). Boiling point: –37°C.	**2453**	
–	–	–	F-D, S-U	Category E. Clear of living quarters.	Flammable, colourless gas. Heavier than air (1.2).	**2454**	
–	–	–	–	–	Transport is prohibited.	**2455**	△

UN No.	Proper Shipping Name (PSN)	Class or division	Subsidiary risk(s)	Packing group	Special provisions	Limited and excepted quantity provisions		Packing		IBC	
						Limited quantities	Excepted quantities	Instructions	Provisions	Instructions	Provisions
(1)	(2) 3.1.2	(3) 2.0	(4) 2.0	(5) 2.0.1.3	(6) 3.3	(7a) 3.4	(7b) 3.5	(8) 4.1.4	(9) 4.1.4	(10) 4.1.4	(11) 4.1.4
2456	2-CHLOROPROPENE	3	–	I	–	0	E3	P001	–	–	–
2457	2,3-DIMETHYLBUTANE	3	–	II	–	1 L	E2	P001	–	IBC02	–
2458	HEXADIENES	3	–	II	–	1 L	E2	P001	–	IBC02	–
2459	2-METHYL-1-BUTENE	3	–	I	–	0	E3	P001	–	–	–
2460	2-METHYL-2-BUTENE	3	–	II	–	1 L	E2	P001	–	IBC02	B8
2461	METHYLPENTADIENES	3	–	II	–	1 L	E2	P001	–	IBC02	–
2463	ALUMINIUM HYDRIDE	4.3	–	I	–	0	E0	P403	PP31	–	–
2464	BERYLLIUM NITRATE	5.1	6.1	II	–	1 kg	E2	P002	–	IBC08	B2 B4
2465	DICHLOROISOCYANURIC ACID, DRY or DICHLOROISOCYANURIC ACID, SALTS	5.1	–	II	135	1 kg	E2	P002	–	IBC08	B2 B4
2466	POTASSIUM SUPEROXIDE	5.1	–	I	–	0	E0	P503	–	IBC06	B1
2468	TRICHLOROISOCYANURIC ACID, DRY	5.1	–	II	–	1 kg	E2	P002	–	IBC08	B2 B4
2469	ZINC BROMATE	5.1	–	III	–	5 kg	E1	P002 LP02	–	IBC08	B3
2470	PHENYLACETONITRILE, LIQUID	6.1	–	III	–	5 L	E1	P001 LP01	–	IBC03	–
2471	OSMIUM TETROXIDE	6.1	– P	I	–	0	E5	P002	PP30 PP31	IBC07	B1
2473	SODIUM ARSANILATE	6.1	–	III	–	5 kg	E1	P002 LP02	–	IBC08	B3
2474	THIOPHOSGENE	6.1	–	I	279 354	0	E0	P602	–	–	–

Portable tanks and bulk containers			EmS	Stowage and segregation	Properties and observations	UN No.
	Tank instructions	Provisions				
(12)	(13) 4.2.5 4.3	(14) 4.2.5	(15) 5.4.3.2 7.3	(16) 7.1 7.2	(17)	(18)
–	T11	TP2	F-E, S-D	Category E.	Colourless liquid. Flashpoint: below –18°C c.c. Explosive limits: 2.5% to 12%. Boiling point: 23°C. Immiscible with water. Harmful if swallowed or by inhalation. Irritating to skin, eyes and mucous membranes.	2456
–	T7	TP1	F-E, S-D	Category E.	Colourless liquid. Flashpoint: –29°C c.c. Explosive limits: 1.2% to 7%. Immiscible with water. Irritating to skin, eyes and mucous membranes. Narcotic in high concentrations.	2457
–	T4	TP1	F-E, S-D	Category B.	Colourless liquids. 1,3-HEXADIENE: flashpoint –3°C c.c. 1,4-HEXADIENE: flashpoint –25°C c.c. 1,5-HEXADIENE: flashpoint –27°C c.c. 2,4-HEXADIENE: flashpoint –7°C c.c. Immiscible with water. Harmful by inhalation. Irritating to skin, eyes and mucous membranes.	2458
–	T11	TP2	F-E, S-D	Category E.	Colourless, volatile liquid with a disagreeable odour. Flashpoint: below –18°C c.c. Immiscible with water. Irritating to skin, eyes and mucous membranes.	2459
–	T7	TP1	F-E, S-D	Category E.	Colourless, volatile liquid with a disagreeable odour. Flashpoint: below –18°C c.c. Immiscible with water. Irritating to skin, eyes and mucous membranes.	2460
–	T4	TP1	F-E, S-D	Category E.	Colourless liquids. Flashpoint: below –18°C c.c. Immiscible with water. Irritating to skin, eyes and mucous membranes.	2461
–	–	–	F-G, S-O	Category E.	White to grey powder. In contact with water, acids or moisture, evolves hydrogen, which may be ignited by the heat of the reaction.	2463
–	T3	TP33	F-A, S-Q	Category A.	White or light yellow deliquescent crystals, or fine dust. Mixtures with combustible material are readily ignited and may burn fiercely. Toxic if swallowed, by skin contact or by dust inhalation.	2464
–	T3	TP33	F-A, S-Q	Category A. Keep as dry as reasonably practicable.	White crystalline powder or granules; slightly hygroscopic. Partially soluble in water. Mixtures with combustible material are sensitive to friction and are liable to ignite. Harmful by inhalation. Irritating to skin, eyes and mucous membranes.	2465
–	–	–	F-G, S-Q	Category E. Keep as dry as reasonably practicable. "Separated from" permanganates, acids and class 4.1.	Yellow flakes. Particularly if wetted with small quantities of water, a mixture with combustible material may ignite, following impact or friction. When involved in a fire, or in contact with water or acids, decomposes evolving oxygen. Highly irritating to skin, eyes and mucous membranes.	2466
–	T3	TP33	F-A, S-Q	Category A. Keep as dry as reasonably practicable.	Colourless powder or granules. Mixtures with combustible material are sensitive to friction and are liable to ignite. On contact with nitrogen compounds, fumes of nitrogen trichloride can be formed, which are very explosive. Harmful by inhalation. Irritating to skin, eyes and mucous membranes.	2468
–	T1	TP33	F-H, S-Q	Category A. "Separated from" ammonium compounds and cyanides.	Colourless powder. Soluble in water. Reacts vigorously with sulphuric acid. Reacts fiercely with cyanides when heated or by friction. May form explosive mixtures with combustible material, powdered metals or ammonium compounds. These mixtures are sensitive to friction and are liable to ignite. When involved in a fire, may cause an explosion.	2469
–	T4	TP1	F-A, S-A	Category A. "Separated from" acids.	Colourless to light brown liquid. Immiscible with water. Toxic if swallowed, by skin contact or by inhalation.	2470
–	T6	TP33	F-A, S-A	Category B. Clear of living quarters.	Pale yellow, crystalline, volatile solid with an irritating odour. Highly toxic if swallowed, by skin contact or by inhalation.	2471
–	T1	TP33	F-A, S-A	Category A.	White, crystalline powder. Soluble in water. Toxic if swallowed, by skin contact or by dust inhalation.	2473
–	T20	TP2 TP13 TP37	F-A, S-A	Category D. Clear of living quarters. "Separated from" acids.	Red fuming liquid with a foul phosgene-like odour. Decomposes slowly in water. Reacts with acids, evolving toxic and corrosive fumes. Highly toxic if swallowed, by skin contact or by inhalation.	2474

UN No.	Proper Shipping Name (PSN)	Class or division	Subsidiary risk(s)	Packing group	Special provisions	Limited and excepted quantity provisions		Packing		IBC	
						Limited quantities	Excepted quantities	Instructions	Provisions	Instructions	Provisions
(1)	(2) 3.1.2	(3) 2.0	(4) 2.0	(5) 2.0.1.3	(6) 3.3	(7a) 3.4	(7b) 3.5	(8) 4.1.4	(9) 4.1.4	(10) 4.1.4	(11) 4.1.4
2475	VANADIUM TRICHLORIDE	8	–	III	–	5 kg	E1	P002 LP02	–	IBC08	B3
2477	METHYL ISOTHIOCYANATE	6.1	3	I	354	0	E0	P602	–	–	–
2478	ISOCYANATES, FLAMMABLE, TOXIC, N.O.S. or ISOCYANATE SOLUTION, FLAMMABLE, TOXIC, N.O.S.	3	6.1	II	274	1 L	E2	P001	PP31	IBC02	–
2478	ISOCYANATES, FLAMMABLE, TOXIC, N.O.S. or ISOCYANATE SOLUTION, FLAMMABLE, TOXIC, N.O.S.	3	6.1	III	223 274	5 L	E1	P001	PP31	IBC03	–
2480	METHYL ISOCYANATE	6.1	3	I	354	0	E0	P601	–	–	–
2481	ETHYL ISOCYANATE	6.1	3	I	354	0	E0	P602	–	–	–
2482	*n*-PROPYL ISOCYANATE	6.1	3	I	354	0	E0	P602	–	–	–
2483	ISOPROPYL ISOCYANATE	6.1	3	I	354	0	E0	P602	–	–	–
2484	*tert*-BUTYL ISOCYANATE	6.1	3	I	354	0	E0	P602	–	–	–
2485	*n*-BUTYL ISOCYANATE	6.1	3	I	354	0	E0	P602	–	–	–
2486	ISOBUTYL ISOCYANATE	6.1	3	I	354	0	E0	P602	–	–	–
2487	PHENYL ISOCYANATE	6.1	3	I	354	0	E0	P602	–	–	–
2488	CYCLOHEXYL ISOCYANATE	6.1	3	I	354	0	E0	P602	–	–	–

Portable tanks and bulk containers			EmS	Stowage and segregation	Properties and observations	UN No.
	Tank instructions	Provisions				
(12)	(13) 4.2.5 4.3	(14) 4.2.5	(15) 5.4.3.2 7.3	(16) 7.1 7.2	(17)	(18)
–	T1	TP33	F-A, S-B	Category A. Clear of living quarters.	Pink, deliquescent crystals. Decomposes in water, evolving hydrogen chloride, a corrosive gas apparent as white fumes. In the presence of moisture, highly corrosive to most metals. Irritating to skin, eyes and mucous membranes.	**2475**
–	T20	TP2 TP13 TP37	F-E, S-D	Category D. Clear of living quarters.	White crystals. Usually shipped as an oily liquid with a flashpoint below 60°C c.c. Melting point: 36°C (pure substance). Flashpoint: 32°C c.c. (pure substance). Insoluble in water. When involved in a fire, evolves toxic gases. Highly toxic if swallowed, by skin contact or by inhalation.	**2477**
–	T11	TP2 TP13 TP27	F-E, S-D	Category D. Clear of living quarters.	Flammable toxic liquids with a pungent odour. Immiscible with water but react with it to form carbon dioxide. Toxic if swallowed, by skin contact or by inhalation. Irritating to skin, eyes and mucous membranes.	**2478**
–	T7	TP1 TP13 TP28	F-E, S-D	Category A.	See entry above.	**2478**
–	T22	TP2 TP13	F-E, S-D	Category D. Clear of living quarters. "Separated from" acids.	Flammable liquid with a pungent odour. Flashpoint: –7°C c.c. (pure product). Boiling point: 38°C (pure product). Vapour heavier than air. Immiscible with water but reacts violently with it. In contact with water or acids, evolves highly toxic nitrous fumes. Highly toxic if swallowed, by skin contact or by inhalation. Irritating to skin, eyes and mucous membranes.	**2480**
–	T20	TP2 TP13 TP37	F-E, S-D	Category D. Clear of living quarters. "Separated from" acids.	Liquid with a pungent odour. Flashpoint: –18°C to 0°C c.c. Boiling point: 60°C. Immiscible with water but reacts violently with it. On contact with water or acids, or when heated above boiling point, evolves highly toxic nitrous fumes. Highly toxic if swallowed, by skin contact or by inhalation. Irritating to skin, eyes and mucous membranes.	**2481**
–	T20	TP2 TP13 TP37	F-E, S-D	Category D. Clear of living quarters.	Flammable liquid with a pungent odour. Immiscible with water but reacts violently with it, evolving gases. Flashpoint: –18°C to 23°C c.c. Highly toxic if swallowed, by skin contact or by inhalation. Irritating to skin, eyes and mucous membranes.	**2482**
–	T20	TP2 TP13 TP37	F-E, S-D	Category D. Clear of living quarters.	Liquid with a pungent odour. Flashpoint: –10°C to 0°C c.c. Immiscible with water but reacts violently with it, evolving gases. Highly toxic if swallowed, by skin contact or by inhalation. Irritating to skin, eyes and mucous membranes.	**2483**
–	T20	TP2 TP13 TP37	F-E, S-D	Category D. Clear of living quarters.	Colourless liquid with a pungent odour. Immiscible with water but reacts violently with it, evolving gases. Flashpoint: 11°C c.c. Highly toxic if swallowed, by skin contact or by inhalation. Irritating to skin, eyes and mucous membranes.	**2484**
–	T20	TP2 TP13 TP37	F-E, S-D	Category D. Clear of living quarters.	Colourless liquid with a pungent odour. Immiscible with water but reacts violently with it, evolving gases. Flashpoint: 19°C c.c. Highly toxic if swallowed, by skin contact or by inhalation. Irritating to skin, eyes and mucous membranes.	**2485**
–	T20	TP2 TP13 TP37	F-E, S-D	Category D. Clear of living quarters.	Liquid with a pungent odour. Immiscible with water but reacts violently with it, evolving gases. Highly toxic if swallowed, by skin contact or by inhalation. Irritating to skin, eyes and mucous membranes.	**2486**
–	T20	TP2 TP13 TP37	F-E, S-D	Category D. Clear of living quarters.	Colourless to yellowish liquid with a pungent odour. Flashpoint: 51°C c.c. Immiscible with water. Reacts with water, evolving carbon dioxide. Highly toxic if swallowed, by skin contact or by inhalation. Irritating to skin, eyes and mucous membranes.	**2487**
–	T20	TP2 TP13 TP37	F-E, S-D	Category D. Clear of living quarters.	Yellowish liquid with an irritating odour. Flashpoint: 53°C c.c. Immiscible with water. Reacts with water, evolving carbon dioxide. Highly toxic if swallowed, by skin contact or by inhalation. Irritating to skin, eyes and mucous membranes.	**2488**

UN No.	Proper Shipping Name (PSN)	Class or division	Subsidiary risk(s)	Packing group	Special provisions	Limited and excepted quantity provisions		Packing		IBC	
						Limited quantities	Excepted quantities	Instructions	Provisions	Instructions	Provisions
(1)	(2) 3.1.2	(3) 2.0	(4) 2.0	(5) 2.0.1.3	(6) 3.3	(7a) 3.4	(7b) 3.5	(8) 4.1.4	(9) 4.1.4	(10) 4.1.4	(11) 4.1.4
2490	DICHLOROISOPROPYL ETHER	6.1	–	II	–	100 mL	E4	P001	–	IBC02	–
2491	ETHANOLAMINE or ETHANOLAMINE SOLUTION	8	–	III	223	5 L	E1	P001 LP01	–	IBC03	–
2493	HEXAMETHYLENEIMINE	3	8	II	–	1 L	E2	P001	–	IBC02	–
△ 2495	IODINE PENTAFLUORIDE	5.1	6.1/8	I	–	0	E0	P200	–	–	–
2496	PROPIONIC ANHYDRIDE	8	–	III	–	5 L	E1	P001 LP01	–	IBC03	–
2498	1,2,3,6-TETRAHYDRO-BENZALDEHYDE	3	–	III	–	5 L	E1	P001 LP01	–	IBC03	–
2501	TRIS-(1-AZIRIDINYL)-PHOSPHINE OXIDE SOLUTION	6.1	–	II	–	100 mL	E4	P001	–	IBC02	–
2501	TRIS-(1-AZIRIDINYL)-PHOSPHINE OXIDE SOLUTION	6.1	–	III	223	5 L	E1	P001 LP01	–	IBC03	–
2502	VALERYL CHLORIDE	8	3	II	–	1 L	E2	P001	–	IBC02	–
2503	ZIRCONIUM TETRACHLORIDE	8	–	III	–	5 kg	E1	P002 LP02	–	IBC08	B3
2504	TETRABROMOETHANE	6.1	– P	III	–	5 L	E1	P001 LP01	–	IBC03	–
2505	AMMONIUM FLUORIDE	6.1	–	III	–	5 kg	E1	P002 LP02	–	IBC08	B3
2506	AMMONIUM HYDROGEN SULPHATE	8	–	II	–	1 kg	E2	P002	–	IBC08	B2 B4
2507	CHLOROPLATINIC ACID, SOLID	8	–	III	–	5 kg	E1	P002 LP02	–	IBC08	B3
2508	MOLYBDENUM PENTACHLORIDE	8	–	III	–	5 kg	E1	P002 LP02	–	IBC08	B3
2509	POTASSIUM HYDROGEN SULPHATE	8	–	II	–	1 kg	E2	P002	–	IBC08	B2 B4
2511	2-CHLOROPROPIONIC ACID	8	–	III	223	5 L	E1	P001 LP01	–	IBC03	–

Portable tanks and bulk containers			EmS	Stowage and segregation	Properties and observations	UN No.
	Tank instructions	Provisions				
(12)	(13) 4.2.5 4.3	(14) 4.2.5	(15) 5.4.3.2 7.3	(16) 7.1 7.2	(17)	(18)
–	T7	TP2	F-A, S-A	Category B.	Colourless liquid. Immiscible with water. Toxic if swallowed, by skin contact or by inhalation.	**2490**
–	T4	TP1	F-A, S-B	Category A. "Separated from" acids.	Colourless. Miscible with water. Corrosive to copper, copper compounds, copper alloys and rubber. Reacts violently with acids. Liquid and vapour cause burns to skin, eyes and mucous membranes.	**2491**
–	T7	TP1	F-E, S-C	Category B. Clear of living quarters.	Yellowish liquid with an ammoniacal odour. Flashpoint: 18°C c.c. Miscible with water. Harmful by inhalation. Absorbed through the skin. Causes burns to skin, eyes and mucous membranes.	**2493**
–	–	–	F-A, S-Q	Category D. Protected from sources of heat. Clear of living quarters. Segregation as for class 5.1 but "separated from" classes 4.1 and 7. "Separated from" acids.	Colourless, fuming liquid (density 3.75). Powerful oxidant; may cause fire in contact with organic material such as wood, cotton or straw. Reacts violently with water, evolving hydrogen fluoride, a toxic, extremely corrosive gas apparent as white fumes. In contact with acids or acid fumes, evolves highly toxic fumes of iodine, fluorine and their compounds. Highly corrosive to most metals. Toxic if swallowed, by skin contact or by inhalation. Causes burns to skin, eyes and mucous membranes.	**2495** △
–	T4	TP1	F-A, S-B	Category A.	Colourless, combustible liquid with a pungent odour. Reacts with water, forming propionic acid. Corrosive to skin, eyes and mucous membranes.	**2496**
–	T2	TP1	F-E, S-D	Category A.	Colourless liquid. Flashpoint: 57°C o.c. Immiscible with water.	**2498**
–	T7	TP2	F-A, S-A	Category A.	Aqueous solution. Miscible with water. Toxic if swallowed, by skin contact or by inhalation.	**2501**
–	T4	TP1	F-A, S-A	Category A.	See entry above.	**2501**
–	T7	TP2	F-E, S-C	Category C. Clear of living quarters.	Liquid with a penetrating odour. Flashpoint: 23°C c.c. or above. Reacts with water, evolving hydrogen chloride, a corrosive gas apparent as white fumes. Corrosive to most metals. Causes burns to skin, eyes and mucous membranes.	**2502**
–	T1	TP33	F-A, S-B	Category A.	White, lustrous crystals. Reacts with water, evolving hydrogen chloride, a corrosive gas apparent as white fumes. In the presence of moisture, corrosive to most metals. Irritating to mucous membranes.	**2503**
–	T4	TP1	F-A, S-A	Category A.	Colourless to yellowish liquid with a camphor-like odour. Toxic if swallowed, by skin contact or by inhalation.	**2504**
–	T1	TP33	F-A, S-A	Category A. "Separated from" acids.	Colourless crystals or powder with an ammonia-like odour. Readily soluble in water. Decomposes in contact with acids, evolving hydrogen fluoride, a corrosive gas. Toxic if swallowed, by skin contact or by dust inhalation.	**2505**
–	T3	TP33	F-A, S-B	Category A. Clear of living quarters.	White, rhombic crystals. Soluble in water. When involved in a fire, evolves extremely irritating and corrosive fumes. In the presence of moisture, corrosive to most metals. Causes burns to skin, eyes and mucous membranes.	**2506**
–	T1	TP33	F-A, S-B	Category A.	Red-brown crystals. Soluble in water.	**2507**
–	T1	TP33	F-A, S-B	Category C. Clear of living quarters.	Black or green-black crystals. Hygroscopic. Reacts violently with water, evolving hydrogen chloride, a corrosive gas apparent as white fumes. Harmful if swallowed. Dust and vapour irritate skin, eyes and mucous membranes.	**2508**
–	T3	TP33	F-A, S-B	Category A.	Colourless crystals. Soluble in water. When involved in a fire, evolves extremely irritating and corrosive fumes. In the presence of moisture, corrosive to most metals. Irritating to skin, eyes and mucous membranes.	**2509**
–	T4	TP2	F-A, S-B	Category A.	Colourless, aqueous solution with a specific odour. Causes burns to skin, eyes and mucous membranes.	**2511**

3 DGL

UN No.	Proper Shipping Name (PSN)	Class or division	Subsidiary risk(s)	Packing group	Special provisions	Limited and excepted quantity provisions		Packing		IBC	
						Limited quantities	Excepted quantities	Instructions	Provisions	Instructions	Provisions
(1)	(2) 3.1.2	(3) 2.0	(4) 2.0	(5) 2.0.1.3	(6) 3.3	(7a) 3.4	(7b) 3.5	(8) 4.1.4	(9) 4.1.4	(10) 4.1.4	(11) 4.1.4
2512	AMINOPHENOLS (*o*-, *m*-, *p*-)	6.1	–	III	279	5 kg	E1	P002 LP02	–	IBC08	B3
2513	BROMOACETYL BROMIDE	8	–	II	–	1 L	E2	P001	–	IBC02	B20
2514	BROMOBENZENE	3	– P	III	–	5 L	E1	P001 LP01	–	IBC03	–
2515	BROMOFORM	6.1	– P	III	–	5 L	E1	P001 LP01	–	IBC03	–
△ 2516	CARBON TETRABROMIDE	6.1	– P	III	–	5 kg	E1	P002 LP02	–	IBC08	B3
2517	1-CHLORO-1,1-DIFLUORO-ETHANE (REFRIGERANT GAS R 142b)	2.1	–	–	–	0	E0	P200	–	–	–
2518	1,5,9-CYCLODODECATRIENE	6.1	– P	III	–	5 L	E1	P001 LP01	–	IBC03	–
2520	CYCLOOCTADIENES	3	–	III	–	5 L	E1	P001 LP01	–	IBC03	–
2521	DIKETENE, STABILIZED	6.1	3	I	354	0	E0	P602	–	–	–
2522	2-DIMETHYLAMINOETHYL METHACRYLATE	6.1	–	II	–	100 mL	E4	P001	–	IBC02	–
2524	ETHYL ORTHOFORMATE	3	–	III	–	5 L	E1	P001 LP01	–	IBC03	–
2525	ETHYL OXALATE	6.1	–	III	–	5 L	E1	P001 LP01	–	IBC03	–
2526	FURFURYLAMINE	3	8	III	–	5 L	E1	P001	–	IBC03	–
2527	ISOBUTYL ACRYLATE, STABILIZED	3	–	III	–	5 L	E1	P001 LP01	–	IBC03	–
2528	ISOBUTYL ISOBUTYRATE	3	–	III	–	5 L	E1	P001 LP01	–	IBC03	–
2529	ISOBUTYRIC ACID	3	8	III	–	5 L	E1	P001	–	IBC03	–
2531	METHACRYLIC ACID, STABILIZED	8	–	II	–	1 L	E2	P001	–	IBC02	–
2533	METHYL TRICHLOROACETATE	6.1	–	III	–	5 L	E1	P001 LP01	–	IBC03	–

Portable tanks and bulk containers			EmS	Stowage and segregation	Properties and observations	UN No.
	Tank instructions	Provisions				
(12)	(13) 4.2.5 4.3	(14) 4 2.5	(15) 5.4.3.2 7.3	(16) 7.1 7.2	(17)	(18)
–	T1	TP33	F-A, S-A	Category A.	White or brownish (*ortho-* and *para-*) or reddish-yellow (*meta-*) crystals. Soluble in water. Toxic if swallowed, by skin contact or by inhalation.	2512
–	T8	TP2	F-A, S-B	Category C. Clear of living quarters. "Separated from" alkalis.	Clear liquid, colourless. Boiling point: 150°C. Reacts violently with water, evolving hydrogen bromide, an irritating and corrosive gas apparent as white fumes. In the presence of moisture, highly corrosive to most metals. Reacts violently with alkalis such as ammonia and hydrazine. Causes very severe burns to skin, eyes and mucous membranes. Vapour causes tears.	2513
–	T2	TP1	F-E, S-D	Category A.	Colourless liquid with a characteristic odour. Flashpoint: 51°C c.c. Explosive limits: 0.5% to 2.8%. Immiscible with water.	2514
–	T4	TP1	F-A, S-A	Category A. Keep as cool as reasonably practicable. Clear of living quarters.	Colourless liquid or crystals (melting point 9°C) with a chloroform-like odour. Toxic if swallowed, by skin contact or by inhalation. Narcotic effect.	2515
–	T1	TP33	F-A, S-A	Category A. Protected from sources of heat.	Colourless crystals. Melting point: 48°C. Insoluble in water. Toxic if swallowed, by skin contact or by inhalation of dust and vapour.	2516 △
–	T50	–	F-D, S-U	Category B. Clear of living quarters.	Flammable gas. Explosive limits: 8.5% to 14%. Much heavier than air (3.5).	2517
–	T4	TP1	F-A, S-A	Category A. Clear of living quarters.	Colourless liquid. Toxic if swallowed, by skin contact or by inhalation.	2518
–	T2	TP1	F-E, S-D	Category A.	Colourless liquids. Immiscible with water. 1,5-CYCLOOCTADIENE: flashpoint 38°C c.c. Irritating to skin, eyes and mucous membranes.	2520
–	T20	TP2 TP13 TP37	F-E, S-D	Category D. "Away from" acids and alkalis. Clear of living quarters.	Colourless flammable liquid with a pungent odour. Flashpoint: 44°C c.c. Immiscible with water, but hydrolyses slowly in contact with it. The presence of acids, bases or amines can initiate explosive polymerization. Highly toxic if swallowed, by skin contact or by inhalation.	2521
–	T7	TP2	F-A, S-A	Category D. Clear of living quarters.	Combustible liquid. Causes tears. Toxic if swallowed, by skin contact or by inhalation.	2522
–	T2	TP1	F-E, S-D	Category A.	Colourless liquid with an ethereal odour. Flashpoint: 30°C c.c. Immiscible with water.	2524
–	T4	TP1	F-A, S-A	Category A.	Colourless, oily, aromatic liquid. Slowly decomposed by water. Toxic if swallowed, by skin contact or by dust inhalation.	2525
–	T4	TP1	F-E, S-C	Category A. Clear of living quarters.	Pale yellow, oily liquid. Flashpoint: 37°C o.c. Miscible with water. Harmful by inhalation. Causes burns to skin and eyes. Irritating to mucous membranes.	2526
–	T2	TP1	F-E, S-D	Category A.	Colourless liquid with a pungent odour. Flashpoint: 29°C o.c. Immiscible with water. Harmful by inhalation. Irritating to skin, eyes and mucous membranes.	2527
–	T2	TP1	F-E, S-D	Category A.	Colourless liquid with a fruity odour. Flashpoint: 37°C c.c. Explosive limits: 0.96% to 7.59%. Immiscible with water.	2528
–	T4	TP1	F-E, S-C	Category A.	Colourless liquid with a pungent odour. Flashpoint: 55°C c.c. Explosive limits: 2% to 9.2%. Miscible with water. Causes burns to skin and eyes. Irritating to skin, eyes and mucous membranes.	2529
–	T7	TP2 TP18 TP30	F-A, S-B	Category C. Clear of living quarters.	Colourless, combustible liquid with a specific odour. Miscible with water. Polymerizes readily above its melting point (15°C), thereby generating heat and possible risk of explosion; should therefore be properly stabilized. Cooling below melting point (15°C) followed by subsequent reheating can release uninhibited monomer that readily polymerizes. Decomposes when heated, evolving toxic gases. Causes burns to skin, eyes and mucous membranes.	2531
–	T4	TP1	F-A, S-A	Category A.	Colourless liquid. Immiscible with water. Toxic if swallowed, by skin contact or by inhalation.	2533

UN No.	Proper Shipping Name (PSN)	Class or division	Subsidiary risk(s)	Packing group	Special provisions	Limited and excepted quantity provisions		Packing		IBC	
						Limited quantities	Excepted quantities	Instructions	Provisions	Instructions	Provisions
(1)	(2) 3.1.2	(3) 2.0	(4) 2.0	(5) 2.0.1.3	(6) 3.3	(7a) 3.4	(7b) 3.5	(8) 4.1.4	(9) 4.1.4	(10) 4.1.4	(11) 4.1.4
2534	METHYLCHLOROSILANE	2.3	2.1/8	–	–	0	E0	P200	–	–	–
2535	4-METHYLMORPHOLINE (*N*-METHYLMORPHOLINE)	3	8	II	–	1 L	E2	P001	–	IBC02	–
2536	METHYLTETRAHYDROFURAN	3	–	II	–	1 L	E2	P001	–	IBC02	–
2538	NITRONAPHTHALENE	4.1	–	III	–	5 kg	E1	P002 LP02	–	IBC08	B3
2541	TERPINOLENE	3	–	III	–	5 L	E1	P001 LP01	–	IBC03	–
2542	TRIBUTYLAMINE	6.1	–	II	–	100 mL	E4	P001	–	IBC02	–
2545	HAFNIUM POWDER, DRY	4.2	–	I	–	0	E0	P404	PP31	–	–
2545	HAFNIUM POWDER, DRY	4.2	–	II	–	0	E2	P410	PP31	IBC06	B2
2545	HAFNIUM POWDER, DRY	4.2	–	III	223	0	E1	P002 LP02	PP31	IBC08	B3
2546	TITANIUM POWDER, DRY	4.2	–	I	–	0	E0	P404	PP31	–	–
2546	TITANIUM POWDER, DRY	4.2	–	II	–	0	E2	P410	PP31	IBC06	B2
2546	TITANIUM POWDER, DRY	4.2	–	III	223	0	E1	P002 LP02	PP31	IBC08	B3
2547	SODIUM SUPEROXIDE	5.1	–	I	–	0	E0	P503	–	IBC06	B1
2548	CHLORINE PENTAFLUORIDE	2.3	5.1/8	–	–	0	E0	P200	–	–	–
2552	HEXAFLUOROACETONE HYDRATE, LIQUID	6.1	–	II	–	100 mL	E4	P001	–	IBC02	–
2554	METHYLALLYL CHLORIDE	3	–	II	–	1 L	E2	P001	–	IBC02	–
2555	NITROCELLULOSE WITH WATER (not less than 25% water, by mass)	4.1	–	II	28	0	E0	P406	PP31	–	–

Portable tanks and bulk containers			EmS	Stowage and segregation	Properties and observations	UN No.
	Tank instructions	Provisions				
(12)	(13) 4.2.5 4.3	(14) 4.2.5	(15) 5.4.3.2 7.3	(16) 7.1 7.2	(17)	(18)
–	–	–	F-D, S-U	Category D. Clear of living quarters. Segregation as for class 2.1 but "away from" class 4.3.	Liquefied, flammable, toxic and corrosive colourless gas with a pungent odour. Reacts with water, evolving hydrogen chloride, an irritating and corrosive gas. Heavier than air. Boiling point: 9°C. Highly irritating to skin, eyes and mucous membranes.	2534
–	T7	TP1	F-E, S-C	Category B. Clear of living quarters.	Colourless liquid with an ammonia-like odour. Flashpoint: 13°C c.c. Miscible with water. Harmful by inhalation. Causes burns to skin and eyes. Irritating to mucous membranes.	2535
–	T4	TP1	F-E, S-D	Category B.	Colourless, volatile liquid with an ether-like odour. Flashpoint: –11°C o.c. Immiscible with water.	2536
–	T1	TP33	F-A, S-G	Category A.	Yellow crystals. Insoluble in water. Harmful if swallowed.	2538
–	T2	TP1	F-E, S-E	Category A.	Colourless to pale amber liquid with a lemon odour. Flashpoint: 37°C c.c. Immiscible with water.	2541
–	T7	TP2	F-A, S-A	Category A.	Colourless, combustible liquid with an amine odour. Immiscible with water. When involved in a fire, evolves toxic gases. Toxic if swallowed, by skin contact or by inhalation.	2542
–	–	–	F-G, S-M	Category D.	Black amorphous powder. Insoluble in water. Liable to ignite spontaneously in air. Forms explosive mixtures with oxidizing substances.	2545
–	T3	TP33	F-G, S-M	Category D.	See entry above.	2545
–	T1	TP33	F-G, S-M	Category D.	See entry above.	2545
–	–	–	F-G, S-M	Category D.	Grey powder. Liable to ignite spontaneously in air. Forms explosive mixtures with oxidizing substances.	2546
–	T3	TP33	F-G, S-M	Category D.	See entry above.	2546
–	T1	TP33	F-G, S-M	Category D.	See entry above.	2546
–	–	–	F-G, S-Q	Category E. Keep as dry as reasonably practicable. "Separated from" permanganates, acids and class 4.1.	Pale yellow coarse powder or granules. Particularly if wetted with small quantities of water, a mixture with combustible material may ignite, following impact or friction. When involved in a fire, or in contact with water or acids, decomposes, evolving oxygen. Highly irritating to skin, eyes and mucous membranes.	2547
–	–	–	F-C, S-W	Category D. Clear of living quarters. Segregation as for class 5.1 but "separated from" class 7.	Non-flammable, toxic and corrosive gas. Forms dense, white, corrosive fumes in moist air. Reacts violently with water, evolving hydrogen fluoride, a toxic, irritating and corrosive gas apparent as white fumes. Corrosive to glass and to most metals. Powerful oxidizing agent which may cause violent fires with combustible materials. Much heavier than air (4.5). Highly irritating to skin, eyes and mucous membranes.	2548
–	T7	TP2	F-A, S-A	Category B. Clear of living quarters.	Toxic if swallowed, by skin contact or by inhalation.	2552
–	T4	TP1 TP13	F-E, S-D	Category E.	Colourless to yellowish, volatile liquid with a penetrating odour. Flashpoint: –12°C c.c. Explosive limits: 2.3% to 9.3%. Immiscible with water. When involved in a fire, may evolve highly toxic phosgene gas. Harmful by inhalation. Irritating to skin, eyes and mucous membranes.	2554
–	–	–	F-B, S-J	Category E. "Away from" class 3 and heavy metals and their salts.	Desensitized explosive. Nitrocellulose may be granular or in flakes, blocks or fibrous form. When involved in a fire, evolves toxic fumes; in closed compartments, these fumes may form an explosive mixture with air. May form extremely sensitive compounds with heavy metals or their salts.	2555

UN No.	Proper Shipping Name (PSN)	Class or division	Subsidiary risk(s)	Packing group	Special provisions	Limited and excepted quantity provisions		Packing		IBC	
						Limited quantities	Excepted quantities	Instructions	Provisions	Instructions	Provisions
(1)	(2) 3.1.2	(3) 2.0	(4) 2.0	(5) 2.0.1.3	(6) 3.3	(7a) 3.4	(7b) 3.5	(8) 4.1.4	(9) 4.1.4	(10) 4.1.4	(11) 4.1.4
2556	NITROCELLULOSE WITH ALCOHOL (not less than 25% alcohol, by mass, and not more than 12.6% nitrogen, by dry mass)	4.1	–	II	28	0	E0	P406	PP31	–	–
2557	NITROCELLULOSE with not more than 12.6% nitrogen, by dry mass, MIXTURE WITH or WITHOUT PLASTICIZER, WITH or WITHOUT PIGMENT	4.1	–	II	241	0	E0	P406	PP31	–	–
2558	EPIBROMOHYDRIN	6.1	3 P	I	–	0	E5	P001	–	–	–
2560	2-METHYLPENTAN-2-OL	3	–	III	–	5 L	E1	P001 LP01	–	IBC03	–
2561	3-METHYL-1-BUTENE	3	–	I	–	0	E3	P001	–	–	–
2564	TRICHLOROACETIC ACID SOLUTION	8	–	II	–	1 L	E2	P001	–	IBC02	–
2564	TRICHLOROACETIC ACID SOLUTION	8	–	III	223	5 L	E1	P001 LP01	–	IBC03	–
2565	DICYCLOHEXYLAMINE	8	–	III	–	5 L	E1	P001 LP01	–	IBC03	–
2567	SODIUM PENTACHLOROPHENATE	6.1	– P	II	–	500 g	E4	P002	–	IBC08	B2 B4
2570	CADMIUM COMPOUND	6.1	–	I	274	0	E5	P002	–	IBC07	B1
2570	CADMIUM COMPOUND	6.1	–	II	274	500 g	E4	P002	–	IBC08	B2 B4
2570	CADMIUM COMPOUND	6.1	–	III	223 274	5 kg	E1	P002 LP02	–	IBC08	B3
△ 2571	ALKYLSULPHURIC ACIDS	8	–	II	–	1 L	E2	P001	–	IBC02	–
2572	PHENYLHYDRAZINE	6.1	–	II	–	100 mL	E4	P001	–	IBC02	–
2573	THALLIUM CHLORATE	5.1	6.1 P	II	–	1 kg	E2	P002	–	IBC06	B2
2574	TRICRESYL PHOSPHATE with more than 3% *ortho*-isomer	6.1	– P	II	–	100 mL	E4	P001	–	IBC02	–

Portable tanks and bulk containers			EmS	Stowage and segregation	Properties and observations	UN No.
	Tank instructions	Provisions				
(12)	(13) 4.2.5 4.3	(14) 4.2.5	(15) 5.4.3.2 7.3	(16) 7.1 7.2	(17)	(18)
–	–	–	F-B, S-J	Category D. "Away from" class 3 and heavy metals and their salts.	Nitrocellulose may be granular or in flakes, blocks or fibrous form. In case of leakage, flammable vapours are evolved which, in closed compartments, may form explosive mixtures with air. When involved in a fire, evolves toxic fumes; in closed compartments, these fumes may form an explosive mixture with air. Highly explosive when dry. May form extremely sensitive compounds with heavy metals or their salts.	2556
–	–	–	F-B, S-J	Category D. "Away from" class 3 and heavy metals and their salts.	Nitrocellulose may be in granular form or in flakes. This product may also contain added pigments. When involved in a fire, evolves toxic fumes; in closed compartments, these fumes may form an explosive mixture with air. Burns extremely rapidly with intense heat radiation. The formulation should be prepared so that it remains homogeneous and does not separate during transport. May form extremely sensitive compounds with heavy metals or their salts.	2557
–	T14	TP2 TP13	F-E, S-D	Category D. Clear of living quarters.	Flammable liquid. Flashpoint: 56°C c.c. Highly toxic if swallowed, by skin contact or by inhalation.	2558
–	T2	TP1	F-E, S-D	Category A.	Colourless liquid. Flashpoint: 30°C c.c. Partially miscible with water. Irritating to skin, eyes and mucous membranes.	2560
–	T11	TP2	F-E, S-D	Category E.	Colourless, volatile liquid with a disagreeable odour. Flashpoint: below −18°C c.c. Immiscible with water. Irritating to skin, eyes and mucous membranes.	2561
–	T7	TP2	F-A, S-B	Category B.	Colourless, clear solution with a pungent odour. Corrosive to most metals. Causes burns to skin, eyes and mucous membranes.	2564
–	T4	TP1	F-A, S-B	Category B.	See entry above.	2564
–	T4	TP1	F-A, S-B	Category A.	Clear, colourless, combustible liquid with a fishy odour which may taint other cargoes. Immiscible with water. Causes burns to skin, eyes and mucous membranes.	2565
–	T3	TP33	F-A, S-A	Category A.	White or light brown powder with a pungent odour. Soluble in water. Toxic if swallowed, by skin contact or by dust inhalation.	2567
–	T6	TP33	F-A, S-A	Category A.	Powder or crystals with various colours. May be soluble or insoluble in water. Toxic if swallowed, by skin contact or by dust inhalation.	2570
–	T3	TP33	F-A, S-A	Category A.	See entry above.	2570
–	T1	TP33	F-A, S-A	Category A.	See entry above.	2570
–	T8	TP2 TP13 TP28	F-A, S-B	Category C. For metal drums, category B.	Colourless oily liquids. React with water, evolving heat. Cause burns to skin, eyes and mucous membranes. Highly corrosive to metal.	2571 △
–	T7	TP2	F-A, S-A	Category A. Clear of living quarters.	Pale yellow oily liquid. Melting point: 20°C. Slightly soluble in water. Toxic if swallowed, by skin contact or by inhalation.	2572
–	T3	TP33	F-H, S-Q	Category A. "Separated from" ammonium compounds and cyanides.	Colourless crystals. Slightly soluble in water. Reacts vigorously with sulphuric acid. Reacts fiercely with cyanides when heated or by friction. May form explosive mixtures with combustible material, powdered metals or ammonium compounds. These mixtures are sensitive to friction and are liable to ignite. When involved in a fire, may cause an explosion. Toxic if swallowed, by skin contact or by dust inhalation.	2573
–	T7	TP2	F-A, S-A	Category A.	Colourless liquid. A mixture of isomers. Immiscible with water. Toxic if swallowed, by skin contact or by inhalation.	2574

UN No.	Proper Shipping Name (PSN)	Class or division	Subsidiary risk(s)	Packing group	Special provisions	Limited and excepted quantity provisions		Packing		IBC	
						Limited quantities	Excepted quantities	Instructions	Provisions	Instructions	Provisions
(1)	(2) 3.1.2	(3) 2.0	(4) 2.0	(5) 2.0.1.3	(6) 3.3	(7a) 3.4	(7b) 3.5	(8) 4.1.4	(9) 4.1.4	(10) 4.1.4	(11) 4.1.4
2576	**PHOSPHORUS OXYBROMIDE, MOLTEN**	8	–	II	–	0	E0	–	–	–	–
2577	**PHENYLACETYL CHLORIDE**	8	–	II	–	1 L	E2	P001	–	IBC02	–
2578	**PHOSPHORUS TRIOXIDE**	8	–	III	–	5 kg	E1	P002 LP02	–	IBC08	B3
2579	**PIPERAZINE**	8	–	III	–	5 kg	E1	P002 LP02	–	IBC08	B3
2580	**ALUMINIUM BROMIDE SOLUTION**	8	–	III	223	5 L	E1	P001 LP01	–	IBC03	–
2581	**ALUMINIUM CHLORIDE SOLUTION**	8	–	III	223	5 L	E1	P001 LP01	–	IBC03	–
2582	**FERRIC CHLORIDE SOLUTION**	8	–	III	223	5 L	E1	P001 LP01	–	IBC03	–
2583	**ALKYLSULPHONIC ACIDS, SOLID or ARYLSULPHONIC ACIDS, SOLID** with more than 5% free sulphuric acid	8	–	II	–	1 kg	E2	P002	–	IBC08	B2 B4
2584	**ALKYLSULPHONIC ACIDS, LIQUID or ARYLSULPHONIC ACIDS, LIQUID** with more than 5% free sulphuric acid	8	–	II	–	1 L	E2	P001	–	IBC02	B20
2585	**ALKYLSULPHONIC ACIDS, SOLID or ARYLSULPHONIC ACIDS, SOLID** with not more than 5% free sulphuric acid	8	–	III	–	5 kg	E1	P002 LP02	–	IBC08	B3
2586	**ALKYLSULPHONIC ACIDS, LIQUID or ARYLSULPHONIC ACIDS, LIQUID** with not more than 5% free sulphuric acid	8	–	III	–	5 L	E1	P001 LP01	–	IBC03	–
2587	**BENZOQUINONE**	6.1	–	II	–	500 g	E4	P002	–	IBC08	B2 B4
2588	**PESTICIDE, SOLID, TOXIC, N.O.S.**	6.1	–	I	61 274	0	E5	P002	–	IBC99	–
2588	**PESTICIDE, SOLID, TOXIC, N.O.S.**	6.1	–	II	61 274	500 g	E4	P002	–	IBC08	B2 B4
2588	**PESTICIDE, SOLID, TOXIC, N.O.S.**	6.1	–	III	61 223 274	5 kg	E1	P002 LP02	–	IBC08	B3

Portable tanks and bulk containers			EmS	Stowage and segregation	Properties and observations	UN No.
	Tank instructions	Provisions				
(12)	(13) 4.2.5 4.3	(14) 4.2.5	(15) 5.4.3.2 7.3	(16) 7.1 7.2	(17)	(18)
–	T7	TP3 TP13	F-A, S-B	Category C. Clear of living quarters.	Colourless liquid with a pungent odour. Melting point: 56°C. Reacts violently with water, evolving hydrogen bromide, a toxic and corrosive gas apparent as white fumes. Reacts violently with organic materials (such as wood, cotton, straw), causing fire. When involved in a fire, evolves highly toxic and corrosive gases. In the presence of moisture, highly corrosive to most metals. Vapours and liquid cause burns to skin, eyes and mucous membranes. Shipped molten above its melting point.	2576
–	T7	TP2	F-A, S-B	Category C. Clear of living quarters.	Colourless liquid with a pungent odour. Reacts with water, evolving hydrogen chloride, an irritating and corrosive gas apparent as white fumes. When involved in a fire, evolves highly toxic fumes. Corrosive to most metals. Vapour irritates eyes and mucous membranes. Liquid is corrosive to skin, eyes and mucous membranes.	2577
–	T1	TP33	F-A, S-B	Category A. Keep as cool as reasonably practicable.	Colourless crystals or white deliquescent powder. Melting point: 23°C. Reacts with water, evolving heat and at normal temperatures phosphoric acid, but at higher temperatures phosphine, a highly toxic gas. In the presence of moisture, corrosive to most metals. Causes burns to skin, eyes and mucous membranes.	2578
–	T1	TP33	F-A, S-B	Category A. Keep as cool as reasonably practicable. "Separated from" acids.	Colourless, deliquescent crystals, turning dark on exposure to light. Soluble in water. Decomposes when heated and when involved in a fire, evolving highly toxic nitrous fumes. The solution in water is a strong base and is highly corrosive. Reacts violently with acids. Irritating to skin, eyes and mucous membranes.	2579
–	T4	TP1	F-A, S-B	Category A.	Colourless to yellowish liquid. Highly corrosive to most metals. Vapour highly irritating to skin, eyes and mucous membranes. Liquid causes severe burns to skin, eyes and mucous membranes.	2580
–	T4	TP1	F-A, S-B	Category A.	Colourless to yellowish liquid. Highly corrosive to most metals. Vapour highly irritating to skin, eyes and mucous membranes. Liquid causes severe burns to skin, eyes and mucous membranes.	2581
–	T4	TP1	F-A, S-B	Category A.	Colourless to light brown liquid. Highly corrosive to most metals.	2582
–	T3	TP33	F-A, S-B	Category A.	When involved in a fire, evolve highly toxic gases. Corrosive to most metals, especially in the presence of moisture. Cause burns to skin, eyes and mucous membranes.	2583
–	T8	TP2 TP13	F-A, S-B	Category B.	Liquids usually with a pungent odour. When involved in a fire, evolve highly toxic gases. Highly corrosive to most metals. Cause burns to skin, eyes and mucous membranes.	2584
–	T1	TP33	F-A, S-B	Category A.	Crystalline solids. When involved in a fire, evolve highly toxic gases. In the presence of moisture, corrosive to most metals. Cause burns to skin, eyes and mucous membranes.	2585
–	T4	TP1	F-A, S-B	Category B.	Liquids usually with a pungent odour. When involved in a fire, evolve highly toxic gases. Corrosive to most metals. Cause burns to skin, eyes and mucous membranes.	2586
–	T3	TP33	F-A, S-A	Category A.	Yellow crystals with an irritating and penetrating odour resembling that of chlorine. Slightly soluble in water. Toxic if swallowed, by skin contact or by dust inhalation.	2587
–	T6	TP33	F-A, S-A	Category A. Clear of living quarters.	Solid pesticides present a very wide range of toxic hazard. Toxic if swallowed, by skin contact or by inhalation.	2588
–	T3	TP33	F-A, S-A	Category A. Clear of living quarters.	See entry above.	2588
–	T1	TP33	F-A, S-A	Category A. Clear of living quarters.	See entry above.	2588

3 DGL

UN No.	Proper Shipping Name (PSN)	Class or division	Subsidiary risk(s)	Packing group	Special provisions	Limited and excepted quantity provisions		Packing		IBC	
						Limited quantities	Excepted quantities	Instructions	Provisions	Instructions	Provisions
(1)	(2) 3.1.2	(3) 2.0	(4) 2.0	(5) 2.0.1.3	(6) 3.3	(7a) 3.4	(7b) 3.5	(8) 4.1.4	(9) 4.1.4	(10) 4.1.4	(11) 4.1.4
2589	VINYL CHLOROACETATE	6.1	3	II	–	100 mL	E4	P001	–	IBC02	–
△ 2590	WHITE ASBESTOS (chrysotile, actinolite, anthophyllite, tremolite)	9	–	III	168	5 kg	E1	P002	PP37	IBC08	B2 B3
2591	XENON, REFRIGERATED LIQUID	2.2	–	–	–	120 mL	E1	P203	–	–	–
2599	CHLOROTRIFLUOROMETHANE AND TRIFLUOROMETHANE AZEOTROPIC MIXTURE with approximately 60% chlorotrifluoromethane (REFRIGERANT GAS R 503)	2.2	–	–	–	120 mL	E1	P200	–	–	–
2601	CYCLOBUTANE	2.1	–	–	–	0	E0	P200	–	–	–
2602	DICHLORODIFLUORO-METHANE AND DIFLUOROETHANE AZEOTROPIC MIXTURE with approximately 74% dichlorodifluoromethane (REFRIGERANT GAS R 500)	2.2	–	–	–	120 mL	E1	P200	–	–	–
2603	CYCLOHEPTATRIENE	3	6.1	II	–	1 L	E2	P001	–	IBC02	–
2604	BORON TRIFLUORIDE DIETHYL ETHERATE	8	3	I	–	0	E0	P001	PP31	–	–
2605	METHOXYMETHYL ISOCYANATE	6.1	3	I	354	0	E0	P602	–	–	–
2606	METHYL ORTHOSILICATE	6.1	3	I	354	0	E0	P602	–	–	–

3

DGL

Portable tanks and bulk containers			EmS	Stowage and segregation	Properties and observations	UN No.
	Tank instructions	Provisions				
(12)	(13) 4.2.5 4.3	(14) 4.2.5	(15) 5.4.3.2 7.3	(16) 7.1 7.2	(17)	(18)
–	T7	TP2	F-E, S-D	Category A.	Flammable liquid. Flashpoint: 50°C c.c. Immiscible with water. Toxic if swallowed, by skin contact or by inhalation.	2589
–	T1	TP33	F-A, S-A	Category A. Clear of living quarters. Segregation from foodstuffs as in 7.3.4.2.2, 7.6.3.1.2 or 7.7.3.7.	Mineral fibres of varying length. Non-combustible. Inhalation of the dust of asbestos fibres is dangerous and therefore exposure should be avoided at all times. Always prevent the generation of asbestos dust. A safe level of airborne concentration of asbestos fibres may be obtained through effective packing. Cargo spaces or freight containers that have contained any type of raw asbestos should be carefully cleaned before discharging any remaining cargo, loading other cargo or carrying out repair or maintenance work. Whenever possible, cleaning of cargo spaces should be carried out whilst the ship is in a port where proper facilities and equipment, including proper respiratory apparatus and protective clothing, are available. Parts of the body that may have been exposed should be immediately and thoroughly washed. All waste material should be collected in impermeable and sealed bags for safe disposal ashore. If cleaning cannot be carried out at the discharge port, arrangements should be made in advance for cleaning to be carried out at the next port where necessary facilities are available. If cleaning of cargo spaces has to be carried out at sea, the safety procedures followed and standard of equipment used should be at least as effective as those which would be employed in a port. Until such cleaning is undertaken, the cargo spaces in which the asbestos has been carried should be closed and access to those spaces should be prohibited.	2590 △
–	T75	TP5	F-C, S-V	Category D.	Liquefied, inert, colourless and odourless gas. Much heavier than air (4.5).	2591
–	–	–	F-C, S-V	Category A.	Non-flammable, colourless gas with a mild ethereal odour. Much heavier than air (3.2).	2599
–	–	–	F-D, S-U	Category B. Clear of living quarters.	Liquefied, flammable, colourless gas. Explosive limits: 1.8% to 10%. Heavier than air (1.9). Boiling point: 13°C.	2601
–	T50	–	F-C, S-V	Category A.	Non-flammable, colourless and odourless gas. Much heavier than air (3.7).	2602
–	T7	TP1 TP13	F-E, S-D	Category E. Clear of living quarters.	Colourless to dark yellow liquid with a characteristic odour. Flashpoint: 0°C to 4°C c.c. Immiscible with water. Reacts vigorously with oxidizing substances. Toxic if swallowed, by skin contact or by inhalation.	2603
–	T10	TP2	F-E, S-C	Category D. Clear of living quarters.	Colourless fuming flammable liquid. Flashpoint: 59°C c.c. The flashpoint will be lower when free ether is present. Reacts vigorously with oxidizing substances. Decomposes in contact with water, evolving toxic, corrosive and flammable vapours. Causes burns to skin, eyes and mucous membranes. Inhalation of small quantities of vapour can cause breathing difficulties.	2604
–	T20	TP2 TP13 TP37	F-E, S-D	Category D. Clear of living quarters.	Colourless liquid with a pungent odour. Flashpoint: 13°C c.c. Immiscible with water. Highly toxic if swallowed, by skin contact or by inhalation. Irritating to skin, eyes and mucous membranes.	2605
–	T20	TP2 TP13 TP37	F-E, S-D	Category D. Clear of living quarters.	Colourless, flammable liquid with an ethereal odour. Immiscible with water. Flashpoint: –18°C to 19°C c.c. Highly toxic if swallowed, by skin contact or by inhalation. May cause blindness.	2606

UN No.	Proper Shipping Name (PSN)	Class or division	Subsidiary risk(s)	Packing group	Special provisions	Limited and excepted quantity provisions		Packing		IBC	
						Limited quantities	Excepted quantities	Instructions	Provisions	Instructions	Provisions
(1)	(2) 3.1.2	(3) 2.0	(4) 2.0	(5) 2.0.1.3	(6) 3.3	(7a) 3.4	(7b) 3.5	(8) 4.1.4	(9) 4.1.4	(10) 4.1.4	(11) 4.1.4
2607	ACROLEIN DIMER, STABILIZED	3	–	III	–	5 L	E1	P001 LP01	–	IBC03	–
2608	NITROPROPANES	3	–	III	–	5 L	E1	P001 LP01	–	IBC03	–
2609	TRIALLYL BORATE	6.1	–	III	–	5 L	E1	P001 LP01	–	IBC03	–
2610	TRIALLYLAMINE	3	8	III	–	5 L	E1	P001	–	IBC03	–
2611	PROPYLENE CHLOROHYDRIN	6.1	3	II	–	100 mL	E4	P001	–	IBC02	–
2612	METHYL PROPYL ETHER	3	–	II	–	1 L	E2	P001	–	IBC02	B8
2614	METHALLYL ALCOHOL	3	–	III	–	5 L	E1	P001 LP01	–	IBC03	–
△ 2615	ETHYL PROPYL ETHER	3	–	II	–	1 L	E2	P001	–	IBC02	–
2616	TRIISOPROPYL BORATE	3	–	II	–	1 L	E2	P001	–	IBC02	–
2616	TRIISOPROPYL BORATE	3	–	III	223	5 L	E1	P001 LP01	–	IBC03	–
2617	METHYLCYCLOHEXANOLS, flammable	3	–	III	–	5 L	E1	P001 LP01	–	IBC03	–
2618	VINYLTOLUENES, STABILIZED	3	–	III	–	5 L	E1	P001 LP01	–	IBC03	–
△ 2619	BENZYLDIMETHYLAMINE	8	3	II	–	1 L	E2	P001	–	IBC02	–
2620	AMYL BUTYRATES	3	–	III	–	5 L	E1	P001 LP01	–	IBC03	–
2621	ACETYL METHYL CARBINOL	3	–	III	–	5 L	E1	P001 LP01	–	IBC03	–
2622	GLYCIDALDEHYDE	3	6.1	II	–	1 L	E2	P001	–	IBC02	B8
2623	FIRELIGHTERS, SOLID with flammable liquid	4.1	–	III	–	5 kg	E1	P002 LP02	PP15	–	–
2624	MAGNESIUM SILICIDE	4.3	–	II	–	500 g	E2	P410	PP31 PP40	IBC07	B2

Portable tanks and bulk containers			EmS	Stowage and segregation	Properties and observations	UN No.
	Tank instructions	Provisions				
(12)	(13) 4.2.5 4.3	(14) 4.2.5	(15) 5.4.3.2 7.3	(16) 7.1 7.2	(17)	(18)
–	T2	TP1	F-E, S-D	Category A. Clear of living quarters.	Colourless liquid with a pungent odour. Flashpoint: 48°C o.c. Miscible with water. Harmful by inhalation. Irritating to skin, eyes and mucous membranes.	2607
–	T2	TP1	F-E, S-D	Category A.	Colourless liquids. Explosive limits: 2.2% to 11%. 1-NITROPROPANE: flashpoint approx. 33°C c.c. 2-NITROPROPANE: flashpoint approx. 28°C c.c. Partially miscible with water. Harmful by inhalation.	2608
–	–	–	F-A, S-A	Category A. Keep as dry as reasonably practicable.	Liquid. Hydrolyses in contact with water, forming allyl alcohol. Toxic if swallowed, by skin contact or by inhalation.	2609
–	T4	TP1	F-E, S-C	Category A. Clear of living quarters.	Colourless liquid with a fishy odour. Flashpoint: 39°C o.c. Corrosive when in contact with water. Harmful by inhalation. Causes burns to skin and eyes. Irritating to mucous membranes.	2610
–	T7	TP2 TP13	F-E, S-D	Category A. Keep as cool as reasonably practicable. Clear of living quarters.	Colourless flammable liquid with a mild odour. Flashpoint: 51°C c.c. Miscible with water. Decomposes when heated, evolving highly toxic fumes. Toxic if swallowed, by skin contact or by inhalation.	2611
–	T7	TP2	F-E, S-D	Category E. Clear of living quarters.	Colourless, volatile liquid with an ethereal odour. Flashpoint: below −18°C c.c. Explosive limits: 2% to .... Boiling point: 39°C. Partially miscible with water. Narcotic. Irritating to skin, eyes and mucous membranes.	2612
–	T2	TP1	F-E, S-D	Category A.	Colourless liquid with a pungent odour. Flashpoint: 34°C c.c. Miscible with water. Irritating to skin, eyes and mucous membranes.	2614
–	T4	TP1	F-E, S-D	Category E.	Colourless, volatile liquids. Flashpoint: below −18°C c.c. Explosive limits: 1.7% to 9.0%. Miscible with water. Irritating to skin, eyes and mucous membranes.	2615 △
–	T4	TP1	F-E, S-D	Category B.	Colourless liquid. Flashpoint: 17°C to 60°C c.c. Reacts with water, evolving flammable vapours.	2616
–	T2	TP1	F-E, S-D	Category A.	See entry above.	2616
–	T2	TP1	F-E, S-D	Category A.	Colourless, viscous liquid with a menthol-like odour. Flashpoint: 58°C c.c. Partially miscible with water.	2617
–	T2	TP1	F-E, S-D	Category A.	Colourless liquids. Flashpoint: 54°C to 60°C c.c. Explosive limits: 0.9% to 6.1%. Partially miscible with water. Harmful by inhalation. Irritating to skin, eyes and mucous membranes.	2618
–	T7	TP2	F-E, S-C	Category A. Protected from sources of heat. Clear of living quarters.	Colourless, flammable liquid with an aromatic odour. Flashpoint: 58°C c.c. Immiscible with water. Harmful if swallowed, by skin contact or by inhalation. Corrosive to skin, eyes and mucous membranes.	2619 △
–	T2	TP1	F-E, S-D	Category A.	Colourless liquids. Flashpoint: 52°C to 58°C c.c. Partially miscible with water.	2620
–	T2	TP1	F-E, S-D	Category A.	Yellow liquid with a pleasant odour. Flashpoint: 44°C to 52°C c.c. Miscible with water. Reacts vigorously with oxidizing substances. Irritating to skin, eyes and mucous membranes.	2621
–	T7	TP1	F-E, S-D	Category A. Clear of living quarters.	Colourless liquid with a pungent odour. Flashpoint: 31°C o.c. Miscible with water. Toxic by inhalation. Irritating to skin, eyes and mucous membranes.	2622
–	–	–	F-A, S-I	Category A. "Separated from" acids.	A porous solid, e.g., cellular urea–formaldehyde resin, compacted wood shavings, etc., impregnated with flammable liquid, usually white spirit or kerosene, and designed to burn in a controlled manner. When heated, evolves flammable vapours.	2623
–	T3	TP33	F-G, S-O	Category B. Only to be loaded under dry weather conditions. Under deck in a mechanically ventilated space.	White powder or crystals. Reacts with water or steam, evolving hydrogen, a flammable gas. In contact with acids, evolves silane, a spontaneously flammable gas.	2624

UN No.	Proper Shipping Name (PSN)	Class or division	Subsidiary risk(s)	Packing group	Special provisions	Limited and excepted quantity provisions		Packing		IBC	
						Limited quantities	Excepted quantities	Instructions	Provisions	Instructions	Provisions
(1)	(2) 3.1.2	(3) 2.0	(4) 2.0	(5) 2.0.1.3	(6) 3.3	(7a) 3.4	(7b) 3.5	(8) 4.1.4	(9) 4.1.4	(10) 4.1.4	(11) 4.1.4
△ 2626	CHLORIC ACID, AQUEOUS SOLUTION with not more than 10% chloric acid	5.1	–	II	900	1 L	E2	P504	PP31	IBC02	–
△ 2627	NITRITES, INORGANIC, N.O.S.	5.1	–	II	274 900	1 kg	E2	P002	–	IBC08	B2 B4
2628	POTASSIUM FLUOROACETATE	6.1	–	I	–	0	E5	P002	–	IBC07	B1
2629	SODIUM FLUOROACETATE	6.1	–	I	–	0	E5	P002	–	IBC07	B1
2630	SELENATES or SELENITES	6.1	–	I	274	0	E5	P002	–	IBC07	B1
2642	FLUOROACETIC ACID	6.1	–	I	–	0	E5	P002	–	IBC07	B1
2643	METHYL BROMOACETATE	6.1	–	II	–	100 mL	E4	P001	–	IBC02	–
2644	METHYL IODIDE	6.1	–	I	354	0	E0	P602	–	–	–
2645	PHENACYL BROMIDE	6.1	–	II	–	500 g	E4	P002	–	IBC08	B2 B4
2646	HEXACHLOROCYCLO-PENTADIENE	6.1	–	I	354	0	E0	P602	–	–	–
2647	MALONONITRILE	6.1	–	II	–	500 g	E4	P002	–	IBC08	B2 B4
2648	1,2-DIBROMOBUTAN-3-ONE	6.1	–	II	–	100 mL	E4	P001	–	IBC02	–
2649	1,3-DICHLOROACETONE	6.1	–	II	–	500 g	E4	P002	–	IBC08	B2 B4
2650	1,1-DICHLORO-1-NITRO-ETHANE	6.1	–	II	–	100 mL	E4	P001	–	IBC02	–
2651	4,4′-DIAMINODIPHENYL-METHANE	6.1	– P	III	–	5 kg	E1	P002 LP02	–	IBC08	B3
2653	BENZYL IODIDE	6.1	–	II	–	100 mL	E4	P001	–	IBC02	–
2655	POTASSIUM FLUOROSILICATE	6.1	–	III	–	5 kg	E1	P002 LP02	–	IBC08	B3
2656	QUINOLINE	6.1	–	III	–	5 L	E1	P001 LP01	–	IBC03	–

Portable tanks and bulk containers			EmS	Stowage and segregation	Properties and observations	UN No.	
	Tank instructions	Provisions					
(12)	(13) 4.2.5 4.3	(14) 4.2.5	(15) 5.4.3.2 7.3	(16) 7.1 7.2	(17)	(18)	
–	–	–	F-A, S-Q	Category D. "Separated from" ammonium compounds and cyanides.	Colourless liquid. May decompose, evolving chlorine and oxygen with toxic, corrosive and oxidizing effects. May form explosive mixtures with ammonium compounds, combustible material or powdered metals. Corrosive to most metals. Transport of CHLORIC ACID, AQUEOUS SOLUTION with more than 10% chloric acid is prohibited.	2626	△
–	T3	TP33	F-A, S-Q	Category A. "Separated from" ammonium compounds, cyanides and sulphur.	Solids. Solid mixtures with combustible material are readily ignited and may burn fiercely. Solid mixtures with ammonium compounds or cyanides may explode. If heated, may decompose, giving off toxic nitrous fumes. Harmful if swallowed. Transport of AMMONIUM NITRITES and mixtures of an inorganic nitrite with an ammonium salt is prohibited.	2627	△
–	T6	TP33	F-A, S-A	Category E.	Solid. Soluble in water. Highly toxic if swallowed, by skin contact or by dust inhalation.	2628	
–	T6	TP33	F-A, S-A	Category E.	White powder. Soluble in water. Highly toxic if swallowed, by skin contact or by dust inhalation.	2629	
–	T6	TP33	F-A, S-A	Category E.	A wide range of toxic solids. Generally soluble in water. Highly toxic if swallowed, by skin contact or by dust inhalation.	2630	
–	T6	TP33	F-A, S-A	Category E.	Colourless crystals. Melting point: 33°C. Soluble in water. Highly toxic if swallowed, by skin contact or by dust inhalation.	2642	
–	T7	TP2	F-A, S-A	Category D. Clear of living quarters.	Colourless to straw-coloured liquid. Slightly miscible with water. Causes tears. Toxic if swallowed, by skin contact or by inhalation.	2643	
–	T20	TP2 TP13 TP37	F-A, S-A	Category D. Keep as cool as reasonably practicable. Clear of living quarters.	Colourless liquid. Boiling point: 42°C to 43°C. Slightly miscible with water. When heated, evolves toxic fumes. Highly toxic if swallowed, by skin contact or by inhalation. Has strong narcotic effects.	2644	
–	T3	TP33	F-A, S-A	Category B. Clear of living quarters.	White crystals changing to a greenish colour under the influence of light. Melting point: 50°C. Insoluble in water. Causes tears. Toxic if swallowed, by skin contact or by inhalation.	2645	
–	T20	TP2 TP13 TP35	F-A, S-A	Category D. Clear of living quarters.	Pale yellow liquid with a pungent odour. Immiscible with water. Causes tears. Highly toxic if swallowed, by skin contact or by inhalation.	2646	
–	T3	TP33	F-A, S-A	Category A. Keep as cool as reasonably practicable.	Colourless crystals. Melting point: 32°C. Soluble in water. When heated, evolves highly toxic cyanogen fumes. Toxic if swallowed, by skin contact or by dust inhalation.	2647	
–	–	–	F-A, S-A	Category B. Clear of living quarters.	Liquid. Immiscible with water. Toxic if swallowed, by skin contact or by inhalation. Causes tears.	2648	
–	T3	TP33	F-A, S-A	Category B. Keep as cool as reasonably practicable. Clear of living quarters.	Crystals. Melting point: 45°C. Soluble in water. Decomposes when heated, evolving highly toxic fumes. Toxic if swallowed, by skin contact or by dust inhalation. Causes tears.	2649	
–	T7	TP2	F-A, S-A	Category A. Keep as cool as reasonably practicable. Clear of living quarters. "Separated from" class 5.1.	Liquid. Immiscible with water. May react vigorously with oxidizing substances. Decomposes when heated, evolving highly toxic fumes (oxides of nitrogen). Toxic if swallowed, by skin contact or by inhalation.	2650	
–	T1	TP33	F-A, S-A	Category A.	Tan-coloured flakes or lumps. Slightly soluble in water. Decomposes when heated, evolving highly toxic fumes. Toxic if swallowed by skin contact or by dust inhalation. May be carried in the molten state.	2651	
–	T7	TP2	F-A, S-A	Category B. Keep as cool as reasonably practicable. Clear of living quarters.	Colourless crystals. Melting point: 24°C. Insoluble in water. Toxic if swallowed by skin contact or by dust inhalation. Causes tears.	2653	
–	T1	TP33	F-A, S-A	Category A. "Separated from" acids.	Solids which react with acids, evolving hydrogen fluoride and silicon tetrafluoride, irritating and corrosive gases. Toxic if swallowed, by skin contact or by dust inhalation.	2655	
–	T4	TP1	F-A, S-A	Category A. Keep as cool as reasonably practicable.	Colourless liquid with a pungent odour. Immiscible with water. When heated, evolves highly toxic fumes (of oxides of nitrogen). Toxic if swallowed by skin contact or by inhalation.	2656	

UN No.	Proper Shipping Name (PSN)	Class or division	Subsidiary risk(s)	Packing group	Special provisions	Limited and excepted quantity provisions		Packing		IBC	
						Limited quantities	Excepted quantities	Instructions	Provisions	Instructions	Provisions
(1)	(2) 3.1.2	(3) 2.0	(4) 2.0	(5) 2.0.1.3	(6) 3.3	(7a) 3.4	(7b) 3.5	(8) 4.1.4	(9) 4.1.4	(10) 4.1.4	(11) 4.1.4
2657	SELENIUM DISULPHIDE	6.1	–	II	–	500 g	E4	P002	–	IBC08	B2 B4
2659	SODIUM CHLOROACETATE	6.1	–	III	–	5 kg	E1	P002 LP02	–	IBC08	B3
2660	NITROTOLUIDINES (MONO)	6.1	–	III	–	5 kg	E1	P002 LP02	–	IBC08	B3
2661	HEXACHLOROACETONE	6.1	–	III	–	5 L	E1	P001 LP01	–	IBC03	–
2664	DIBROMOMETHANE	6.1	–	III	–	5 L	E1	P001 LP01	–	IBC03	–
2667	BUTYLTOLUENES	6.1	–	III	–	5 L	E1	P001 LP01	–	IBC03	–
2668	CHLOROACETONITRILE	6.1	3	I	354	0	E0	P602	–	–	–
2669	CHLOROCRESOLS SOLUTION	6.1	–	II	–	100 mL	E4	P001	–	IBC02	–
2669	CHLOROCRESOLS SOLUTION	6.1	–	III	223	5 L	E1	P001 LP01	–	IBC03	–
2670	CYANURIC CHLORIDE	8	–	II	–	1 kg	E2	P002	–	IBC08	B2 B4
2671	AMINOPYRIDINES (*o*-, *m*-, *p*-)	6.1	–	II	–	500 g	E4	P002	–	IBC08	B2 B4
△ 2672	AMMONIA SOLUTION relative density between 0.880 and 0.957 at 15°C in water, with more than 10% but not more than 35% ammonia	8	–	III	–	5 L	E1	P001 LP01	–	IBC03	B11
2673	2-AMINO-4-CHLOROPHENOL	6.1	–	II	–	500 g	E4	P002	–	IBC08	B2 B4
2674	SODIUM FLUOROSILICATE	6.1	–	III	–	5 kg	E1	P002 LP02	–	IBC08	B3
2676	STIBINE	2.3	2.1	–	–	0	E0	P200	–	–	–
2677	RUBIDIUM HYDROXIDE SOLUTION	8	–	II	–	1 L	E2	P001	–	IBC02	–
2677	RUBIDIUM HYDROXIDE SOLUTION	8	–	III	223	5 L	E1	P001 LP01	–	IBC03	–
△ 2678	RUBIDIUM HYDROXIDE	8	–	II	–	1 kg	E2	P002	–	IBC08	B2 B4
2679	LITHIUM HYDROXIDE SOLUTION	8	–	II	–	1 L	E2	P001	–	IBC02	–

Portable tanks and bulk containers			EmS	Stowage and segregation	Properties and observations	UN No.
	Tank instructions	Provisions				
(12)	(13) 4.2.5 4.3	(14) 4.2.5	(15) 5.4.3.2 7.3	(16) 7.1 7.2	(17)	(18)
–	T3	TP33	F-A, S-A	Category A.	Bright red-yellow crystals with a faint odour. Insoluble in water. Toxic if swallowed, by skin contact or by inhalation.	2657
–	T1	TP33	F-A, S-A	Category A.	White powder. Soluble in water. Toxic if swallowed, by skin contact or by dust inhalation.	2659
–	T1	TP33	F-A, S-A	Category A.	Yellow to orange-red crystalline solids. Insoluble in water. Toxic if swallowed, by skin contact or by dust inhalation.	2660
–	T4	TP1	F-A, S-A	Category B. Keep as cool as reasonably practicable. Clear of living quarters.	Colourless to yellowish liquid. Slightly miscible with water. When heated, evolves extremely toxic fumes (phosgene). Causes tears. Toxic if swallowed, by skin contact or by inhalation.	2661
–	T4	TP1	F-A, S-A	Category A.	Clear, colourless liquid. Immiscible with water. Toxic if swallowed, by skin contact or by inhalation.	2664
–	T4	TP1	F-A, S-A	Category A.	Colourless liquids. Immiscible with water. Toxic if swallowed, by skin contact or by inhalation.	2667
–	T20	TP2 TP13 TP37	F-A, S-A	Category D. Keep as cool as reasonably practicable. Clear of living quarters. "Separated from" acids.	Colourless flammable liquid with a pungent odour. Flashpoint: 56°C c.c. Immiscible with water. Decomposes when heated, evolving highly toxic fumes of cyanides. Reacts with steam and acids, evolving toxic and flammable vapours. Highly toxic if swallowed, by skin contact or by inhalation.	2668
–	T7	TP2	F-A, S-A	Category A. Keep as cool as reasonably practicable.	Solutions with a phenol-like odour. Slightly miscible with water. Decompose when heated, evolving extremely toxic fumes (phosgene). Toxic if swallowed, by skin contact or by inhalation.	2669
–	T7	TP2	F-A, S-A	Category A. Keep as cool as reasonably practicable.	See entry above.	2669
–	T3	TP33	F-A, S-B	Category A. Keep as cool as reasonably practicable. Clear of living quarters.	Colourless crystals with a pungent odour. Reacts with water, forming toxic and corrosive acids. Decomposes when heated, evolving toxic and corrosive gases. Causes burns to skin, eyes and mucous membranes.	2670
–	T3	TP33	F-A, S-A	Category B. Keep as cool as reasonably practicable. Clear of living quarters. "Separated from" acids.	White powder or crystals. Melting points: 58°C to 64°C. Soluble in water. Reacts violently with acids. Toxic if swallowed, by skin contact or by dust inhalation.	2671
–	T7	TP1	F-A, S-B	Category A. Under deck in a mechanically ventilated space. Clear of living quarters. "Separated from" acids.	Colourless liquid with a pungent odour. Corrosive to copper, nickel, zinc and tin and their alloys such as brass. Not significantly corrosive to iron and steel. Reacts violently with acids. Liquid and vapour cause burns to skin, eyes and mucous membranes.	2672 △
–	T3	TP33	F-A, S-A	Category A.	Light brown crystals. Slightly soluble in water. Toxic if swallowed, by skin contact or by dust inhalation.	2673
–	T1	TP33	F-A, S-A	Category A. "Separated from" acids.	Solids which react with acids, evolving hydrogen fluoride and silicon tetrafluoride, irritating and corrosive gases. Toxic if swallowed, by skin contact or by dust inhalation.	2674
–	–	–	F-D, S-U	Category D. Clear of living quarters.	Flammable, toxic, colourless gas with a foul odour. Decomposes violently in the presence of water. Much heavier than air (4.3).	2676
–	T7	TP2	F-A, S-B	Category A. "Away from" ammonium salts. "Separated from" acids.	Liquid. Reacts violently with acids. Reacts with ammonium salts, evolving ammonia gas. Corrosive to aluminium, zinc and tin. Causes burns to skin, eyes and mucous membranes.	2677
–	T4	TP1	F-A, S-B	Category A. "Away from" ammonium salts. "Separated from" acids.	See entry above.	2677
–	T3	TP33	F-A, S-B	Category A. "Away from" ammonium salts. "Separated from" acids.	Greyish-white solid, very hygroscopic. Reacts violently with acids. Reacts with ammonium salts, evolving ammonia gas. In the presence of moisture, corrosive to aluminium, zinc and tin. Causes burns to skin, eyes and mucous membranes.	2678 △
–	T7	TP2	F-A, S-B	Category A. "Away from" ammonium salts. "Separated from" acids.	Colourless liquid. Corrosive to aluminium, zinc and tin. Causes burns to skin, eyes and mucous membranes.	2679

UN No.	Proper Shipping Name (PSN)	Class or division	Subsidiary risk(s)	Packing group	Special provisions	Limited and excepted quantity provisions		Packing		IBC	
						Limited quantities	Excepted quantities	Instructions	Provisions	Instructions	Provisions
(1)	(2) 3.1.2	(3) 2.0	(4) 2.0	(5) 2.0.1.3	(6) 3.3	(7a) 3.4	(7b) 3.5	(8) 4.1.4	(9) 4.1.4	(10) 4.1.4	(11) 4.1.4
2679	LITHIUM HYDROXIDE SOLUTION	8	–	III	223	5 L	E1	P001 LP01	–	IBC03	–
2680	LITHIUM HYDROXIDE	8	–	II	–	1 kg	E2	P002	–	IBC08	B2 B4
2681	CAESIUM HYDROXIDE SOLUTION	8	–	II	–	1 L	E2	P001	–	IBC02	–
2681	CAESIUM HYDROXIDE SOLUTION	8	–	III	223	5 L	E1	P001 LP01	–	IBC03	–
2682	CAESIUM HYDROXIDE	8	–	II	–	1 kg	E2	P002	–	IBC08	B2 B4
2683	AMMONIUM SULPHIDE SOLUTION	8	3/6.1	II	–	1 L	E2	P001	–	IBC01	–
2684	3-DIETHYLAMINO-PROPYLAMINE	3	8	III	–	5 L	E1	P001	–	IBC03	–
2685	*N,N*-DIETHYLETHYLENE-DIAMINE	8	3	II	–	1 L	E2	P001	–	IBC02	–
2686	2-DIETHYLAMINOETHANOL	8	3	II	–	1 L	E2	P001	–	IBC02	–
2687	DICYCLOHEXYLAMMONIUM NITRITE	4.1	–	III	–	5 kg	E1	P002 LP02	–	IBC08	B3
2688	1-BROMO-3-CHLOROPROPANE	6.1	–	III	–	5 L	E1	P001 LP01	–	IBC03	–
2689	GLYCEROL-*alpha*-MONOCHLOROHYDRIN	6.1	–	III	–	5 L	E1	P001 LP01	–	IBC03	–
2690	*N-n*-BUTYLIMIDAZOLE	6.1	–	II	–	100 mL	E4	P001	–	IBC02	–
2691	PHOSPHORUS PENTABROMIDE	8	–	II	–	1 kg	E2	P002	–	IBC08	B2 B4
2692	BORON TRIBROMIDE	8	–	I	–	0	E0	P602	–	–	–
2693	BISULPHITES, AQUEOUS SOLUTION, N.O.S.	8	–	III	274	5 L	E1	P001 LP01	–	IBC03	–

Portable tanks and bulk containers			EmS	Stowage and segregation	Properties and observations	UN No.
	Tank instructions	Provisions				
(12)	(13) 4.2.5 4.3	(14) 4.2.5	(15) 5.4.3.2 7.3	(16) 7.1 7.2	(17)	(18)
–	T4	TP2	F-A, S-B	Category A. "Away from" ammonium salts. "Separated from" acids.	Colourless liquid. Corrosive to aluminium, zinc and tin. Causes burns to skin, eyes and mucous membranes.	2679
–	T3	TP33	F-A, S-B	Category A. "Separated from" acids.	Colourless crystals. Soluble in water. Reacts violently with acids. Causes burns to skin, eyes and mucous membranes.	2680
–	T7	TP2	F-A, S-B	Category A. "Away from" ammonium salts. "Separated from" acids.	Colourless liquid. Reacts violently with acids. Reacts with ammonium salts, evolving ammonia gas. Corrosive to glass, aluminium, zinc and tin. Causes burns to skin, eyes and mucous membranes.	2681
–	T4	TP1	F-A, S-B	Category A. "Away from" ammonium salts. "Separated from" acids.	See entry above.	2681
–	T3	TP33	F-A, S-B	Category A. "Away from" ammonium salts. "Separated from" acids.	Colourless or yellowish hygroscopic crystals. Reacts violently with acids. Reacts with ammonium salts, evolving ammonia gas. In the presence of moisture, corrosive to glass, aluminium, zinc and tin. Causes burns to skin, eyes and mucous membranes.	2682
–	T7	TP2 TP13	F-E, S-C	Category B. Keep as cool as reasonably practicable. "Separated from" acids. If flashpoint 60°C c.c. or below, segregation as for class 3, but "away from" class 4.1.	Yellow liquid with a foul odour (of rotten eggs). When heated, evolves toxic and flammable vapours. Reacts violently with acids, evolving hydrogen sulphide, a toxic and flammable gas. Toxic if swallowed, by skin contact or by inhalation. Corrosive to skin, eyes and mucous membranes.	2683
–	T4	TP1	F-E, S-C	Category A.	Colourless liquid with a fishy odour. Flashpoint: 59°C o.c. Miscible with water. Irritating to skin, eyes and mucous membranes.	2684
–	T7	TP2	F-E, S-C	Category A.	Colourless, flammable liquid with a fishy odour. Flashpoint: 46°C o.c. Miscible with water. Harmful by skin contact. Irritating to eyes and mucous membranes.	2685
–	T7	TP2	F-E, S-C	Category A.	Colourless liquid. Miscible with water. Reacts violently with oxidizing substances. Explosive limits: 1.8% to 28%. Flashpoint: 46°C to 60°C c.c. Causes burns to skin, eyes and mucous membranes.	2686
–	T1	TP33	F-A, S-G	Category A.	White powder. Insoluble in water. Harmful if swallowed.	2687
–	T4	TP1	F-A, S-A	Category A.	Colourless liquid. Immiscible with water. Decomposes when heated, evolving highly toxic fumes. Toxic if swallowed, by skin contact or by inhalation.	2688
–	T4	TP1	F-A, S-A	Category A.	Colourless liquid. Miscible with water. Toxic if swallowed, by skin contact or by inhalation.	2689
–	T7	TP2	F-A, S-A	Category A.	Colourless to amber mobile liquid. Miscible with water. Toxic if swallowed, by skin contact or by inhalation.	2690
–	T3	TP33	F-A, S-B	Category B. Keep as cool as reasonably practicable. Clear of living quarters. "Separated from" alkalis and ammonia.	Yellow hygroscopic crystals, evolving fumes in the air which are corrosive and heavier than air. Reacts violently with water, evolving hydrogen bromide, an irritating and corrosive gas apparent as white fumes. Reacts violently with ammonia, bases and many other substances and may cause fire and explosion. Decomposes when heated, evolving corrosive and toxic gases. In the presence of moisture, highly corrosive to most metals. Causes burns to skin, eyes and mucous membranes.	2691
–	T20	TP2 TP13	F-A, S-B	Category C. Keep as cool as reasonably practicable.	Colourless fuming liquid. Reacts violently with water, evolving toxic and corrosive fumes. Decomposes when heated, evolving toxic fumes. In the presence of moisture, highly corrosive to most metals. Liquid and vapour cause severe burns to skin, eyes and mucous membranes.	2692
–	T7	TP1 TP28	F-A, S-B	Category A. Clear of living quarters. "Separated from" acids.	Liquid with a pungent odour. Reacts with acids, evolving sulphur dioxide, a toxic gas. Causes burns to skin, eyes and mucous membranes.	2693

UN No.	Proper Shipping Name (PSN)	Class or division	Subsidiary risk(s)	Packing group	Special provisions	Limited and excepted quantity provisions		Packing		IBC	
						Limited quantities	Excepted quantities	Instructions	Provisions	Instructions	Provisions
(1)	(2) 3.1.2	(3) 2.0	(4) 2.0	(5) 2.0.1.3	(6) 3.3	(7a) 3.4	(7b) 3.5	(8) 4.1.4	(9) 4.1.4	(10) 4.1.4	(11) 4.1.4
2698	TETRAHYDROPHTHALIC ANHYDRIDES with more than 0.05% maleic anhydride	8	–	III	29 169 939	5 kg	E1	P002 LP02	PP14	IBC08	B3
2699	TRIFLUOROACETIC ACID	8	–	I	–	0	E0	P001	–	–	–
2705	1-PENTOL	8	–	II	–	1 L	E2	P001	–	IBC02	–
2707	DIMETHYLDIOXANES	3	–	II	–	1 L	E2	P001	–	IBC02	–
2707	DIMETHYLDIOXANES	3	–	III	223	5 L	E1	P001 LP01	–	IBC03	–
2709	BUTYLBENZENES	3	–	III	–	5 L	E1	P001 LP01	–	IBC03	–
2710	DIPROPYL KETONE	3	–	III	–	5 L	E1	P001 LP01	–	IBC03	–
2713	ACRIDINE	6.1	–	III	–	5 kg	E1	P002 LP02	–	IBC08	B3
2714	ZINC RESINATE	4.1	–	III	–	5 kg	E1	P002	–	IBC06	–
2715	ALUMINIUM RESINATE	4.1	–	III	–	5 kg	E1	P002	–	IBC06	–
2716	1,4-BUTYNEDIOL	6.1	–	III	–	5 kg	E1	P002 LP02	–	IBC08	B3
2717	CAMPHOR, synthetic	4.1	–	III	–	5 kg	E1	P002 LP02	–	IBC08	B3
2719	BARIUM BROMATE	5.1	6.1	II	–	1 kg	E2	P002	–	IBC08	B2 B4
2720	CHROMIUM NITRATE	5.1	–	III	–	5 kg	E1	P002 LP02	–	IBC08	B3
2721	COPPER CHLORATE	5.1	–	II	–	1 kg	E2	P002	–	IBC08	B2 B4
2722	LITHIUM NITRATE	5.1	–	III	–	5 kg	E1	P002 LP02	–	IBC08	B3

Portable tanks and bulk containers			EmS	Stowage and segregation	Properties and observations	UN No.
	Tank instructions	Provisions				
(12)	(13) 4.2.5 4.3	(14) 4.2.5	(15) 5.4.3.2 7.3	(16) 7.1 7.2	(17)	(18)
–	T1	TP33	F-A, S-B	Category A.	White crystalline powders. React with water, evolving heat and forming tetrahydrophthalic acid. Cause burns to skin, eyes and mucous membranes. When heated, evolve acrid fumes which are irritating to skin, eyes and mucous membranes.	2698
–	T10	TP2	F-A, S-B	Category B. Keep as cool as reasonably practicable. Clear of living quarters.	Colourless, fuming, hygroscopic liquid with a pungent odour. Miscible with water. When heated to decomposition or in contact with acids, evolves toxic gases. In the presence of moisture, highly corrosive to most metals. Vapours are highly irritating to skin, eyes and mucous membranes. Liquid causes severe burns to skin, eyes and mucous membranes.	2699
–	T7	TP2	F-A, S-B	Category B. "Away from" acids and alkalis.	Colourless liquid with a perceptible odour. May react in contact with acids and alkalis. Causes burns to skin, eyes and mucous membranes.	2705
–	T4	TP1	F-E, S-D	Category B.	Colourless liquids with a pungent odour. Partially miscible with water. React vigorously with oxidizing substances. Harmful by inhalation. Irritating to skin, eyes and mucous membranes.	2707
–	T2	TP1	F-E, S-D	Category A.	See entry above.	2707
–	T2	TP1	F-E, S-D	Category A.	Colourless liquids with an unpleasant odour. Flashpoint: 34°C to 60°C c.c. Explosive limits: 0.7% to 6.9%. Immiscible with water. Irritating to skin, eyes and mucous membranes.	2709
–	T2	TP1	F-E, S-D	Category A.	Colourless liquid. Flashpoint: 49°C c.c. Immiscible with water.	2710
–	T1	TP33	F-A, S-A	Category A.	Small colourless to yellowish crystals or needles. Sublimes at 100°C. Practically insoluble in water. Toxic if swallowed, by skin contact or by inhalation.	2713
–	T1	TP33	F-A, S-I	Category A.	Powder or clear amber lumps. Insoluble in water. Liable to spontaneous heating. Irritating to skin and mucous membranes.	2714
–	T1	TP33	F-A, S-I	Category A.	Cream to brown coloured mass. Insoluble in water. Liable to spontaneous heating. Irritating to skin and mucous membranes.	2715
–	T1	TP33	F-A, S-A	Category A. "Separated from" acids, alkalis and mercury salts.	White crystals. Melting point: 58°C. Soluble in water. Forms explosive mixtures with mercury salts, strong acids, alkaline compounds and halides. Toxic if swallowed, by skin contact or by inhalation.	2716
–	T1	TP33	F-A, S-I	Category A.	Colourless or white crystals, granules or easily broken masses with a penetrating, pungent and aromatic odour. Slightly soluble in water. When heated, evolves flammable and explosive vapours. Harmful if swallowed.	2717
–	T3	TP33	F-H, S-Q	Category A. "Separated from" ammonium compounds and cyanides.	White crystals or powder. Slightly soluble in water. Reacts vigorously with sulphuric acid. Reacts fiercely with cyanides when heated or by friction. May form explosive mixtures with combustible material, powdered metals or ammonium compounds. These mixtures are sensitive to friction and are liable to ignite. When involved in a fire, may cause an explosion. Toxic if swallowed, by skin contact or by dust inhalation.	2719
–	T1	TP33	F-A, S-Q	Category A.	Purple crystals. Mixtures with combustible material are readily ignited and may burn fiercely. Solutions in water are slightly corrosive. Harmful if swallowed.	2720
–	T3	TP33	F-H, S-Q	Category A. "Separated from" ammonium compounds and cyanides.	Blue-green deliquescent crystals or powder. Soluble in water. Reacts vigorously with sulphuric acid. Reacts fiercely with cyanides when heated or by friction. May form explosive mixtures with combustible material, powdered metals or ammonium compounds. These mixtures are sensitive to friction and are liable to ignite. When involved in a fire, may cause an explosion.	2721
–	T1	TP33	F-A, S-Q	Category A.	Colourless deliquescent crystals. Soluble in water. Mixtures with combustible material are readily ignited and burn fiercely. Harmful if swallowed.	2722

UN No. (1)	Proper Shipping Name (PSN) (2) 3.1.2	Class or division (3) 2.0	Subsidiary risk(s) (4) 2.0	Packing group (5) 2.0.1.3	Special provisions (6) 3.3	Limited and excepted quantity provisions: Limited quantities (7a) 3.4	Limited and excepted quantity provisions: Excepted quantities (7b) 3.5	Packing: Instructions (8) 4.1.4	Packing: Provisions (9) 4.1.4	IBC: Instructions (10) 4.1.4	IBC: Provisions (11) 4.1.4
2723	MAGNESIUM CHLORATE	5.1	–	II	–	1 kg	E2	P002	–	IBC08	B2 B4
2724	MANGANESE NITRATE	5.1	–	III	–	5 kg	E1	P002 LP02	–	IBC08	B3
2725	NICKEL NITRATE	5.1	–	III	–	5 kg	E1	P002 LP02	–	IBC08	B3
2726	NICKEL NITRITE	5.1	–	III	–	5 kg	E1	P002 LP02	–	IBC08	B3
2727	THALLIUM NITRATE	6.1	5.1 P	II	–	500 g	E4	P002	–	IBC06	B2
2728	ZIRCONIUM NITRATE	5.1	–	III	–	5 kg	E1	P002 LP02	–	IBC08	B3
2729	HEXACHLOROBENZENE	6.1	–	III	–	5 kg	E1	P002 LP02	–	IBC08	B3
2730	NITROANISOLES, LIQUID	6.1	–	III	279	5 L	E1	P001 LP01	–	IBC03	–
2732	NITROBROMOBENZENES, LIQUID	6.1	–	III	–	5 L	E1	P001 LP01	–	IBC03	–
2733	AMINES, FLAMMABLE, CORROSIVE, N.O.S. or POLYAMINES, FLAMMABLE, CORROSIVE, N.O.S.	3	8	I	274	0	E0	P001	–	–	–
2733	AMINES, FLAMMABLE, CORROSIVE, N.O.S. or POLYAMINES, FLAMMABLE, CORROSIVE, N.O.S.	3	8	II	274	1 L	E2	P001	–	IBC02	–
2733	AMINES, FLAMMABLE, CORROSIVE, N.O.S. or POLYAMINES, FLAMMABLE, CORROSIVE, N.O.S.	3	8	III	223 274	5 L	E1	P001	–	IBC03	–
2734	AMINES, LIQUID, CORROSIVE, FLAMMABLE, N.O.S. or POLYAMINES, LIQUID, CORROSIVE, FLAMMABLE, N.O.S.	8	3	I	274	0	E0	P001	–	–	–
2734	AMINES, LIQUID, CORROSIVE, FLAMMABLE, N.O.S. or POLYAMINES, LIQUID, CORROSIVE, FLAMMABLE, N.O.S.	8	3	II	274	1 L	E2	P001	–	IBC02	–

Portable tanks and bulk containers			EmS	Stowage and segregation	Properties and observations	UN No.
	Tank instructions	Provisions				
(12)	(13) 4.2.5 4.3	(14) 4.2.5	(15) 5.4.3.2 7.3	(16) 7.1 7.2	(17)	(18)
–	T3	TP33	F-H, S-Q	Category A. "Separated from" ammonium compounds and cyanides.	White deliquescent crystals or powder. Soluble in water. Melting point: 35°C. Reacts vigorously with sulphuric acid. Reacts fiercely with cyanides when heated or by friction. May form explosive mixtures with combustible material, powdered metals or ammonium compounds. These mixtures are sensitive to friction and are liable to ignite. When involved in a fire, may cause an explosion. The cargoes should be protected from moisture prior to and after loading. If weather is inclement, hatches should be closed.	2723
–	T1	TP33	F-A, S-Q	Category A.	Pale pink deliquescent crystals. Soluble in water. Melting point between 26°C and 35°C. Mixtures with combustible material are readily ignited and may burn fiercely. Solutions in water are slightly corrosive. Harmful if swallowed	2724
–	T1	TP33	F-A, S-Q	Category A.	Green deliquescent crystals. Soluble in water. Melting point: 55°C. Mixtures with combustible material are readily ignited and may burn fiercely. Solutions in water are slightly corrosive. Harmful if swallowed.	2725
–	T1	TP33	F-A, S-Q	Category A. "Separated from" ammonium compounds and cyanides.	Reddish-yellow crystals. Decomposes if heated, giving off toxic nitrous fumes. Mixtures with combustible material are readily ignited and may burn fiercely. Mixtures with ammonium compounds or cyanides may explode. Harmful if swallowed.	2726
–	T3	TP33	F-A, S-Q	Category A.	Colourless crystals. Soluble in water. Mixtures with combustible material are readily ignited and may burn fiercely. Toxic if swallowed, by skin contact or by dust inhalation.	2727
–	T1	TP33	F-A, S-Q	Category A.	White crystals, flakes or powder. Soluble in water. Solutions in water are slightly corrosive. Harmful if swallowed	2728
–	T1	TP33	F-A, S-A	Category A.	White needle-like crystals. Insoluble in water. Decomposes when heated, evolving highly toxic fumes. Toxic if swallowed, by skin contact or by dust inhalation.	2729
–	T4	TP1	F-A, S-A	Category A.	Light reddish or amber liquid. Immiscible with water. Toxic if swallowed, by skin contact or by inhalation.	2730
–	T4	TP1	F-A, S-A	Category A.	Colourless to pale yellow liquids. Melting point of 1-BROMO-3-NITROBENZENE: 17°C. Immiscible with water. Toxic if swallowed, by skin contact or by inhalation.	2732
–	T14	TP1 TP27	F-E, S-C	Category D. Clear of living quarters. "Separated from" acids.	Colourless to yellowish liquids with an unpleasant odour. Some are very volatile. Miscible with water. Corrosive to most metals, especially to copper and its alloys. When involved in a fire, evolve toxic gases. React violently with acids. Harmful by inhalation. Cause burns to skin, eyes and mucous membranes.	2733
–	T11	TP1 TP27	F-E, S-C	Category B. Clear of living quarters. "Separated from" acids.	See entry above.	2733
–	T7	TP1 TP23	F-E, S-C	Category A. Clear of living quarters. "Separated from" acids.	See entry above.	2733
–	T14	TP2 TP27	F-E, S-C	Category A. "Separated from" acids.	Colourless to yellowish flammable liquids or solutions with a pungent odour. Miscible with water. When involved in a fire, evolve toxic gasses. Corrosive to most metals, especially to copper and its alloys. React violently with acids. Cause burns to skin, eyes and mucous membranes.	2734
–	T11	TP2 TP27	F-E, S-C	Category A. "Separated from" acids.	See entry above.	2734

UN No.	Proper Shipping Name (PSN)	Class or division	Subsidiary risk(s)	Packing group	Special provisions	Limited and excepted quantity provisions		Packing		IBC	
						Limited quantities	Excepted quantities	Instructions	Provisions	Instructions	Provisions
(1)	(2) 3.1.2	(3) 2.0	(4) 2.0	(5) 2.0.1.3	(6) 3.3	(7a) 3.4	(7b) 3.5	(8) 4.1.4	(9) 4.1.4	(10) 4.1.4	(11) 4.1.4
2735	AMINES, LIQUID, CORROSIVE, N.O.S. or POLYAMINES, LIQUID, CORROSIVE, N.O.S.	8	–	I	274	0	E0	P001	–	–	–
2735	AMINES, LIQUID, CORROSIVE, N.O.S. or POLYAMINES, LIQUID, CORROSIVE, N.O.S.	8	–	II	274	1 L	E2	P001	–	IBC02	–
2735	AMINES, LIQUID, CORROSIVE, N.O.S. or POLYAMINES, LIQUID, CORROSIVE, N.O.S.	8	–	III	223 274	5 L	E1	P001 LP01	–	IBC03	–
2738	*N*-BUTYLANILINE	6.1	–	II	–	100 mL	E4	P001	–	IBC02	–
2739	BUTYRIC ANHYDRIDE	8	–	III	–	5 L	E1	P001 LP01	–	IBC03	–
2740	*n*-PROPYL CHLOROFORMATE	6.1	3/8	I	–	0	E5	P602	–	–	–
2741	BARIUM HYPOCHLORITE with more than 22% available chlorine	5.1	6.1	II	–	1 kg	E2	P002	–	IBC08	B2 B4
△ 2742	CHLOROFORMATES, TOXIC, CORROSIVE, FLAMMABLE, N.O.S.	6.1	3/8	II	274	100 mL	E4	P001	–	IBC01	–
△ 2743	*n*-BUTYL CHLOROFORMATE	6.1	3/8	II	–	100 mL	E4	P001	–	–	–
△ 2744	CYCLOBUTYL CHLOROFORMATE	6.1	3/8	II	–	100 mL	E4	P001	–	IBC01	–
△ 2745	CHLOROMETHYL CHLOROFORMATE	6.1	8	II	–	100 mL	E4	P001	–	IBC02	–
△ 2746	PHENYL CHLOROFORMATE	6.1	8	II	–	100 mL	E4	P001	–	IBC02	–

Portable tanks and bulk containers			EmS	Stowage and segregation	Properties and observations	UN No.
	Tank instructions	Provisions				
(12)	(13) 4.2.5 4.3	(14) 4.2.5	(15) 5.4.3.2 7.3	(16) 7.1 7.2	(17)	(18)
–	T14	TP2 TP27	F-E, S-B	Category A. "Separated from" acids.	Colourless to yellowish liquids or solutions with a pungent odour. Miscible with or soluble in water. When involved in a fire, evolve toxic gases. Corrosive to most metals, especially to copper and its alloys. React violently with acids. Cause burns to skin, eyes and mucous membranes.	**2735**
–	T11	TP1 TP27	F-A, S-B	Category A. "Separated from" acids.	See entry above.	2735
–	T7	TP1 TP28	F-A, S-B	Category A. "Separated from" acids.	See entry above.	**2735**
–	T7	TP2	F-A, S-A	Category A. "Separated from" class 5.1.	Amber liquid with a perceptible odour. Immiscible with water. May react vigorously with oxidizing substances. Toxic if swallowed, by skin contact or by inhalation.	2738
–	T4	TP1	F-A, S-B	Category A.	Colourless liquid. Decomposes in water to form butyric acid.	**2739**
–	T20	TP2 TP13	F-E, S-C	Category B. Segregation as for class 3, but "away from" class 4.1. Clear of living quarters.	Colourless flammable liquid. Flashpoint: 28°C c.c. Decomposed by water, generating propyl alcohol. Highly toxic if swallowed, by skin contact or by inhalation. Causes burns to skin, eyes and mucous membranes.	2740
–	T3	TP33	F-H, S-Q	Category B. "Separated from" ammonium compounds, acids, cyanides, hydrogen peroxide and liquid organic substances.	White powder with pungent odour. Reacts with acids, evolving chlorine, an irritating, corrosive and toxic gas. Reacts fiercely with cyanides when heated or by friction. May form explosive mixtures with combustible material, powdered metals or ammonium compounds. These mixtures are sensitive to friction and are liable to ignite. When involved in a fire, may cause an explosion. Toxic if swallowed, by skin contact or by dust inhalation. Dust irritates mucous membranes. Contact with eyes will cause serious injury to the cornea (blindness) if not treated immediately by using copious amounts of water followed by medical attention.	**2741**
–	–	–	F-E, S-C	Category A. Keep as cool and dry as reasonably practicable. Protected from sources of heat. Clear of living quarters. Segregation as for class 3, but "away from" class 4.1.	A wide range of colourless to yellowish flammable liquids. React and decompose with water or heat, evolving hydrogen chloride, an irritating and corrosive gas apparent as white fumes. Flashpoint: cyclohexyl chloroformate: 53°C c.c. Toxic if swallowed, by skin contact or by inhalation. Cause burns to skin, eyes and mucous membranes.	2742 △
–	T20	TP2 TP13	F-E, S-C	Category A. Keep as cool and dry as reasonably practicable. Protected from sources of heat. Clear of living quarters. Segregation as for class 3, but "away from" class 4.1.	A wide range of colourless to yellowish flammable liquids. React and decompose with water or heat, evolving hydrogen chloride, an irritating and corrosive gas apparent as white fumes. Flashpoint: 32°C c.c. to 39°C c.c. Toxic if swallowed, by skin contact or by inhalation. Cause burns to skin, eyes and mucous membranes.	**2743** △
–	T7	TP2 TP13	F-E, S-C	Category A. Keep as cool and dry as reasonably practicable. Protected from sources of heat. Clear of living quarters. Segregation as for class 3, but "away from" class 4.1.	A wide range of colourless to yellowish flammable liquids. React and decompose with water or heat, evolving hydrogen chloride, an irritating and corrosive gas apparent as white fumes. Flashpoint: 38°C c.c. Toxic if swallowed, by skin contact or by inhalation. Cause burns to skin, eyes and mucous membranes.	2744 △
–	T7	TP2 TP13	F-A, S-B	Category A. Keep as cool and dry as reasonably practicable. Protected from sources of heat. Clear of living quarters.	A wide range of colourless to yellowish liquids. React and decompose with water or heat, evolving hydrogen chloride, an irritating and corrosive gas apparent as white fumes. Toxic if swallowed, by skin contact or by inhalation. Cause burns to skin, eyes and mucous membranes.	**2745** △
–	T7	TP2 TP13	F-A, S-B	Category A. Keep as cool and dry as reasonably practicable. Protected from sources of heat. Clear of living quarters.	A wide range of colourless to yellowish liquids. React and decompose with water or heat, evolving hydrogen chloride, an irritating and corrosive gas apparent as white fumes. Toxic if swallowed, by skin contact or by inhalation. Cause burns to skin, eyes and mucous membranes.	2746 △

UN No.	Proper Shipping Name (PSN)	Class or division	Subsidiary risk(s)	Packing group	Special provisions	Limited and excepted quantity provisions		Packing		IBC	
						Limited quantities	Excepted quantities	Instructions	Provisions	Instructions	Provisions
(1)	(2) 3.1.2	(3) 2.0	(4) 2.0	(5) 2.0.1.3	(6) 3.3	(7a) 3.4	(7b) 3.5	(8) 4.1.4	(9) 4.1.4	(10) 4.1.4	(11) 4.1.4
△ 2747	*tert*-BUTYLCYCLOHEXYL CHLOROFORMATE	6.1	–	III	–	5 L	E1	P001 LP01	–	IBC03	–
△ 2748	2-ETHYLHEXYL CHLOROFORMATE	6.1	8	II	–	100 mL	E4	P001	–	IBC02	–
2749	TETRAMETHYLSILANE	3	–	I	–	0	E3	P001	–	–	–
2750	1,3-DICHLOROPROPANOL-2	6.1	–	II	–	100 mL	E4	P001	–	IBC02	–
2751	DIETHYLTHIOPHOSPHORYL CHLORIDE	8	–	II	–	1 L	E2	P001	–	IBC02	–
2752	1,2-EPOXY-3-ETHOXY-PROPANE	3	–	III	–	5 L	E1	P001 LP01	–	IBC03	–
2753	*N*-ETHYLBENZYLTOLUIDINES, LIQUID	6.1	–	III	–	5 L	E1	P001 LP01	–	IBC03	–
2754	*N*-ETHYLTOLUIDINES	6.1	–	II	–	100 mL	E4	P001	–	IBC02	–
2757	CARBAMATE PESTICIDE, SOLID, TOXIC	6.1	–	I	61 274	0	E5	P002	–	IBC07	B1
2757	CARBAMATE PESTICIDE, SOLID, TOXIC	6.1	–	II	61 274	500 g	E4	P002	–	IBC08	B2 B4
2757	CARBAMATE PESTICIDE, SOLID, TOXIC	6.1	–	III	61 223 274	5 kg	E1	P002 LP02	–	IBC08	B3
2758	CARBAMATE PESTICIDE, LIQUID, FLAMMABLE, TOXIC, flashpoint less than 23°C	3	6.1	I	61 274	0	E0	P001	–	–	–
2758	CARBAMATE PESTICIDE, LIQUID, FLAMMABLE, TOXIC, flashpoint less than 23°C	3	6.1	II	61 274	1 L	E2	P001	–	IBC02	–
2759	ARSENICAL PESTICIDE, SOLID, TOXIC	6.1	–	I	61 274	0	E5	P002	–	IBC07	B1
2759	ARSENICAL PESTICIDE, SOLID, TOXIC	6.1	–	II	61 274	500 g	E4	P002	–	IBC08	B2 B4
2759	ARSENICAL PESTICIDE, SOLID, TOXIC	6.1	–	III	61 223 274	5 kg	E1	P002 LP02	–	IBC08	B3
2760	ARSENICAL PESTICIDE, LIQUID, FLAMMABLE, TOXIC, flashpoint less than 23°C	3	6.1	I	61 274	0	E0	P001	–	–	–
2760	ARSENICAL PESTICIDE, LIQUID, FLAMMABLE, TOXIC, flashpoint less than 23°C	3	6.1	II	61 274	1 L	E2	P001	–	IBC02	–
2761	ORGANOCHLORINE PESTICIDE, SOLID, TOXIC	6.1	–	I	61 274	0	E5	P002	–	IBC07	B1

Portable tanks and bulk containers			EmS	Stowage and segregation	Properties and observations	UN No.	
	Tank instructions	Provisions					
(12)	(13) 4.2.5 4.3	(14) 4.2.5	(15) 5.4.3.2 7.3	(16) 7.1 7.2	(17)	(18)	
–	T4	TP1	F-A, S-A	Category A. Keep as cool and dry as reasonably practicable. Protected from sources of heat.	Colourless to yellowish liquid. Reacts with water or decomposes if heated, evolving hydrogen chloride, an irritating and corrosive gas apparent as white fumes. Toxic if swallowed, by skin contact or by inhalation.	**2747**	△
–	T7	TP2 TP13	F-A, S-B	Category A. Keep as cool and dry as reasonably practicable. Protected from sources of heat. Clear of living quarters.	A wide range of colourless to yellowish liquids. React and decompose with water or heat, evolving hydrogen chloride, an irritating and corrosive gas apparent as white fumes. Toxic if swallowed, by skin contact or by inhalation. Cause burns to skin, eyes and mucous membranes.	**2748**	△
–	T14	TP2	F-E, S-D	Category D.	Colourless, volatile liquid. Flashpoint: below −18°C c.c. Boiling point: 27°C. Immiscible with water. Harmful if swallowed or by inhalation. Irritating to skin, eyes and mucous membranes.	**2749**	
–	T7	TP2	F-A, S-A	Category A. Keep as cool as reasonably practicable. Clear of living quarters.	Colourless, slightly viscous liquid with a chloroform-like odour. Immiscible with water. Decomposes when heated, evolving extremely toxic fumes (phosgene). Toxic if swallowed, by skin contact or by inhalation.	**2750**	
–	T7	TP2	F-A, S-B	Category D. Keep as cool as reasonably practicable. Clear of living quarters.	Colourless liquid with a perceptible odour. Reacts slowly with water, forming hydrochloric acid. When involved in a fire, evolves toxic gases (hydrogen chloride and sulphur dioxide). Vapour highly irritating to eyes and mucous membranes. Liquid causes burns to skin, eyes and mucous membranes.	**2751**	
–	T2	TP1	F-E, S-D	Category A.	Immiscible with water. Flashpoint: 47°C c.c. Irritating to skin, eyes and mucous membranes.	**2752**	
–	T7	TP1	F-A, S-A	Category A.	Liquids with a strong odour. Immiscible with water. Toxic if swallowed, by skin contact or by inhalation.	**2753**	
–	T7	TP2	F-A, S-A	Category A.	Colourless to light amber flammable liquids. Immiscible with water. Toxic if swallowed, by skin contact or by inhalation.	**2754**	
–	T6	TP33	F-A, S-A	Category A. Clear of living quarters.	Solid pesticides present a very wide range of toxic hazard. Toxic if swallowed, by skin contact or by inhalation.	**2757**	
–	T3	TP33	F-A, S-A	Category A. Clear of living quarters.	See entry above.	**2757**	
–	T1	TP33	F-A, S-A	Category A. Clear of living quarters.	See entry above.	**2757**	
–	T14	TP2 TP13 TP27	F-E, S-D	Category B. Clear of living quarters.	Pesticides frequently contain petroleum or coal tar distillates, or other flammable liquids. Miscibility with water depends upon the composition. Toxic if swallowed, by skin contact or by inhalation.	**2758**	
–	T11	TP2 TP13 TP27	F-E, S-D	Category B. Clear of living quarters.	See entry above.	**2758**	
–	T6	TP33	F-A, S-A	Category A. Clear of living quarters.	Solid pesticides present a very wide range of toxic hazard. Toxic if swallowed, by skin contact or by inhalation.	**2759**	
–	T3	TP33	F-A, S-A	Category A. Clear of living quarters.	See entry above.	**2759**	
–	T1	TP33	F-A, S-A	Category A. Clear of living quarters.	See entry above.	**2759**	
–	T14	TP2 TP13 TP27	F-E, S-D	Category B. Clear of living quarters.	Pesticides frequently contain petroleum or coal tar distillates, or other flammable liquids. Miscibility with water depends upon the composition. Toxic if swallowed, by skin contact or by inhalation.	**2760**	
–	T11	TP2 TP13 TP27	F-E, S-D	Category B. Clear of living quarters.	See entry above.	**2760**	
–	T6	TP33	F-A, S-A	Category A. Clear of living quarters.	Solid pesticides present a very wide range of toxic hazard. Toxic if swallowed, by skin contact or by inhalation.	**2761**	

UN No.	Proper Shipping Name (PSN)	Class or division	Subsidiary risk(s)	Packing group	Special provisions	Limited and excepted quantity provisions		Packing		IBC	
						Limited quantities	Excepted quantities	Instructions	Provisions	Instructions	Provisions
(1)	(2) 3.1.2	(3) 2.0	(4) 2.0	(5) 2.0.1.3	(6) 3.3	(7a) 3.4	(7b) 3.5	(8) 4.1.4	(9) 4.1.4	(10) 4.1.4	(11) 4.1.4
2761	ORGANOCHLORINE PESTICIDE, SOLID, TOXIC	6.1	–	II	61 274	500 g	E4	P002	–	IBC08	B2 B4
2761	ORGANOCHLORINE PESTICIDE, SOLID, TOXIC	6.1	–	III	61 223 274	5 kg	E1	P002 LP02	–	IBC08	B3
2762	ORGANOCHLORINE PESTICIDE, LIQUID, FLAMMABLE, TOXIC, flashpoint less than 23°C	3	6.1	I	61 274	0	E0	P001	–	–	–
2762	ORGANOCHLORINE PESTICIDE, LIQUID, FLAMMABLE, TOXIC, flashpoint less than 23°C	3	6.1	II	61 274	1 L	E2	P001	–	IBC02	–
2763	TRIAZINE PESTICIDE, SOLID, TOXIC	6.1	–	I	61 274	0	E5	P002	–	IBC07	B1
2763	TRIAZINE PESTICIDE, SOLID, TOXIC	6.1	–	II	61 274	500 g	E4	P002	–	IBC08	B2 B4
2763	TRIAZINE PESTICIDE, SOLID, TOXIC	6.1	–	III	61 223 274	5 kg	E1	P002	–	IBC08	B3
2764	TRIAZINE PESTICIDE, LIQUID, FLAMMABLE, TOXIC, flashpoint less than 23°C	3	6.1	I	61 274	0	E0	P001	–	–	–
2764	TRIAZINE PESTICIDE, LIQUID, FLAMMABLE, TOXIC, flashpoint less than 23°C	3	6.1	II	61 274	1 L	E2	P001	–	IBC02	–
2771	THIOCARBAMATE PESTICIDE, SOLID, TOXIC	6.1	–	I	61 274	0	E5	P002	–	IBC07	B1
2771	THIOCARBAMATE PESTICIDE, SOLID, TOXIC	6.1	–	II	61 274	500 g	E4	P002	–	IBC08	B2 B4
2771	THIOCARBAMATE PESTICIDE, SOLID, TOXIC	6.1	–	III	61 223 274	5 kg	E1	P002 LP02	–	IBC08	B3
2772	THIOCARBAMATE PESTICIDE, LIQUID, FLAMMABLE, TOXIC, flashpoint less than 23°C	3	6.1	I	61 274	0	E0	P001	–	–	–
2772	THIOCARBAMATE PESTICIDE, LIQUID, FLAMMABLE, TOXIC, flashpoint less than 23°C	3	6.1	II	61 274	1 L	E2	P001	–	IBC02	–
2775	COPPER BASED PESTICIDE, SOLID, TOXIC	6.1	–	I	61 274	0	E5	P002	–	IBC07	B1
2775	COPPER BASED PESTICIDE, SOLID, TOXIC	6.1	–	II	61 274	500 g	E4	P002	–	IBC08	B2 B4
2775	COPPER BASED PESTICIDE, SOLID, TOXIC	6.1	–	III	61 223 274	5 kg	E1	P002 LP02	–	IBC08	B3
2776	COPPER BASED PESTICIDE, LIQUID, FLAMMABLE, TOXIC, flashpoint less than 23°C	3	6.1	I	61 274	0	E0	P001	–	–	–
2776	COPPER BASED PESTICIDE, LIQUID, FLAMMABLE, TOXIC, flashpoint less than 23°C	3	6.1	II	61 274	1 L	E2	P001	–	IBC02	–
2777	MERCURY BASED PESTICIDE, SOLID, TOXIC	6.1	– P	I	61 274	0	E5	P002	–	IBC07	B1
2777	MERCURY BASED PESTICIDE, SOLID, TOXIC	6.1	– P	II	61 274	500 g	E4	P002	–	IBC08	B2 B4

Portable tanks and bulk containers			EmS	Stowage and segregation	Properties and observations	UN No.
	Tank instructions	Provisions				
(12)	(13) 4.2.5 4.3	(14) 4.2.5	(15) 5.4.3.2 7.3	(16) 7.1 7.2	(17)	(18)
–	T3	TP33	F-A, S-A	Category A. Clear of living quarters.	Solid pesticides present a very wide range of toxic hazard. Toxic if swallowed, by skin contact or by inhalation.	2761
–	T1	TP33	F-A, S-A	Category A. Clear of living quarters.	See entry above.	2761
–	T14	TP2 TP13 TP27	F-E, S-D	Category B. Clear of living quarters.	Pesticides frequently contain petroleum or coal tar distillates, or other flammable liquids. Miscibility with water depends upon the composition. Toxic if swallowed, by skin contact or by inhalation.	2762
–	T11	TP2 TP13 TP27	F-E, S-D	Category B. Clear of living quarters.	See entry above.	2762
–	T6	TP33	F-A, S-A	Category A. Clear of living quarters.	Solid pesticides present a very wide range of toxic hazard. Toxic if swallowed, by skin contact or by inhalation.	2763
–	T3	TP33	F-A, S-A	Category A. Clear of living quarters.	See entry above.	2763
–	T3	TP33	F-A, S-A	Category A. Clear of living quarters.	See entry above.	2763
–	T14	TP2 TP13 TP27	F-E, S-D	Category B. Clear of living quarters.	Pesticides frequently contain petroleum or coal tar distillates, or other flammable liquids. Miscibility with water depends upon the composition. Toxic if swallowed, by skin contact or by inhalation.	2764
–	T11	TP2 TP13 TP27	F-E, S-D	Category B. Clear of living quarters.	See entry above.	2764
–	T6	TP33	F-A, S-A	Category A. Clear of living quarters.	Solid pesticides present a very wide range of toxic hazard. Toxic if swallowed, by skin contact or by inhalation.	2771
–	T3	TP33	F-A, S-A	Category A. Clear of living quarters.	See entry above.	2771
–	T1	TP33	F-A, S-A	Category A. Clear of living quarters.	See entry above.	2771
–	T14	TP2 TP13 TP27	F-E, S-D	Category B. Clear of living quarters.	Pesticides frequently contain petroleum or coal tar distillates, or other flammable liquids. Miscibility with water depends upon the composition. Toxic if swallowed, by skin contact or by inhalation.	2772
–	T11	TP2 TP13 TP27	F-E, S-D	Category B. Clear of living quarters.	See entry above.	2772
–	T6	TP33	F-A, S-A	Category A. Clear of living quarters.	Solid pesticides present a very wide range of toxic hazard. Toxic if swallowed, by skin contact or by inhalation.	2775
–	T3	TP33	F-A, S-A	Category A. Clear of living quarters.	See entry above.	2775
–	T1	TP33	F-A, S-A	Category A. Clear of living quarters.	See entry above.	2775
–	T14	TP2 TP13 TP27	F-E, S-D	Category B. Clear of living quarters.	Pesticides frequently contain petroleum or coal tar distillates, or other flammable liquids. Miscibility with water depends upon the composition. Toxic if swallowed, by skin contact or by inhalation.	2776
–	T11	TP2 TP13 TP27	F-E, S-D	Category B. Clear of living quarters.	See entry above.	2776
–	T6	TP33	F-A, S-A	Category A. Clear of living quarters.	Solid pesticides present a very wide range of toxic hazard. Toxic if swallowed, by skin contact or by inhalation.	2777
–	T3	TP33	F-A, S-A	Category A. Clear of living quarters.	See entry above.	2777

UN No.	Proper Shipping Name (PSN)	Class or division	Subsidiary risk(s)	Packing group	Special provisions	Limited and excepted quantity provisions		Packing		IBC	
						Limited quantities	Excepted quantities	Instructions	Provisions	Instructions	Provisions
(1)	(2) 3.1.2	(3) 2.0	(4) 2.0	(5) 2.0.1.3	(6) 3.3	(7a) 3.4	(7b) 3.5	(8) 4.1.4	(9) 4.1.4	(10) 4.1.4	(11) 4.1.4
2777	MERCURY BASED PESTICIDE, SOLID, TOXIC	6.1	– P	III	61 223 274	5 kg	E1	P002 LP02	–	IBC08	B3
2778	MERCURY BASED PESTICIDE, LIQUID, FLAMMABLE, TOXIC, flashpoint less than 23°C	3	6.1 P	I	61 274	0	E0	P001	–	–	–
2778	MERCURY BASED PESTICIDE, LIQUID, FLAMMABLE, TOXIC, flashpoint less than 23°C	3	6.1 P	II	61 274	1 L	E2	P001	–	IBC02	–
2779	SUBSTITUTED NITROPHENOL PESTICIDE, SOLID, TOXIC	6.1	–	I	61 274	0	E5	P002	–	IBC07	B1
2779	SUBSTITUTED NITROPHENOL PESTICIDE, SOLID, TOXIC	6.1	–	II	61 274	500 g	E4	P002	–	IBC08	B2 B4
2779	SUBSTITUTED NITROPHENOL PESTICIDE, SOLID, TOXIC	6.1	–	III	61 223 274	5 kg	E1	P002 LP02	–	IBC08	B3
2780	SUBSTITUTED NITROPHENOL PESTICIDE, LIQUID, FLAMMABLE, TOXIC, flashpoint less than 23°C	3	6.1	I	61 274	0	E0	P001	–	–	–
2780	SUBSTITUTED NITROPHENOL PESTICIDE, LIQUID, FLAMMABLE, TOXIC, flashpoint less than 23°C	3	6.1	II	61 274	1 L	E2	P001	–	IBC02	–
2781	BIPYRIDILIUM PESTICIDE, SOLID, TOXIC	6.1	–	I	61 274	0	E5	P002	–	IBC07	B1
2781	BIPYRIDILIUM PESTICIDE, SOLID, TOXIC	6.1	–	II	61 274	500 g	E4	P002	–	IBC08	B2 B4
2781	BIPYRIDILIUM PESTICIDE, SOLID, TOXIC	6.1	–	III	61 223 274	5 kg	E1	P002 LP02	–	IBC08	B3
2782	BIPYRIDILIUM PESTICIDE, LIQUID, FLAMMABLE, TOXIC, flashpoint less than 23°C	3	6.1	I	61 274	0	E0	P001	–	–	–
2782	BIPYRIDILIUM PESTICIDE, LIQUID, FLAMMABLE, TOXIC, flashpoint less than 23°C	3	6.1	II	61 274	1 L	E2	P001	–	IBC02	–
2783	ORGANOPHOSPHORUS PESTICIDE, SOLID, TOXIC	6.1	–	I	61 274	0	E5	P002	–	IBC07	B1
2783	ORGANOPHOSPHORUS PESTICIDE, SOLID, TOXIC	6.1	–	II	61 274	500 g	E4	P002	–	IBC08	B2 B4
2783	ORGANOPHOSPHORUS PESTICIDE, SOLID, TOXIC	6.1	–	III	61 223 274	5 kg	E1	P002 LP02	–	IBC08	B3
2784	ORGANOPHOSPHORUS PESTICIDE, LIQUID, FLAMMABLE, TOXIC, flashpoint less than 23°C	3	6.1	I	61 274	0	E0	P001	–	–	–
2784	ORGANOPHOSPHORUS PESTICIDE, LIQUID, FLAMMABLE, TOXIC, flashpoint less than 23°C	3	6.1	II	61 274	1 L	E2	P001	–	IBC02	–
△ 2785	4-THIAPENTANAL	6.1	–	III	–	5 L	E1	P001 LP01	PP31	IBC03	–

Portable tanks and bulk containers			EmS	Stowage and segregation	Properties and observations	UN No.
	Tank instructions	Provisions				
(12)	(13) 4.2.5 4.3	(14) 4.2.5	(15) 5.4.3.2 7.3	(16) 7.1 7.2	(17)	(18)
–	T1	TP33	F-A, S-A	Category A. Clear of living quarters.	Solid pesticides present a very wide range of toxic hazard. Toxic if swallowed, by skin contact or by inhalation.	2777
–	T14	TP2 TP13 TP27	F-E, S-D	Category B. Clear of living quarters.	Pesticides frequently contain petroleum or coal tar distillates, or other flammable liquids. Miscibility with water depends upon the composition. Toxic if swallowed, by skin contact or by inhalation.	2778
–	T11	TP2 TP13 TP27	F-E, S-D	Category B. Clear of living quarters.	See entry above.	2778
–	T6	TP33	F-A, S-A	Category A. Clear of living quarters.	Solid pesticides present a very wide range of toxic hazard. Toxic if swallowed, by skin contact or by inhalation.	2779
–	T3	TP33	F-A, S-A	Category A. Clear of living quarters.	See entry above.	2779
–	T1	TP33	F-A, S-A	Category A. Clear of living quarters.	See entry above.	2779
–	T14	TP2 TP13 TP27	F-E, S-D	Category B. Clear of living quarters.	Pesticides frequently contain petroleum or coal tar distillates, or other flammable liquids. Miscibility with water depends upon the composition. Toxic if swallowed, by skin contact or by inhalation.	2780
–	T11	TP2 TP13 TP27	F-E, S-D	Category B. Clear of living quarters.	See entry above.	2780
–	T6	TP33	F-A, S-A	Category A. Clear of living quarters.	Solid pesticides present a very wide range of toxic hazard. Toxic if swallowed, by skin contact or by inhalation.	2781
–	T3	TP33	F-A, S-A	Category A. Clear of living quarters.	See entry above.	2781
–	T1	TP33	F-A, S-A	Category A. Clear of living quarters.	See entry above.	2781
–	T14	TP2 TP13 TP27	F-E, S-D	Category B. Clear of living quarters.	Pesticides frequently contain petroleum or coal tar distillates, or other flammable liquids. Miscibility with water depends upon the composition. Toxic if swallowed, by skin contact or by inhalation.	2782
–	T11	TP2 TP13 TP27	F-E, S-D	Category B. Clear of living quarters.	See entry above.	2782
–	T6	TP33	F-A, S-A	Category A. Clear of living quarters.	Solid pesticides present a very wide range of toxic hazard. Toxic if swallowed, by skin contact or by inhalation.	2783
–	T3	TP33	F-A, S-A	Category A. Clear of living quarters.	See entry above.	2783
–	T1	TP33	F-A, S-A	Category A. Clear of living quarters.	See entry above.	2783
–	T14	TP2 TP13 TP27	F-E, S-D	Category B. Clear of living quarters.	Pesticides frequently contain petroleum or coal tar distillates, or other flammable liquids. Miscibility with water depends upon the composition. Toxic if swallowed, by skin contact or by inhalation.	2784
–	T11	TP2 TP13 TP27	F-E, S-D	Category B. Clear of living quarters.	See entry above.	2784
–	T4	TP1	F-A, S-A	Category D. Protected from sources of heat. "Away from" acids and alkalis.	Colourless liquid with an extremely foul and persistent odour. Miscible with water. Decomposes rapidly in contact with acids and bases. Toxic if swallowed, by skin contact or by inhalation.	2785 △

UN No.	Proper Shipping Name (PSN)	Class or division	Subsidiary risk(s)	Packing group	Special provisions	Limited and excepted quantity provisions		Packing		IBC	
						Limited quantities	Excepted quantities	Instructions	Provisions	Instructions	Provisions
(1)	(2) 3.1.2	(3) 2.0	(4) 2.0	(5) 2.0.1.3	(6) 3.3	(7a) 3.4	(7b) 3.5	(8) 4.1.4	(9) 4.1.4	(10) 4.1.4	(11) 4.1.4
2786	ORGANOTIN PESTICIDE, SOLID, TOXIC	6.1	– P	I	61 274	0	E5	P002	–	IBC07	B1
2786	ORGANOTIN PESTICIDE, SOLID, TOXIC	6.1	– P	II	61 274	500 g	E4	P002	–	IBC08	B2 B4
2786	ORGANOTIN PESTICIDE, SOLID, TOXIC	6.1	– P	III	61 223 274	5 kg	E1	P002 LP02	–	IBC08	B3
2787	ORGANOTIN PESTICIDE, LIQUID, FLAMMABLE, TOXIC, flashpoint less than 23°C	3	6.1 P	I	61 274	0	E0	P001	–	–	–
2787	ORGANOTIN PESTICIDE, LIQUID, FLAMMABLE, TOXIC, flashpoint less than 23°C	3	6.1 P	II	61 274	1 L	E2	P001	–	IBC02	–
2788	ORGANOTIN COMPOUND, LIQUID, N.O.S.	6.1	– P	I	43 274	0	E5	P001	–	–	–
2788	ORGANOTIN COMPOUND, LIQUID, N.O.S.	6.1	– P	II	43 274	100 mL	E4	P001	–	IBC02	–
2788	ORGANOTIN COMPOUND, LIQUID, N.O.S.	6.1	– P	III	43 223 274	5 L	E1	P001 LP01	–	IBC03	–
2789	ACETIC ACID, GLACIAL or ACETIC ACID SOLUTION, more than 80% acid, by mass	8	3	II	–	1 L	E2	P001	–	IBC02	–
2790	ACETIC ACID SOLUTION not less than 50% but not more than 80% acid, by mass	8	–	II	–	1 L	E2	P001	–	IBC02	–
2790	ACETIC ACID SOLUTION more than 10% and less than 50% acid, by mass	8	–	III	–	5 L	E1	P001 LP01	–	IBC03	–
2793	FERROUS METAL BORINGS, SHAVINGS, TURNINGS, or CUTTINGS in a form liable to self-heating	4.2	–	III	223 931	0	E1	P003 LP02	PP20	IBC08	B3 B6
2794	BATTERIES, WET, FILLED WITH ACID electric storage	8	–	–	295	1 L	E0	P801	–	–	–
2795	BATTERIES, WET, FILLED WITH ALKALI electric storage	8	–	–	295	1 L	E0	P801	–	–	–

Portable tanks and bulk containers			EmS	Stowage and segregation	Properties and observations	UN No.
	Tank instructions	Provisions				
(12)	(13) 4.2.5 4.3	(14) 4.2.5	(15) 5.4.3.2 7.3	(16) 7.1 7.2	(17)	(18)
–	T6	TP33	F-A, S-A	Category A. Clear of living quarters.	Solid pesticides present a very wide range of toxic hazard. Toxic if swallowed, by skin contact or by inhalation.	2786
–	T3	TP33	F-A, S-A	Category A. Clear of living quarters.	See entry above.	2786
–	T1	TP33	F-A, S-A	Category A. Clear of living quarters.	See entry above.	2786
–	T14	TP2 TP13 TP27	F-E, S-D	Category B. Clear of living quarters.	Pesticides frequently contain petroleum or coal tar distillates, or other flammable liquids. Miscibility with water depends upon the composition. Toxic if swallowed, by skin contact or by inhalation.	2787
–	T11	TP2 TP13 TP27	F-E, S-D	Category B. Clear of living quarters.	See entry above.	2787
–	T14	TP2 TP13 TP27	F-A, S-A	Category A. Clear of living quarters.	A wide variety of toxic liquids. Toxic if swallowed, by skin contact or by inhalation.	2788
–	T11	TP2 TP13 TP27	F-A, S-A	Category A. Clear of living quarters.	See entry above.	2788
–	T7	TP2 TP28	F-A, S-A	Category A. Clear of living quarters.	See entry above.	2788
–	T7	TP2	F-E, S-C	Category A.	Colourless flammable liquid with a pungent odour. When pure, crystallizes below 16°C. Flashpoint: 40°C c.c. (pure product), 60°C c.c. (80% solution). Explosive limits: 4% to 17%. Miscible with water. Corrosive to lead and most other metals. Corrosive to skin, eyes and mucous membranes.	2789
–	T7	TP2	F-A, S-B	Category A.	Colourless liquid with a pungent odour. Miscible with water. Corrosive to lead and most other metals. Corrosive to skin, eyes and mucous membranes.	2790
–	T4	TP1	F-A, S-B	Category A.	See entry above.	2790
–	BK2	–	F-G, S-J	Category A.	These cargoes are liable to self-heating and to ignite spontaneously, particularly when in a finely divided form, wet or contaminated with such materials as unsaturated cutting oil, oily rags and other combustible matter. Self-heating or inadequate ventilation may cause dangerous depletion of oxygen in the stowage spaces. Excessive amounts of cast iron borings or organic materials may encourage heating. The swarf should be protected from moisture prior to and after loading. If, during loading, the weather is inclement, hatches should be closed or otherwise protected to keep the material dry.	2793
–	–	–	F-A, S-B	Category A. For unit loads in open cargo transport units, category B. "Separated from" acids.	Metal plates immersed in acid electrolyte in a glass, hard rubber or plastics receptacle. When electrically charged, may cause fire through short-circuiting of terminals. Acid electrolyte is corrosive to most metals. Causes burns to skin, eyes and mucous membranes. Used batteries being transported for disposal or reclamation should be carefully checked prior to shipment to ensure the integrity of each battery and its suitability for transport.	2794
–	–	–	F-A, S-B	Category A. For unit loads in open cargo transport units, category B.	Metal plates immersed in alkaline electrolyte in a glass, hard rubber or plastics receptacle. When electrically charged, may cause fire through short-circuiting of terminals. Alkaline electrolyte is corrosive to aluminium, zinc and tin. Reacts violently with acids. Causes burns to skin, eyes and mucous membranes. Used batteries being transported for disposal or reclamation should be carefully checked prior to shipment to ensure the integrity of each battery and its suitability for transport.	2795

UN No.	Proper Shipping Name (PSN)	Class or division	Subsidiary risk(s)	Packing group	Special provisions	Limited and excepted quantity provisions		Packing		IBC	
						Limited quantities	Excepted quantities	Instructions	Provisions	Instructions	Provisions
(1)	(2) 3.1.2	(3) 2.0	(4) 2.0	(5) 2.0.1.3	(6) 3.3	(7a) 3.4	(7b) 3.5	(8) 4.1.4	(9) 4.1.4	(10) 4.1.4	(11) 4.1.4
2796	SULPHURIC ACID with not more than 51% acid or BATTERY FLUID, ACID	8	–	II	–	1 L	E2	P001	–	IBC02	B20
2797	BATTERY FLUID, ALKALI	8	–	II	–	1 L	E2	P001	–	IBC02	–
2798	PHENYLPHOSPHORUS DICHLORIDE	8	–	II	–	1 L	E2	P001	–	IBC02	–
2799	PHENYLPHOSPHORUS THIODICHLORIDE	8	–	II	–	1 L	E2	P001	–	IBC02	–
2800	BATTERIES, WET, NON-SPILLABLE electric storage	8	–	–	29 238	1 L	E0	P003	PP16	–	–
2801	DYE, LIQUID, CORROSIVE, N.O.S. or DYE INTERMEDIATE, LIQUID, CORROSIVE, N.O.S.	8	–	I	274	0	E0	P001	–	–	–
2801	DYE, LIQUID, CORROSIVE, N.O.S. or DYE INTERMEDIATE, LIQUID, CORROSIVE, N.O.S.	8	–	II	274	1 L	E2	P001	–	IBC02	–
2801	DYE, LIQUID, CORROSIVE, N.O.S. or DYE INTERMEDIATE, LIQUID, CORROSIVE, N.O.S.	8	–	III	223 274	5 L	E1	P001 LP01	–	IBC03	–
2802	COPPER CHLORIDE	8	– P	III	–	500 g	E1	P002 LP02	–	IBC08	B3
△ 2803	GALLIUM	8	–	III	–	5 kg	E0	P800	PP41	–	–
2805	LITHIUM HYDRIDE, FUSED SOLID	4.3	–	II	–	500 g	E2	P410	PP31 PP40	IBC04	–
2806	LITHIUM NITRIDE	4.3	–	I	–	0	E0	P403	PP31	IBC04	B1
2807	MAGNETIZED MATERIAL	9	–	–	960	–	–	–	–	–	–
△ 2809	MERCURY	8	6.1	III	365	5 kg	E0	P800	–	–	–
2810	TOXIC LIQUID, ORGANIC, N.O.S.	6.1	–	I	274 315	0	E5	P001	–	–	–
2810	TOXIC LIQUID, ORGANIC, N.O.S.	6.1	–	II	274	100 mL	E4	P001	–	IBC02	–
2810	TOXIC LIQUID, ORGANIC, N.O.S.	6.1	–	III	223 274	5 L	E1	P001 LP01	–	IBC03	–
2811	TOXIC SOLID, ORGANIC, N.O.S.	6.1	–	I	274	0	E5	P002	–	IBC99	–

Portable tanks and bulk containers			EmS	Stowage and segregation	Properties and observations	UN No.
	Tank instructions	Provisions				
(12)	(13) 4.2.5 4.3	(14) 4.2.5	(15) 5.4.3.2 7.3	(16) 7.1 7.2	(17)	(18)
–	T8	TP2	F-A, S-B	Category B.	Colourless liquid, mixture not exceeding 1.405 relative density. Highly corrosive to most metals. Causes burns to skin, eyes and mucous membranes.	2796
–	T7	TP2 TP28	F-A, S-B	Category A. "Away from" ammonium salts. "Separated from" acids.	Reacts violently with acids. Reacts with ammonium salts, evolving ammonia gas. Corrosive to aluminium, zinc and tin.	2797
–	T7	TP2	F-A, S-B	Category B. Clear of living quarters.	Colourless liquid. Causes burns to skin, eyes and mucous membranes.	2798
–	T7	TP2	F-A, S-B	Category B. Clear of living quarters.	Colourless liquid which fumes slightly in air. Reacts with water or steam, evolving toxic and flammable vapours. Causes burns to skin, eyes and mucous membranes.	2799
–	–	–	F-A, S-B	Category A.	Metal plates immersed in gelled alkaline or acid electrolyte in a glass, hard rubber or plastics receptacle of a non-spillable type. When electrically charged, may cause fire through short-circuiting of terminals. Cause burns to skin, eyes and mucous membranes.	2800
–	T14	TP2 TP27	F-A, S-B	Category A.	A wide range of corrosive liquids. Cause burns to skin, eyes and mucous membranes.	2801
–	T11	TP2 TP27	F-A, S-B	Category A.	See entry above.	2801
–	T7	TP1 TP28	F-A, S-B	Category A.	See entry above.	2801
–	T1	TP33	F-A, S-B	Category A.	White to yellow-brown crystals or powder. Partially to fully soluble in water. Corrosive to steel. Causes burns to skin, eyes and mucous membranes.	2802
–	T1	TP33	F-A, S-B	Category B. Protected from sources of heat.	Silvery-white metallic element that melts at 29°C, becoming a bright, shiny liquid. Insoluble in water. Highly corrosive to aluminium. Harmful if swallowed, by skin contact or by inhalation. Special care should be taken if a leakage occurs when carried in aluminium freight containers. Carriage should be prohibited in hovercraft and other ships constructed from aluminium.	2803 △
–	T3	TP33	F-G, S-N	Category E. "Separated from" acids.	White, crystalline mass. Reacts with water, moisture or acids, evolving hydrogen which may be ignited by the heat of the reaction.	2805
–	–	–	F-A, S-O	Category E.	Brownish-red crystals or fine, free-flowing powder. Reacts slowly with water to form lithium hydroxide and ammonia.	2806
	–	–	–	–	Not subject to the provisions of this Code but may be subject to provisions governing the transport of dangerous goods by other modes.	2807
–	–	–	F-A, S-B	Category B. Clear of living quarters. "Away from" azides.	A silvery metallic element occurring in the liquid state at normal temperatures. Relative density: 13.546. Melting point: −39°C . Highly corrosive to aluminium. Toxic if swallowed, by skin contact or by inhalation. Special care should be taken if a leakage occurs during transport, especially when carried in breakable packages and in aluminium freight containers. Carriage should be prohibited in hovercraft and other ships constructed from aluminium.	2809 △
–	T14	TP2 TP13 TP27	F-A, S-A	Category B. Clear of living quarters.	Toxic if swallowed, by skin contact or by inhalation.	2810
–	T11	TP2 TP13 TP27	F-A, S-A	Category B. Clear of living quarters.	See entry above.	2810
–	T7	TP1 TP28	F-A, S-A	Category A. Clear of living quarters.	See entry above.	2810
–	T6	TP33	F-A, S-A	Category B.	Toxic if swallowed, by skin contact or by inhalation.	2811

UN No.	Proper Shipping Name (PSN)	Class or division	Subsidiary risk(s)	Packing group	Special provisions	Limited and excepted quantity provisions		Packing		IBC	
						Limited quantities	Excepted quantities	Instructions	Provisions	Instructions	Provisions
(1)	(2) 3.1.2	(3) 2.0	(4) 2.0	(5) 2.0.1.3	(6) 3.3	(7a) 3.4	(7b) 3.5	(8) 4.1.4	(9) 4.1.4	(10) 4.1.4	(11) 4.1.4
2811	TOXIC SOLID, ORGANIC, N.O.S.	6.1	–	II	274	500 g	E4	P002	–	IBC08	B2 B4
2811	TOXIC SOLID, ORGANIC, N.O.S.	6.1	–	III	223 274	5 kg	E1	P002	–	IBC08	B3
2812	SODIUM ALUMINATE, SOLID	8	–	–	960	–	–	–	–	–	–
2813	WATER-REACTIVE SOLID, N.O.S.	4.3	–	I	274	0	E0	P403	PP31 PP83	IBC99	–
2813	WATER-REACTIVE SOLID, N.O.S.	4.3	–	II	274	500 g	E2	P410	PP31 PP40 PP83	IBC07	B2
2813	WATER-REACTIVE SOLID, N.O.S.	4.3	–	III	223 274	1 kg	E1	P410	PP31 PP83	IBC08	B4
2814	INFECTIOUS SUBSTANCE, AFFECTING HUMANS	6.2	–	–	318 341	0	E0	P620	–	–	–
2815	*N*-AMINOETHYLPIPERAZINE	8	–	III	–	5 L	E1	P001 LP01	–	IBC03	–
2817	AMMONIUM HYDROGEN-DIFLUORIDE SOLUTION	8	6.1	II	–	1 L	E2	P001	–	IBC02	B20
2817	AMMONIUM HYDROGEN-DIFLUORIDE SOLUTION	8	6.1	III	223	5 L	E1	P001	–	IBC03	–
2818	AMMONIUM POLYSULPHIDE SOLUTION	8	6.1	II	–	1 L	E2	P001	–	IBC02	–
2818	AMMONIUM POLYSULPHIDE SOLUTION	8	6.1	III	223	5 L	E1	P001	–	IBC03	–
2819	AMYL ACID PHOSPHATE	8	–	III	–	5 L	E1	P001 LP01	–	IBC03	–
2820	BUTYRIC ACID	8	–	III	–	5 L	E1	P001 LP01	–	IBC03	–
2821	PHENOL SOLUTION	6.1	–	II	–	100 mL	E4	P001	–	IBC02	–
2821	PHENOL SOLUTION	6.1	–	III	223	5 L	E1	P001 LP01	–	IBC03	–
2822	2-CHLOROPYRIDINE	6.1	–	II	–	100 mL	E4	P001	–	IBC02	–
2823	CROTONIC ACID, SOLID	8	–	III	–	5 kg	E1	P002 LP02	–	IBC08	B2 B3
2826	ETHYL CHLOROTHIOFORMATE	8	3 P	II	–	0	E2	P001	–	–	–
2829	CAPROIC ACID	8	–	III	–	5 L	E1	P001 LP01	–	IBC03	–

Portable tanks and bulk containers			EmS	Stowage and segregation	Properties and observations	UN No.
	Tank instructions	Provisions				
(12)	(13) 4.2.5 4.3	(14) 4.2.5	(15) 5.4.3.2 7.3	(16) 7.1 7.2	(17)	(18)
–	T3	TP33	F-A, S-A	Category B.	Toxic if swallowed, by skin contact or by inhalation.	2811
–	T1	TP33	F-A, S-A	Category A.	See entry above.	2811
	–	–	–	–	Not subject to the provisions of this Code but may be subject to provisions governing the transport of dangerous goods by other modes.	2812
–	T9	TP7 TP33	F-G, S-N	Category E. Clear of living quarters.	–	2813
–	T3	TP33	F-G, S-N	Category E. Clear of living quarters.	–	2813
–	T1	TP33	F-G, S-N	Category E. Clear of living quarters.	–	2813
–	BK2	–	F-A, S-T	As approved by the competent authorities of the countries involved in the shipment.	Substances which are dangerous to humans or to humans and animals.	2814
–	T4	TP1	F-A, S-B	Category A. Keep as cool as reasonably practicable.	Yellow liquid. Miscible with water. Corrosive to skin, eyes and mucous membranes.	2815
–	T8	TP2 TP13	F-A, S-B	Category B. Clear of living quarters.	Colourless liquid. Miscible with water. Highly corrosive to most metals and glass. Toxic if swallowed, by skin contact or by inhalation. Causes burns to skin, eyes and mucous membranes.	2817
–	T4	TP1 TP13	F-A, S-B	Category B. Clear of living quarters.	See entry above.	2817
–	T7	TP2 TP13	F-A, S-B	Category B. Keep as cool as reasonably practicable. Clear of living quarters. "Separated from" acids.	Unstable yellowish liquid with a foul odour (of rotten eggs). Miscible with water. Reacts violently with acids. Decomposes in contact with acids, evolving hydrogen sulphide, a toxic and flammable gas. Toxic if swallowed, by skin contact or by inhalation. Causes burns to skin, eyes and mucous membranes.	2818
–	T4	TP1 TP13	F-A, S-B	Category B. Keep as cool as reasonably practicable. Clear of living quarters. "Separated from" acids.	See entry above.	2818
–	T4	TP1	F-A, S-B	Category A.	Clear colourless liquid. A mixture of primary and amyl isomers. Immiscible with water. Corrosive to skin, eyes and mucous membranes.	2819
–	T4	TP1	F-A, S-B	Category A. Keep as cool as reasonably practicable.	Colourless liquid with a penetrating and unpleasant odour. Freezing point: −5°C to −8°C. Miscible with water. Corrosive to most metals. Harmful if swallowed or by inhalation. Corrosive to skin, eyes and mucous membranes.	2820
–	T7	TP2	F-A, S-A	Category A.	Yellowish solutions with a perceptible odour. Toxic if swallowed, by skin contact or by inhalation. Rapidly absorbed through the skin.	2821
–	T4	TP1	F-A, S-A	Category A.	See entry above.	2821
–	T7	TP2	F-A, S-A	Category A. Clear of living quarters.	Colourless oily liquid. Slightly miscible with water. Toxic if swallowed, by skin contact or by inhalation.	2822
–	T1	TP33	F-A, S-B	Category A. Keep as cool as reasonably practicable.	White crystalline solid. Soluble in water. Decomposes when heated, evolving toxic fumes. Causes burns to skin, eyes and mucous membranes.	2823
–	T7	TP2	F-E, S-C	Category A. Clear of living quarters.	Colourless, flammable liquid. Flashpoint: 29°C c.c. Causes burns to skin, eyes and mucous membranes.	2826
–	T4	TP1	F-A, S-B	Category A.	Oily, colourless or yellowish liquid. Melting point: −4°C. Partially miscible with water. Corrosive to mild steel. Causes burns to skin, eyes and mucous membranes.	2829

UN No.	Proper Shipping Name (PSN)	Class or division	Subsidiary risk(s)	Packing group	Special provisions	Limited and excepted quantity provisions		Packing		IBC	
						Limited quantities	Excepted quantities	Instructions	Provisions	Instructions	Provisions
(1)	(2) 3.1.2	(3) 2.0	(4) 2.0	(5) 2.0.1.3	(6) 3.3	(7a) 3.4	(7b) 3.5	(8) 4.1.4	(9) 4.1.4	(10) 4.1.4	(11) 4.1.4
2830	LITHIUM FERROSILICON	4.3	–	II	–	500 g	E2	P410	PP31 PP40	IBC07	B2
2831	1,1,1-TRICHLOROETHANE	6.1	–	III	–	5 L	E1	P001 LP01	–	IBC03	–
△ 2834	PHOSPHOROUS ACID	8	–	III	–	5 kg	E1	P002 LP02	–	IBC08	B3
2835	SODIUM ALUMINIUM HYDRIDE	4.3	–	II	–	500 g	E2	P410	PP31 PP40	IBC04	–
2837	BISULPHATES, AQUEOUS SOLUTION	8	–	II	–	1 L	E2	P001	–	IBC02	–
2837	BISULPHATES, AQUEOUS SOLUTION	8	–	III	223	5 L	E1	P001 LP01	–	IBC03	–
2838	VINYL BUTYRATE, STABILIZED	3	–	II	–	1 L	E2	P001	–	IBC02	–
2839	ALDOL	6.1	–	II	–	100 mL	E4	P001	–	IBC02	–
2840	BUTYRALDOXIME	3	–	III	–	5 L	E1	P001 LP01	–	IBC03	–
2841	DI-*n*-AMYLAMINE	3	6.1	III	–	5 L	E1	P001	–	IBC03	–
2842	NITROETHANE	3	–	III	–	5 L	E1	P001 LP01	–	IBC03	–
2844	CALCIUM MANGANESE SILICON	4.3	–	III	–	1 kg	E1	P410	PP31	IBC08	B4
2845	PYROPHORIC LIQUID, ORGANIC, N.O.S.	4.2	–	I	274	0	E0	P400	PP31	–	–
2846	PYROPHORIC SOLID, ORGANIC, N.O.S.	4.2	–	I	274	0	E0	P404	PP31	–	–
2849	3-CHLOROPROPANOL-1	6.1	–	III	–	5 L	E1	P001 LP01	–	IBC03	–
2850	PROPYLENE TETRAMER	3	–	III	–	5 L	E1	P001 LP01	–	IBC03	–
2851	BORON TRIFLUORIDE DIHYDRATE	8	–	II	–	1 L	E2	P001	–	IBC02	–
2852	DIPICRYL SULPHIDE, WETTED with not less than 10% water, by mass	4.1	–	I	28	0	E0	P406	PP24 PP31	–	–

Portable tanks and bulk containers			EmS	Stowage and segregation	Properties and observations	UN No.
	Tank instructions	Provisions				
(12)	(13) 4.2.5 4.3	(14) 4.2.5	(15) 5.4.3.2 7.3	(16) 7.1 7.2	(17)	(18)
–	T3	TP33	F-G, S-N	Category E. Only to be loaded under dry weather conditions. Under deck in a mechanically ventilated space. Clear of living quarters.	Dark, crystalline, metal-like powder or brittle lumps. In contact with moisture, evolves flammable and toxic gases.	2830
–	T4	TP1	F-A, S-A	Category A. Clear of living quarters.	Colourless liquid. Immiscible with water. Decomposes when heated, evolving highly toxic fumes (phosgene and hydrogen chloride). Toxic if swallowed, by skin contact or by inhalation. Narcotic in high concentrations.	2831
–	T1	TP33	F-A, S-B	Category A. Protected from sources of heat.	Colourless to yellow deliquescent crystals. Soluble in water. Mildly corrosive to most metals. Causes burns to skin, eyes and mucous membranes.	2834 △
–	T3	TP33	F-G, S-O	Category E. "Separated from" acids.	White, crystalline solid. Reacts with water, moisture or acids, evolving hydrogen, which may be ignited by the heat of the reaction.	2835
–	T7	TP2	F-A, S-B	Category A.	Colourless to white liquid. Miscible with water. Corrosive to most metals. Causes burns to skin, eyes and mucous membranes.	2837
–	T4	TP1	F-A, S-B	Category A.	See entry above.	2837
–	T4	TP1	F-E, S-D	Category B.	Colourless liquid with a pungent odour. Flashpoint: 12°C c.c. Explosive limits: 1.4% to 8.8%. Immiscible with water. Irritating to skin, eyes and mucous membranes.	2838
–	T7	TP2	F-A, S-A	Category A. Keep as cool as reasonably practicable.	Clear, colourless to yellow viscous liquid. Miscible with water. Decomposes at 85°C, evolving toxic fumes. May react vigorously with oxidizing substances. Toxic if swallowed, by skin contact or by inhalation.	2839
–	T2	TP1	F-E, S-D	Category A.	Colourless liquid. Immiscible with water. Flashpoint: 58°C c.c. Harmful by inhalation. Irritating to skin, eyes and mucous membranes.	2840
–	T4	TP1	F-E, S-D	Category A.	Colourless liquid with an ammoniacal odour. Flashpoint: 52°C c.c. Slightly miscible with water. Toxic if swallowed, by skin contact or by inhalation.	2841
–	T2	TP1	F-E, S-D	Category A.	Colourless, oily liquid. Flashpoint: 28°C c.c. Explosive limits: 3.4% to .... When involved in a fire, evolves nitrous toxic fumes. Slightly soluble in water. Irritating to skin, eyes and mucous membranes.	2842
–	T1	TP33	F-G, S-N	Category A. Only to be loaded under dry weather conditions. Under deck in a mechanically ventilated space. "Separated from" acids.	In contact with water, evolves hydrogen, a flammable gas. In contact with acid, evolves silane, a spontaneously flammable gas.	2844
–	T22	TP2 TP7	F-G, S-M	Category D. Separated longitudinally by an intervening complete compartment or hold from class 1.	Highly flammable liquids, may ignite spontaneously in moist air. In contact with air, evolve irritating and slightly toxic fumes.	2845
–	–	–	F-G, S-M	Category D.	Liable to ignite spontaneously in air. If shaken, may produce sparks. In contact with water, evolve hydrogen, a flammable gas.	2846
–	T4	TP1	F-A, S-A	Category A.	Colourless to light-yellow liquid. Miscible with water. Mildly corrosive to steel. Toxic if swallowed, by skin contact or by inhalation.	2849
–	T2	TP1	F-E, S-E	Category A.	Colourless liquid. Immiscible with water. Irritating to skin, eyes and mucous membranes.	2850
–	T7	TP2	F-A, S-B	Category B. Keep as cool as reasonably practicable. Clear of living quarters.	Colourless, non-fuming liquid. Boiling range: 58°C to 60°C. Reacts with water, evolving corrosive and toxic fumes. Corrosive to mild steel. Causes burns to skin, eyes and mucous membranes.	2851
–	–	–	F-B, S-J	Category D. "Away from" class 3 and heavy metals and their salts.	Desensitized explosive. Golden-yellow, crystalline leaflets. Explosive and sensitive to shock and heat in the dry state. May form extremely sensitive compounds with heavy metals or their salts.	2852

UN No.	Proper Shipping Name (PSN)	Class or division	Subsidiary risk(s)	Packing group	Special provisions	Limited and excepted quantity provisions		Packing		IBC	
						Limited quantities	Excepted quantities	Instructions	Provisions	Instructions	Provisions
(1)	(2) 3.1.2	(3) 2.0	(4) 2.0	(5) 2.0.1.3	(6) 3.3	(7a) 3.4	(7b) 3.5	(8) 4.1.4	(9) 4.1.4	(10) 4.1.4	(11) 4.1.4
2853	MAGNESIUM FLUOROSILICATE	6.1	–	III	–	5 kg	E1	P002 LP02	–	IBC08	B3
2854	AMMONIUM FLUOROSILICATE	6.1	–	III	–	5 kg	E1	P002 LP02	–	IBC08	B3
2855	ZINC FLUOROSILICATE	6.1	–	III	–	5 kg	E1	P002 LP02	–	IBC08	B3
2856	FLUOROSILICATES, N.O.S.	6.1	–	III	274	5 kg	E1	P002 LP02	–	IBC08	B3
△ 2857	REFRIGERATING MACHINES containing non-flammable, non-toxic gases or ammonia solutions (UN 2672)	2.2	–	–	119	0	E0	P003	PP32	–	–
2858	ZIRCONIUM, DRY coiled wire, finished metal sheets, strip (thinner than 254 microns but not thinner than 18 microns)	4.1	–	III	921	5 kg	E1	P002 LP02	–	–	–
2859	AMMONIUM METAVANADATE	6.1	–	II	–	500 g	E4	P002	–	IBC08	B2 B4
2861	AMMONIUM POLYVANADATE	6.1	–	II	–	500 g	E4	P002	–	IBC08	B2 B4
2862	VANADIUM PENTOXIDE, non-fused form	6.1	–	III	–	5 kg	E1	P002 LP02	–	IBC08	B3
2863	SODIUM AMMONIUM VANADATE	6.1	–	II	–	500 g	E4	P002	–	IBC08	B2 B4
2864	POTASSIUM METAVANADATE	6.1	–	II	–	500 g	E4	P002	–	IBC08	B2 B4
2865	HYDROXYLAMINE SULPHATE	8	–	III	–	5 kg	E1	P002 LP02	–	IBC08	B3
2869	TITANIUM TRICHLORIDE MIXTURE	8	–	II	–	1 kg	E2	P002	–	IBC08	B2 B4
2869	TITANIUM TRICHLORIDE MIXTURE	8	–	III	223	5 kg	E1	P002 LP02	–	IBC08	B3
△ 2870	ALUMINIUM BOROHYDRIDE	4.2	4.3	I	–	0	E0	P400	–	–	–
2870	ALUMINIUM BOROHYDRIDE IN DEVICES	4.2	4.3	I	–	0	E0	P002	PP13	–	–
2871	ANTIMONY POWDER	6.1	–	III	–	5 kg	E1	P002 LP02	–	IBC08	B3
2872	DIBROMOCHLOROPROPANES	6.1	–	II	–	100 mL	E4	P001	–	IBC02	–
2872	DIBROMOCHLOROPROPANES	6.1	–	III	223	5 L	E1	P001 LP01	–	IBC03	–
△ 2873	DIBUTYLAMINOETHANOL	6.1	–	III	–	5 L	E1	P001 LP01	–	IBC03	–

Portable tanks and bulk containers			EmS	Stowage and segregation	Properties and observations	UN No.
	Tank instructions	Provisions				
(12)	(13) 4.2.5 4.3	(14) 4.2.5	(15) 5.4.3.2 7.3	(16) 7.1 7.2	(17)	(18)
–	T1	TP33	F-A, S-A	Category A. "Separated from" acids.	Solids which react with acids, evolving hydrogen fluoride and silicon tetrafluoride, irritating and corrosive gases. Toxic if swallowed, by skin contact or by dust inhalation.	2853
–	T1	TP33	F-A, S-A	Category A. "Separated from" acids.	Solids which react with acids, evolving hydrogen fluoride and silicon tetrafluoride, irritating and corrosive gases. Toxic if swallowed, by skin contact or by dust inhalation.	2854
–	T1	TP33	F-A, S-A	Category A. "Separated from" acids.	Solids which react with acids, evolving hydrogen fluoride and silicon tetrafluoride, irritating and corrosive gases. Toxic if swallowed, by skin contact or by dust inhalation.	2855
–	T1	TP33	F-A, S-A	Category A. "Separated from" acids.	Solids which react with acids, evolving hydrogen fluoride and silicon tetrafluoride, irritating and corrosive gases. Toxic if swallowed, by skin contact or by dust inhalation.	2856
–	–	–	F-C, S-V	Category A.	–	2857 △
–	–	–	F-G, S-G	Category A.	Hard silvery metal.	2858
–	T3	TP33	F-A, S-A	Category A. Segregation as for class 5.1 but "away from" classes 4.1, 5.1 and 7.	White crystalline powder. Slightly soluble in water. May act as an oxidizing substance. Toxic if swallowed, by skin contact or by inhalation.	2859
–	T3	TP33	F-A, S-A	Category A. Segregation as for class 5.1 but "away from" classes 4.1, 5.1 and 7.	Orange powder. Slightly soluble in water. May act as an oxidizing substance. Toxic if swallowed, by skin contact or by inhalation.	2861
–	T1	TP33	F-A, S-A	Category A.	Brownish powder. Slightly soluble in water. Toxic if swallowed, by skin contact or by inhalation.	2862
–	T3	TP33	F-A, S-A	Category A.	Orange wet cake (with 10% to 15% water). Soluble in water. Toxic if swallowed, by skin contact or by dust inhalation.	2863
–	T3	TP33	F-A, S-A	Category A.	White crystalline powder. Slightly soluble in water. Toxic if swallowed, by skin contact or by inhalation.	2864
–	T1	TP33	F-A, S-B	Category A.	Colourless to white crystalline powder. Soluble in water. May decompose explosively when heated. Causes burns to skin, eyes and mucous membranes.	2865
–	T3	TP33	F-A, S-B	Category A. Clear of living quarters.	Violet crystalline solid. Reacts in moist air or in water, evolving heat and hydrogen chloride, an irritating and corrosive gas apparent as white fumes. In the presence of moisture, corrosive to most metals. Causes burns to skin, eyes and mucous membranes.	2869
–	T1	TP33	F-A, S-B	Category A. Clear of living quarters.	See entry above.	2869
–	T21	TP7 TP33	F-G, S-M	Category D.	Liquid. Ignites spontaneously in air. Reacts with water or steam to produce heat or hydrogen, which may form explosive mixtures with air.	2870 △
–	–	–	F-G, S-M	Category D.	–	2870
–	T1	TP33	F-A, S-A	Category A.	Metallic antimony in the form of a fine grey powder. Insoluble in water. Toxic if swallowed, by skin contact or by dust inhalation.	2871
–	T7	TP2	F-A, S-A	Category A.	Colourless liquid with a perceptible odour. Immiscible with water. Toxic if swallowed, by skin contact or by inhalation.	2872
–	T4	TP1	F-A, S-A	Category A.	See entry above.	2872
–	T4	TP1	F-A, S-A	Category A.	Colourless liquid with a perceptible odour. Miscible with water. Toxic if swallowed, by skin contact or by inhalation.	2873 △

3

DGL

UN No.	Proper Shipping Name (PSN)	Class or division	Subsidiary risk(s)	Packing group	Special provisions	Limited and excepted quantity provisions		Packing		IBC	
						Limited quantities	Excepted quantities	Instructions	Provisions	Instructions	Provisions
(1)	(2) 3.1.2	(3) 2.0	(4) 2.0	(5) 2.0.1.3	(6) 3.3	(7a) 3.4	(7b) 3.5	(8) 4.1.4	(9) 4.1.4	(10) 4.1.4	(11) 4.1.4
2874	FURFURYL ALCOHOL	6.1	–	III	–	5 L	E1	P001 LP01	–	IBC03	–
2875	HEXACHLOROPHENE	6.1	–	III	–	5 kg	E1	P002 LP02	–	IBC08	B3
2876	RESORCINOL	6.1	–	III	–	5 kg	E1	P002 LP02	–	IBC08	B3
△ 2878	TITANIUM SPONGE GRANULES or TITANIUM SPONGE POWDERS	4.1	–	III	223	5 kg	E1	P002 LP02	–	IBC08	B3
2879	SELENIUM OXYCHLORIDE	8	6.1	I	–	0	E0	P001	–	–	–
△ 2880	CALCIUM HYPOCHLORITE, HYDRATED or CALCIUM HYPOCHLORITE, HYDRATED MIXTURE with not less than 5.5% but not more than 16% water	5.1	–	II	314 322	1 kg	E2	P002	PP85	–	–
△ 2880	CALCIUM HYPOCHLORITE, HYDRATED or CALCIUM HYPOCHLORITE, HYDRATED MIXTURE with not less than 5.5% but not more than 16% water	5.1	–	III	223 314	5 kg	E1	P002	PP85	–	–
2881	METAL CATALYST, DRY	4.2	–	I	274	0	E0	P404	PP31	–	–
2881	METAL CATALYST, DRY	4.2	–	II	274	0	E2	P410	PP31	IBC06	B2
2881	METAL CATALYST, DRY	4.2	–	III	223 274	0	E1	P002 LP02	PP31	IBC08	B3
△ 2900	INFECTIOUS SUBSTANCE, AFFECTING ANIMALS only	6.2	–	–	318 341	0	E0	P620	–	–	–
2901	BROMINE CHLORIDE	2.3	5.1/8	–	–	0	E0	P200	–	–	–
2902	PESTICIDE, LIQUID, TOXIC, N.O.S.	6.1	–	I	61 274	0	E5	P001	–	–	–

Portable tanks and bulk containers			EmS	Stowage and segregation	Properties and observations	UN No.
	Tank instructions	Provisions				
(12)	(13) 4.2.5 4.3	(14) 4.2.5	(15) 5.4.3.2 7.3	(16) 7.1 7.2	(17)	(18)
–	T4	TP1	F-A, S-A	Category A. "Separated from" acids and class 5.1.	Clear, colourless, mobile liquid, becoming brown to dark-red upon exposure to light and air. Miscible with water. Reacts explosively with oxidizing substances. Toxic if swallowed, by skin contact or by inhalation.	2874
–	T1	TP33	F-A, S-A	Category A.	White, odourless powder or crystals. Insoluble in water. Toxic if swallowed, by skin contact or by dust inhalation.	2875
–	T1	TP33	F-A, S-A	Category A.	White to pink crystals. Soluble in water. Toxic if swallowed, by skin contact or by dust inhalation.	2876
–	T1	TP33	F-G, S-G	Category D. "Separated from" class 5.1.	Silvery grey granules or dark grey, amorphous powder. May react with carbon dioxide, evolving oxygen. Forms explosive mixtures with oxidizing substances.	2878 △
–	T10	TP2 TP13	F-A, S-B	Category E. Clear of living quarters.	Colourless, yellowish liquid. Reacts violently with water, evolving hydrogen chloride, an irritating and corrosive gas apparent as white fumes. In the presence of moisture, highly corrosive to most metals. Toxic if swallowed, by skin contact or by inhalation. Causes severe burns to skin, eyes and mucous membranes.	2879
–	–	–	F-H, S-Q	Category D. Protected from sources of heat. Cargo transport units shall be protected from direct sunlight. Packages in cargo transport units shall be stowed so as to allow for adequate air circulation throughout the cargo. "Separated from" ammonium compounds, acids, cyanides, hydrogen peroxides and liquid organic substances.	White or yellowish solid (powder, granules or tablets) with chlorine-like odour. Soluble in water. May cause fire in contact with organic material or ammonium compounds. Substances are liable to exothermic decomposition at elevated temperatures. This condition may lead to fire or explosion. Decomposition can be initiated by heat or by impurities (e.g., powdered metals (iron, manganese, cobalt, magnesium) and their compounds). Liable to heat slowly. Reacts with acids, evolving chlorine, an irritating, corrosive and toxic gas. In the presence of moisture, corrosive to most metals. Dust irritates mucous membranes.	2880 △
–	–	–	F-H, S-Q	Category D. Protected from sources of heat. Cargo transport units shall be protected from direct sunlight. Packages in cargo transport units shall be stowed so as to allow for adequate air circulation throughout the cargo. "Separated from" ammonium compounds, acids, cyanides, hydrogen peroxides and liquid organic substances.	See entry above.	2880 △
–	T21	TP7 TP33	F-G, S-M	Category C.	Liable to ignite spontaneously in air.	2881
–	T3	TP33	F-G, S-M	Category C.	See entry above	2881
–	T1	TP33	F-G, S-M	Category C.	See entry above.	2881
–	BK2	–	F-A, S-T	As approved by the competent authorities of the countries involved in the shipment.	Substances which are dangerous to animals only. For action to be taken in the event of damage to, or leaking from, a package containing infectious substances, refer to 7.8.3.	2900 △
–	–	–	F-C, S-W	Category D. Clear of living quarters. Segregation as for class 5.1 but "separated from" class 7.	Reddish-yellow non-flammable, toxic and corrosive gas. When heated to decomposition, emits highly toxic and corrosive fumes of bromine and chlorine. Reacts with water, evolving toxic and corrosive fumes. Powerful oxidizing agent which may cause violent fires with combustible materials. Much heavier than air. Highly irritating to skin, eyes and mucous membranes.	2901
–	T14	TP2 TP13 TP27	F-A, S-A	Category B. Clear of living quarters.	Liquid pesticides which present a very wide range of toxic hazard. Miscibility with water depends upon the composition. Toxic if swallowed, by skin contact or by inhalation.	2902

UN No.	Proper Shipping Name (PSN)	Class or division	Subsidiary risk(s)	Packing group	Special provisions	Limited and excepted quantity provisions		Packing		IBC	
						Limited quantities	Excepted quantities	Instructions	Provisions	Instructions	Provisions
(1)	(2) 3.1.2	(3) 2.0	(4) 2.0	(5) 2.0.1.3	(6) 3.3	(7a) 3.4	(7b) 3.5	(8) 4.1.4	(9) 4.1.4	(10) 4.1.4	(11) 4.1.4
2902	PESTICIDE, LIQUID, TOXIC, N.O.S.	6.1	–	II	61 274	100 mL	E4	P001	–	IBC02	–
2902	PESTICIDE, LIQUID, TOXIC, N.O.S.	6.1	–	III	61 223 274	5 L	E1	P001 LP01	–	IBC03	–
2903	PESTICIDE, LIQUID, TOXIC, FLAMMABLE, N.O.S. flashpoint not less than 23°C	6.1	3	I	61 274	0	E5	P001	–	–	–
2903	PESTICIDE, LIQUID, TOXIC, FLAMMABLE, N.O.S. flashpoint not less than 23°C	6.1	3	II	61 274	100 mL	E4	P001	–	IBC02	–
2903	PESTICIDE, LIQUID, TOXIC, FLAMMABLE, N.O.S. flashpoint not less than 23°C	6.1	3	III	61 223 274	5 L	E1	P001	–	IBC03	–
2904	CHLOROPHENOLATES, LIQUID or PHENOLATES, LIQUID	8	–	III	–	5 L	E1	P001 LP01	–	IBC03	–
2905	CHLOROPHENOLATES, SOLID or PHENOLATES, SOLID	8	–	III	–	5 kg	E1	P002 LP02	–	IBC08	B3
2907	ISOSORBIDE DINITRATE MIXTURE with not less than 60% lactose, mannose, starch, or calcium hydrogen phosphate	4.1	–	II	127	0	E0	P406	PP26 PP80	IBC06	B2 B12
△ 2908	RADIOACTIVE MATERIAL, EXCEPTED PACKAGE – EMPTY PACKAGING	7	See SP290	–	290	0	E0	See 4.1.9	See 4.1.9	See 4.1.9	See 4.1.9
△ 2909	RADIOACTIVE MATERIAL, EXCEPTED PACKAGE – ARTICLES MANUFACTURED FROM NATURAL URANIUM or DEPLETED URANIUM or NATURAL THORIUM	7	See SP290	–	290	0	E0	See 4.1.9	See 4.1.9	See 4.1.9	See 4.1.9
△ 2910	RADIOACTIVE MATERIAL, EXCEPTED PACKAGE – LIMITED QUANTITY OF MATERIAL	7	See SP290	–	290 325	0	E0	See 4.1.9	See 4.1.9	See 4.1.9	See 4.1.9
△ 2911	RADIOACTIVE MATERIAL, EXCEPTED PACKAGE – INSTRUMENTS or ARTICLES	7	See SP290	–	290	0	E0	See 4.1.9	See 4.1.9	See 4.1.9	See 4.1.9
△ 2912	RADIOACTIVE MATERIAL, LOW SPECIFIC ACTIVITY (LSA-I), non fissile or fissile – excepted	7	See SP172	–	172 317 325	0	E0	See 4.1.9	See 4.1.9	See 4.1.9	See 4.1.9
△ 2913	RADIOACTIVE MATERIAL, SURFACE CONTAMINATED OBJECTS (SCO-I or SCO-II), non fissile or fissile – excepted	7	See SP172	–	172 317	0	E0	See 4.1.9	See 4.1.9	See 4.1.9	See 4.1.9
△ 2915	RADIOACTIVE MATERIAL, TYPE A PACKAGE, non-special form, non fissile or fissile – excepted	7	See SP172	–	172 317 325	0	E0	See 4.1.9	See 4.1.9	See 4.1.9	See 4.1.9
△ 2916	RADIOACTIVE MATERIAL, TYPE B(U) PACKAGE, non fissile or fissile – excepted	7	See SP172	–	172 317 325	0	E0	See 4.1.9	See 4.1.9	See 4.1.9	See 4.1.9

Portable tanks and bulk containers			EmS	Stowage and segregation	Properties and observations	UN No.
	Tank instructions	Provisions				
(12)	(13) 4.2.5 4.3	(14) 4.2.5	(15) 5.4.3.2 7.3	(16) 7.1 7.2	(17)	(18)
–	T11	TP2 TP13 TP27	F-A, S-A	Category B. Clear of living quarters.	Liquid pesticides which present a very wide range of toxic hazard. Miscibility with water depends upon the composition. Toxic if swallowed, by skin contact or by inhalation.	**2902**
–	T7	TP2 TP28	F-A, S-A	Category A. Clear of living quarters.	See entry above.	2902
–	T14	TP2 TP13 TP27	F-E, S-D	Category B. Clear of living quarters.	Liquid flammable pesticides having a flashpoint between 23°C and 60°C c.c., presenting a very wide range of toxic hazard. They frequently contain petroleum or coal tar distillates, or other flammable liquids. Flashpoint and miscibility with water depend upon the composition. Toxic if swallowed, by skin contact or by inhalation.	**2903**
–	T11	TP2 TP13 TP27	F-E, S-D	Category B. Clear of living quarters.	See entry above.	2903
–	T7	TP2	F-E, S-D	Category A. Clear of living quarters.	See entry above.	**2903**
–	–	–	F-A, S-B	Category A.	A wide range of corrosive liquids. Cause burns to skin, eyes and mucous membranes.	2904
–	T1	TP33	F-A, S-B	Category A.	A wide range of corrosive solids. Soluble in water. Cause burns to skin, eyes and mucous membranes.	**2905**
–	–	–	F-A, S-J	Category E. "Away from" class 3 and heavy metals and their salts.	Desensitized explosive. Pure isosorbide dinitrate is explosive. May form extremely sensitive compounds with heavy metals or their salts.	2907
–	–	–	F-I, S-S	Category A.	See 1.5.1 and 5.1.5.4.2.	**2908** △
–	–	–	F-I, S-S	Category A.	See 1.5.1 and 5.1.5.4.2.	2909 △
–	–	–	F-I, S-S	Category A.	See 1.5.1 and 5.1.5.4.2.	**2910** △
–	–	–	F-I, S-S	Category A.	See 1.5.1 and 5.1.5.4.2.	2911 △
–	T5	TP4	F-I, S-S	Category A, except for uranyl nitrate hexahydrate solution for which category D applies.	See 1.5.1.	**2912** △
–	T5	TP4	F-I, S-S	Category A.	See 1.5.1.	2913 △
–	–	–	F-I, S-S	Category A, except for uranyl nitrate hexahydrate solution, uranium metal pyrophoric and thorium metal pyrophoric for which category D applies.	See 1.5.1.	**2915** △
–	–	–	F-I, S-S	Category A, taking account of any supplementary requirements specified in the transport documents.	See 1.5.1. For ships transporting an INF cargo as defined in regulation VII/14 of the SOLAS Convention, 1974, as amended, refer also to the INF Code.	2916 △

	UN No.	Proper Shipping Name (PSN)	Class or division	Subsidiary risk(s)	Packing group	Special provisions	Limited and excepted quantity provisions		Packing		IBC	
							Limited quantities	Excepted quantities	Instructions	Provisions	Instructions	Provisions
	(1)	(2) 3.1.2	(3) 2.0	(4) 2.0	(5) 2.0.1.3	(6) 3.3	(7a) 3.4	(7b) 3.5	(8) 4.1.4	(9) 4.1.4	(10) 4.1.4	(11) 4.1.4
△	2917	RADIOACTIVE MATERIAL, TYPE B(M) PACKAGE, non fissile or fissile – excepted	7	See SP172	–	172 317 325	0	E0	See 4.1.9	See 4.1.9	See 4.1.9	See 4.1.9
△	2919	RADIOACTIVE MATERIAL TRANSPORTED UNDER SPECIAL ARRANGEMENT, non fissile or fissile – excepted	7	See SP172	–	172 317 325	0	E0	See 4.1.9	See 4.1.9	See 4.1.9	See 4.1.9
△	2920	CORROSIVE LIQUID, FLAMMABLE, N.O.S.	8	3	I	274	0	E0	P001	–	–	–
△	2920	CORROSIVE LIQUID, FLAMMABLE, N.O.S.	8	3	II	274	1 L	E2	P001	–	IBC02	–
△	2921	CORROSIVE SOLID, FLAMMABLE, N.O.S.	8	4.1	I	274	0	E0	P002	–	IBC99	–
△	2921	CORROSIVE SOLID, FLAMMABLE, N.O.S.	8	4.1	II	274	1 kg	E2	P002	–	IBC08	B2 B4
	2922	CORROSIVE LIQUID, TOXIC, N.O.S.	8	6.1	I	274	0	E0	P001	–	–	–
	2922	CORROSIVE LIQUID, TOXIC, N.O.S.	8	6.1	II	274	1 L	E2	P001	–	IBC02	–
	2922	CORROSIVE LIQUID, TOXIC, N.O.S.	8	6.1	III	223 274	5 L	E1	P001	–	IBC03	–
	2923	CORROSIVE SOLID, TOXIC, N.O.S.	8	6.1	I	274	0	E0	P002	–	IBC99	–
	2923	CORROSIVE SOLID, TOXIC, N.O.S.	8	6.1	II	274	1 kg	E2	P002	–	IBC08	B2 B4
	2923	CORROSIVE SOLID, TOXIC, N.O.S.	8	6.1	III	223 274	5 kg	E1	P002	–	IBC08	B3
	2924	FLAMMABLE LIQUID, CORROSIVE, N.O.S.	3	8	I	274	0	E0	P001	–	–	–
	2924	FLAMMABLE LIQUID, CORROSIVE, N.O.S.	3	8	II	274	1 L	E2	P001	–	IBC02	–
	2924	FLAMMABLE LIQUID, CORROSIVE, N.O.S.	3	8	III	223 274	5 L	E1	P001	–	IBC03	–
	2925	FLAMMABLE SOLID, CORROSIVE, ORGANIC, N.O.S.	4.1	8	II	274 915	1 kg	E2	P002	–	IBC06	B2
	2925	FLAMMABLE SOLID, CORROSIVE, ORGANIC, N.O.S.	4.1	8	III	223 274 915	5 kg	E1	P002	–	IBC06	–
	2926	FLAMMABLE SOLID, TOXIC, ORGANIC, N.O.S.	4.1	6.1	II	274 915	1 kg	E2	P002	–	IBC06	B2
	2926	FLAMMABLE SOLID, TOXIC, ORGANIC, N.O.S.	4.1	6.1	III	223 274 915	5 kg	E1	P002	–	IBC06	–
	2927	TOXIC LIQUID, CORROSIVE, ORGANIC, N.O.S.	6.1	8	I	274 315	0	E5	P001	–	–	–
	2927	TOXIC LIQUID, CORROSIVE, ORGANIC, N.O.S.	6.1	8	II	274	100 mL	E4	P001	–	IBC02	–

3

DGL

Portable tanks and bulk containers			EmS	Stowage and segregation	Properties and observations	UN No.
	Tank instructions	Provisions				
(12)	(13) 4.2.5 4.3	(14) 4.2.5	(15) 5.4.3.2 7.3	(16) 7.1 7.2	(17)	(18)
–	–	–	F-I, S-S	Category A, taking account of any supplementary requirements specified in the transport documents.	See 1.5.1. For ships transporting an INF cargo as defined in regulation VII/14 of the SOLAS Convention, 1974, as amended, refer also to the INF Code.	**2917** △
–	–	–	F-I, S-S	Category A, taking account of any supplementary requirements specified in the competent authority approval certificate(s).	See 1.5.1. For ships transporting an INF cargo as defined in regulation VII/14 of the SOLAS Convention, 1974, as amended, refer also to the INF Code.	2919 △
–	T14	TP2 TP27	F-E, S-C	Category C. Protected from sources of heat. Clear of living quarters.	Causes burns to skin, eyes and mucous membranes.	**2920** △
–	T11	TP2 TP27	F-E, S-C	Category C. Protected from sources of heat. Clear of living quarters.	See entry above.	2920 △
–	T6	TP33	F-A, S-G	Category B. Keep as cool as reasonably practicable. Protected from sources of heat.	Causes burns to skin, eyes and mucous membranes.	**2921** △
–	T3	TP33	F-A, S-G	Category B. Keep as cool as reasonably practicable. Protected from sources of heat.	See entry above.	2921 △
–	T14	TP2 TP13 TP27	F-A, S-B	Category B. Clear of living quarters.	Causes burns to skin, eyes and mucous membranes. Toxic if swallowed, by skin contact or by inhalation.	**2922**
–	T7	TP2	F-A, S-B	Category B. Clear of living quarters.	See entry above.	2922
–	T7	TP1 TP28	F-A, S-B	Category B. Clear of living quarters.	See entry above.	**2922**
–	T6	TP33	F-A, S-B	Category B. Clear of living quarters.	Causes burns to skin, eyes and mucous membranes. Toxic if swallowed, by skin contact or by inhalation.	2923
–	T3	TP33	F-A, S-B	Category B. Clear of living quarters.	See entry above.	**2923**
–	T1	TP33	F-A, S-B	Category B. Clear of living quarters.	See entry above.	2923
–	T14	TP2	F-E, S-C	Category E. Clear of living quarters.	Causes burns to skin, eyes and mucous membranes.	**2924**
–	T11	TP2 TP27	F-E, S-C	Category B. Clear of living quarters.	See entry above.	2924
–	T7	TP1 TP28	F-E, S-C	Category A. Clear of living quarters.	See entry above.	**2924**
–	T3	TP33	F-A, S-G	Category D. Clear of living quarters.	Causes burns to skin, eyes and mucous membranes.	2925
–	T1	TP33	F-A, S-G	Category D. Clear of living quarters.	See entry above.	**2925**
–	T3	TP33	F-A, S-G	Category B. Clear of living quarters.	Toxic if swallowed, by skin contact or by dust inhalation. Should be handled with care to minimize exposure, particularly to dust.	2926
–	T1	TP33	F-A, S-G	Category B. Clear of living quarters.	See entry above.	**2926**
–	T14	TP2 TP13 TP27	F-A, S-B	Category B. Clear of living quarters.	Toxic if swallowed, by skin contact or by inhalation. Causes burns to skin, eyes and mucous membranes.	2927
–	T11	TP2 TP27	F-A, S-B	Category B. Clear of living quarters.	See entry above.	**2927**

UN No.	Proper Shipping Name (PSN)	Class or division	Subsidiary risk(s)	Packing group	Special provisions	Limited and excepted quantity provisions		Packing		IBC	
						Limited quantities	Excepted quantities	Instructions	Provisions	Instructions	Provisions
(1)	(2) 3.1.2	(3) 2.0	(4) 2.0	(5) 2.0.1.3	(6) 3.3	(7a) 3.4	(7b) 3.5	(8) 4.1.4	(9) 4.1.4	(10) 4.1.4	(11) 4.1.4
2928	TOXIC SOLID, CORROSIVE, ORGANIC, N.O.S.	6.1	8	I	274	0	E5	P002	–	IBC99	–
2928	TOXIC SOLID, CORROSIVE, ORGANIC, N.O.S.	6.1	8	II	274	500 g	E4	P002	–	IBC06	B2
2929	TOXIC LIQUID, FLAMMABLE, ORGANIC, N.O.S.	6.1	3	I	274 315	0	E5	P001	–	–	–
2929	TOXIC LIQUID, FLAMMABLE, ORGANIC, N.O.S.	6.1	3	II	274	100 mL	E4	P001	–	IBC02	–
2930	TOXIC SOLID, FLAMMABLE, ORGANIC, N.O.S.	6.1	4.1	I	274	0	E5	P002	–	IBC99	–
2930	TOXIC SOLID, FLAMMABLE, ORGANIC, N.O.S.	6.1	4.1	II	274	500 g	E4	P002	–	IBC08	B2 B4
2931	VANADYL SULPHATE	6.1	–	II	–	500 g	E4	P002	–	IBC08	B2 B4
2933	METHYL 2-CHLORO-PROPIONATE	3	–	III	–	5 L	E1	P001 LP01	–	IBC03	–
2934	ISOPROPYL 2-CHLORO-PROPIONATE	3	–	III	–	5 L	E1	P001 LP01	–	IBC03	–
2935	ETHYL 2-CHLORO-PROPIONATE	3	–	III	–	5 L	E1	P001 LP01	–	IBC03	–
2936	THIOLACTIC ACID	6.1	–	II	–	100 mL	E4	P001	–	IBC02	–
2937	*alpha*-METHYLBENZYL ALCOHOL, LIQUID	6.1	–	III	–	5 L	E1	P001 LP01	–	IBC03	–
2940	9-PHOSPHABICYCLONONANES (CYCLOOCTADIENE-PHOSPHINES)	4.2	–	II	–	0	E2	P410	PP31	IBC06	B2
2941	FLUOROANILINES	6.1	–	III	–	5 L	E1	P001 LP01	–	IBC03	–
2942	2-TRIFLUOROMETHYLANILINE	6.1	–	III	–	5 L	E1	P001 LP01	–	IBC03	–
2943	TETRAHYDRO-FURFURYLAMINE	3	–	III	–	5 L	E1	P001 LP01	–	IBC03	–
2945	*N*-METHYLBUTYLAMINE	3	8	II	–	1 L	E2	P001	–	IBC02	–
2946	2-AMINO-5-DIETHYLAMINO-PENTANE	6.1	–	III	–	5 L	E1	P001 LP01	–	IBC03	–
2947	ISOPROPYL CHLOROACETATE	3	–	III	–	5 L	E1	P001 LP01	–	IBC03	–
2948	3-TRIFLUOROMETHYLANILINE	6.1	–	II	–	100 mL	E4	P001	–	IBC02	–
2949	SODIUM HYDROSULPHIDE, HYDRATED with not less than 25% water of crystallization	8	–	II	–	1 kg	E2	P002	–	IBC08	B2 B4

Portable tanks and bulk containers			EmS	Stowage and segregation	Properties and observations	UN No.
	Tank instructions	Provisions				
(12)	(13) 4.2.5 4.3	(14) 4.2.5	(15) 5.4.3.2 7.3	(16) 7.1 7.2	(17)	(18)
–	T6	TP33	F-A, S-B	Category B. Clear of living quarters.	Toxic if swallowed, by skin contact or by inhalation. Causes burns to skin, eyes and mucous membranes.	2928
–	T3	TP33	F-A, S-B	Category B. Clear of living quarters.	See entry above.	2928
–	T14	TP2 TP13 TP27	F-E, S-D	Category B. Clear of living quarters.	Toxic if swallowed, by skin contact or by inhalation.	2929
–	T11	TP2 TP13 TP27	F-E, S-D	Category B. Clear of living quarters.	See entry above.	2929
–	T6	TP33	F-A, S-G	Category B.	Toxic if swallowed, by skin contact or by inhalation.	2930
–	T3	TP33	F-A, S-G	Category B.	See entry above.	2930
–	T3	TP33	F-A, S-A	Category A.	Blue, crystalline powder. Soluble in water. Toxic if swallowed, by skin contact or by dust inhalation.	2931
–	T2	TP1	F-E, S-D	Category A.	Colourless liquid with an ether-like odour. Flashpoint: 32°C c.c. Slightly soluble in water. Irritating to skin, eyes and mucous membranes.	2933
–	T2	TP1	F-E, S-D	Category A.	Colourless liquid with a sweetish odour. Flashpoint: 50°C c.c. Immiscible with water. Irritating to skin, eyes and mucous membranes.	2934
–	T2	TP1	F-E, S-D	Category A.	Colourless liquid with a pungent odour. Flashpoint: 38°C c.c. Immiscible with water. Irritating to skin, eyes and mucous membranes.	2935
–	T7	TP2	F-A, S-A	Category A.	Oily liquid with a foul odour. Melting point: 10°C. Miscible with water. Toxic if swallowed, by skin contact or by inhalation.	2936
–	T4	TP1	F-A, S-A	Category A.	Colourless liquid. Slightly miscible with water. Melting point: 21°C (pure substance). Toxic if swallowed, by skin contact or by inhalation.	2937
–	T3	TP33	F-A, S-J	Category A.	Colourless, waxy solids. Melting point: 40°C to 60°C. React in contact with materials such as sawdust or other cellulose-based materials, resulting in charring and evolution of toxic fumes. Irritating to skin, eyes and mucous membranes.	2940
–	T4	TP1	F-A, S-A	Category A.	Liquids. Freezing points: −28°C to −2°C. Immiscible with water. Toxic if swallowed, by skin contact or by inhalation.	2941
–	–	–	F-A, S-A	Category A.	Liquid. Immiscible with water. Toxic if swallowed, by skin contact or by inhalation.	2942
–	T2	TP1	F-E, S-D	Category A.	Colourless to yellowish liquid with an ammoniacal odour. Flashpoint: 45°C c.c. Miscible with water. Harmful by inhalation. Irritating to skin, eyes and mucous membranes.	2943
–	T7	TP1	F-E, S-C	Category B. Clear of living quarters.	Colourless liquid. Flashpoint: 0°C c.c. Miscible with water. Harmful by inhalation. Causes burns to skin and eyes. Irritating to mucous membranes.	2945
–	T4	TP1	F-A, S-A	Category A.	Liquid with an acrid odour. Miscible with water. Toxic if swallowed, by skin contact or by inhalation.	2946
–	T2	TP1	F-E, S-D	Category A.	Colourless liquid with a pungent odour. Flashpoint: 56°C c.c. Slightly soluble in water. Harmful by inhalation. Irritating to skin, eyes and mucous membranes.	2947
–	T7	TP2	F-A, S-A	Category A. Clear of living quarters.	Colourless to yellowish liquid. Melting point: 5°C. Slightly miscible with water. Toxic if swallowed, by skin contact or by inhalation.	2948
–	T7	TP2	F-A, S-B	Category A. "Separated from" acids.	Colourless needles or yellow flakes. Soluble in water with a foul odour. Melting point: 52°C. Reacts violently with acids, evolving hydrogen sulphide a toxic and flammable gas. Causes burns to skin, eyes and mucous membranes.	2949

UN No.	Proper Shipping Name (PSN)	Class or division	Subsidiary risk(s)	Packing group	Special provisions	Limited and excepted quantity provisions		Packing		IBC	
						Limited quantities	Excepted quantities	Instructions	Provisions	Instructions	Provisions
(1)	(2) 3.1.2	(3) 2.0	(4) 2.0	(5) 2.0.1.3	(6) 3.3	(7a) 3.4	(7b) 3.5	(8) 4.1.4	(9) 4.1.4	(10) 4.1.4	(11) 4.1.4
2950	MAGNESIUM GRANULES, COATED particle size not less than 149 microns	4.3	–	III	920	1 kg	E1	P410	–	IBC08	B4
△ 2956	5-*tert*-BUTYL-2,4,6-TRINITRO-*m*-XYLENE (MUSK XYLENE)	4.1	–	III	132 133	0	E1	P409	–	–	–
△ 2965	BORON TRIFLUORIDE DIMETHYL ETHERATE	4.3	3/8	I	–	0	E0	P401	PP31	–	–
2966	THIOGLYCOL	6.1	–	II	–	100 mL	E4	P001	–	IBC02	–
2967	SULPHAMIC ACID	8	–	III	–	5 kg	E1	P002 LP02	–	IBC08	B3
△ 2968	MANEB, STABILIZED or MANEB PREPARATION, STABILIZED against self-heating	4.3	– P	III	223 946	1 kg	E1	P002	–	IBC08	B4
△ 2969	CASTOR BEANS or CASTOR MEAL or CASTOR POMACE or CASTOR FLAKE	9	–	II	141	5 kg	E2	P002	PP34	IBC08	B2 B4
△ 2977	RADIOACTIVE MATERIAL, URANIUM HEXAFLUORIDE, FISSILE	7	8	–	172	0	E0	See 4.1.9	See 4.1.9	See 4.1.9	See 4.1.9
△ 2978	RADIOACTIVE MATERIAL, URANIUM HEXAFLUORIDE non fissile or fissile – excepted	7	8	–	172 317	0	E0	See 4.1.9	See 4.1.9	See 4.1.9	See 4.1.9
2983	ETHYLENE OXIDE AND PROPYLENE OXIDE MIXTURE with not more than 30% ethylene oxide	3	6.1	I	–	0	E0	P200	–	–	–
△ 2984	HYDROGEN PEROXIDE, AQUEOUS SOLUTION with not less than 8% but less than 20% hydrogen peroxide (stabilized as necessary)	5.1	–	III	65	5 L	E1	P504	–	IBC02	B5
△ 2985	CHLOROSILANES, FLAMMABLE, CORROSIVE, N.O.S.	3	8	II	–	0	E0	P010	–	–	–

Portable tanks and bulk containers			EmS	Stowage and segregation	Properties and observations	UN No.	
	Tank instructions	Provisions					
(12)	(13) 4.2.5 4.3	(14) 4.2.5	(15) 5.4.3.2 7.3	(16) 7.1 7.2	(17)	(18)	
–	T1 BK2	TP33	F-G, S-O	Category A. "Separated from" acids.	Coated granules with particle size ranging from 149 to 2000 microns. In contact with water or acids, evolve hydrogen, a flammable gas.	2950	
–	–	–	F-B, S-G	Category D. Keep as cool as reasonably practicable. Protected from sources of heat. Clear of living quarters. During transport it shall be stored (or kept) in a cool and well-ventilated place. For packages carrying a subsidiary risk of class 1, segregation as for class 1, division 1.3.	Insoluble in water. May explode if involved in a fire under confined conditions. Sensitive to strong detonation shock. Harmful if swallowed or by skin contact.	2956	△
–	T10	TP2 TP7 TP13	F-G, S-O	Category D. Clear of living quarters. Segregation as for class 3, but "away from" classes 3, 4.1 and 8.	Colourless, flammable liquid. Flashpoint: 20°C c.c. but widely variable, depending upon free ether content. Freezing point: –14°C. Decomposes in contact with water, forming dimethyl ether, a flammable gas. Causes burns to skin, eyes and mucous membranes.	2965	△
–	T7	TP2	F-A, S-A	Category A.	Colourless liquid with a foul odour. Miscible with water. Decomposes when heated, evolving sulphur dioxide. Toxic if swallowed, by skin contact or by inhalation.	2966	
–	T1	TP33	F-A, S-B	Category A.	White crystalline powder. Soluble in water. Decomposes when heated, evolving toxic fumes. Causes burns to skin, eyes and mucous membranes.	2967	
–	T1	TP33	F-G, S-L	Category B. Segregation from foodstuffs as in 7.3.4.2.2, 7.6.3.1.2 or 7.7.3.7. "Separated from" acids.	Yellow powder. May evolve toxic, irritating or flammable fumes when wet, when involved in a fire or in contact with acids. Requires certification from the shipper that the substance is not class 4.2.	2968	△
–	T3 BK2	TP33	F-A, S-A	Category E. Clear of living quarters. Segregation from foodstuffs as in 7.3.4.2.2, 7.6.3.1.2 or 7.7.3.7. "Away from" class 5.1. "Separated from" class 6.2.	Whole beans or meal. The latter is the residue remaining after the oil has been extracted from the seeds. Castor beans contain a powerful allergen which, by inhalation of dust or by skin contact with crushed bean products, can give rise to severe irritation of the skin, eyes and mucous membranes in some persons. They are also toxic by ingestion. When handling these products, wear at least a dust mask and goggles. Avoid unnecessary skin contact.	2969	△
–	–	–	F-I, S-S	Category A, taking account of any supplementary requirements specified in the transport documents.	See 1.5.1.	2977	△
–	–	–	F-I, S-S	Category A, taking account of any supplementary requirements specified in the transport documents.	See 1.5.1.	2978	△
–	T14	TP2 TP7 TP13	F-E, S-D	Category E. Clear of living quarters.	Colourless, volatile liquid with an ethereal odour. Flashpoint: below –18°C c.c. Explosive limits: 2.2% to 55%. Boiling point: 23°C to 28°C. Miscible with water. Corrosive to aluminium. Toxic if swallowed, by skin contact or by inhalation. Irritating to eyes and mucous membranes.	2983	
–	T4	TP1 TP6 TP24	F-H, S-Q	Category B. Protected from sources of heat. "Separated from" permanganates and class 4.1. See 7.2.6.3.2.	Colourless liquid. Slowly decomposes, evolving oxygen; the rate of decomposition increases in contact with metals, except aluminium.	2984	△
–	T14	TP2 TP7 TP13 TP27	F-E, S-C	Category B. Clear of living quarters.	Colourless liquids with a pungent odour. When involved in a fire, evolve toxic gases. React violently with water, evolving hydrogen chloride, an irritating and corrosive gas. In the presence of moisture, highly corrosive to most metals. Cause burns to skin, eyes and mucous membranes.	2985	△

3

DGL

	UN No.	Proper Shipping Name (PSN)	Class or division	Subsidiary risk(s)	Packing group	Special provisions	Limited and excepted quantity provisions		Packing		IBC	
							Limited quantities	Excepted quantities	Instructions	Provisions	Instructions	Provisions
	(1)	(2) 3.1.2	(3) 2.0	(4) 2.0	(5) 2.0.1.3	(6) 3.3	(7a) 3.4	(7b) 3.5	(8) 4.1.4	(9) 4.1.4	(10) 4.1.4	(11) 4.1.4
△	2986	CHLOROSILANES, CORROSIVE, FLAMMABLE, N.O.S.	8	3	II	–	0	E0	P010	–	–	–
△	2987	CHLOROSILANES, CORROSIVE, N.O.S.	8	–	II	–	0	E0	P010	–	–	–
	2988	CHLOROSILANES, WATER-REACTIVE, FLAMMABLE, CORROSIVE, N.O.S.	4.3	3/8	I	–	0	E0	P401	PP31	–	–
△	2989	LEAD PHOSPHITE, DIBASIC	4.1	–	II	922	1 kg	E2	P002	–	IBC08	B2 B4
△	2989	LEAD PHOSPHITE, DIBASIC	4.1	–	III	922	5 kg	E1	P002 LP02	–	IBC08	B3
△	2990	LIFE-SAVING APPLIANCES, SELF-INFLATING	9	–	–	296	0	E0	P905	–	–	–
	2991	CARBAMATE PESTICIDE, LIQUID, TOXIC, FLAMMABLE, flashpoint not less than 23°C	6.1	3	I	61 274	0	E5	P001	–	–	–
	2991	CARBAMATE PESTICIDE, LIQUID, TOXIC, FLAMMABLE, flashpoint not less than 23°C	6.1	3	II	61 274	100 mL	E4	P001	–	IBC02	–
	2991	CARBAMATE PESTICIDE, LIQUID, TOXIC, FLAMMABLE, flashpoint not less than 23°C	6.1	3	III	61 223 274	5 L	E1	P001 LP01	–	IBC03	–
	2992	CARBAMATE PESTICIDE, LIQUID, TOXIC	6.1	–	I	61 274	0	E5	P001	–	–	–
	2992	CARBAMATE PESTICIDE, LIQUID, TOXIC	6.1	–	II	61 274	100 mL	E4	P001	–	IBC02	–
	2992	CARBAMATE PESTICIDE, LIQUID, TOXIC	6.1	–	III	61 223 274	5 L	E1	P001 LP01	–	IBC03	–
	2993	ARSENICAL PESTICIDE, LIQUID, TOXIC, FLAMMABLE, flashpoint not less than 23°C	6.1	3	I	61 274	0	E5	P001	–	–	–

Portable tanks and bulk containers			EmS	Stowage and segregation	Properties and observations	UN No.	
	Tank instructions	Provisions					
(12)	(13) 4.2.5 4.3	(14) 4.2.5	(15) 5.4.3.2 7.3	(16) 7.1 7.2	(17)	(18)	
–	T14	TP2 TP7 TP13 TP27	F-E, S-C	Category C. Clear of living quarters.	Colourless, flammable liquids with a pungent odour. Immiscible with water. React violently with water or steam, evolving hydrogen chloride, an irritating and corrosive gas apparent as white fumes. When involved in a fire, evolve toxic gas. In the presence of moisture, highly corrosive to most metals. Cause burns to skin, eyes and mucous membranes.	**2986**	△
–	T14	TP2 TP7 TP13 TP27	F-A, S-B	Category C. Clear of living quarters.	Colourless liquids with a pungent odour. Immiscible with water. React violently with water or steam, evolving hydrogen chloride, an irritating and corrosive gas apparent as white fumes. When involved in a fire, evolve toxic gases. In the presence of moisture, highly corrosive to most metals. Cause burns to skin, eyes and mucous membranes.	**2987**	△
–	T14	TP2 TP7 TP13	F-G, S-N	Category D. Clear of living quarters. Segregation as for class 3 but "away from" classes 3, 4.1 and 8.	Colourless, very volatile liquids, flammable and corrosive, with a pungent odour. Immiscible with water. React violently with water or steam to produce heat which may lead to self-ignition; toxic and corrosive fumes will be evolved. May react vigorously in contact with oxidizing substances. Cause burns to skin, eyes and mucous membranes.	**2988**	
–	T3	TP33	F-A, S-G	Category B. Segregation from foodstuffs as in 7.3.4.2.2, 7.6.3.1.2 or 7.7.3.7.	Fine white crystals or powder. Insoluble in water. Combustion can be sustained, even in the absence of air. Harmful if swallowed.	**2989**	△
–	T1	TP33	F-A, S-G	Category B. Segregation from foodstuffs as in 7.3.4.2.2, 7.6.3.1.2 or 7.7.3.7.	See entry above.	**2989**	△
–	–	–	F-A, S-V	Category A. "Separated from" class 6.2. Within the appliance, to the extent that the dangerous goods are packaged as integral parts of the complete life-saving appliance, there is no need to apply the provisions on segregation of substances in chapter 7.2.	These articles may contain: (a) Class 2.2 compressed gases; (b) signal devices (class 1) which may include smoke and illumination signal flares; signal devices must be packed in plastic or fibreboard inner packagings; (c) electric storage batteries; (d) first aid kit; or (e) "strike anywhere" matches.	2990	△
–	T14	TP2 TP13 TP27	F-E, S-D	Category B. Clear of living quarters.	Liquid flammable pesticides having a flashpoint between 23°C and 60°C c.c., presenting a very wide range of toxic hazard. They frequently contain petroleum or coal tar distillates, or other flammable liquids. Flashpoint and miscibility with water depend upon the composition. Toxic if swallowed, by skin contact or by inhalation.	**2991**	
–	T11	TP2 TP13 TP27	F-E, S-D	Category B. Clear of living quarters.	See entry above.	2991	
–	T7	TP2 TP28	F-E, S-D	Category A. Clear of living quarters.	See entry above.	**2991**	
–	T14	TP2 TP13 TP27	F-A, S-A	Category B. Clear of living quarters.	Miscibility with water depends upon the composition. Toxic if swallowed, by skin contact or by inhalation.	2992	
–	T11	TP2 TP13 TP27	F-A, S-A	Category B. Clear of living quarters.	See entry above.	**2992**	
–	T7	TP2 TP28	F-A, S-A	Category A. Clear of living quarters.	See entry above.	2992	
–	T14	TP2 TP13 TP27	F-E, S-D	Category B. Clear of living quarters.	Liquid flammable pesticides having a flashpoint between 23°C and 60°C c.c., presenting a very wide range of toxic hazard. They frequently contain petroleum or coal tar distillates, or other flammable liquids. Flashpoint and miscibility with water depend upon the composition. Toxic if swallowed, by skin contact or by inhalation.	**2993**	

3

DGL

UN No.	Proper Shipping Name (PSN)	Class or division	Subsidiary risk(s)	Packing group	Special provisions	Limited and excepted quantity provisions		Packing		IBC	
						Limited quantities	Excepted quantities	Instructions	Provisions	Instructions	Provisions
(1)	(2) 3.1.2	(3) 2.0	(4) 2.0	(5) 2.0.1.3	(6) 3.3	(7a) 3.4	(7b) 3.5	(8) 4.1.4	(9) 4.1.4	(10) 4.1.4	(11) 4.1.4
2993	ARSENICAL PESTICIDE, LIQUID, TOXIC, FLAMMABLE, flashpoint not less than 23°C	6.1	3	II	61 274	100 mL	E4	P001	–	IBC02	–
2993	ARSENICAL PESTICIDE, LIQUID, TOXIC, FLAMMABLE, flashpoint not less than 23°C	6.1	3	III	61 223 274	5 L	E1	P001	–	IBC03	–
2994	ARSENICAL PESTICIDE, LIQUID, TOXIC	6.1	–	I	61 274	0	E5	P001	–	–	–
2994	ARSENICAL PESTICIDE, LIQUID, TOXIC	6.1	–	II	61 274	100 mL	E4	P001	–	IBC02	–
2994	ARSENICAL PESTICIDE, LIQUID, TOXIC	6.1	–	III	61 223 274	5 L	E1	P001 LP01	–	IBC03	–
2995	ORGANOCHLORINE PESTICIDE, LIQUID, TOXIC, FLAMMABLE, flashpoint not less than 23°C	6.1	3	I	61 274	0	E5	P001	–	–	–
2995	ORGANOCHLORINE PESTICIDE, LIQUID, TOXIC, FLAMMABLE, flashpoint not less than 23°C	6.1	3	II	61 274	100 mL	E4	P001	–	IBC02	–
2995	ORGANOCHLORINE PESTICIDE, LIQUID, TOXIC, FLAMMABLE, flashpoint not less than 23°C	6.1	3	III	61 223 274	5 L	E1	P001	–	IBC03	–
2996	ORGANOCHLORINE PESTICIDE, LIQUID, TOXIC	6.1	–	I	61 274	0	E5	P001	–	–	–
2996	ORGANOCHLORINE PESTICIDE, LIQUID, TOXIC	6.1	–	II	61 274	100 mL	E4	P001	–	IBC02	–
2996	ORGANOCHLORINE PESTICIDE, LIQUID, TOXIC	6.1	–	III	61 223 274	5 L	E1	P001 LP01	–	IBC03	–
2997	TRIAZINE PESTICIDE, LIQUID, TOXIC, FLAMMABLE, flashpoint not less than 23°C	6.1	3	I	61 274	0	E5	P001	–	–	–
2997	TRIAZINE PESTICIDE, LIQUID, TOXIC, FLAMMABLE, flashpoint not less than 23°C	6.1	3	II	61 274	100 mL	E4	P001	–	IBC02	–
2997	TRIAZINE PESTICIDE, LIQUID, TOXIC, FLAMMABLE, flashpoint not less than 23°C	6.1	3	III	61 223 274	5 L	E1	P001	–	IBC03	–
2998	TRIAZINE PESTICIDE, LIQUID, TOXIC	6.1	–	I	61 274	0	E5	P001	–	–	–
2998	TRIAZINE PESTICIDE, LIQUID, TOXIC	6.1	–	II	61 274	100 mL	E4	P001	–	IBC02	–
2998	TRIAZINE PESTICIDE, LIQUID, TOXIC	6.1	–	III	61 223 274	5 L	E1	P001 LP01	–	IBC03	–

Portable tanks and bulk containers			EmS	Stowage and segregation	Properties and observations	UN No.
	Tank instructions	Provisions				
(12)	(13) 4.2.5 4.3	(14) 4.2.5	(15) 5.4.3.2 7.3	(16) 7.1 7.2	(17)	(18)
–	T11	TP2 TP13 TP27	F-E, S-D	Category B. Clear of living quarters.	Liquid flammable pesticides having a flashpoint between 23°C and 60°C c.c., presenting a very wide range of toxic hazard. They frequently contain petroleum or coal tar distillates, or other flammable liquids. Flashpoint and miscibility with water depend upon the composition. Toxic if swallowed, by skin contact or by inhalation.	2993
–	T7	TP2 TP28	F-E, S-D	Category A. Clear of living quarters.	See entry above.	2993
–	T14	TP2 TP13 TP27	F-A, S-A	Category B. Clear of living quarters.	Liquid pesticides which present a very wide range of toxic hazard. Miscibility with water depends upon the composition. Toxic if swallowed, by skin contact or by inhalation.	2994
–	T11	TP2 TP13 TP27	F-A, S-A	Category B. Clear of living quarters.	See entry above.	2994
–	T7	TP2 TP28	F-A, S-A	Category A. Clear of living quarters.	See entry above.	2994
–	T14	TP2 TP13 TP27	F-E, S-D	Category B. Clear of living quarters.	It frequently contains petroleum or coal tar distillates, or other flammable liquids. Flashpoint and miscibility with water depend upon the composition. Toxic if swallowed, by skin contact or by inhalation.	2995
–	T11	TP2 TP13 TP27	F-E, S-D	Category B. Clear of living quarters.	See entry above.	2995
–	T7	TP2 TP28	F-E, S-D	Category A. Clear of living quarters.	See entry above.	2995
–	T14	TP2 TP13 TP27	F-A, S-A	Category B. Clear of living quarters.	Miscibility with water depends upon the composition. Toxic if swallowed, by skin contact or by inhalation.	2996
–	T11	TP2 TP13 TP27	F-A, S-A	Category B. Clear of living quarters.	See entry above.	2996
–	T7	TP2 TP28	F-A, S-A	Category A. Clear of living quarters.	See entry above.	2996
–	T14	TP2 TP13 TP27	F-E, S-D	Category B. Clear of living quarters.	It frequently contains petroleum or coal tar distillates, or other flammable liquids. Flashpoint and miscibility with water depend upon the composition. Toxic if swallowed, by skin contact or by inhalation.	2997
–	T11	TP2 TP13 TP27	F-E, S-D	Category B. Clear of living quarters.	See entry above.	2997
–	T7	TP2 TP28	F-E, S-D	Category A. Clear of living quarters.	See entry above.	2997
–	T14	TP2 TP13 TP27	F-A, S-A	Category B. Clear of living quarters.	See alphabetical index to identify those pesticides which are marine pollutants. Liquid pesticides which present a very wide range of toxic hazard. Miscibility with water depends upon the composition. Toxic if swallowed, by skin contact or by inhalation.	2998
–	T11	TP2 TP13 TP27	F-A, S-A	Category B. Clear of living quarters.	See entry above.	2998
–	T7	TP2 TP28	F-A, S-A	Category A. Clear of living quarters.	See entry above.	2998

UN No.	Proper Shipping Name (PSN)	Class or division	Subsidiary risk(s)	Packing group	Special provisions	Limited and excepted quantity provisions		Packing		IBC	
						Limited quantities	Excepted quantities	Instructions	Provisions	Instructions	Provisions
(1)	(2) 3.1.2	(3) 2.0	(4) 2.0	(5) 2.0.1.3	(6) 3.3	(7a) 3.4	(7b) 3.5	(8) 4.1.4	(9) 4.1.4	(10) 4.1.4	(11) 4.1.4
3005	THIOCARBAMATE PESTICIDE, LIQUID, TOXIC, FLAMMABLE, flashpoint not less than 23°C	6.1	3	I	61 274	0	E5	P001	–	–	–
3005	THIOCARBAMATE PESTICIDE, LIQUID, TOXIC, FLAMMABLE, flashpoint not less than 23°C	6.1	3	II	61 274	100 mL	E4	P001	–	IBC02	–
3005	THIOCARBAMATE PESTICIDE, LIQUID, TOXIC, FLAMMABLE, flashpoint not less than 23°C	6.1	3	III	61 223 274	5 L	E1	P001	–	IBC03	–
3006	THIOCARBAMATE PESTICIDE, LIQUID, TOXIC	6.1	–	I	61 274	0	E5	P001	–	–	–
3006	THIOCARBAMATE PESTICIDE, LIQUID, TOXIC	6.1	–	II	61 274	100 mL	E4	P001	–	IBC02	–
3006	THIOCARBAMATE PESTICIDE, LIQUID, TOXIC	6.1	–	III	61 223 274	5 L	E1	P001 LP01	–	IBC03	–
3009	COPPER BASED PESTICIDE, LIQUID, TOXIC, FLAMMABLE, flashpoint not less than 23°C	6.1	3	I	61 274	0	E5	P001	–	–	–
3009	COPPER BASED PESTICIDE, LIQUID, TOXIC, FLAMMABLE, flashpoint not less than 23°C	6.1	3	II	61 274	100 mL	E4	P001	–	IBC02	–
3009	COPPER BASED PESTICIDE, LIQUID, TOXIC, FLAMMABLE, flashpoint not less than 23°C	6.1	3	III	61 223 274	5 L	E1	P001	–	IBC03	–
3010	COPPER BASED PESTICIDE, LIQUID, TOXIC	6.1	–	I	61 274	0	E5	P001	–	–	–
3010	COPPER BASED PESTICIDE, LIQUID, TOXIC	6.1	–	II	61 274	100 mL	E4	P001	–	IBC02	–
3010	COPPER BASED PESTICIDE, LIQUID, TOXIC	6.1	–	III	61 223 274	5 L	E1	P001 LP01	–	IBC03	–
3011	MERCURY BASED PESTICIDE, LIQUID, TOXIC, FLAMMABLE, flashpoint not less than 23°C	6.1	3 P	I	61 274	0	E5	P001	–	–	–
3011	MERCURY BASED PESTICIDE, LIQUID, TOXIC, FLAMMABLE, flashpoint not less than 23°C	6.1	3 P	II	61 274	100 mL	E4	P001	–	IBC02	–
3011	MERCURY BASED PESTICIDE, LIQUID, TOXIC, FLAMMABLE, flashpoint not less than 23°C	6.1	3 P	III	61 223 274	5 L	E1	P001	–	IBC03	–
3012	MERCURY BASED PESTICIDE, LIQUID, TOXIC	6.1	– P	I	61 274	0	E5	P001	–	–	–
3012	MERCURY BASED PESTICIDE, LIQUID, TOXIC	6.1	– P	II	61 274	100 mL	E4	P001	–	IBC02	–

Portable tanks and bulk containers			EmS	Stowage and segregation	Properties and observations	UN No.
	Tank instructions	Provisions				
(12)	(13) 4.2.5 4.3	(14) 4.2.5	(15) 5.4.3.2 7.3	(16) 7.1 7.2	(17)	(18)
–	T14	TP2 TP13	F-E, S-D	Category B. Clear of living quarters.	Liquid flammable pesticides having a flashpoint between 23°C and 60°C c.c., presenting a very wide range of toxic hazard. They frequently contain petroleum or coal tar distillates, or other flammable liquids. Flashpoint and miscibility with water depend upon the composition. Toxic if swallowed, by skin contact or by inhalation.	3005
–	T11	TP2 TP13 TP27	F-E, S-D	Category B. Clear of living quarters.	See entry above.	3005
–	T7	TP2 TP28	F-E, S-D	Category A. Clear of living quarters.	See entry above.	3005
–	T14	TP2 TP13	F-A, S-A	Category B. Clear of living quarters.	Liquid pesticides which present a very wide range of toxic hazard. Miscibility with water depends upon the composition. Toxic if swallowed, by skin contact or by inhalation.	3006
–	T11	TP2 TP13 TP27	F-A, S-A	Category B. Clear of living quarters.	See entry above.	3006
–	T7	TP2 TP28	F-A, S-A	Category A. Clear of living quarters.	See entry above.	3006
–	T14	TP2 TP13 TP27	F-E, S-D	Category B. Clear of living quarters.	Liquid flammable pesticides having a flashpoint between 23°C and 60°C c.c., presenting a very wide range of toxic hazard. They frequently contain petroleum or coal tar distillates, or other flammable liquids. Flashpoint and miscibility with water depend upon the composition. Toxic if swallowed, by skin contact or by inhalation.	3009
–	T11	TP2 TP13 TP27	F-E, S-D	Category B. Clear of living quarters.	See entry above.	3009
–	T7	TP2 TP28	F-E, S-D	Category A. Clear of living quarters.	See entry above.	3009
–	T14	TP2 TP13 TP27	F-A, S-A	Category B. Clear of living quarters.	Liquid pesticides which present a very wide range of toxic hazard. Miscibility with water depends upon the composition. Toxic if swallowed, by skin contact or by inhalation.	3010
–	T11	TP2 TP13 TP27	F-A, S-A	Category B. Clear of living quarters.	See entry above.	3010
–	T7	TP2 TP28	F-A, S-A	Category A. Clear of living quarters.	See entry above.	3010
–	T14	TP2 TP13 TP27	F-E, S-D	Category B. Clear of living quarters.	Liquid flammable pesticides having a flashpoint between 23°C and 60°C c.c., presenting a very wide range of toxic hazard. They frequently contain petroleum or coal tar distillates, or other flammable liquids. Flashpoint and miscibility with water depend upon the composition. Toxic if swallowed, by skin contact or by inhalation.	3011
–	T11	TP2 TP13 TP27	F-E, S-D	Category B. Clear of living quarters.	See entry above.	3011
–	T7	TP2 TP28	F-E, S-D	Category A. Clear of living quarters.	See entry above.	3011
–	T14	TP2 TP13 TP27	F-A, S-A	Category B. Clear of living quarters.	Liquid pesticides which present a very wide range of toxic hazard. Miscibility with water depends upon the composition. Toxic if swallowed, by skin contact or by inhalation.	3012
–	T11	TP2 TP13 TP27	F-A, S-A	Category B. Clear of living quarters.	See entry above.	3012

UN No.	Proper Shipping Name (PSN)	Class or division	Subsidiary risk(s)	Packing group	Special provisions	Limited and excepted quantity provisions		Packing		IBC	
						Limited quantities	Excepted quantities	Instructions	Provisions	Instructions	Provisions
(1)	(2) 3.1.2	(3) 2.0	(4) 2.0	(5) 2.0.1.3	(6) 3.3	(7a) 3.4	(7b) 3.5	(8) 4.1.4	(9) 4.1.4	(10) 4.1.4	(11) 4.1.4
3012	MERCURY BASED PESTICIDE, LIQUID, TOXIC	6.1	– P	III	61 223 274	5 L	E1	P001 LP01	–	IBC03	–
3013	SUBSTITUTED NITROPHENOL PESTICIDE, LIQUID, TOXIC, FLAMMABLE, flashpoint not less than 23°C	6.1	3	I	61 274	0	E5	P001	–	–	–
3013	SUBSTITUTED NITROPHENOL PESTICIDE, LIQUID, TOXIC, FLAMMABLE, flashpoint not less than 23°C	6.1	3	II	61 274	100 mL	E4	P001	–	IBC02	–
3013	SUBSTITUTED NITROPHENOL PESTICIDE, LIQUID, TOXIC, FLAMMABLE, flashpoint not less than 23°C	6.1	3	III	61 223 274	5 L	E1	P001	–	IBC03	–
3014	SUBSTITUTED NITROPHENOL PESTICIDE, LIQUID, TOXIC	6.1	–	I	61 274	0	E5	P001	–	–	–
3014	SUBSTITUTED NITROPHENOL PESTICIDE, LIQUID, TOXIC	6.1	–	II	61 274	100 mL	E4	P001	–	IBC02	–
3014	SUBSTITUTED NITROPHENOL PESTICIDE, LIQUID, TOXIC	6.1	–	III	61 223 274	5 L	E1	P001 LP01	–	IBC03	–
3015	BIPYRIDILIUM PESTICIDE, LIQUID, TOXIC, FLAMMABLE, flashpoint not less than 23°C	6.1	3	I	61 274	0	E5	P001	–	–	–
3015	BIPYRIDILIUM PESTICIDE, LIQUID, TOXIC, FLAMMABLE, flashpoint not less than 23°C	6.1	3	II	61 274	100 mL	E4	P001	–	IBC02	–
3015	BIPYRIDILIUM PESTICIDE, LIQUID, TOXIC, FLAMMABLE, flashpoint not less than 23°C	6.1	3	III	61 223 274	5 L	E1	P001	–	IBC03	–
3016	BIPYRIDILIUM PESTICIDE, LIQUID, TOXIC	6.1	–	I	61 274	0	E5	P001	–	–	–
3016	BIPYRIDILIUM PESTICIDE, LIQUID, TOXIC	6.1	–	II	61 274	100 mL	E4	P001	–	IBC02	–
3016	BIPYRIDILIUM PESTICIDE, LIQUID, TOXIC	6.1	–	III	61 223 274	5 L	E1	P001 LP01	–	IBC03	–
3017	ORGANOPHOSPHORUS PESTICIDE, LIQUID, TOXIC, FLAMMABLE, flashpoint not less than 23°C	6.1	3	I	61 274	0	E5	P001	–	–	–
3017	ORGANOPHOSPHORUS PESTICIDE, LIQUID, TOXIC, FLAMMABLE, flashpoint not less than 23°C	6.1	3	II	61 274	100 mL	E4	P001	–	IBC02	–
3017	ORGANOPHOSPHORUS PESTICIDE, LIQUID, TOXIC, FLAMMABLE, flashpoint not less than 23°C	6.1	3	III	61 223 274	5 L	E1	P001	–	IBC03	–

Portable tanks and bulk containers			EmS	Stowage and segregation	Properties and observations	UN No.
	Tank instructions	Provisions				
(12)	(13) 4.2.5 4.3	(14) 4.2.5	(15) 5.4.3.2 7.3	(16) 7.1 7.2	(17)	(18)
–	T7	TP2 TP28	F-A, S-A	Category A. Clear of living quarters.	Liquid pesticides which present a very wide range of toxic hazard. Miscibility with water depends upon the composition. Toxic if swallowed, by skin contact or by inhalation.	3012
–	T14	TP2 TP13 TP27	F-E, S-D	Category B. Clear of living quarters.	Liquid flammable pesticides having a flashpoint between 23°C and 60°C c.c., presenting a very wide range of toxic hazard. They frequently contain petroleum or coal tar distillates, or other flammable liquids. Flashpoint and miscibility with water depend upon the composition. Toxic if swallowed, by skin contact or by inhalation.	3013
–	T11	TP2 TP13 TP27	F-E, S-D	Category B. Clear of living quarters.	See entry above.	3013
–	T7	TP2 TP28	F-E, S-D	Category A. Clear of living quarters.	See entry above.	3013
–	T14	TP2 TP13 TP27	F-A, S-A	Category B. Clear of living quarters.	Liquid pesticides which present a very wide range of toxic hazard. Miscibility with water depends upon the composition. Toxic if swallowed, by skin contact or by inhalation.	3014
–	T11	TP2 TP13 TP27	F-A, S-A	Category B. Clear of living quarters.	See entry above.	3014
–	T7	TP2 TP28	F-A, S-A	Category A. Clear of living quarters.	See entry above.	3014
–	T14	TP2 TP13 TP27	F-E, S-D	Category B. Clear of living quarters.	Liquid flammable pesticides having a flashpoint between 23°C and 60°C c.c., presenting a very wide range of toxic hazard. They frequently contain petroleum or coal tar distillates, or other flammable liquids. Flashpoint and miscibility with water depend upon the composition. Toxic if swallowed, by skin contact or by inhalation.	3015
–	T11	TP2 TP13 TP27	F-E, S-D	Category B. Clear of living quarters.	See entry above.	3015
–	T7	TP2 TP28	F-E, S-D	Category A. Clear of living quarters.	See entry above.	3015
–	T14	TP2 TP13 TP27	F-A, S-A	Category B. Clear of living quarters.	Liquid pesticides which present a very wide range of toxic hazard. Miscibility with water depends upon the composition. Toxic if swallowed, by skin contact or by inhalation.	3016
–	T11	TP2 TP13 TP27	F-A, S-A	Category B. Clear of living quarters.	See entry above.	3016
–	T7	TP2 TP28	F-A, S-A	Category A. Clear of living quarters.	See entry above.	3016
–	T14	TP2 TP13 TP27	F-E, S-D	Category B. Clear of living quarters.	Liquid flammable pesticides having a flashpoint between 23°C and 60°C c.c., presenting a very wide range of toxic hazard. They frequently contain petroleum or coal tar distillates, or other flammable liquids. Flashpoint and miscibility with water depend upon the composition. Toxic if swallowed, by skin contact or by inhalation.	3017
–	T11	TP2 TP13 TP27	F-E, S-D	Category B. Clear of living quarters.	See entry above.	3017
–	T7	TP2 TP28	F-E, S-D	Category A. Clear of living quarters.	See entry above.	3017

UN No.	Proper Shipping Name (PSN)	Class or division	Subsidiary risk(s)	Packing group	Special provisions	Limited and excepted quantity provisions		Packing		IBC	
						Limited quantities	Excepted quantities	Instructions	Provisions	Instructions	Provisions
(1)	(2) 3.1.2	(3) 2.0	(4) 2.0	(5) 2.0.1.3	(6) 3.3	(7a) 3.4	(7b) 3.5	(8) 4.1.4	(9) 4.1.4	(10) 4.1.4	(11) 4.1.4
3018	ORGANOPHOSPHORUS PESTICIDE, LIQUID, TOXIC	6.1	–	I	61 274	0	E5	P001	–	–	–
3018	ORGANOPHOSPHORUS PESTICIDE, LIQUID, TOXIC	6.1	–	II	61 274	100 mL	E4	P001	–	IBC02	–
3018	ORGANOPHOSPHORUS PESTICIDE, LIQUID, TOXIC	6.1	–	III	61 223 274	5 L	E1	P001 LP01	–	IBC03	–
3019	ORGANOTIN PESTICIDE, LIQUID, TOXIC, FLAMMABLE, flashpoint not less than 23°C	6.1	3 P	I	61 274	0	E5	P001	–	–	–
3019	ORGANOTIN PESTICIDE, LIQUID, TOXIC, FLAMMABLE, flashpoint not less than 23°C	6.1	3 P	II	61 274	100 mL	E4	P001	–	IBC02	–
3019	ORGANOTIN PESTICIDE, LIQUID, TOXIC, FLAMMABLE, flashpoint not less than 23°C	6.1	3 P	III	61 223 274	5 L	E1	P001	–	IBC03	–
3020	ORGANOTIN PESTICIDE, LIQUID, TOXIC	6.1	– P	I	61 274	0	E5	P001	–	–	–
3020	ORGANOTIN PESTICIDE, LIQUID, TOXIC	6.1	– P	II	61 274	100 mL	E4	P001	–	IBC02	–
3020	ORGANOTIN PESTICIDE, LIQUID, TOXIC	6.1	– P	III	61 223 274	5 L	E1	P001 LP01	–	IBC03	–
3021	PESTICIDE, LIQUID, FLAMMABLE, TOXIC, N.O.S., flashpoint less than 23°C	3	6.1	I	61 274	0	E0	P001	–	–	–
3021	PESTICIDE, LIQUID, FLAMMABLE, TOXIC, N.O.S., flashpoint less than 23°C	3	6.1	II	61 274	1 L	E2	P001	–	IBC02	–
3022	1,2-BUTYLENE OXIDE, STABILIZED	3	–	II	–	1 L	E2	P001	–	IBC02	–
3023	2-METHYL-2-HEPTANETHIOL	6.1	3	I	354	0	E0	P602	–	–	–
3024	COUMARIN DERIVATIVE PESTICIDE, LIQUID, FLAMMABLE, TOXIC, flashpoint less than 23°C	3	6.1	I	61 274	0	E0	P001	–	–	–
3024	COUMARIN DERIVATIVE PESTICIDE, LIQUID, FLAMMABLE, TOXIC, flashpoint less than 23°C	3	6.1	II	61 274	1 L	E2	P001	–	IBC02	–
3025	COUMARIN DERIVATIVE PESTICIDE, LIQUID, TOXIC, FLAMMABLE, flashpoint not less than 23°C	6.1	3	I	61 274	0	E5	P001	–	–	–

Portable tanks and bulk containers			EmS	Stowage and segregation	Properties and observations	UN No.
	Tank instructions	Provisions				
(12)	(13) 4.2.5 4.3	(14) 4.2.5	(15) 5.4.3.2 7.3	(16) 7.1 7.2	(17)	(18)
–	T14	TP2 TP13 TP27	F-A, S-A	Category B. Clear of living quarters.	Liquid pesticides which present a very wide range of toxic hazard. Miscibility with water depends upon the composition. Toxic if swallowed, by skin contact or by inhalation.	3018
–	T11	TP2 TP13 TP27	F-A, S-A	Category B. Clear of living quarters.	See entry above.	3018
–	T7	TP2 TP28	F-A, S-A	Category A. Clear of living quarters.	See entry above.	3018
–	T14	TP2 TP13 TP27	F-E, S-D	Category B. Clear of living quarters.	Liquid flammable pesticides having a flashpoint between 23°C and 60°C c.c., presenting a very wide range of toxic hazard. They frequently contain petroleum or coal tar distillates, or other flammable liquids. Flashpoint and miscibility with water depend upon the composition. Toxic if swallowed, by skin contact or by inhalation.	3019
–	T11	TP2 TP13 TP27	F-E, S-D	Category B. Clear of living quarters.	See entry above.	3019
–	T7	TP2 TP28	F-E, S-D	Category A. Clear of living quarters.	See entry above.	3019
–	T14	TP2 TP13 TP27	F-A, S-A	Category B. Clear of living quarters.	Liquid pesticides which present a very wide range of toxic hazard. Miscibility with water depends upon the composition. Toxic if swallowed, by skin contact or by inhalation.	3020
–	T11	TP2 TP13 TP27	F-A, S-A	Category B. Clear of living quarters.	See entry above.	3020
–	T7	TP2 TP28	F-A, S-A	Category A. Clear of living quarters.	See entry above.	3020
–	T14	TP2 TP13 TP27	F-E, S-D	Category B. Clear of living quarters.	Pesticides frequently contain petroleum or coal tar distillates, or other flammable liquids. Miscibility with water depends upon the composition. Toxic if swallowed, by skin contact or by inhalation.	3021
–	T11	TP2 TP13 TP27	F-E, S-D	Category B. Clear of living quarters.	See entry above.	3021
–	T4	TP1	F-E, S-D	Category B. "Away from" acids and alkalis.	Colourless liquid. Flashpoint: −15°C c.c. Explosive limits: 1.5% to 18.3%. Reacts violently with acids, alkalis and oxidizers. Miscible with water. Harmful if swallowed or by inhalation. Irritating to skin, eyes and mucous membranes.	3022
–	T20	TP2 TP13 TP35	F-E, S-D	Category D. Clear of living quarters. "Separated from" odour-absorbing cargoes.	Colourless flammable liquid with a foul odour. Flashpoint: 31°C c.c. Miscible with water. Highly toxic if swallowed, by skin contact or by inhalation.	3023
–	T14	TP2 TP13 TP27	F-E, S-D	Category B. Clear of living quarters.	Pesticides frequently contain petroleum or coal tar distillates, or other flammable liquids. Miscibility with water depends upon the composition. Toxic if swallowed, by skin contact or by inhalation.	3024
–	T11	TP2 TP13 TP27	F-E, S-D	Category B. Clear of living quarters.	See entry above.	3024
–	T14	TP2 TP13 TP27	F-E, S-D	Category B. Clear of living quarters.	Liquid flammable pesticides having a flashpoint between 23°C and 60°C c.c., presenting a very wide range of toxic hazard. They frequently contain petroleum or coal tar distillates, or other flammable liquids. Flashpoint and miscibility with water depend upon the composition. Toxic if swallowed, by skin contact or by inhalation.	3025

UN No. (1)	Proper Shipping Name (PSN) (2) 3.1.2	Class or division (3) 2.0	Subsidiary risk(s) (4) 2.0	Packing group (5) 2.0.1.3	Special provisions (6) 3.3	Limited and excepted quantity provisions: Limited quantities (7a) 3.4	Limited and excepted quantity provisions: Excepted quantities (7b) 3.5	Packing: Instructions (8) 4.1.4	Packing: Provisions (9) 4.1.4	IBC: Instructions (10) 4.1.4	IBC: Provisions (11) 4.1.4
3025	COUMARIN DERIVATIVE PESTICIDE, LIQUID, TOXIC, FLAMMABLE, flashpoint not less than 23°C	6.1	3	II	61 274	100 mL	E4	P001	–	IBC02	–
3025	COUMARIN DERIVATIVE PESTICIDE, LIQUID, TOXIC, FLAMMABLE, flashpoint not less than 23°C	6.1	3	III	61 223 274	5 L	E1	P001	–	IBC03	–
3026	COUMARIN DERIVATIVE PESTICIDE, LIQUID, TOXIC	6.1	–	I	61 274	0	E5	P001	–	–	–
3026	COUMARIN DERIVATIVE PESTICIDE, LIQUID, TOXIC	6.1	–	II	61 274	100 mL	E4	P001	–	IBC02	–
3026	COUMARIN DERIVATIVE PESTICIDE, LIQUID, TOXIC	6.1	–	III	61 223 274	5 L	E1	P001 LP01	–	IBC03	–
3027	COUMARIN DERIVATIVE PESTICIDE, SOLID, TOXIC	6.1	–	I	61 274	0	E5	P002	–	IBC07	B1
3027	COUMARIN DERIVATIVE PESTICIDE, SOLID, TOXIC	6.1	–	II	61 274	500 g	E4	P002	–	IBC08	B2 B4
3027	COUMARIN DERIVATIVE PESTICIDE, SOLID, TOXIC	6.1	–	III	61 223 274	5 kg	E1	P002 LP02	–	IBC08	B3
3028	BATTERIES, DRY, CONTAINING POTASSIUM HYDROXIDE, SOLID electric storage	8	–	III	295 304	5 kg	E0	P801	–	–	–
3048	ALUMINIUM PHOSPHIDE PESTICIDE	6.1	–	I	153 930	0	E5	P002	PP31	IBC07	B1
△ 3054	CYCLOHEXYL MERCAPTAN	3	–	III	–	5 L	E1	P001 LP01	–	IBC03	–
3055	2-(2-AMINOETHOXY)ETHANOL	8	–	III	–	5 L	E1	P001 LP01	–	IBC03	–
3056	*n*-HEPTALDEHYDE	3	–	III	–	5 L	E1	P001 LP01	–	IBC03	–
3057	TRIFLUOROACETYL CHLORIDE	2.3	8	–	–	0	E0	P200	–	–	–
△ 3064	NITROGLYCERIN SOLUTION IN ALCOHOL with more than 1% but not more than 5% nitroglycerin	3	–	II	359	0	E0	P300	–	–	–
3065	ALCOHOLIC BEVERAGES, with more than 70% alcohol by volume	3	–	II	–	5 L	E2	P001	PP2	IBC02	–

Portable tanks and bulk containers			EmS	Stowage and segregation	Properties and observations	UN No.
	Tank instructions	Provisions				
(12)	(13) 4.2.5 4.3	(14) 4.2.5	(15) 5.4.3.2 7.3	(16) 7.1 7.2	(17)	(18)
–	T11	TP2 TP13 TP27	F-E, S-D	Category B. Clear of living quarters.	Liquid flammable pesticides having a flashpoint between 23°C and 60°C c.c., presenting a very wide range of toxic hazard. They frequently contain petroleum or coal tar distillates, or other flammable liquids. Flashpoint and miscibility with water depend upon the composition. Toxic if swallowed, by skin contact or by inhalation.	3025
–	T7	TP1 TP28	F-E, S-D	Category A. Clear of living quarters.	See entry above.	3025
–	T14	TP2 TP13 TP27	F-A, S-A	Category B. Clear of living quarters.	Liquid pesticides which present a very wide range of toxic hazard. Miscibility with water depends upon the composition. Toxic if swallowed, by skin contact or by inhalation.	3026
–	T11	TP2 TP27	F-A, S-A	Category B. Clear of living quarters.	See entry above.	3026
–	T7	TP1 TP28	F-A, S-A	Category A. Clear of living quarters.	See entry above.	3026
–	T6	TP33	F-A, S-A	Category A. Clear of living quarters.	Solid pesticides present a very wide range of toxic hazard. Toxic if swallowed, by skin contact or by inhalation.	3027
–	T3	TP33	F-A, S-A	Category A. Clear of living quarters.	See entry above.	3027
–	T1	TP33	F-A, S-A	Category A. Clear of living quarters.	See entry above.	3027
–	–	–	F-A, S-B	Category A. "Separated from" acids.	Series of metal plates immersed in dry potassium hydroxide in a closed receptacle. When electrically charged, may cause fire through short-circuiting of terminals. Batteries need not be individually marked and labelled if the pallet bears the appropriate mark and label. Used batteries being transported for disposal or reclamation should be carefully checked prior to shipment to ensure the integrity of each battery and its suitability for transport. React violently with acids.	3028
–	T6	TP33	F-A, S-A	Category E, in a mechanically ventilated space. Clear of living quarters.	Waxed pellets, adequately stabilized powder, tablets or crystals. Highly toxic if swallowed, by skin contact or by inhalation.	3048
–	T2	TP1	F-E, S-D	Category A. Clear of living quarters. Segregation from foodstuffs as in 7.3.4.2, 7.6.3.1.2 or 7.7.3.6. "Separated from" odour-absorbing cargoes.	Colourless liquid with a garlic-like odour. Flashpoint: 49°C c.c. Immiscible with water. Harmful by inhalation. Irritating to skin, eyes and mucous membranes.	3054 △
–	T4	TP1	F-A, S-B	Category A.	Colourless, slightly viscous liquid with a mild odour. Miscible with water. Harmful if swallowed or by inhalation. Corrosive to skin, eyes and mucous membranes.	3055
–	T2	TP1	F-E, S-D	Category A.	Colourless or pale yellow, oily liquid with a pungent odour. Flashpoint: 35°C to 45°C c.c. Explosive limits: 1.1% to 5.2%. Slightly soluble in water. Irritating to skin, eyes and mucous membranes.	3056
–	T50	TP21	F-C, S-U	Category D. Clear of living quarters.	Liquefied, non-flammable, toxic and corrosive gas. Reacts with water. Corrosive to glass and to most metals, including steel. Heavier than air (1.4 at 20°C). Highly irritating to skin, eyes and mucous membranes.	3057
–	–	–	F-E, S-D	Category E.	Immiscible with water. Ignites readily. When involved in a fire, evolves toxic nitrous fumes. Not explosive in this state but damage to, or leakage from, a package may allow solvent to evaporate and thus leave the nitroglycerin in an explosive state.	3064 △
–	T4	TP1	F-E, S-D	Category A.	Aqueous solutions of ethanol produced and supplied as alcoholic beverages. Miscible with water. Flashpoint: –13°C c.c. or greater.	3065

3

DGL

UN No.	Proper Shipping Name (PSN)	Class or division	Subsidiary risk(s)	Packing group	Special provisions	Limited and excepted quantity provisions		Packing		IBC	
						Limited quantities	Excepted quantities	Instructions	Provisions	Instructions	Provisions
(1)	(2) 3.1.2	(3) 2.0	(4) 2.0	(5) 2.0.1.3	(6) 3.3	(7a) 3.4	(7b) 3.5	(8) 4.1.4	(9) 4.1.4	(10) 4.1.4	(11) 4.1.4
3065	ALCOHOLIC BEVERAGES, with more than 24% but not more than 70% alcohol by volume	3	–	III	144 145 247	5 L	E1	P001	PP2	IBC03	–
3066	PAINT (including paint, lacquer, enamel, stain, shellac, varnish, polish, liquid filler and liquid lacquer base) or PAINT RELATED MATERIAL (including paint thinning or reducing compound)	8	–	II	163	1 L	E2	P001	–	IBC02	–
3066	PAINT (including paint, lacquer, enamel, stain, shellac, varnish, polish, liquid filler and liquid lacquer base) or PAINT RELATED MATERIAL (including paint thinning or reducing compound)	8	–	III	163 223	5 L	E1	P001	–	IBC03	–
3070	ETHYLENE OXIDE AND DICHLORODIFLUORO-METHANE MIXTURE with not more than 12.5% ethylene oxide	2.2	–	–	–	120 mL	E1	P200	–	–	–
3071	MERCAPTANS, LIQUID, TOXIC, FLAMMABLE, N.O.S. or MERCAPTAN MIXTURE, LIQUID, TOXIC, FLAMMABLE, N.O.S.	6.1	3	II	274	100 mL	E4	P001	–	IBC02	–
△ 3072	LIFE-SAVING APPLIANCES, NOT SELF-INFLATING containing dangerous goods as equipment	9	–	–	296	0	E0	P905	–	–	–
3073	VINYLPYRIDINES, STABILIZED	6.1	3/8	II	–	100 mL	E4	P001	–	IBC01	–
△ 3077	ENVIRONMENTALLY HAZARDOUS SUBSTANCE, SOLID, N.O.S.	9	–	III	274 335 966 967	5 kg	E1	P002 LP02	PP12	IBC08	B3

Portable tanks and bulk containers			EmS	Stowage and segregation	Properties and observations	UN No.	
	Tank instructions	Provisions					
(12)	(13) 4.2.5 4.3	(14) 4.2.5	(15) 5.4.3.2 7.3	(16) 7.1 7.2	(17)	(18)	
–	T2	TP1	F-E, S-D	Category A.	Alcoholic beverages containing more than 24% alcohol but not more than 70% by volume, when transported as part of the manufacturing process, may be transported in wooden barrels with a capacity of more than 250 litres and not more than 500 litres meeting the general requirements of 4.1.1, as appropriate, on the following conditions: 1. the wooden barrels should be checked and tightened before filling; 2. sufficient ullage (not less than 3%) should be left to allow for the expansion of the liquid; 3. the wooden barrels should be transported with the bungholes pointing upwards; 4. the wooden barrels should be transported in containers meeting the requirements of the International Convention for Safe Containers (CSC), as amended. Each wooden barrel should be secured in custom-made cradles and should be wedged by appropriate means to prevent them from being displaced in any way during transport; and 5. when carried on board ships, the containers should be stowed in open cargo spaces or in enclosed cargo spaces complying with the requirements for class 3 flammable liquids with a flashpoint of 23°C c.c. or less in regulation II-2/19 of SOLAS, 74, as amended.	3065	
–	T7	TP2 TP27 TP28	F-A, S-B	Category B. Clear of living quarters.	Corrosive content. Causes burns to skin, eyes and mucous membranes.	3066	
–	T4	TP1 TP27 TP29	F-A, S-B	Category A. Clear of living quarters.	See entry above.	3066	
–	T50	–	F-C, S-V	Category A.	Liquefied, non-flammable gas. Much heavier than air.	3070	
–	T11	TP2 TP13 TP27	F-E, S-D	Category C. Clear of living quarters. "Separated from" all odour-absorbing cargoes.	Colourless to yellow flammable liquids with a garlic odour. Immiscible with water. Toxic if swallowed, by skin contact or by inhalation.	3071	
–	–	–	F-A, S-V	Category A. "Separated from" class 6.2. Within the appliance, to the extent that the dangerous goods are packaged as integral parts of the complete life-saving appliance, there is no need to apply the provisions on segregation of substances in chapter 7.2.	These articles may contain: (a) Class 2.2 compressed gases; (b) signal devices (class 1) which may include smoke and illumination signal flares; signal devices must be packed in plastic or fibreboard inner packagings; (c) electric storage batteries; (d) first aid kit; or (e) "strike anywhere" matches.	3072	△
–	T7	TP2 TP13	F-E, S-C	Category C. Clear of living quarters. Segregation as for class 3 but "away from" class 4.1. "Separated from" acids.	Colourless to straw-coloured flammable liquids. Flashpoint: 42°C to 51°C c.c. Toxic if swallowed, by skin contact or by inhalation. Causes burns to skin, eyes and mucous membranes. React violently with acids.	3073	
–	T1 BK1 BK2 BK3	TP33	F-A, S-F	Category A. When transported in BK3 bulk container, see 7.6.2.12 and 7.7.3.9.	–	3077	△

UN No.	Proper Shipping Name (PSN)	Class or division	Subsidiary risk(s)	Packing group	Special provisions	Limited and excepted quantity provisions		Packing		IBC	
						Limited quantities	Excepted quantities	Instructions	Provisions	Instructions	Provisions
(1)	(2) 3.1.2	(3) 2.0	(4) 2.0	(5) 2.0.1.3	(6) 3.3	(7a) 3.4	(7b) 3.5	(8) 4.1.4	(9) 4.1.4	(10) 4.1.4	(11) 4.1.4
3078	CERIUM turnings or gritty powder	4.3	–	II	–	500 g	E2	P410	PP31 PP40	IBC07	B2
3079	METHACRYLONITRILE, STABILIZED	6.1	3	I	354	0	E0	P602	–	–	–
△ 3080	ISOCYANATES, TOXIC, FLAMMABLE, N.O.S or ISOCYANATE SOLUTION, TOXIC, FLAMMABLE, N.O.S.	6.1	3	II	274	100 mL	E4	P001	–	IBC02	–
3082	ENVIRONMENTALLY HAZARDOUS SUBSTANCE, LIQUID, N.O.S.	9	–	III	274 335	5 L	E1	P001 LP01	PP1	IBC03	–
3083	PERCHLORYL FLUORIDE	2.3	5.1	–	–	0	E0	P200	–	–	–
3084	CORROSIVE SOLID, OXIDIZING, N.O.S.	8	5.1	I	274	0	E0	P002	–	–	–
3084	CORROSIVE SOLID, OXIDIZING, N.O.S.	8	5.1	II	274	1 kg	E2	P002	–	IBC06	B2
3085	OXIDIZING SOLID, CORROSIVE, N.O.S.	5.1	8	I	274	0	E0	P503	–	–	–
3085	OXIDIZING SOLID, CORROSIVE, N.O.S.	5.1	8	II	274	1 kg	E2	P002	–	IBC06	B2
3085	OXIDIZING SOLID, CORROSIVE, N.O.S.	5.1	8	III	223 274	5 kg	E1	P002	–	IBC08	B3
3086	TOXIC SOLID, OXIDIZING, N.O.S.	6.1	5.1	I	274	0	E5	P002	–	–	–
3086	TOXIC SOLID, OXIDIZING, N.O.S.	6.1	5.1	II	274	500 g	E4	P002	–	IBC06	B2
△ 3087	OXIDIZING SOLID, TOXIC, N.O.S.	5.1	6.1	I	274 900	0	E0	P503	–	–	–
△ 3087	OXIDIZING SOLID, TOXIC, N.O.S.	5.1	6.1	II	274 900	1 kg	E2	P002	–	IBC06	B2
△ 3087	OXIDIZING SOLID, TOXIC, N.O.S.	5.1	6.1	III	223 274 900	5 kg	E1	P002	–	IBC08	B3
3088	SELF-HEATING SOLID, ORGANIC, N.O.S.	4.2	–	II	274	0	E2	P410	PP31	IBC06	B2
3088	SELF-HEATING SOLID, ORGANIC, N.O.S.	4.2	–	III	223 274	0	E1	P002 LP02	PP31	IBC08	B3

Portable tanks and bulk containers			EmS	Stowage and segregation	Properties and observations	UN No.
	Tank instructions	Provisions				
(12)	(13) 4.2.5 4.3	(14) 4.2.5	(15) 5.4.3.2 7.3	(16) 7.1 7.2	(17)	(18)
–	T3	TP33	F-G, S-O	Category E. "Separated from" acids.	Grey, ductile metal or powder. Decomposes in water and reacts violently with acids, evolving hydrogen, which may be ignited by the heat of the reaction.	3078
–	T20	TP2 TP13 TP37	F-E, S-D	Category D. Clear of living quarters.	Colourless, mobile liquid with a pungent odour. Flashpoint: 4°C c.c. Explosive limits: 3% to 17%. Partially miscible with water. Highly toxic if swallowed, by skin contact or by inhalation. Practice has shown that this substance may leak from packagings that ordinarily are leakproof to other chemicals.	3079
–	T11	TP2 TP13 TP27	F-E, S-D	Category D. Protected from sources of heat. Clear of living quarters.	Flammable liquids or solutions with a pungent odour. Immiscible with or insoluble in water, but react with it to form carbon dioxide. Toxic if swallowed, by skin contact or by inhalation. Irritating to skin, eyes and mucous membranes.	3080 △
–	T4	TP2 TP29	F-A, S-F	Category A.	–	3082
–	–	–	F-C, S-W	Category D. Clear of living quarters.	Non-flammable, toxic, colourless gas with a characteristic sweet odour. Strong oxidizing agent; may cause fire in contact with organic materials. Reacts with water or moist air to produce toxic and corrosive fumes. Mixtures with oils or combustible materials may explode. Much heavier than air (3.6). Irritating to skin, eyes and mucous membranes.	3083
–	T6	TP33	F-A, S-Q	Category C.	Causes burns to skin, eyes and mucous membranes.	3084
–	T3	TP33	F-A, S-Q	Category C.	See entry above.	3084
–	–	–	F-A, S-Q	Category D. Keep as dry as reasonably practicable. "Separated from" ammonium compounds, cyanides and peroxides.	Causes burns to skin, eyes and mucous membranes. Particular care in handling should be exercised if packages have become wetted.	3085
–	T3	TP33	F-A, S-Q	Category B. Keep as dry as reasonably practicable. "Separated from" ammonium compounds, cyanides and peroxides.	See entry above.	3085
–	T1	TP33	F-A, S-Q	Category B. Keep as dry as reasonably practicable. "Separated from" ammonium compounds, cyanides and peroxides.	See entry above.	3085
–	T6	TP33	F-A, S-Q	Category C.	Toxic if swallowed, by skin contact or by inhalation.	3086
–	T3	TP33	F-A, S-Q	Category C.	See entry above.	3086
–	–	–	F-A, S-Q	Category D."Separated from" ammonium compounds, cyanides and peroxides.	Toxic if swallowed, by skin contact or by dust inhalation. Should be handled with care to minimize exposure, particularly to dust.	3087 △
–	T3	TP33	F-A, S-Q	Category B. "Separated from" ammonium compounds, cyanides and peroxides.	See entry above.	3087 △
–	T1	TP33	F-A, S-Q	Category B. "Separated from" ammonium compounds, cyanides and peroxides.	See entry above.	3087 △
–	T3	TP33	F-A, S-J	Category C.	Liable to self-heating or spontaneous combustion.	3088
–	T1	TP33	F-A, S-J	Category C.	See entry above.	3088

UN No.	Proper Shipping Name (PSN)	Class or division	Subsidiary risk(s)	Packing group	Special provisions	Limited and excepted quantity provisions		Packing		IBC	
						Limited quantities	Excepted quantities	Instructions	Provisions	Instructions	Provisions
(1)	(2) 3.1.2	(3) 2.0	(4) 2.0	(5) 2.0.1.3	(6) 3.3	(7a) 3.4	(7b) 3.5	(8) 4.1.4	(9) 4.1.4	(10) 4.1.4	(11) 4.1.4
3089	METAL POWDER, FLAMMABLE, N.O.S.	4.1	–	II	–	1 kg	E2	P002	–	IBC08	B2 B4
3089	METAL POWDER, FLAMMABLE, N.O.S.	4.1	–	III	223	5 kg	E1	P002	–	IBC06	–
3090	LITHIUM METAL BATTERIES (including lithium alloy batteries)	9	–	II	188 230 310 957	0	E0	P903	–	–	–
△ 3091	LITHIUM METAL BATTERIES CONTAINED IN EQUIPMENT or LITHIUM METAL BATTERIES PACKED WITH EQUIPMENT (including lithium alloy batteries)	9	–	II	188 230 360 957	0	E0	P903	–	–	–
3092	1-METHOXY-2-PROPANOL	3	–	III	–	5 L	E1	P001 LP01	–	IBC03	–
3093	CORROSIVE LIQUID, OXIDIZING, N.O.S.	8	5.1	I	274	0	E0	P001	–	–	–
3093	CORROSIVE LIQUID, OXIDIZING, N.O.S.	8	5.1	II	274	1 L	E2	P001	–	IBC02	–
3094	CORROSIVE LIQUID, WATER-REACTIVE, N.O.S.	8	4.3	I	274	0	E0	P001	–	–	–
3094	CORROSIVE LIQUID, WATER-REACTIVE, N.O.S.	8	4.3	II	274	500 mL	E2	P001	–	–	–
3095	CORROSIVE SOLID, SELF-HEATING, N.O.S.	8	4.2	I	274	0	E0	P002	–	–	–
3095	CORROSIVE SOLID, SELF-HEATING, N.O.S.	8	4.2	II	274	1 kg	E2	P002	–	IBC06	B2
3096	CORROSIVE SOLID, WATER-REACTIVE, N.O.S.	8	4.3	I	274	0	E0	P002	–	–	–
3096	CORROSIVE SOLID, WATER-REACTIVE, N.O.S.	8	4.3	II	274	1 kg	E2	P002	–	IBC06	B2
3097	FLAMMABLE SOLID, OXIDIZING, N.O.S.	4.1	5.1	II	76 274	0	E2	P099	–	–	–
3097	FLAMMABLE SOLID, OXIDIZING, N.O.S.	4.1	5.1	III	76 274	0	E1	P099	–	–	–
3098	OXIDIZING LIQUID, CORROSIVE, N.O.S.	5.1	8	I	274	0	E0	P502	–	–	–
3098	OXIDIZING LIQUID, CORROSIVE, N.O.S.	5.1	8	II	274	1 L	E2	P504	–	IBC01	–
3098	OXIDIZING LIQUID, CORROSIVE, N.O.S.	5.1	8	III	223 274	5 L	E1	P504	–	IBC02	–
3099	OXIDIZING LIQUID, TOXIC, N.O.S.	5.1	6.1	I	274	0	E0	P502	–	–	–

Portable tanks and bulk containers			EmS	Stowage and segregation	Properties and observations	UN No.
	Tank instructions	Provisions				
(12)	(13) 4.2.5 4.3	(14) 4.2.5	(15) 5.4.3.2 7.3	(16) 7.1 7.2	(17)	(18)
–	T3	TP33	F-G, S-G	Category B. "Separated from" class 5.1.	–	3089
–	T1	TP33	F-G, S-G	Category B. "Separated from" class 5.1.	–	3089
–	–	–	F-A, S-I	Category A.	Electrical batteries containing lithium encased in a rigid metallic body. Lithium batteries may also be shipped in, or packed with, equipment. Electrical lithium batteries may cause fire due to an explosive rupture of the body caused by improper construction or reaction with contaminants.	3090
–	–	–	F-A, S-I	Category A.	See above.	3091 △
–	T2	TP1	F-E, S-D	Category A.	Colourless liquid. Flashpoint: 29°C to 35°C c.c. Explosive limits: 1.7% to 11.5%. Miscible with water. Reacts with strong oxidizing substances. Irritating to skin, eyes and mucous membranes.	3092
–	–	–	F-A, S-Q	Category C.	Causes burns to skin, eyes and mucous membranes.	3093
–	–	–	F-A, S-Q	Category C.	See entry above.	3093
–	–	–	F-G, S-L	Category D.	Causes burns to skin, eyes and mucous membranes.	3094
–	–	–	F-G, S-L	Category D.	See entry above.	3094
–	T6	TP33	F-A, S-N	Category D.	Causes burns to skin, eyes and mucous membranes.	3095
–	T3	TP33	F-A, S-N	Category D.	See entry above.	3095
–	T6	TP33	F-G, S-L	Category D.	Causes burns to skin, eyes and mucous membranes.	3096
–	T3	TP33	F-G, S-L	Category D.	See entry above.	3096
–	–	–	F-A, S-Q	–	–	3097
–	T1	TP33	F-A, S-Q	–	–	3097
–	–	–	F-A, S-Q	Category D. Keep as dry as reasonably practicable. "Separated from" ammonium compounds, cyanides and peroxides.	Causes burns to skin, eyes and mucous membranes. Particular care in handling should be exercised if packages have become wetted.	3098
–	–	–	F-A, S-Q	Category B. Keep as dry as reasonably practicable. "Separated from" ammonium compounds, cyanides and peroxides.	See entry above.	3098
–	–	–	F-A, S-Q	Category B. Keep as dry as reasonably practicable. "Separated from" ammonium compounds, cyanides and peroxides.	See entry above.	3098
–	–	–	F-A, S-Q	Category D. "Separated from" ammonium compounds, cyanides and peroxides.	Toxic if swallowed, by skin contact or by dust inhalation. Should be handled with care to minimize exposure, particularly to dust.	3099

UN No.	Proper Shipping Name (PSN)	Class or division	Subsidiary risk(s)	Packing group	Special provisions	Limited and excepted quantity provisions		Packing		IBC	
						Limited quantities	Excepted quantities	Instructions	Provisions	Instructions	Provisions
(1)	(2) 3.1.2	(3) 2.0	(4) 2.0	(5) 2.0.1.3	(6) 3.3	(7a) 3.4	(7b) 3.5	(8) 4.1.4	(9) 4.1.4	(10) 4.1.4	(11) 4.1.4
3099	OXIDIZING LIQUID, TOXIC, N.O.S.	5.1	6.1	II	274	1 L	E2	P504	–	IBC01	–
3099	OXIDIZING LIQUID, TOXIC, N.O.S.	5.1	6.1	III	223 274	5 L	E1	P504	–	IBC02	–
3100	OXIDIZING SOLID, SELF-HEATING, N.O.S.	5.1	4.2	I	76 274	0	E0	P099	–	–	–
3100	OXIDIZING SOLID, SELF-HEATING, N.O.S.	5.1	4.2	II	76 274	0	E2	P099	–	–	–
△ 3101	ORGANIC PEROXIDE TYPE B, LIQUID	5.2	See SP181	–	122 181 195 274	25 mL	E0	P520	–	–	–
△ 3102	ORGANIC PEROXIDE TYPE B, SOLID	5.2	See SP181	–	122 181 195 274	100 g	E0	P520	–	–	–
△ 3103	ORGANIC PEROXIDE TYPE C, LIQUID	5.2	–	–	122 195 274	25 mL	E0	P520	–	–	–
△ 3104	ORGANIC PEROXIDE TYPE C, SOLID	5.2	–	–	122 195 274	100 g	E0	P520	–	–	–
△ 3105	ORGANIC PEROXIDE TYPE D, LIQUID	5.2	–	–	122 274	125 mL	E0	P520	–	–	–
△ 3106	ORGANIC PEROXIDE TYPE D, SOLID	5.2	–	–	122 274	500 g	E0	P520	–	–	–
△ 3107	ORGANIC PEROXIDE TYPE E, LIQUID	5.2	–	–	122 274	125 mL	E0	P520	–	–	–
△ 3108	ORGANIC PEROXIDE TYPE E, SOLID	5.2	–	–	122 274	500 g	E0	P520	–	–	–
△ 3109	ORGANIC PEROXIDE TYPE F, LIQUID	5.2	–	–	122 274	125 mL	E0	P520	–	IBC520	–
△ 3110	ORGANIC PEROXIDE TYPE F, SOLID	5.2	–	–	122 274	500 g	E0	P520	–	IBC520	–

Portable tanks and bulk containers			EmS	Stowage and segregation	Properties and observations	UN No.	
	Tank instructions	Provisions					
(12)	(13) 4.2.5 4.3	(14) 4.2.5	(15) 5.4.3.2 7.3	(16) 7.1 7.2	(17)	(18)	
–	–	–	F-A, S-Q	Category B. "Separated from" ammonium compounds, cyanides and peroxides.	Toxic if swallowed, by skin contact or by dust inhalation. Should be handled with care to minimize exposure, particularly to dust.	**3099**	
–	–	–	F-A, S-Q	Category B. "Separated from" ammonium compounds, cyanides and peroxides.	See entry above.	**3099**	
–	–	–	F-A, S-Q	–	–	**3100**	
–	–	–	F-A, S-Q	–	–	**3100**	
–	–	–	F-J, S-R	Category D. Protected from sources of heat. For packages carrying a subsidiary risk label of class 1, segregation as for class 1, division 1.3. "Separated from" acids and alkalis.	May explode at elevated temperatures or in a fire. Burns vigorously. Immiscible with water. Contact with the eyes and skin should be avoided. May evolve irritant or toxic fumes.	**3101**	△
–	–	–	F-J, S-R	Category D. Protected from sources of heat. For packages carrying a subsidiary risk label of class 1, segregation as for class 1, division 1.3. "Separated from" acids and alkalis.	May explode at elevated temperatures or in a fire. Burns vigorously. Insoluble in water. Contact with the eyes and skin should be avoided. Addition of water to disuccinic acid peroxide will decrease its thermal stability. May evolve irritant or toxic fumes.	**3102**	△
–	–	–	F-J, S-R	Category D. Protected from sources of heat. "Separated from" acids and alkalis.	May decompose violently at elevated temperatures or in a fire. Burns vigorously. Immiscible with water except for *tert*-butyl hydroperoxide. Contact with the eyes and skin should be avoided. May evolve irritant or toxic fumes.	**3103**	△
–	–	–	F-J, S-R	Category D. Protected from sources of heat. "Separated from" acids and alkalis.	May decompose violently at elevated temperatures or in a fire. Burns vigorously. Insoluble in water. Contact with the eyes and skin should be avoided. May evolve irritant or toxic fumes.	**3104**	△
–	–	–	F-J, S-R	Category D. Protected from sources of heat. "Separated from" acids and alkalis. See 7.2.6.3.2.	Decomposes at elevated temperatures or in a fire. Burns vigorously. Immiscible with water except for acetylacetone peroxide, *tert*-butyl hydroperoxide and peroxyacetic acid, type D, stabilized. Contact with the eyes and skin should be avoided. May evolve irritant or toxic fumes.	**3105**	△
–	–	–	F-J, S-R	Category D. Protected from sources of heat. "Separated from" acids and alkalis.	Decomposes at elevated temperatures or in a fire. Burns vigorously. Insoluble in water except for 3-chloroperoxybenzoic acid. Contact with the eyes and skin should be avoided. May evolve irritant or toxic fumes.	**3106**	△
–	–	–	F-J, S-R	Category D. Protected from sources of heat. "Separated from" acids and alkalis. See 7.2.6.3.2.	Decomposes at elevated temperatures or in a fire. Burns vigorously. Immiscible with water except for *tert*-amyl hydroperoxide, and *tert*-butyl hydroperoxide and peroxyacetic acid, type E, stabilized. Contact with the eyes and skin should be avoided. May evolve irritant or toxic fumes.	**3107**	△
–	–	–	F-J, S-R	Category D. Protected from sources of heat. "Separated from" acids and alkalis.	Decomposes at elevated temperatures or in a fire. Burns vigorously. Insoluble in water. Contact with the eyes and skin should be avoided. May evolve irritant or toxic fumes.	**3108**	△
–	T23	–	F-J, S-R	Category D. Protected from sources of heat. "Separated from" acids and alkalis. See 7.2.6.3.2.	Decomposes at elevated temperatures or in a fire. Burns vigorously. Immiscible with water except for *tert*-butyl hydroperoxide; dibenzoyl peroxide; dilauroyl peroxide and peroxyacetic acid, type F, stabilized. Contact with the eyes and skin should be avoided. May evolve irritant or toxic fumes.	**3109**	△
–	T23	TP33	F-J, S-R	Category D. Protected from sources of heat. "Separated from" acids and alkalis.	Decomposes at elevated temperatures or in a fire. Burns vigorously. Insoluble in water. Contact with the eyes and skin should be avoided. May evolve irritant or toxic fumes.	**3110**	△

UN No.	Proper Shipping Name (PSN)	Class or division	Subsidiary risk(s)	Packing group	Special provisions	Limited and excepted quantity provisions		Packing		IBC	
						Limited quantities	Excepted quantities	Instructions	Provisions	Instructions	Provisions
(1)	(2) 3.1.2	(3) 2.0	(4) 2.0	(5) 2.0.1.3	(6) 3.3	(7a) 3.4	(7b) 3.5	(8) 4.1.4	(9) 4.1.4	(10) 4.1.4	(11) 4.1.4
△ 3111	ORGANIC PEROXIDE TYPE B, LIQUID, TEMPERATURE CONTROLLED	5.2	See SP181	–	122 181 195 274 923	0	E0	P520	–	–	–
△ 3112	ORGANIC PEROXIDE TYPE B, SOLID, TEMPERATURE CONTROLLED	5.2	See SP181	–	122 181 195 274 923	0	E0	P520	–	–	–
△ 3113	ORGANIC PEROXIDE TYPE C, LIQUID, TEMPERATURE CONTROLLED	5.2	–	–	122 195 274 923	0	E0	P520	–	–	–
△ 3114	ORGANIC PEROXIDE TYPE C, SOLID, TEMPERATURE CONTROLLED	5.2	–	–	122 195 274 923	0	E0	P520	–	–	–
△ 3115	ORGANIC PEROXIDE TYPE D, LIQUID, TEMPERATURE CONTROLLED	5.2	–	–	122 274 923	0	E0	P520	–	–	–
△ 3116	ORGANIC PEROXIDE TYPE D, SOLID, TEMPERATURE CONTROLLED	5.2	–	–	122 274 923	0	E0	P520	–	–	–
△ 3117	ORGANIC PEROXIDE TYPE E, LIQUID, TEMPERATURE CONTROLLED	5.2	–	–	122 274 923	0	E0	P520	–	–	–
△ 3118	ORGANIC PEROXIDE TYPE E, SOLID, TEMPERATURE CONTROLLED	5.2	–	–	122 274 923	0	E0	P520	–	–	–
△ 3119	ORGANIC PEROXIDE TYPE F, LIQUID, TEMPERATURE CONTROLLED	5.2	–	–	122 274 923	0	E0	P520	–	IBC520	–

Portable tanks and bulk containers			EmS	Stowage and segregation	Properties and observations	UN No.	
	Tank instructions	Provisions					
(12)	(13) 4.2.5 4.3	(14) 4.2.5	(15) 5.4.3.2 7.3	(16) 7.1 7.2	(17)	(18)	
–	–	–	F-F, S-R	Category D. Protected from sources of heat. For packages carrying a subsidiary risk label of class 1, segregation as for class 1, division 1.3. Shall be transported under temperature control. "Separated from" acids and alkalis.	May explode at temperatures higher than the emergency temperature or in a fire. Burns vigorously. Immiscible with water. Contact with the eyes and skin should be avoided. Control and emergency temperatures for each formulation are given in table 2.5.3.2.4. The temperature should be checked regularly. May evolve irritant or toxic fumes.	3111	△
–	–	–	F-F, S-R	Category D. Protected from sources of heat. For packages carrying a subsidiary risk label of class 1, segregation as for class 1, division 1.3. Shall be transported under temperature control. "Separated from" acids and alkalis.	May explode at temperatures higher than the emergency temperature or in a fire. Burns vigorously. Immiscible with water. Contact with the eyes and skin should be avoided. Control and emergency temperatures for each formulation are given in table 2.5.3.2.4. The temperature should be checked regularly. May evolve irritant or toxic fumes.	3112	△
–	–	–	F-F, S-R	Category D. Protected from sources of heat. Shall be transported under temperature control. "Separated from" acids and alkalis.	May decompose violently at temperatures higher than the emergency temperature or in a fire. Burns vigorously. Immiscible with water. Contact with the eyes and skin should be avoided. Control and emergency temperatures for each formulation are given in table 2.5.3.2.4. The temperature should be checked regularly. May evolve irritant or toxic fumes.	3113	△
–	–	–	F-F, S-R	Category D. Protected from sources of heat. Shall be transported under temperature control. "Separated from" acids and alkalis.	May decompose violently at temperatures higher than the emergency temperature or in a fire. Burns vigorously. Insoluble in water. Contact with the eyes and skin should be avoided. Control and emergency temperatures for each formulation are given in table 2.5.3.2.4. The temperature should be checked regularly. May evolve irritant or toxic fumes.	3114	△
–	–	–	F-F, S-R	Category D. Protected from sources of heat. Shall be transported under temperature control. "Separated from" acids and alkalis.	Decomposes at temperatures higher than the emergency temperature or in a fire. Burns vigorously. Immiscible with water. Contact with the eyes and skin should be avoided. Control and emergency temperatures for each formulation are given in table 2.5.3.2.4. The temperature should be checked regularly. May evolve irritant or toxic fumes.	3115	△
–	–	–	F-F, S-R	Category D. Protected from sources of heat. Shall be transported under temperature control. "Separated from" acids and alkalis.	Decomposes at temperatures higher than the emergency temperature or in a fire. Burns vigorously. Insoluble in water except for diperoxyazelaic acid. Contact with the eyes and skin should be avoided. Control and emergency temperatures for each formulation are given in table 2.5.3.2.4. The temperature should be checked regularly. May evolve irritant or toxic fumes.	3116	△
–	–	–	F-F, S-R	Category D. Protected from sources of heat. Shall be transported under temperature control. "Separated from" acids and alkalis.	Decomposes at temperatures higher than the emergency temperature or in a fire. Burns vigorously. Immiscible with water. Contact with the eyes and skin should be avoided. Control and emergency temperatures for each formulation are given in table 2.5.3.2.4. The temperature should be checked regularly. May evolve irritant or toxic fumes.	3117	△
–	–	–	F-F, S-R	Category D. Protected from sources of heat. Shall be transported under temperature control. "Separated from" acids and alkalis.	Decomposes at temperatures higher than the emergency temperature or in a fire. Burns vigorously. Insoluble in water except for di-(2-ethylhexyl) peroxydicarbonate. Contact with the eyes and skin should be avoided. Control and emergency temperatures for each formulation are given in table 2.5.3.2.4. The temperature should be checked regularly. May evolve irritant or toxic fumes.	3118	△
–	T23	–	F-F, S-R	Category D. Protected from sources of heat. Shall be transported under temperature control. "Separated from" acids and alkalis.	Decomposes at temperatures higher than the emergency temperature or in a fire. Burns vigorously. Immiscible with water except for di-(4-*tert*-butylcyclohexyl) peroxydicarbonate, dicetyl peroxydicarbonate and dimyristyl peroxydicarbonate. Contact with the eyes and skin should be avoided. Control and emergency temperatures for each formulation are given in table 2.5.3.2.4. The temperature should be checked regularly. May evolve irritant or toxic fumes.	3119	△

UN No.	Proper Shipping Name (PSN)	Class or division	Subsidiary risk(s)	Packing group	Special provisions	Limited and excepted quantity provisions		Packing		IBC	
						Limited quantities	Excepted quantities	Instructions	Provisions	Instructions	Provisions
(1)	(2) 3.1.2	(3) 2.0	(4) 2.0	(5) 2.0.1.3	(6) 3.3	(7a) 3.4	(7b) 3.5	(8) 4.1.4	(9) 4.1.4	(10) 4.1.4	(11) 4.1.4
△ 3120	ORGANIC PEROXIDE TYPE F, SOLID, TEMPERATURE CONTROLLED	5.2	–	–	122 274 923	0	E0	P520	–	IBC520	–
3121	OXIDIZING SOLID, WATER-REACTIVE, N.O.S.	5.1	4.3	I	76 274	0	E0	P099	–	–	–
3121	OXIDIZING SOLID, WATER-REACTIVE, N.O.S.	5.1	4.3	II	76 274	0	E2	P099	–	–	–
3122	TOXIC LIQUID, OXIDIZING, N.O.S.	6.1	5.1	I	274 315	0	E5	P001	–	–	–
3122	TOXIC LIQUID, OXIDIZING, N.O.S.	6.1	5.1	II	274	100 mL	E4	P001	–	IBC02	–
3123	TOXIC LIQUID, WATER-REACTIVE, N.O.S.	6.1	4.3	I	274 315	0	E5	P099	–	–	–
3123	TOXIC LIQUID, WATER-REACTIVE, N.O.S.	6.1	4.3	II	274	100 mL	E4	P001	–	IBC02	–
3124	TOXIC SOLID, SELF-HEATING, N.O.S.	6.1	4.2	I	274	0	E5	P002	–	–	–
3124	TOXIC SOLID, SELF-HEATING, N.O.S.	6.1	4.2	II	274	0	E4	P002	–	IBC06	B2
3125	TOXIC SOLID, WATER-REACTIVE, N.O.S.	6.1	4.3	I	274	0	E5	P099	–	–	–
3125	TOXIC SOLID, WATER-REACTIVE, N.O.S.	6.1	4.3	II	274	500 g	E4	P002	–	IBC06	B2
3126	SELF-HEATING SOLID, CORROSIVE, ORGANIC, N.O.S.	4.2	8	II	76 274	0	E2	P410	–	IBC05	B2
3126	SELF-HEATING SOLID, CORROSIVE, ORGANIC, N.O.S.	4.2	8	III	76 223 274	0	E1	P002	–	IBC08	B3
3127	SELF-HEATING SOLID, OXIDIZING, N.O.S.	4.2	5.1	II	76 274	0	E2	P099	–	–	–
3127	SELF-HEATING SOLID, OXIDIZING, N.O.S.	4.2	5.1	III	76 223 274	0	E1	P099	–	–	–
3128	SELF-HEATING SOLID, TOXIC, ORGANIC, N.O.S.	4.2	6.1	II	76 274	0	E2	P410	–	IBC05	B2
3128	SELF-HEATING SOLID, TOXIC, ORGANIC, N.O.S.	4.2	6.1	III	76 223 274	0	E1	P002	–	IBC08	B3
△ 3129	WATER-REACTIVE LIQUID, CORROSIVE, N.O.S.	4.3	8	I	76 274	0	E0	P402	–	–	–
△ 3129	WATER-REACTIVE LIQUID, CORROSIVE, N.O.S.	4.3	8	II	76 274	0	E2	P402	–	IBC01	–
△ 3129	WATER-REACTIVE LIQUID, CORROSIVE, N.O.S.	4.3	8	III	76 223 274	0	E1	P001	–	IBC02	–
3130	WATER-REACTIVE LIQUID, TOXIC, N.O.S.	4.3	6.1	I	76 274	0	E0	P402	–	–	–
3130	WATER-REACTIVE LIQUID, TOXIC, N.O.S.	4.3	6.1	II	76 274	0	E2	P402	–	IBC01	–

Portable tanks and bulk containers			EmS	Stowage and segregation	Properties and observations	UN No.
	Tank instructions	Provisions				
(12)	(13) 4.2.5 4.3	(14) 4.2.5	(15) 5.4.3.2 7.3	(16) 7.1 7.2	(17)	(18)
–	T23	TP33	F-F, S-R	Category D. Protected from sources of heat. Shall be transported under temperature control. "Separated from" acids and alkalis.	Decomposes at temperatures higher than the emergency temperature or in a fire. Burns vigorously. Insoluble in water. Contact with the eyes and skin should be avoided. Control and emergency temperatures for each formulation are given in the table 2.5.3.2.4. The temperature should be checked regularly. May evolve irritant or toxic fumes.	3120 △
–	–	–	F-G, S-L	–	–	3121
–	–	–	F-G, S-L	–	–	3121
–	–	–	F-A, S-Q	Category C.	Toxic if swallowed, by skin contact or by inhalation.	3122
–	–	–	F-A, S-Q	Category C.	See entry above.	3122
–	–	–	F-G, S-N	Category D. Clear of living quarters.	Toxic if swallowed, by skin contact or by inhalation.	3123
–	–	–	F-G, S-N	Category D. Clear of living quarters.	See entry above.	3123
–	T6	TP33	F-A, S-J	Category D. Clear of living quarters.	Highly toxic if swallowed, by skin contact or by inhalation.	3124
–	T3	TP33	F-A, S-J	Category D. Clear of living quarters.	See entry above.	3124
–	T6	TP33	F-G, S-N	Category D. Clear of living quarters.	Toxic if swallowed, by skin contact or by inhalation.	3125
–	T3	TP33	F-G, S-N	Category D. Clear of living quarters.	See entry above.	3125
–	T3	TP33	F-A, S-J	Category C.	–	3126
–	T1	TP33	F-A, S-J	Category C.	–	3126
–	T3	TP33	F-A, S-J	–	–	3127
–	T1	TP33	F-A, S-J	–	–	3127
–	T3	TP33	F-A, S-J	Category C.	–	3128
–	T1	TP33	F-A, S-J	Category C.	–	3128
–	T14	TP2 TP7 TP13	F-G, S-N	Category D.	–	3129 △
–	T11	TP2 TP7	F-G, S-N	Category E. If under deck, in a mechanically ventilated space.	–	3129 △
–	T7	TP2 TP7	F-G, S-N	Category E.	–	3129 △
–	–	–	F-G, S-N	Category D.	–	3130
–	–	–	F-G, S-N	Category E. If under deck, in a mechanically ventilated space.	–	3130

3

DGL

UN No.	Proper Shipping Name (PSN)	Class or division	Subsidiary risk(s)	Packing group	Special provisions	Limited and excepted quantity provisions		Packing		IBC	
						Limited quantities	Excepted quantities	Instructions	Provisions	Instructions	Provisions
(1)	(2) 3.1.2	(3) 2.0	(4) 2.0	(5) 2.0.1.3	(6) 3.3	(7a) 3.4	(7b) 3.5	(8) 4.1.4	(9) 4.1.4	(10) 4.1.4	(11) 4.1.4
3130	WATER-REACTIVE LIQUID, TOXIC, N.O.S.	4.3	6.1	III	76 223 274	0	E1	P001	–	IBC02	–
3131	WATER-REACTIVE SOLID, CORROSIVE, N.O.S.	4.3	8	I	76 274	0	E0	P403	–	–	–
3131	WATER-REACTIVE SOLID, CORROSIVE, N.O.S.	4.3	8	II	76 274	0	E2	P410	PP40	IBC06	B2
3131	WATER-REACTIVE SOLID, CORROSIVE, N.O.S.	4.3	8	III	76 223 274	0	E1	P410	–	IBC08	B4
3132	WATER-REACTIVE SOLID, FLAMMABLE, N.O.S.	4.3	4.1	I	76 274	0	E0	P403	–	IBC99	–
3132	WATER-REACTIVE SOLID, FLAMMABLE, N.O.S.	4.3	4.1	II	76 274	0	E2	P410	PP40	IBC04	–
3132	WATER-REACTIVE SOLID, FLAMMABLE, N.O.S.	4.3	4.1	III	76 223 274	0	E1	P410	PP40	IBC06	–
3133	WATER-REACTIVE SOLID, OXIDIZING, N.O.S.	4.3	5.1	II	76 274	0	E2	P099	–	–	–
3133	WATER-REACTIVE SOLID, OXIDIZING, N.O.S.	4.3	5.1	III	76 223 274	0	E1	P099	–	–	–
3134	WATER-REACTIVE SOLID, TOXIC, N.O.S.	4.3	6.1	I	274	0	E0	P403	–	–	–
3134	WATER-REACTIVE SOLID, TOXIC, N.O.S.	4.3	6.1	II	274	500 g	E2	P410	PP40	IBC05	B2
3134	WATER-REACTIVE SOLID, TOXIC, N.O.S.	4.3	6.1	III	223 274	1 kg	E1	P410	–	IBC08	B4
3135	WATER-REACTIVE SOLID, SELF-HEATING, N.O.S.	4.3	4.2	I	76 274	0	E0	P403	–	–	–
3135	WATER-REACTIVE SOLID, SELF-HEATING, N.O.S.	4.3	4.2	II	76 274	0	E2	P410	–	IBC05	B2
3135	WATER-REACTIVE SOLID, SELF-HEATING, N.O.S.	4.3	4.2	III	76 223 274	0	E1	P410	–	IBC08	B4
3136	TRIFLUOROMETHANE, REFRIGERATED LIQUID	2.2	–	–	–	120 mL	E1	P203	–	–	–
3137	OXIDIZING SOLID, FLAMMABLE, N.O.S.	5.1	4.1	I	76 274	0	E0	P099	–	–	–
3138	ETHYLENE, ACETYLENE AND PROPYLENE MIXTURE, REFRIGERATED LIQUID containing at least 71.5% ethylene, with not more than 22.5% acetylene and not more than 6% propylene	2.1	–	–	–	0	E0	P203	–	–	–
3139	OXIDIZING LIQUID, N.O.S.	5.1	–	I	274	0	E0	P502	–	–	–
3139	OXIDIZING LIQUID, N.O.S.	5.1	–	II	274	1 L	E2	P504	–	IBC02	–
3139	OXIDIZING LIQUID, N.O.S.	5.1	–	III	223 274	5 L	E1	P504	–	IBC02	–

Portable tanks and bulk containers			EmS	Stowage and segregation	Properties and observations	UN No.
	Tank instructions	Provisions				
(12)	(13) 4.2.5 4.3	(14) 4.2.5	(15) 5.4.3.2 7.3	(16) 7.1 7.2	(17)	(18)
–	–	–	F-G, S-N	Category E. If under deck, in a mechanically ventilated space.	–	3130
–	T9	TP7 TP33	F-G, S-L	Category D.	–	3131
–	T3	TP33	F-G, S-L	Category E. If under deck, in a mechanically ventilated space.	–	3131
–	T1	TP33	F-G, S-L	Category E. If under deck, in a mechanically ventilated space.	–	3131
–	–	–	F-G, S-N	–	–	3132
–	T3	TP33	F-G, S-N	–	–	3132
–	T1	TP33	F-G, S-N	–	–	3132
–	–	–	F-G, S-L	–	–	3133
–	–	–	F-G, S-L	–	–	3133
–	–	–	F-G, S-N	Category D.	–	3134
–	T3	TP33	F-G, S-N	Category E. If under deck, in a mechanically ventilated space.	–	3134
–	T1	TP33	F-G, S-N	Category E. If under deck, in a mechanically ventilated space.	–	3134
–	–	–	F-G, S-N	–	–	3135
–	T3	TP33	F-G, S-N	–	–	3135
–	T1	TP33	F-G, S-N	–	–	3135
–	T75	TP5	F-C, S-V	Category D.	Liquefied, non-flammable gas. Much heavier than air (2.4).	3136
–	–	–	F-G, S-Q	–	–	3137
–	T75	TP5	F-D, S-U	Category D. Clear of living quarters. "Separated from" chlorine.	Liquefied, flammable, colourless mixture of gases with a garlic odour. Explosive limits: 2.7% to 36%. Lighter than air (0.96).	3138
–	–	–	F-A, S-Q	Category D. "Separated from" ammonium compounds, cyanides and peroxides.	–	3139
–	–	–	F-A, S-Q	Category B. "Separated from" ammonium compounds, cyanides and peroxides.	–	3139
–	–	–	F-A, S-Q	Category B. "Separated from" ammonium compounds, cyanides and peroxides.	–	3139

UN No.	Proper Shipping Name (PSN)	Class or division	Subsidiary risk(s)	Packing group	Special provisions	Limited and excepted quantity provisions		Packing		IBC	
						Limited quantities	Excepted quantities	Instructions	Provisions	Instructions	Provisions
(1)	(2) 3.1.2	(3) 2.0	(4) 2.0	(5) 2.0.1.3	(6) 3.3	(7a) 3.4	(7b) 3.5	(8) 4.1.4	(9) 4.1.4	(10) 4.1.4	(11) 4.1.4
3140	ALKALOIDS, LIQUID, N.O.S. or ALKALOIDS SALTS, LIQUID, N.O.S.	6.1	–	I	43 274	0	E5	P001	–	–	–
3140	ALKALOIDS, LIQUID, N.O.S. or ALKALOIDS SALTS, LIQUID, N.O.S.	6.1	–	II	43 274	100 mL	E4	P001	–	IBC02	–
3140	ALKALOIDS, LIQUID, N.O.S. or ALKALOIDS SALTS, LIQUID, N.O.S.	6.1	–	II	43 223 274	5 L	E1	P001 LP01	–	IBC03	–
3141	ANTIMONY COMPOUND, INORGANIC, LIQUID, N.O.S.	6.1	–	III	45 274	5 L	E1	P001 LP01	–	IBC03	–
3142	DISINFECTANT, LIQUID, TOXIC, N.O.S.	6.1	–	I	274	0	E5	P001	–	–	–
3142	DISINFECTANT, LIQUID, TOXIC, N.O.S.	6.1	–	II	274	100 mL	E4	P001	–	IBC02	–
3142	DISINFECTANT, LIQUID, TOXIC, N.O.S.	6.1	–	III	223 274	5 L	E1	P001 LP01	–	IBC03	–
3143	DYE, SOLID, TOXIC, N.O.S. or DYE INTERMEDIATE, SOLID, TOXIC, N.O.S.	6.1	–	I	274	0	E5	P002	–	IBC07	B1
3143	DYE, SOLID, TOXIC, N.O.S. or DYE INTERMEDIATE, SOLID, TOXIC, N.O.S.	6.1	–	II	274	500 g	E4	P002	–	IBC08	B2 B4
3143	DYE, SOLID, TOXIC, N.O.S. or DYE INTERMEDIATE, SOLID, TOXIC, N.O.S.	6.1	–	III	223 274	5 kg	E1	P002 LP02	–	IBC08	B3
3144	NICOTINE COMPOUND, LIQUID, N.O.S. or NICOTINE PREPARATION, LIQUID, N.O.S.	6.1	–	I	43 274	0	E5	P001	–	–	–
3144	NICOTINE COMPOUND, LIQUID, N.O.S. or NICOTINE PREPARATION, LIQUID, N.O.S.	6.1	–	II	43 274	100 mL	E4	P001	–	IBC02	–
3144	NICOTINE COMPOUND, LIQUID, N.O.S. or NICOTINE PREPARATION, LIQUID, N.O.S.	6.1	–	III	43 223 274	5 L	E1	P001 LP01	–	IBC03	–
3145	ALKYLPHENOLS, LIQUID, N.O.S. (including C_2–C_{12} homologues)	8	–	I	–	0	E0	P001	–	–	–
3145	ALKYLPHENOLS, LIQUID, N.O.S. (including C_2–C_{12} homologues)	8	–	II	–	1 L	E2	P001	–	IBC02	–
3145	ALKYLPHENOLS, LIQUID, N.O.S. (including C_2–C_{12} homologues)	8	–	III	223	5 L	E1	P001 LP01	–	IBC03	–
3146	ORGANOTIN COMPOUND, SOLID, N.O.S.	6.1	– P	I	43 274	0	E5	P002	–	IBC07	B1
3146	ORGANOTIN COMPOUND, SOLID, N.O.S.	6.1	– P	II	43 274	500 g	E4	P002	–	IBC08	B2 B4
3146	ORGANOTIN COMPOUND, SOLID, N.O.S.	6.1	– P	III	43 223 274	5 kg	E1	P002 LP02	–	IBC08	B3
3147	DYE, SOLID, CORROSIVE, N.O.S. or DYE INTERMEDIATE, SOLID, CORROSIVE, N.O.S.	8	–	I	274	0	E0	P002	–	IBC07	B1
3147	DYE, SOLID, CORROSIVE, N.O.S. or DYE INTERMEDIATE, SOLID, CORROSIVE, N.O.S.	8	–	II	274	1 kg	E2	P002	–	IBC08	B2 B4

Portable tanks and bulk containers			EmS	Stowage and segregation	Properties and observations	UN No.
	Tank instructions	Provisions				
(12)	(13) 4.2.5 4.3	(14) 4.2.5	(15) 5.4.3.2 7.3	(16) 7.1 7.2	(17)	(18)
–	–	–	F-A, S-A	Category A.	A wide range of toxic liquids, generally of vegetable origin. Toxic if swallowed, by skin contact or by inhalation.	3140
–	–	–	F-A, S-A	Category A.	See entry above.	3140
–	–	–	F-A, S-A	Category A.	See entry above.	3140
–	–	–	F-A, S-A	Category A.	A wide range of toxic liquids. Toxic if swallowed, by skin contact or by inhalation.	3141
–	–	–	F-A, S-A	Category A. Clear of living quarters.	A wide range of toxic liquids. Toxic if swallowed, by skin contact or by inhalation.	3142
–	–	–	F-A, S-A	Category A. Clear of living quarters.	See entry above.	3142
–	–	–	F-A, S-A	Category A. Clear of living quarters.	See entry above.	3142
–	T6	TP33	F-A, S-A	Category A.	A wide range of toxic solids. Toxic if swallowed, by skin contact or by inhalation.	3143
–	T3	TP33	F-A, S-A	Category A.	See entry above.	3143
–	T1	TP33	F-A, S-A	Category A.	See entry above.	3143
–	–	–	F-A, S-A	Category B. Clear of living quarters.	A wide variety of toxic liquids. Toxic if swallowed, by skin contact or by inhalation.	3144
–	–	–	F-A, S-A	Category B. Clear of living quarters.	See entry above.	3144
–	–	–	F-A, S-A	Category B. Clear of living quarters.	See entry above.	3144
–	T14	TP2	F-A, S-B	Category B.	A wide range of colourless to pale straw-coloured liquids with penetrating odours (sometimes camphor-like). Liquids slightly miscible with water. Cause burns to skin, eyes and mucous membranes.	3145
–	T11	TP2 TP27	F-A, S-B	Category B.	See entry above.	3145
–	T7	TP1 TP28	F-A, S-B	Category A.	See entry above.	3145
–	T6	TP33	F-A, S-A	Category B. Clear of living quarters.	A wide variety of toxic solids. Toxic if swallowed, by skin contact or by inhalation.	3146
–	T3	TP33	F-A, S-A	Category A. Clear of living quarters.	See entry above.	3146
–	T1	TP33	F-A, S-A	Category A. Clear of living quarters.	See entry above.	3146
–	T6	TP33	F-A, S-B	Category A.	A wide range of corrosive solids or pastes. Cause burns to skin, eyes and mucous membranes.	3147
–	T3	TP33	F-A, S-B	Category A.	See entry above.	3147

3

DGL

UN No. (1)	Proper Shipping Name (PSN) (2) 3.1.2	Class or division (3) 2.0	Subsidiary risk(s) (4) 2.0	Packing group (5) 2.0.1.3	Special provisions (6) 3.3	Limited and excepted quantity provisions: Limited quantities (7a) 3.4	Limited and excepted quantity provisions: Excepted quantities (7b) 3.5	Packing: Instructions (8) 4.1.4	Packing: Provisions (9) 4.1.4	IBC: Instructions (10) 4.1.4	IBC: Provisions (11) 4.1.4
3147	DYE, SOLID, CORROSIVE, N.O.S. or DYE INTERMEDIATE, SOLID, CORROSIVE, N.O.S.	8	–	III	223 274	5 kg	E1	P002 LP02	–	IBC08	B3
△ 3148	WATER-REACTIVE LIQUID, N.O.S.	4.3	–	I	274	0	E0	P402	PP31	–	–
△ 3148	WATER-REACTIVE LIQUID, N.O.S.	4.3	–	II	274	500 mL	E2	P402	PP31	IBC01	–
△ 3148	WATER-REACTIVE LIQUID, N.O.S.	4.3	–	III	223 274	1 L	E1	P001	PP31	IBC02	–
△ 3149	HYDROGEN PEROXIDE AND PEROXYACETIC ACID MIXTURE with acid(s), water and not more than 5% peroxyacetic acid, STABILIZED	5.1	8	II	196	1 L	E2	P504	PP10	IBC02	B5
3150	DEVICES, SMALL, HYDROCARBON GAS POWERED or HYDROCARBON GAS REFILLS FOR SMALL DEVICES with release device	2.1	–	–	–	0	E0	P003	–	–	–
△ 3151	POLYHALOGENATED BIPHENYLS, LIQUID or POLYHALOGENATED TERPHENYLS, LIQUID	9	– P	II	203 305	1 L	E2	P906	–	IBC02	–
△ 3152	POLYHALOGENATED BIPHENYLS, SOLID or POLYHALOGENATED TERPHENYLS, SOLID	9	– P	II	203 305 958	1 kg	E2	P906	–	IBC08	B2 B4
3153	PERFLUORO(METHYL VINYL ETHER)	2.1	–	–	–	0	E0	P200	–	–	–
3154	PERFLUORO(ETHYL VINYL ETHER)	2.1	–	–	–	0	E0	P200	–	–	–
3155	PENTACHLOROPHENOL	6.1	– P	II	43	500 g	E4	P002	–	IBC08	B2 B4
3156	COMPRESSED GAS, OXIDIZING, N.O.S.	2.2	5.1	–	274	0	E0	P200	–	–	–
3157	LIQUEFIED GAS, OXIDIZING, N.O.S.	2.2	5.1	–	274	0	E0	P200	–	–	–
3158	GAS, REFRIGERATED LIQUID, N.O.S.	2.2	–	–	274	120 mL	E1	P203	–	–	–
3159	1,1,1,2-TETRAFLUOROETHANE (REFRIGERANT GAS R 134a)	2.2	–	–	–	120 mL	E1	P200	–	–	–
3160	LIQUEFIED GAS, TOXIC, FLAMMABLE, N.O.S.	2.3	2.1	–	274	0	E0	P200	–	–	–
3161	LIQUEFIED GAS, FLAMMABLE, N.O.S.	2.1	–	–	274	0	E0	P200	–	–	–
3162	LIQUEFIED GAS, TOXIC, N.O.S.	2.3	–	–	274	0	E0	P200	–	–	–
3163	LIQUEFIED GAS, N.O.S.	2.2	–	–	274	120 mL	E1	P200	–	–	–
3164	ARTICLES, PRESSURIZED, PNEUMATIC or HYDRAULIC (containing non-flammable gas)	2.2	–	–	283	120 mL	E0	P003	–	–	–

Portable tanks and bulk containers			EmS	Stowage and segregation	Properties and observations	UN No.	
	Tank instructions	Provisions					
(12)	(13) 4.2.5 4.3	(14) 4.2.5	(15) 5.4.3.2 7.3	(16) 7.1 7.2	(17)	(18)	
–	T1	TP33	F-A, S-B	Category A.	A wide range of corrosive solids or pastes. Cause burns to skin, eyes and mucous membranes.	3147	
–	T13	TP2 TP7 TP38	F-G, S-N	Category E. Clear of living quarters.	–	3148	△
–	T7	TP2 TP7	F-G, S-N	Category E. Clear of living quarters.	–	3148	△
–	T7	TP2 TP7	F-G, S-N	Category E. Clear of living quarters.	–	3148	△
–	T7	TP2 TP6 TP24	F-H, S-Q	Category D. Protected from sources of heat. "Separated from" permanganates and class 4.1. See 7.2.6.3.2.	Colourless liquid. Carried as an aqueous solution. Slowly decomposes, evolving oxygen; the rate of decomposition increases on contact with most metals. In contact with combustible material, may cause fire. Causes burns to skin, eyes and mucous membranes. Even though stabilized, these solutions may evolve oxygen.	3149	△
–	–	–	F-D, S-U	Category B. Clear of living quarters.	Various small devices used for cosmetic and other purposes, and their refills.	3150	
–	–	–	F-A, S-A	Category A. Segregation from foodstuffs as in 7.3.4.2.1, 7.6.3.1.2 or 7.7.3.6.	Viscous liquids with a perceptible odour. Harmful by ingestion or by skin contact. This entry also covers articles, such as transformers and condensers, containing free liquid polyhalogenated biphenyls or polyhalogenated terphenyls.	3151	△
–	T3	TP33	F-A, S-A	Category A. Segregation from foodstuffs as in 7.3.4.2.1, 7.6.3.1.2 or 7.7.3.6.	Solid with a perceptible odour. Melting point of solids varies from 2°C to 164°C. Harmful by ingestion or by skin contact. This entry covers articles, such as rags, cotton waste, clothing or sawdust, containing polyhalogenated biphenyls or polyhalogenated terphenyls where no free visible liquid is present.	3152	△
–	T50	–	F-D, S-U	Category E. Clear of living quarters.	Explosive limits: 7% to 73%. Much heavier than air (4.8). Boiling point: −27°C.	3153	
–	–	–	F-D, S-U	Category E. Clear of living quarters.	Explosive limits: 7% to 73%. Much heavier than air (6.4). Boiling point: 12°C.	3154	
–	T3	TP33	F-A, S-A	Category A.	Toxic if swallowed, by skin contact or by dust inhalation.	3155	
–	–	–	F-C, S-W	Category D.	–	3156	
–	–	–	F-C, S-W	Category D.	–	3157	
–	T75	TP5	F-C, S-V	Category D.	–	3158	
–	T50	–	F-C, S-V	Category A.	Non-flammable gas with a mild ether-like odour. Much heavier than air (3.5).	3159	
–	–	–	F-D, S-U	Category D. Clear of living quarters.	–	3160	
–	T50	–	F-D, S-U	Category D. Clear of living quarters.	–	3161	
–	–	–	F-C, S-U	Category D. Clear of living quarters.	–	3162	
–	T50	–	F-C, S-V	Category A.	–	3163	
–	–	–	F-C, S-V	Category A.	Articles containing non-flammable, non-toxic gas necessary for their operation.	3164	

	UN No.	Proper Shipping Name (PSN)	Class or division	Subsidiary risk(s)	Packing group	Special provisions	Limited and excepted quantity provisions		Packing		IBC	
							Limited quantities	Excepted quantities	Instructions	Provisions	Instructions	Provisions
	(1)	(2) 3.1.2	(3) 2.0	(4) 2.0	(5) 2.0.1.3	(6) 3.3	(7a) 3.4	(7b) 3.5	(8) 4.1.4	(9) 4.1.4	(10) 4.1.4	(11) 4.1.4
△	3165	AIRCRAFT HYDRAULIC POWER UNIT FUEL TANK (containing a mixture of anhydrous hydrazine and methylhydrazine) (M86 fuel)	3	6.1/8	I	–	0	E0	P301	–	–	–
	3166	ENGINE, INTERNAL COMBUSTION or VEHICLE, FLAMMABLE GAS POWERED or VEHICLE, FLAMMABLE LIQUID POWERED or ENGINE, FUEL CELL, FLAMMABLE GAS POWERED or ENGINE, FUEL CELL, FLAMMABLE LIQUID POWERED or VEHICLE, FUEL CELL, FLAMMABLE GAS POWERED or VEHICLE, FUEL CELL, FLAMMABLE LIQUID POWERED	9	–	–	312 356 961 962	–	–	–	–	–	–
	3167	GAS SAMPLE, NON-PRESSURIZED, FLAMMABLE, N.O.S., not refrigerated liquid	2.1	–	–	209	0	E0	P201	–	–	–
	3168	GAS SAMPLE, NON-PRESSURIZED, TOXIC, FLAMMABLE, N.O.S., not refrigerated liquid	2.3	2.1	–	209	0	E0	P201	–	–	–
	3169	GAS SAMPLE, NON-PRESSURIZED, TOXIC, N.O.S., not refrigerated liquid	2.3	–	–	209	0	E0	P201	–	–	–
	3170	ALUMINIUM SMELTING BY-PRODUCTS or ALUMINIUM REMELTING BY-PRODUCTS	4.3	–	II	244	500 g	E2	P410	PP31 PP40	IBC07	B2
	3170	ALUMINIUM SMELTING BY-PRODUCTS or ALUMINIUM REMELTING BY-PRODUCTS	4.3	–	III	223 244	1 kg	E1	P002	PP31	IBC08	B4
△	3171	BATTERY-POWERED VEHICLE or BATTERY-POWERED EQUIPMENT	9	–	–	240 961 962	–	–	–	–	–	–
	3172	TOXINS, EXTRACTED FROM LIVING SOURCES, LIQUID, N.O.S.	6.1	–	I	210 274	0	E5	P001	–	–	–
	3172	TOXINS, EXTRACTED FROM LIVING SOURCES, LIQUID, N.O.S.	6.1	–	II	210 274	100 mL	E4	P001	–	IBC02	–
	3172	TOXINS, EXTRACTED FROM LIVING SOURCES, LIQUID, N.O.S.	6.1	–	III	210 223 274	5 L	E1	P001 LP01	–	IBC03	–
	3174	TITANIUM DISULPHIDE	4.2	–	III	–	0	E1	P002 LP02	PP31	IBC08	B3
	3175	SOLIDS CONTAINING FLAMMABLE LIQUID, N.O.S.	4.1	–	II	216 274	1 kg	E2	P002	PP9	IBC06	B2
	3176	FLAMMABLE SOLID, ORGANIC, MOLTEN, N.O.S.	4.1	–	II	274	0	E0	–	–	–	–

Portable tanks and bulk containers			EmS	Stowage and segregation	Properties and observations	UN No.
	Tank instructions	Provisions				
(12)	(13) 4.2.5 4.3	(14) 4.2.5	(15) 5.4.3.2 7.3	(16) 7.1 7.2	(17)	(18)
–	–	–	F-E, S-C	Category D. Clear of living quarters. Segregation as for class 3 but "away from" class 4.1 and class 8.	The mixture is miscible with water and may react dangerously with oxidizing substances. The mixture is highly toxic if swallowed, by skin contact or by inhalation. Causes burns to skin, eyes and mucous membranes.	3165 △
–	–	–	*	Category A.	Types of articles transported under this entry include internal combustion engines, compression/ignition engines, fuel cell powered engines, motor vehicles, hybrid vehicles, motorcycles and boats. *F-D, S-U for gases or F-E, S-E for liquids.	3166
–	–	–	F-D, S-U	Category D.	–	3167
–	–	–	F-D, S-U	Category D.	–	3168
–	–	–	F-C, S-U	Category D.	–	3169
–	T3 BK2	TP33	F-G, S-P	Category B. Under deck in a mechanically ventilated space. Only to be loaded under dry weather conditions.	Grey powder or lumps with some metallic inclusions. Contact with water may cause heating with possible evolution of flammable and toxic gases such as hydrogen and ammonia. This entry includes e.g., aluminium dross, aluminium skimmings, spent cathodes, spent pot iner and aluminium salt slags.	3170
–	T1 BK2	TP33	F-G, S-P	Category B. Under deck in a mechanically ventilated space. Only to be loaded under dry weather conditions.	See entry above.	3170
–	–	–	F-A, S-I	Category A.	Types of articles transported under this entry include vehicles or equipment powered by wet batteries, sodium batteries or lithium batteries with the batteries installed, such as electrically-powered cars, lawnmowers, wheelchairs and other mobility aids.	3171 △
–	–	–	F-A, S-A	Category B.	Toxins from plant, animal or bacterial sources which contain infectious substances or toxins that are contained in infectious substances should be classified in class 6.2. Toxic if swallowed, by skin contact or by inhalation.	3172
–	–	–	F-A, S-A	Category B.	See entry above.	3172
–	–	–	F-A, S-A	Category A.	See entry above.	3172
–	T1	TP33	F-A, S-J	Category A.	Yellow or grey powder with an unpleasant odour. In contact with water slowly evolves hydrogen sulphide gas.	3174
–	T3 BK2	TP33	F-A, S-I	Category B.	Mixtures of non-dangerous solids (such as soil, sand, production materials, etc.) and flammable liquids.	3175
–	T3	TP3 TP26	F-A, S-H	Category C.	Shipped molten above its melting point.	3176

UN No.	Proper Shipping Name (PSN)	Class or division	Subsidiary risk(s)	Packing group	Special provisions	Limited and excepted quantity provisions		Packing		IBC	
						Limited quantities	Excepted quantities	Instructions	Provisions	Instructions	Provisions
(1)	(2) 3.1.2	(3) 2.0	(4) 2.0	(5) 2.0.1.3	(6) 3.3	(7a) 3.4	(7b) 3.5	(8) 4.1.4	(9) 4.1.4	(10) 4.1.4	(11) 4.1.4
3176	FLAMMABLE SOLID, ORGANIC, MOLTEN, N.O.S.	4.1	–	III	223 274	0	E0	–	–	–	–
3178	FLAMMABLE SOLID, INORGANIC, N.O.S.	4.1	–	II	274 915	1 kg	E2	P002	–	IBC08	B2 B4
3178	FLAMMABLE SOLID, INORGANIC, N.O.S.	4.1	–	III	223 274 915	5 kg	E1	P002 LP02	–	IBC08	B3
3179	FLAMMABLE SOLID, TOXIC, INORGANIC, N.O.S.	4.1	6.1	II	274 915	1 kg	E2	P002	–	IBC06	B2
3179	FLAMMABLE SOLID, TOXIC, INORGANIC, N.O.S.	4.1	6.1	III	223 274 915	5 kg	E1	P002	–	IBC06	–
3180	FLAMMABLE SOLID, CORROSIVE, INORGANIC, N.O.S.	4.1	8	II	274 915	1 kg	E2	P002	–	IBC06	B2
3180	FLAMMABLE SOLID, CORROSIVE, INORGANIC, N.O.S.	4.1	8	III	223 274 915	5 kg	E1	P002	–	IBC06	–
3181	METAL SALTS OF ORGANIC COMPOUNDS, FLAMMABLE, N.O.S.	4.1	–	II	274	1 kg	E2	P002	PP31	IBC08	B2 B4
3181	METAL SALTS OF ORGANIC COMPOUNDS, FLAMMABLE, N.O.S.	4.1	–	III	223 274	5 kg	E1	P002 LP02	PP31	IBC08	B3
3182	METAL HYDRIDES, FLAMMABLE, N.O.S.	4.1	–	II	274	1 kg	E2	P410	PP31 PP40	IBC04	–
3182	METAL HYDRIDES, FLAMMABLE, N.O.S.	4.1	–	III	223 274	5 kg	E1	P002	PP31	IBC04	–
3183	SELF-HEATING LIQUID, ORGANIC, N.O.S.	4.2	–	II	274	0	E2	P001	PP31	IBC02	–
3183	SELF-HEATING LIQUID, ORGANIC, N.O.S.	4.2	–	III	223 274	0	E1	P001	PP31	IBC02	–
3184	SELF-HEATING LIQUID, TOXIC, ORGANIC, N.O.S.	4.2	6.1	II	274	0	E2	P402	PP31	IBC02	–
3184	SELF-HEATING LIQUID, TOXIC, ORGANIC, N.O.S.	4.2	6.1	III	223 274	0	E1	P001	PP31	IBC02	–
3185	SELF-HEATING LIQUID, CORROSIVE, ORGANIC, N.O.S.	4.2	8	II	274	0	E2	P402	PP31	IBC02	–
3185	SELF-HEATING LIQUID, CORROSIVE, ORGANIC, N.O.S.	4.2	8	III	223 274	0	E1	P001	PP31	IBC02	–
3186	SELF-HEATING LIQUID, INORGANIC, N.O.S.	4.2	–	II	274	0	E2	P001	PP31	IBC02	–
3186	SELF-HEATING LIQUID, INORGANIC, N.O.S.	4.2	–	III	223 274	0	E1	P001	PP31	IBC02	–
3187	SELF-HEATING LIQUID, TOXIC, INORGANIC, N.O.S.	4.2	6.1	II	274	0	E2	P402	PP31	IBC02	–
3187	SELF-HEATING LIQUID, TOXIC, INORGANIC, N.O.S.	4.2	6.1	III	223 274	0	E1	P001	PP31	IBC02	–
3188	SELF-HEATING LIQUID, CORROSIVE, INORGANIC, N.O.S.	4.2	8	II	274	0	E2	P402	PP31	IBC02	–
3188	SELF-HEATING LIQUID, CORROSIVE, INORGANIC, N.O.S.	4.2	8	III	223 274	0	E1	P001	PP31	IBC02	–

Portable tanks and bulk containers			EmS	Stowage and segregation	Properties and observations	UN No.
	Tank instructions	Provisions				
(12)	(13) 4.2.5 4.3	(14) 4.2.5	(15) 5.4.3.2 7.3	(16) 7.1 7.2	(17)	(18)
–	T1	TP3 TP26	F-A, S-H	Category C.	Shipped molten above its melting point.	3176
–	T3	TP33	F-A, S-G	Category B.	–	3178
–	T1	TP33	F-A, S-G	Category B.	–	3178
–	T3	TP33	F-A, S-G	Category B. Clear of living quarters.	Toxic if swallowed, by skin contact or by dust inhalation. Should be handled with care to minimize exposure, particularly to dust.	3179
–	T1	TP33	F-A, S-G	Category B. Clear of living quarters.	See entry above.	3179
–	T3	TP33	F-A, S-G	Category D. Clear of living quarters.	Causes burns to skin, eyes and mucous membranes.	3180
–	T1	TP33	F-A, S-G	Category D. Clear of living quarters.	See entry above.	3180
–	T3	TP33	F-A, S-I	Category B. Clear of living quarters.	Decomposes in water. Liable to spontaneous heating. Irritating to skin and mucous membranes.	3181
–	T1	TP33	F-A, S-I	Category B. Clear of living quarters.	See entry above.	3181
–	T3	TP33	F-A, S-G	Category E.	–	3182
–	T1	TP33	F-A, S-G	Category E.	–	3182
–	–	–	F-A, S-J	Category C.	–	3183
–	–	–	F-A, S-J	Category C.	–	3183
–	–	–	F-A, S-J	Category C.	–	3184
–	–	–	F-A, S-J	Category C.	–	3184
–	–	–	F-A, S-J	Category C.	–	3185
–	–	–	F-A, S-J	Category C.	–	3185
–	–	–	F-A, S-J	Category C.	–	3186
–	–	–	F-A, S-J	Category C.	–	3186
–	–	–	F-A, S-J	Category C.	–	3187
–	–	–	F-A, S-J	Category C.	–	3187
–	–	–	F-A, S-J	Category C.	–	3188
–	–	–	F-A, S-J	Category C.	–	3188

UN No.	Proper Shipping Name (PSN)	Class or division	Subsidiary risk(s)	Packing group	Special provisions	Limited and excepted quantity provisions		Packing		IBC	
						Limited quantities	Excepted quantities	Instructions	Provisions	Instructions	Provisions
(1)	(2) 3.1.2	(3) 2.0	(4) 2.0	(5) 2.0.1.3	(6) 3.3	(7a) 3.4	(7b) 3.5	(8) 4.1.4	(9) 4.1.4	(10) 4.1.4	(11) 4.1.4
3189	METAL POWDER, SELF-HEATING, N.O.S.	4.2	–	II	274	0	E2	P410	PP31	IBC06	B2
3189	METAL POWDER, SELF-HEATING, N.O.S.	4.2	–	III	223 274	0	E1	P002 LP02	PP31	IBC08	B3
3190	SELF-HEATING SOLID, INORGANIC, N.O.S.	4.2	–	II	274	0	E2	P410	PP31	IBC06	B2
3190	SELF-HEATING SOLID, INORGANIC, N.O.S.	4.2	–	III	223 274	0	E1	P002 LP02	PP31	IBC08	B3
3191	SELF-HEATING SOLID, TOXIC, INORGANIC, N.O.S.	4.2	6.1	II	274	0	E2	P410	–	IBC05	B2
3191	SELF-HEATING SOLID, TOXIC, INORGANIC, N.O.S.	4.2	6.1	III	223 274	0	E1	P002	–	IBC08	B3
3192	SELF-HEATING SOLID, CORROSIVE, INORGANIC, N.O.S.	4.2	8	II	274	0	E2	P410	–	IBC05	B2
3192	SELF-HEATING SOLID, CORROSIVE, INORGANIC, N.O.S.	4.2	8	III	274	0	E1	P002	–	IBC08	B3
3194	PYROPHORIC LIQUID, INORGANIC, N.O.S.	4.2	–	I	274	0	E0	P400	–	–	–
3200	PYROPHORIC SOLID, INORGANIC, N.O.S.	4.2	–	I	274	0	E0	P404	PP31	–	–
3205	ALKALINE EARTH METAL ALCOHOLATES, N.O.S.	4.2	–	II	183 274	0	E2	P410	PP31	IBC06	B2
3205	ALKALINE EARTH METAL ALCOHOLATES, N.O.S.	4.2	–	III	183 223 274	0	E1	P002 LP02	PP31	IBC08	B3
3206	ALKALI METAL ALCOHOLATES, SELF-HEATING, CORROSIVE, N.O.S.	4.2	8	II	182 274	0	E2	P410	PP31	IBC05	B2
3206	ALKALI METAL ALCOHOLATES, SELF-HEATING, CORROSIVE, N.O.S.	4.2	8	III	182 223 274	0	E1	P002	PP31	IBC08	B3
3208	METALLIC SUBSTANCE, WATER-REACTIVE, N.O.S.	4.3	–	I	274	0	E0	P403	PP31	IBC99	–
3208	METALLIC SUBSTANCE, WATER-REACTIVE, N.O.S.	4.3	–	II	274	500 g	E2	P410	PP31	IBC07	B2
3208	METALLIC SUBSTANCE, WATER-REACTIVE, N.O.S.	4.3	–	III	223 274	1 kg	E1	P410	PP31 PP40	IBC08	B4
3209	METALLIC SUBSTANCE, WATER-REACTIVE, SELF-HEATING, N.O.S.	4.3	4.2	I	274	0	E0	P403	PP31	–	–
3209	METALLIC SUBSTANCE, WATER-REACTIVE, SELF-HEATING, N.O.S.	4.3	4.2	II	274	0	E2	P410	PP31 PP40	IBC05	B2
3209	METALLIC SUBSTANCE, WATER-REACTIVE, SELF-HEATING, N.O.S.	4.3	4.2	III	223 274	0	E1	P410	PP31	IBC08	B4

Portable tanks and bulk containers			EmS	Stowage and segregation	Properties and observations	UN No.
	Tank instructions	Provisions				
(12)	(13) 4.2.5 4.3	(14) 4.2.5	(15) 5.4.3.2 7.3	(16) 7.1 7.2	(17)	(18)
–	T3	TP33	F-G, S-J	Category C.	Forms explosive mixtures with oxidizing substances.	3189
–	T1	TP33	F-G, S-J	Category C.	See entry above.	3189
–	T3	TP33	F-A, S-J	Category C.	Liable to self-heating or spontaneous combustion.	3190
–	T1	TP33	F-A, S-J	Category C.	See entry above.	3190
–	T3	TP33	F-A, S-J	Category C.	–	3191
–	T1	TP33	F-A, S-J	Category C.	–	3191
–	T3	TP33	F-A, S-J	Category C.	–	3192
–	T1	TP33	F-A, S-J	Category C.	–	3192
–	–	–	F-G, S-M	Category D. "Separated longitudinally by an intervening complete compartment or hold from" class 1.	Highly flammable liquids, may ignite spontaneously in moist air. In contact with air, evolve irritating and slightly toxic fumes.	3194
–	T21	TP7 TP33	F-G, S-M	Category D.	Liable to ignite spontaneously in air. If shaken, may produce sparks. In contact with water, evolve hydrogen, a flammable gas.	3200
–	T3	TP33	F-A, S-J	Category B.	Free-flowing hygroscopic powders. Irritating to skin, eyes and mucous membranes.	3205
–	T1	TP33	F-A, S-J	Category B.	See entry above.	3205
–	T3	TP33	F-A, S-J	Category B.	Free-flowing hygroscopic powder. Cause burns to skin, eyes and mucous membranes.	3206
–	T1	TP33	F-A, S-J	Category B.	See entry above.	3206
–	–	–	F-G, S-N	Category E. Clear of living quarters.	–	3208
–	T3	TP33	F-G, S-N	Category E. Clear of living quarters.	–	3208
–	T1	TP33	F-G, S-N	Category E. Clear of living quarters.	–	3208
–	–	–	F-G, S-N	Category E. Clear of living quarters.	–	3209
–	T3	TP33	F-G, S-N	Category E. Clear of living quarters.	–	3209
–	T1	TP33	F-G, S-N	Category E. Clear of living quarters.	–	3209

	UN No. (1)	Proper Shipping Name (PSN) (2) 3.1.2	Class or division (3) 2.0	Subsidiary risk(s) (4) 2.0	Packing group (5) 2.0.1.3	Special provisions (6) 3.3	Limited and excepted quantity provisions: Limited quantities (7a) 3.4	Excepted quantities (7b) 3.5	Packing: Instructions (8) 4.1.4	Provisions (9) 4.1.4	IBC: Instructions (10) 4.1.4	Provisions (11) 4.1.4
△	3210	**CHLORATES, INORGANIC, AQUEOUS SOLUTION, N.O.S.**	5.1	–	II	274 351	1 L	E2	P504	–	IBC02	–
	3210	**CHLORATES, INORGANIC, AQUEOUS SOLUTION, N.O.S.**	5.1	–	III	223 274 351	5 L	E1	P504	–	IBC02	–
	3211	**PERCHLORATES, INORGANIC, AQUEOUS SOLUTION, N.O.S.**	5.1	–	II	–	1 L	E2	P504	–	IBC02	–
	3211	**PERCHLORATES, INORGANIC, AQUEOUS SOLUTION, N.O.S.**	5.1	–	III	223	5 L	E1	P504	–	IBC02	–
△	3212	**HYPOCHLORITES, INORGANIC, N.O.S.**	5.1	–	II	274 349 900 903	1 kg	E2	P002	–	IBC08	B2 B4
△	3213	**BROMATES, INORGANIC, AQUEOUS SOLUTION, N.O.S.**	5.1	–	II	274 350	1 L	E2	P504	–	IBC02	–
	3213	**BROMATES, INORGANIC, AQUEOUS SOLUTION, N.O.S.**	5.1	–	III	223 274 350	5 L	E1	P504	–	IBC02	–

Portable tanks and bulk containers			EmS	Stowage and segregation	Properties and observations	UN No.
	Tank instructions	Provisions				
(12)	(13) 4.2.5 4.3	(14) 4.2.5	(15) 5.4.3.2 7.3	(16) 7.1 7.2	(17)	(18)
–	T4	TP1	F-H, S-Q	Category B. "Separated from" ammonium compounds, cyanides and sulphur.	When involved in a fire, may cause an explosion. Leakage and subsequent evaporation of the water of the solutions may present increased dangers as follows: 1. in contact with combustible material (particularly with fibrous material such as jute, cotton or sisal) or sulphur, danger of spontaneous combustion; 2. in contact with ammonium compounds, powdered metals or oils, danger of explosion Transport of ammonium chlorate, aqueous solution is **prohibited**.	3210 △
–	T4	TP1	F-H, S-Q	Category B. "Separated from" ammonium compounds, cyanides and sulphur.	See entry above.	3210
–	T4	TP1	F-H, S-Q	Category B. "Separated from" ammonium compounds, cyanides and sulphur.	When involved in a fire, may cause an explosion. Leakage and subsequent evaporation of the water of the solutions may present increased dangers as follows: 1. in contact with combustible material (particularly with fibrous material such as jute, cotton or sisal) or sulphur, danger of spontaneous combustion; 2. in contact with ammonium compounds, powdered metals or oils, danger of explosion	3211
–	T4	TP1	F-H, S-Q	Category B. "Separated from" ammonium compounds, cyanides and sulphur.	See entry above.	3211
–	T3	TP33	F-H, S-Q	Category D. Category E, Closed Cargo Transport Unit and pallet boxes only. Ventilation may be required. The possible need to open hatches in case of fire to provide maximum ventilation and to apply water in an emergency, and the consequent risk to the stability of the ship through flooding of the cargo space, shall be considered before loading. "Separated from" ammonium compounds, acids, cyanides, hydrogen peroxide and liquid organic substances. Protected from sources of heat.	Solids. Critical ambient temperature of decomposition may be as low as 60°C. May cause fire in contact with organic material or ammonium compounds. React with acids, evolving chlorine, an irritating, corrosive and toxic gas. In the presence of moisture, corrosive to most metals. Dust irritates mucous membranes. Transport of ammonium hypochlorite and mixtures of a hypochlorite with an ammonium salt is **prohibited**.	3212 △
–	T4	TP1	F-H, S-Q	Category B. "Separated from" ammonium compounds, cyanides and sulphur.	When involved in a fire, may cause an explosion. Leakage and subsequent evaporation of the water of the solutions may present increased dangers as follows: 1. in contact with combustible material (particularly with fibrous material such as jute, cotton or sisal) or sulphur, danger of spontaneous combustion; 2. in contact with ammonium compounds, powdered metals or oils, danger of explosion Transport of ammonium bromate, aqueous solution is **prohibited**.	3213 △
–	T4	TP1	F-H, S-Q	Category B. "Separated from" ammonium compounds, cyanides and sulphur.	See entry above.	3213

3

DGL

	UN No.	Proper Shipping Name (PSN)	Class or division	Subsidiary risk(s)	Packing group	Special provisions	Limited and excepted quantity provisions		Packing		IBC	
							Limited quantities	Excepted quantities	Instructions	Provisions	Instructions	Provisions
	(1)	(2) 3.1.2	(3) 2.0	(4) 2.0	(5) 2.0.1.3	(6) 3.3	(7a) 3.4	(7b) 3.5	(8) 4.1.4	(9) 4.1.4	(10) 4.1.4	(11) 4.1.4
△	3214	PERMANGANATES, INORGANIC, AQUEOUS SOLUTION, N.O.S.	5.1	–	II	274 353	1 L	E2	P504	–	IBC02	–
	3215	PERSULPHATES, INORGANIC, N.O.S.	5.1	–	III	–	5 kg	E1	P002 LP02	–	IBC08	B3
	3216	PERSULPHATES, INORGANIC, AQUEOUS SOLUTION, N.O.S.	5.1	–	III	–	5 L	E1	P504	–	IBC02	–
	3218	NITRATES, INORGANIC, AQUEOUS SOLUTION, N.O.S.	5.1	–	II	270	1 L	E2	P504	–	IBC02	–
	3218	NITRATES, INORGANIC, AQUEOUS SOLUTION, N.O.S.	5.1	–	III	223 270	5 L	E1	P504	–	IBC02	–
△	3219	NITRITES, INORGANIC, AQUEOUS SOLUTION, N.O.S.	5.1	–	II	274	1 L	E2	P504	–	IBC01	–
	3219	NITRITES, INORGANIC, AQUEOUS SOLUTION, N.O.S.	5.1	–	III	223 274 900	5 L	E1	P504	–	IBC02	–
	3220	PENTAFLUOROETHANE (REFRIGERANT GAS R 125)	2.2	–	–	–	120 mL	E1	P200	–	–	–
△	3221	SELF-REACTIVE LIQUID TYPE B	4.1	See SP181	–	181 274	25 mL	E0	P520	PP21	–	–

Portable tanks and bulk containers			EmS	Stowage and segregation	Properties and observations	UN No.
	Tank instructions	Provisions				
(12)	(13) 4.2.5 4.3	(14) 4.2.5	(15) 5.4.3.2 7.3	(16) 7.1 7.2	(17)	(18)
–	T4	TP1	F-H, S-Q	Category D. "Separated from" ammonium compounds, cyanides, peroxides and sulphur.	When involved in a fire, may cause an explosion. Leakage and subsequent evaporation of the water of the solutions may present increased dangers as follows: 1. in contact with combustible material (particularly with fibrous material such as jute, cotton or sisal) or sulphur, danger of spontaneous combustion; 2. in contact with ammonium compounds, powdered metals or oils, danger of explosion. Transport of ammonium permanganate, aqueous solution is **prohibited**.	**3214** △
–	T1	TP33	F-A, S-Q	Category A. "Separated from" ammonium compounds and cyanides. However, the segregation provisions concerning ammonium compounds do not apply to mixtures of ammonium persulphates and/or potassium persulphates and/or sodium persulphates.	Solids. Solid mixtures with combustible material are sensitive to friction and are liable to ignite. React fiercely with cyanides when heated or by friction. May form explosive mixture with powdered metals or ammonium compounds.	3215
–	T4	TP1 TP29	F-A, S-Q	Category A. "Separated from" ammonium compounds, cyanides and sulphur.	When involved in a fire, may cause an explosion. Leakage and subsequent evaporation of the water of the solutions may present increased dangers as follows: 1. in contact with combustible material (particularly with fibrous material such as jute, cotton or sisal) or sulphur, danger of spontaneous combustion; 2. in contact with ammonium compounds, powdered metals or oils, danger of explosion.	**3216**
–	T4	TP1	F-A, S-Q	Category B. "Separated from" ammonium compounds, cyanides and sulphur.	When involved in a fire, may cause an explosion. Leakage and subsequent evaporation of the water of the solutions may present increased dangers as follows: 1. in contact with combustible material (particularly with fibrous material such as jute, cotton or sisal) or sulphur, danger of spontaneous combustion; 2. in contact with ammonium compounds, powdered metals or oils, danger of explosion.	3218
–	T4	TP1	F-A, S-Q	Category B. "Separated from" ammonium compounds, cyanides and sulphur.	See entry above.	**3218**
–	T4	TP1	F-A, S-Q	Category B. "Separated from" ammonium compounds, cyanides and sulphur.	When involved in a fire, may cause an explosion. Leakage and subsequent evaporation of the water of the solutions may present increased dangers as follows: 1. in contact with combustible material (particularly with fibrous material such as jute, cotton or sisal) or sulphur, danger of spontaneous combustion; 2. in contact with ammonium compounds, powdered metals or oils, danger of explosion. Transport of ammonium nitrites, aqueous solution is **prohibited**.	3219 △
–	T4	TP1	F-A, S-Q	Category B. "Separated from" ammonium compounds, cyanides and sulphur.	See entry above.	**3219**
–	T50	–	F-C, S-V	Category A.	Liquefied, non-flammable gas with a mild ether-like odour. Much heavier than air (4.2).	3220
–	–	–	F-J, S-G	Category D. Protected from sources of heat. For packages carrying a subsidiary risk label of class 1, segregation as for class 1, division 1.3. "Separated from" acids and alkalis.	May explode at elevated temperatures or in a fire. Burns vigorously. Immiscible with water. Contact with alkalis or acids may cause dangerous decomposition. The products of combustion or self-accelerating decomposition may be toxic by inhalation.	**3221** △

	UN No.	Proper Shipping Name (PSN)	Class or division	Subsidiary risk(s)	Packing group	Special provisions	Limited and excepted quantity provisions		Packing		IBC	
							Limited quantities	Excepted quantities	Instructions	Provisions	Instructions	Provisions
	(1)	(2) 3.1.2	(3) 2.0	(4) 2.0	(5) 2.0.1.3	(6) 3.3	(7a) 3.4	(7b) 3.5	(8) 4.1.4	(9) 4.1.4	(10) 4.1.4	(11) 4.1.4
△	3222	SELF-REACTIVE SOLID TYPE B	4.1	See SP181	–	181 274	100 g	E0	P520	PP21	–	–
△	3223	SELF-REACTIVE LIQUID TYPE C	4.1	–	–	274	25 mL	E0	P520	PP21	–	–
△	3224	SELF-REACTIVE SOLID TYPE C	4.1	–	–	274	100 g	E0	P520	PP21	–	–
△	3225	SELF-REACTIVE LIQUID TYPE D	4.1	–	–	274	125 mL	E0	P520	–	–	–
△	3226	SELF-REACTIVE SOLID TYPE D	4.1	–	–	274	500 g	E0	P520	–	–	–
△	3227	SELF-REACTIVE LIQUID TYPE E	4.1	–	–	274	125 mL	E0	P520	–	–	–
△	3228	SELF-REACTIVE SOLID TYPE E	4.1	–	–	274	500 g	E0	P520	–	–	–
△	3229	SELF-REACTIVE LIQUID TYPE F	4.1	–	–	274	125 mL	E0	P520	–	IBC99	–
△	3230	SELF-REACTIVE SOLID TYPE F	4.1	–	–	274	500 g	E0	P520	–	IBC99	–
△	3231	SELF-REACTIVE LIQUID TYPE B, TEMPERATURE CONTROLLED	4.1	See SP181	–	181 194 274 923	0	E0	P520	PP21	–	–

Portable tanks and bulk containers			EmS	Stowage and segregation	Properties and observations	UN No.	
	Tank instructions	Provisions					
(12)	(13) 4.2.5 4.3	(14) 4.2.5	(15) 5.4.3.2 7.3	(16) 7.1 7.2	(17)	(18)	
–	–	–	F-J, S-G	Category D. Protected from sources of heat. For packages carrying a subsidiary risk label of class 1, segregation as for class 1, division 1.3. "Separated from" acids and alkalis.	May explode at elevated temperatures or in a fire. Burns vigorously. Insoluble in water. Contact with alkalis or acids may cause dangerous decomposition. The products of combustion or self-accelerating decomposition may be toxic by inhalation.	3222	△
–	–	–	F-J, S-G	Category D. Protected from sources of heat. "Separated from" acids and alkalis.	May decompose violently at elevated temperatures or in a fire. Burns vigorously. Immiscible with water. Contact with alkalis or acids may cause dangerous decomposition. The products of combustion or self-accelerating decomposition may be toxic by inhalation.	3223	△
–	–	–	F-J, S-G	Category D. Protected from sources of heat. "Separated from" acids and alkalis.	May decompose violently at elevated temperatures or in a fire. Burns vigorously. Insoluble in water. Contact with alkalis or acids may cause dangerous decomposition. The products of combustion or self-accelerating decomposition may be toxic by inhalation.	3224	△
–	–	–	F-J, S-G	Category D. Protected from sources of heat. "Separated from" acids and alkalis.	Decomposes at elevated temperatures or in a fire. Burns vigorously. Immiscible with water. Contact with alkalis or acids may cause dangerous decomposition. The products of combustion or self-accelerating decomposition may be toxic by inhalation.	3225	△
–	–	–	F-J, S-G	Category D. Protected from sources of heat. "Separated from" acids and alkalis.	Decomposes at elevated temperatures or in a fire. Burns vigorously. Contact with alkalis or acids may cause dangerous decomposition. The products of combustion or self-accelerating decomposition may be toxic by inhalation. Insoluble in water except: 4-(BENZYL(ETHYL)AMINO)-3-ETHOXYBENZENEDIAZONIUM ZINC CHLORIDE 3-CHLORO-4-DIETHYLAMINOBENZENEDIAZONIUM ZINC CHLORIDE 4-DIPROPYLAMINOBENZENEDIAZONIUM ZINC CHLORIDE SODIUM 2-DIAZO-1-NAPHTHOL-4-SULPHONATE SODIUM 2-DIAZO-1-NAPHTHOL-5-SULPHONATE	3226	△
–	–	–	F-J, S-G	Category D. Protected from sources of heat. "Separated from" acids and alkalis.	Decomposes at elevated temperatures or in a fire. Burns vigorously. Immiscible with water. Contact with alkalis or acids may cause dangerous decomposition. The products of combustion or self-accelerating decomposition may be toxic by inhalation.	3227	△
–	–	–	F-J, S-G	Category D. Protected from sources of heat. "Separated from" acids and alkalis.	Decomposes at elevated temperatures or in a fire. Burns vigorously. Insoluble in water. Contact with alkalis or acids may cause dangerous decomposition. The products of combustion or self-accelerating decomposition may be toxic by inhalation.	3228	△
–	T23	–	F-J, S-G	Category D. Protected from sources of heat. "Separated from" acids and alkalis.	Decomposes at elevated temperatures or in a fire. Burns vigorously. Immiscible with water. Contact with alkalis or acids may cause dangerous decomposition. The products of combustion or self-accelerating decomposition may be toxic by inhalation.	3229	△
–	T23	–	F-J, S-G	Category D. Protected from sources of heat. "Separated from" acids and alkalis.	Decomposes at elevated temperatures or in a fire. Burns vigorously. Insoluble in water. Contact with alkalis or acids may cause dangerous decomposition. The products of combustion or self-accelerating decomposition may be toxic by inhalation.	3230	△
–	–	–	F-F, S-K	Category D. Protected from sources of heat. For packages carrying a subsidiary risk label of class 1, segregation as for class 1, division 1.3. Shall be transported under temperature control. "Separated from" acids and alkalis.	May explode at temperatures higher than the emergency temperature or in a fire. Burns vigorously. Immiscible with water. Contact with alkalis or acids may cause dangerous decomposition. The products of combustion or self-accelerating decomposition may be toxic by inhalation. Control and emergency temperatures for each formulation can be found in 2.4.2.3.2.3. The temperature should be checked regularly.	3231	△

	UN No.	Proper Shipping Name (PSN)	Class or division	Subsidiary risk(s)	Packing group	Special provisions	Limited and excepted quantity provisions		Packing		IBC	
							Limited quantities	Excepted quantities	Instructions	Provisions	Instructions	Provisions
	(1)	(2) 3.1.2	(3) 2.0	(4) 2.0	(5) 2.0.1.3	(6) 3.3	(7a) 3.4	(7b) 3.5	(8) 4.1.4	(9) 4.1.4	(10) 4.1.4	(11) 4.1.4
△	3232	SELF-REACTIVE SOLID TYPE B, TEMPERATURE CONTROLLED	4.1	See SP181	–	181 194 274 923	0	E0	P520	PP21	–	–
△	3233	SELF-REACTIVE LIQUID TYPE C, TEMPERATURE CONTROLLED	4.1	–	–	194 274 923	0	E0	P520	PP21	–	–
△	3234	SELF-REACTIVE SOLID TYPE C, TEMPERATURE CONTROLLED	4.1	–	–	194 274 923	0	E0	P520	PP21	–	–
△	3235	SELF-REACTIVE LIQUID TYPE D, TEMPERATURE CONTROLLED	4.1	–	–	194 274 923	0	E0	P520	–	–	–
△	3236	SELF-REACTIVE SOLID TYPE D, TEMPERATURE CONTROLLED	4.1	–	–	194 274 923	0	E0	P520	–	–	–
△	3237	SELF-REACTIVE LIQUID TYPE E, TEMPERATURE CONTROLLED	4.1	–	–	194 274 923	0	E0	P520	–	–	–
△	3238	SELF-REACTIVE SOLID TYPE E, TEMPERATURE CONTROLLED	4.1	–	–	194 274 923	0	E0	P520	–	–	–

Portable tanks and bulk containers			EmS	Stowage and segregation	Properties and observations	UN No.	
	Tank instructions	Provisions					
(12)	(13) 4.2.5 4.3	(14) 4.2.5	(15) 5.4.3.2 7.3	(16) 7.1 7.2	(17)	(18)	
–	–	–	F-F, S-K	Category D. Protected from sources of heat. For packages carrying a subsidiary risk label of class 1, segregation as for class 1, division 1.3. Shall be transported under temperature control. "Separated from" acids and alkalis.	May explode at temperatures higher than the emergency temperature or in a fire. Burns vigorously. Insoluble in water. Contact with alkalis or acids may cause dangerous decomposition. The products of combustion or self-accelerating decomposition may be toxic by inhalation. Control and emergency temperatures for each formulation can be found in 2.4.2.3.2.3. The temperature should be checked regularly.	3232	△
–	–	–	F-F, S-K	Category D. Protected from sources of heat. Shall be transported under temperature control. "Separated from" acids and alkalis.	May explode at temperatures higher than the emergency temperature or in a fire. Burns vigorously. Immiscible with water. Contact with alkalis or acids may cause dangerous decomposition. The products of combustion or self-accelerating decomposition may be toxic by inhalation. Control and emergency temperatures for each formulation can be found in 2.4.2.3.2.3. The temperature should be checked regularly.	3233	△
–	–	–	F-F, S-K	Category D. Protected from sources of heat. Shall be transported under temperature control. "Separated from" acids and alkalis.	May explode at temperatures higher than the emergency temperature or in a fire. Burns vigorously. Insoluble in water except: 3-METHYL-4-(PYRROLIDIN-1-YL)BENZENEDIAZONIUM TETRAFLUOROBORATE TETRAMINEPALLADIUM(II) NITRATE Contact with alkalis or acids may cause dangerous decomposition. The products of combustion or self-accelerating decomposition may be toxic by inhalation. Control and emergency temperatures for each formulation can be found in 2.4.2.3.2.3. The temperature should be checked regularly.	3234	△
–	–	–	F-F, S-K	Category D. Protected from sources of heat. Shall be transported under temperature control. "Separated from" acids and alkalis.	Decomposes at temperatures higher than the emergency temperature or in a fire. Burns vigorously. Immiscible with water. Contact with alkalis or acids may cause dangerous decomposition. The products of combustion or self-accelerating decomposition may be toxic by inhalation.	3235	△
–	–	–	F-F, S-K	Category D. Protected from sources of heat. Shall be transported under temperature control. "Separated from" acids and alkalis.	Decomposes at temperatures higher than the emergency temperature or in a fire. Burns vigorously. Soluble in water except: AZODICARBONAMIDE FORMULATION TYPE D 2,2'-AZODI(2,4-DIMETHYL-4-METHOXYVALERONITRILE) 2,2'-AZODI(2,4-DIMETHYLVALERONITRILE) 2,2'-AZODI(2-METHYLBUTYRONITRILE) *N*-FORMYL-2-(NITROMETHYLENE)-1,3-PERHYDROTHIAZINE 4-NITROSOPHENOL Contact with alkalis or acids may cause dangerous decomposition. The products of combustion or self-accelerating decomposition may be toxic by inhalation.	3236	△
–	–	–	F-F, S-K	Category D. Protected from sources of heat. Shall be transported under temperature control. "Separated from" acids and alkalis.	Decomposes at temperatures higher than the emergency temperature or in a fire. Burns vigorously. Immiscible with water. Contact with alkalis or acids may cause dangerous decomposition. The products of combustion or self-accelerating decomposition may be toxic by inhalation. Control and emergency temperatures for each formulation can be found in 2.4.2.3.2.3. The temperature should be checked regularly.	3237	△
–	–	–	F-F, S-K	Category D. Protected from sources of heat. Shall be transported under temperature control. "Separated from" acids and alkalis.	Decomposes at temperatures higher than the emergency temperature or in a fire. Burns vigorously. Insoluble in water. Contact with alkalis or acids may cause dangerous decomposition. The products of combustion or self-accelerating decomposition may be toxic by inhalation. Control and emergency temperatures for each formulation can be found in 2.4.2.3.2.3. The temperature should be checked regularly.	3238	△

UN No.	Proper Shipping Name (PSN)	Class or division	Subsidiary risk(s)	Packing group	Special provisions	Limited and excepted quantity provisions		Packing		IBC	
						Limited quantities	Excepted quantities	Instructions	Provisions	Instructions	Provisions
(1)	(2) 3.1.2	(3) 2.0	(4) 2.0	(5) 2.0.1.3	(6) 3.3	(7a) 3.4	(7b) 3.5	(8) 4.1.4	(9) 4.1.4	(10) 4.1.4	(11) 4.1.4
△ 3239	SELF-REACTIVE LIQUID TYPE F, TEMPERATURE CONTROLLED	4.1	–	–	194 274 923	0	E0	P520	–	–	–
△ 3240	SELF-REACTIVE SOLID TYPE F, TEMPERATURE CONTROLLED	4.1	–	–	194 274 923	0	E0	P520	–	–	–
△ 3241	2-BROMO-2-NITROPROPANE-1,3-DIOL	4.1	–	III	–	5 kg	E1	P520	PP22	IBC08	B3
3242	AZODICARBONAMIDE	4.1	–	II	215	500 g	E2	P409	–	–	–
3243	SOLIDS CONTAINING TOXIC LIQUID, N.O.S.	6.1	–	II	217 274	500 g	E4	P002	PP9	IBC02	–
3244	SOLIDS CONTAINING CORROSIVE LIQUID, N.O.S.	8	–	II	218 274	1 kg	E2	P002	PP9	IBC05	–
△ 3245	GENETICALLY MODIFIED MICROORGANISMS or GENETICALLY MODIFIED ORGANISMS	9	–	–	219	0	E0	P904	–	IBC99	–
3246	METHANESULPHONYL CHLORIDE	6.1	8	I	354	0	E0	P602	–	–	–
△ 3247	SODIUM PEROXOBORATE, ANHYDROUS	5.1	–	II	–	1 kg	E2	P002	–	IBC08	B2 B4
3248	MEDICINE, LIQUID, FLAMMABLE, TOXIC, N.O.S.	3	6.1	II	220 221	1 L	E2	P001	–	–	–
3248	MEDICINE, LIQUID, FLAMMABLE, TOXIC, N.O.S.	3	6.1	III	220 221 223	5 L	E1	P001	–	–	–
3249	MEDICINE, SOLID, TOXIC, N.O.S.	6.1	–	II	221	500 g	E4	P002	–	–	–
3249	MEDICINE, SOLID, TOXIC, N.O.S.	6.1	–	III	221 223	5 kg	E1	P002 LP02	–	–	–
3250	CHLOROACETIC ACID, MOLTEN	6.1	8	II	–	0	E0	–	–	–	–

Portable tanks and bulk containers			EmS	Stowage and segregation	Properties and observations	UN No.	
	Tank instructions	Provisions					
(12)	(13) 4.2.5 4.3	(14) 4.2.5	(15) 5.4.3.2 7.3	(16) 7.1 7.2	(17)	(18)	
–	T23	–	F-F, S-K	Category D. Protected from sources of heat. Shall be transported under temperature control. "Separated from" acids and alkalis.	Decomposes at temperatures higher than the emergency temperature or in a fire. Burns vigorously. Immiscible with water. Contact with alkalis or acids may cause dangerous decomposition. The products of combustion or self-accelerating decomposition may be toxic by inhalation. Control and emergency temperatures for each formulation can be found in 2.4.2.3.2.3. The temperature must be checked regularly.	3239	△
–	T23	–	F-F, S-K	Category D. Protected from sources of heat. Shall be transported under temperature control. "Separated from" acids and alkalis.	Decomposes at temperatures higher than the emergency temperature or in a fire. Burns vigorously. Insoluble in water. Contact with alkalis or acids may cause dangerous decomposition. The products of combustion or self-accelerating decomposition may be toxic by inhalation. Control and emergency temperatures for each formulation can be found in 2.4.2.3.2.3. The temperature must be checked regularly.	3240	△
–	–	–	F-J, S-G	Category C. Protected from sources of heat. Clear of living quarters. Keep as cool as reasonably practicable. During transport, it shall be stored (or kept) in a cool ventilated place.	White crystals. Soluble in water. Decomposes when heated, evolving toxic gases. Sensitive to strong detonation shock. This substance shall be packed in accordance with packing method OP6 (see applicable packing instruction).	3241	△
–	T3	TP33	F-J, S-G	Category D. "Separated from" class 5.1, acids and alkalis.	Yellow or orange powder. Insoluble in water. Heat may cause exothermic decomposition, producing carbon monoxide (a toxic and flammable gas) and nitrogen. May explode if involved in a fire under confined conditions. Addition of activators (e.g., zinc compounds) may result in a decrease of thermal stability and/or a change in explosive properties.	3242	
–	T3 BK2	TP33	F-A, S-A	Category B. Clear of living quarters.	Mixtures of non-dangerous solids (such as soil, sand, production materials, etc.) and toxic liquids. Toxic if swallowed, by skin contact or by inhalation.	3243	
–	T3 BK2	TP33	F-A, S-B	Category B. Clear of living quarters.	Mixtures of non-dangerous solids (such as soil, sand, production materials, etc.) and corrosive liquids. Cause burns to skin, eyes and mucous membranes.	3244	
–	–	–	F-A, S-T	As approved by the competent authorities of the countries involved in the transport. Segregation from foodstuffs as in 7.3.4.2.1, 7.6.3.1.2 or 7.7.3.6.	–	3245	△
–	T20	TP2 TP13 TP37	F-A, S-B	Category D. Clear of living quarters.	Pale yellow liquid. Highly toxic if swallowed, by skin contact or by inhalation. Causes burns to skin, eyes and mucous membranes.	3246	
–	T3	TP33	F-A, S-Q	Category A. Keep as dry as reasonably practicable. Protected from sources of heat.	Yellowish, odourless crystals. Soluble in water. Mixtures with combustible material are readily ignited and may burn fiercely. Harmful if swallowed.	3247	△
–	–	–	F-E, S-D	Category B. Clear of living quarters..	Toxic if swallowed, by skin contact or by inhalation.	3248	
–	–	–	F-E, S-D	Category A.	See entry above.	3248	
–	T3	TP33	F-A, S-A	Category C. Clear of living quarters.	Toxic if swallowed, by skin contact or by dust inhalation.	3249	
–	T1	TP33	F-A, S-A	Category C. Clear of living quarters.	See entry above.	3249	
–	T7	TP3 TP28	F-A, S-B	Category C. Clear of living quarters.	Molten liquid. Melting point may be as low as 50°C. Toxic if swallowed, by skin contact or by inhalation. Causes burns to skin, eyes and mucous membranes.	3250	

UN No.	Proper Shipping Name (PSN)	Class or division	Subsidiary risk(s)	Packing group	Special provisions	Limited and excepted quantity provisions		Packing		IBC	
						Limited quantities	Excepted quantities	Instructions	Provisions	Instructions	Provisions
(1)	(2) 3.1.2	(3) 2.0	(4) 2.0	(5) 2.0.1.3	(6) 3.3	(7a) 3.4	(7b) 3.5	(8) 4.1.4	(9) 4.1.4	(10) 4.1.4	(11) 4.1.4
△ 3251	ISOSORBIDE-5-MONONITRATE	4.1	–	III	226	5 kg	E1	P409	–	–	–
3252	DIFLUOROMETHANE (REFRIGERANT GAS R 32)	2.1	–	–	–	0	E0	P200	–	–	–
3253	DISODIUM TRIOXOSILICATE	8	–	III	–	5 kg	E1	P002 LP02	–	IBC08	B3
3254	TRIBUTYLPHOSPHANE	4.2	–	I	–	0	E0	P400	–	–	–
3255	*tert*-BUTYL HYPOCHLORITE	4.2	8	I	76	0	E0	P099	–	–	–
3256	ELEVATED TEMPERATURE LIQUID, FLAMMABLE, N.O.S. with flashpoint above 60°C, at or above its flashpoint	3	–	III	274	0	E0	P099	–	IBC01	–
3257	ELEVATED TEMPERATURE LIQUID, N.O.S. at or above 100°C and below its flashpoint (including molten metals, molten salts, etc.)	9	–	III	232 274	0	E0	P099	–	IBC01	–
3258	ELEVATED TEMPERATURE SOLID, N.O.S. at or above 240°C	9	–	III	232 274	0	E0	P099	–	–	–
3259	AMINES, SOLID, CORROSIVE, N.O.S. or POLYAMINES, SOLID, CORROSIVE, N.O.S.	8	–	I	274	0	E0	P002	–	IBC07	B1
3259	AMINES, SOLID, CORROSIVE, N.O.S. or POLYAMINES, SOLID, CORROSIVE, N.O.S.	8	–	II	274	1 kg	E2	P002	–	IBC08	B2 B4
3259	AMINES, SOLID, CORROSIVE, N.O.S. or POLYAMINES, SOLID, CORROSIVE, N.O.S.	8	–	III	223 274	5 kg	E1	P002 LP02	–	IBC08	B3
3260	CORROSIVE SOLID, ACIDIC, INORGANIC, N.O.S.	8	–	I	274	0	E0	P002	–	IBC07	B1
3260	CORROSIVE SOLID, ACIDIC, INORGANIC, N.O.S.	8	–	II	274	1 kg	E2	P002	–	IBC08	B2 B4
3260	CORROSIVE SOLID, ACIDIC, INORGANIC, N.O.S.	8	–	III	223 274	5 kg	E1	P002 LP02	–	IBC08	B3
3261	CORROSIVE SOLID, ACIDIC, ORGANIC, N.O.S.	8	–	I	274	0	E0	P002	–	IBC07	B1
3261	CORROSIVE SOLID, ACIDIC, ORGANIC, N.O.S.	8	–	II	274	1 kg	E2	P002	–	IBC08	B2 B4
3261	CORROSIVE SOLID, ACIDIC, ORGANIC, N.O.S.	8	–	III	223 274	5 kg	E1	P002 LP02	–	IBC08	B3

Portable tanks and bulk containers			EmS	Stowage and segregation	Properties and observations	UN No.
	Tank instructions	Provisions				
(12)	(13) 4.2.5 4.3	(14) 4.2.5	(15) 5.4.3.2 7.3	(16) 7.1 7.2	(17)	(18)
–	–	–	F-F, S-G	Category D. Protected from sources of heat. Clear of living quarters. Keep as cool as reasonably practicable. During transport, it shall be stored (or kept) in a cool ventilated place.	May explode if involved in a fire under confined conditions. Sensitive to strong detonation shock.	3251 △
–	T50	–	F-D, S-U	Category D. Clear of living quarters.	Flammable colourless gas. Heavier than air (1.8).	3252
–	T1	TP33	F-A, S-B	Category A. "Separated from" acids.	Colourless hygroscopic solid. Dangerous reaction with oxidizers. In the presence of moisture, reacts with aluminium, zinc, tin and their compounds, evolving hydrogen, a flammable gas. Causes burns to skin, eyes and mucous membranes. Reacts violently with acids.	3253
–	T21	TP2 TP7	F-A, S-M	Category D. "Separated from" CARBON TETRACHLORIDE (UN 1846).	Colourless yellowish liquid. Insoluble in water. Strong garlic odour (phosphine). Liable to heat and ignite spontaneously in air. If involved in a fire, evolves phosphine, a flammable and highly toxic gas. Reacts violently with oxidizing substances (peroxides, halogens, nitric oxides and carbon tetrachloride). Irritating to mucous membranes.	3254
–	–	–	F-A, S-M	Category D.	Volatile flammable slightly yellow liquid with a pungent odour. Immiscible with water. Boiling point: 77°C to 79°C. Flashpoint between –15°C and –10°C. Exposure to light causes immediate dangerous decomposition. Causes burns to skin, eyes and mucous membranes.	3255
–	T3	TP3 TP29	F-E, S-D	Category A.	–	3256
–	T3	TP3 TP29	F-A, <u>S-P</u>	Category A. If under deck, in a mechanically ventilated space.	Any liquid which is transported at or above 100°C but below its flashpoint. May cause fire if in contact with combustible material due to extreme temperature.	3257
–	–	–	F-A, <u>S-P</u>	Category A. If under deck, in a mechanically ventilated space.	Any solid which is transported at or above 240°C. May cause fire if in contact with combustible material due to extreme temperature.	3258
–	T6	TP33	F-A, S-B	Category A. "Separated from" acids.	Colourless to yellowish solids with a pungent odour. Miscible with or soluble in water. When involved in a fire, evolve toxic gases. Corrosive to most metals, especially to copper and its alloys. Cause burns to skin, eyes and mucous membranes. React violently with acids.	3259
–	T3	TP33	F-A, S-B	Category A. "Separated from" acids.	See entry above.	3259
–	T1	TP33	F-A, S-B	Category A. "Separated from" acids.	See entry above.	3259
–	T6	TP33	F-A, S-B	Category B.	Causes burns to skin, eyes and mucous membranes.	3260
–	T3	TP33	F-A, S-B	Category B.	See entry above.	3260
–	T1	TP33	F-A, S-B	Category A.	See entry above.	3260
–	T6	TP33	F-A, S-B	Category B.	Causes burns to skin, eyes and mucous membranes.	3261
–	T3	TP33	F-A, S-B	Category B.	See entry above.	3261
–	T1	TP33	F-A, S-B	Category A.	See entry above.	3261

UN No.	Proper Shipping Name (PSN)	Class or division	Subsidiary risk(s)	Packing group	Special provisions	Limited and excepted quantity provisions		Packing		IBC	
						Limited quantities	Excepted quantities	Instructions	Provisions	Instructions	Provisions
(1)	(2) 3.1.2	(3) 2.0	(4) 2.0	(5) 2.0.1.3	(6) 3.3	(7a) 3.4	(7b) 3.5	(8) 4.1.4	(9) 4.1.4	(10) 4.1.4	(11) 4.1.4
3262	CORROSIVE SOLID, BASIC, INORGANIC, N.O.S.	8	–	I	274	0	E0	P002	–	IBC07	B1
3262	CORROSIVE SOLID, BASIC, INORGANIC, N.O.S.	8	–	II	274	1 kg	E2	P002	–	IBC08	B2 B4
3262	CORROSIVE SOLID, BASIC, INORGANIC, N.O.S.	8	–	III	223 274	5 kg	E1	P002 LP02	–	IBC08	B3
3263	CORROSIVE SOLID, BASIC, ORGANIC, N.O.S.	8	–	I	274	0	E0	P002	–	IBC07	B1
3263	CORROSIVE SOLID, BASIC, ORGANIC, N.O.S.	8	–	II	274	1 kg	E2	P002	–	IBC08	B2 B4
3263	CORROSIVE SOLID, BASIC, ORGANIC, N.O.S.	8	–	III	223 274	5 kg	E1	P002 LP02	–	IBC08	B3
3264	CORROSIVE LIQUID, ACIDIC, INORGANIC, N.O.S.	8	–	I	274	0	E0	P001	–	–	–
3264	CORROSIVE LIQUID, ACIDIC, INORGANIC, N.O.S.	8	–	II	274	1 L	E2	P001	–	IBC02	–
3264	CORROSIVE LIQUID, ACIDIC, INORGANIC, N.O.S.	8	–	III	223 274	5 L	E1	P001 LP01	–	IBC03	–
3265	CORROSIVE LIQUID, ACIDIC, ORGANIC, N.O.S.	8	–	I	274	0	E0	P001	–	–	–
3265	CORROSIVE LIQUID, ACIDIC, ORGANIC, N.O.S.	8	–	II	274	1 L	E2	P001	–	IBC02	–
3265	CORROSIVE LIQUID, ACIDIC, ORGANIC, N.O.S.	8	–	III	223 274	5 L	E1	P001 LP01	–	IBC03	–
3266	CORROSIVE LIQUID, BASIC, INORGANIC, N.O.S.	8	–	I	274	0	E0	P001	–	–	–
3266	CORROSIVE LIQUID, BASIC, INORGANIC, N.O.S.	8	–	II	274	1 L	E2	P001	–	IBC02	–
3266	CORROSIVE LIQUID, BASIC, INORGANIC, N.O.S.	8	–	III	223 274	5 L	E1	P001 LP01	–	IBC03	–
3267	CORROSIVE LIQUID, BASIC, ORGANIC, N.O.S.	8	–	I	274	0	E0	P001	–	–	–
3267	CORROSIVE LIQUID, BASIC, ORGANIC, N.O.S.	8	–	II	274	1 L	E2	P001	–	IBC02	–
3267	CORROSIVE LIQUID, BASIC, ORGANIC, N.O.S.	8	–	III	223 274	5 L	E1	P001 LP01	–	IBC03	–
3268	AIR BAG INFLATORS or AIR BAG MODULES or SEAT-BELT PRETENSIONERS	9	–	III	280 289	0	E0	P902 LP902	–	–	–
3269	POLYESTER RESIN KIT	3	–	II	236 340	5 L	See SP340	P302	–	–	–
3269	POLYESTER RESIN KIT	3	–	III	236 340	5 L	See SP340	P302	–	–	–
3270	NITROCELLULOSE MEMBRANE FILTERS with not more than 12.6% nitrogen, by dry mass	4.1	–	II	237 286	1 kg	E2	P411	–	–	–

Portable tanks and bulk containers			EmS	Stowage and segregation	Properties and observations	UN No.
	Tank instructions	Provisions				
(12)	(13) 4.2.5 4.3	(14) 4.2.5	(15) 5.4.3.2 7.3	(16) 7.1 7.2	(17)	(18)
–	T6	TP33	F-A, S-B	Category B. "Separated from" acids.	Reacts violently with acids. Causes burns to skin, eyes and mucous membranes.	3262
–	T3	TP33	F-A, S-B	Category B. "Separated from" acids.	See entry above.	3262
–	T1	TP33	F-A, S-B	Category A. "Separated from" acids.	See entry above.	3262
–	T6	TP33	F-A, S-B	Category B. "Separated from" acids.	Reacts violently with acids. Causes burns to skin, eyes and mucous membranes.	3263
–	T3	TP33	F-A, S-B	Category B. "Separated from" acids.	See entry above.	3263
–	T1	TP33	F-A, S-B	Category A. "Separated from" acids.	See entry above.	3263
–	T14	TP2 TP27	F-A, S-B	Category B. Clear of living quarters.	Causes burns to skin, eyes and mucous membranes.	3264
–	T11	TP2 TP27	F-A, S-B	Category B. Clear of living quarters.	See entry above.	3264
–	T7	TP1 TP28	F-A, S-B	Category A. Clear of living quarters.	See entry above.	3264
–	T14	TP2 TP27	F-A, S-B	Category B. Clear of living quarters.	Causes burns to skin, eyes and mucous membranes.	3265
–	T11	TP2 TP27	F-A, S-B	Category B. Clear of living quarters.	See entry above.	3265
–	T7	TP1 TP28	F-A, S-B	Category A. Clear of living quarters.	See entry above.	3265
–	T14	TP2 TP27	F-A, S-B	Category B. Clear of living quarters. "Separated from" acids.	Reacts violently with acids. Causes burns to skin, eyes and mucous membranes.	3266
–	T11	TP2 TP27	F-A, S-B	Category B. Clear of living quarters. "Separated from" acids.	See entry above.	3266
–	T7	TP1 TP28	F-A, S-B	Category A. Clear of living quarters. "Separated from" acids.	See entry above.	3266
–	T14	TP2 TP27	F-A, S-B	Category B. Clear of living quarters. "Separated from" acids.	Reacts violently with acids. Causes burns to skin, eyes and mucous membranes.	3267
–	T11	TP2 TP27	F-A, S-B	Category B. Clear of living quarters. "Separated from" acids.	See entry above.	3267
–	T7	TP1 TP28	F-A, S-B	Category A. Clear of living quarters. "Separated from" acids.	See entry above.	3267
–	–	–	F-B, S-X	Category A.	–	3268
–	–	–	F-E, S-D	Category B.	Polyester resin kits consist of two components: a base material (flammable liquid, packaging group II) and an activator (organic peroxide), each separately packed in an inner packaging.	3269
–	–	–	F-E, S-D	Category A.	See entry above.	3269
–	–	–	F-A, S-I	Category D.	Filters may be small round pieces or large sheets. When involved in a fire, evolve toxic fumes; in closed compartments, these fumes may form an explosive mixture with air. Burn rapidly with intense heat radiation.	3270

UN No.	Proper Shipping Name (PSN)	Class or division	Subsidiary risk(s)	Packing group	Special provisions	Limited and excepted quantity provisions		Packing		IBC	
						Limited quantities	Excepted quantities	Instructions	Provisions	Instructions	Provisions
(1)	(2) 3.1.2	(3) 2.0	(4) 2.0	(5) 2.0.1.3	(6) 3.3	(7a) 3.4	(7b) 3.5	(8) 4.1.4	(9) 4.1.4	(10) 4.1.4	(11) 4.1.4
3271	ETHERS, N.O.S.	3	–	II	274	1 L	E2	P001	–	IBC02	–
3271	ETHERS, N.O.S.	3	–	III	223 274	5 L	E1	P001 LP01	–	IBC03	–
3272	ESTERS, N.O.S.	3	–	II	274	1 L	E2	P001	–	IBC02	–
3272	ESTERS, N.O.S.	3	–	III	223 274	5 L	E1	P001 LP01	–	IBC03	–
3273	NITRILES, FLAMMABLE, TOXIC, N.O.S.	3	6.1	I	274	0	E0	P001	–	–	–
3273	NITRILES, FLAMMABLE, TOXIC, N.O.S.	3	6.1	II	274	1 L	E2	P001	–	IBC02	–
3274	ALCOHOLATES SOLUTION, N.O.S. in alcohol	3	8	II	274	1 L	E2	P001	–	IBC02	–
3275	NITRILES, TOXIC, FLAMMABLE, N.O.S.	6.1	3	I	274 315	0	E5	P001	–	–	–
3275	NITRILES, TOXIC, FLAMMABLE, N.O.S.	6.1	3	II	274	100 mL	E4	P001	–	IBC02	–
△ 3276	NITRILES, LIQUID, TOXIC, N.O.S.	6.1	–	I	274 315	0	E5	P001	–	–	–
△ 3276	NITRILES, LIQUID, TOXIC, N.O.S.	6.1	–	II	274	100 mL	E4	P001	–	IBC02	–
△ 3276	NITRILES, LIQUID, TOXIC, N.O.S.	6.1	–	III	223 274	5 L	E1	P001 LP01	–	IBC03	–
△ 3277	CHLOROFORMATES, TOXIC, CORROSIVE, N.O.S.	6.1	8	II	274	100 mL	E4	P001	–	IBC02	–
△ 3278	ORGANOPHOSPHORUS COMPOUND, LIQUID, TOXIC, N.O.S.	6.1	–	I	43 274 315	0	E5	P001	–	–	–
△ 3278	ORGANOPHOSPHORUS COMPOUND, LIQUID, TOXIC, N.O.S.	6.1	–	II	43 274	100 mL	E4	P001	–	IBC02	–
△ 3278	ORGANOPHOSPHORUS COMPOUND, LIQUID, TOXIC, N.O.S.	6.1	–	III	43 223 274	5 L	E1	P001 LP01	–	IBC03	–
3279	ORGANOPHOSPHORUS COMPOUND, TOXIC, FLAMMABLE, N.O.S.	6.1	3	I	43 274 315	0	E5	P001	–	–	–
3279	ORGANOPHOSPHORUS COMPOUND, TOXIC, FLAMMABLE, N.O.S.	6.1	3	II	43 274	100 mL	E4	P001	–	–	–
3280	ORGANOARSENIC COMPOUND, LIQUID, N.O.S.	6.1	–	I	274 315	0	E5	P001	–	–	–

Portable tanks and bulk containers			EmS	Stowage and segregation	Properties and observations	UN No.
	Tank instructions	Provisions				
(12)	(13) 4.2.5 4.3	(14) 4.2.5	(15) 5.4.3.2 7.3	(16) 7.1 7.2	(17)	(18)
–	T7	TP1 TP8 TP23	F-E, S-D	Category B.	–	3271
–	T4	TP1 TP29	F-E, S-D	Category A.	–	3271
–	T7	TP1 TP8 TP23	F-E, S-D	Category B.	–	3272
–	T4	TP1 TP29	F-E, S-D	Category A.	–	3272
–	T14	TP2 TP13 TP27	F-E, S-D	Category E. Clear of living quarters. "Separated from" acids.	Liquids evolving toxic vapour. React with acids or acid fumes, evolving hydrogen cyanide, a highly toxic and flammable gas. Toxic if swallowed, by skin contact or by inhalation.	3273
–	T11	TP2 TP13 TP27	F-E, S-D	Category B. Clear of living quarters. "Separated from" acids.	See entry above.	3273
–	–	–	F-E, S-C	Category B.	Colourless solution. Reacts violently with water. Causes burns to skin, eyes and mucous membranes.	3274
–	T14	TP2 TP13 TP27	F-E, S-D	Category B. Clear of living quarters. "Separated from" acids.	Flammable liquids, evolving toxic vapour. React with acids or acid fumes, evolving hydrogen cyanide, a highly toxic and flammable gas. Miscible with water. Toxic if swallowed, by skin contact or by inhalation.	3275
–	T11	TP2 TP13 TP27	F-E, S-D	Category B. Clear of living quarters. "Separated from" acids.	See entry above.	3275
–	T14	TP2 TP13 TP27	F-A, S-A	Category B. "Separated from" acids.	Liquids, evolving toxic vapour. React with acids or acid fumes, evolving hydrogen cyanide, a highly toxic and flammable gas. Miscible with water. Toxic if swallowed, by skin contact or by inhalation.	3276 △
–	T11	TP2 TP27	F-A, S-A	Category B. "Separated from" acids.	See entry above.	3276 △
–	T7	TP1 TP28	F-A, S-A	Category A. "Separated from" acids.	See entry above.	3276 △
–	T8	TP2 TP13 TP28	F-A, S-B	Category A. Keep as cool and dry as reasonably practicable. Protected from sources of heat. Clear of living quarters.	React and decompose with water or heat, evolving hydrogen chloride, an irritating and corrosive gas apparent as white fumes. Toxic if swallowed, by skin contact or by inhalation. Cause burns to skin, eyes and mucous membranes.	3277 △
–	T14	TP2 TP13 TP27	F-A, S-A	Category B.	Toxic if swallowed, by skin contact or by inhalation.	3278 △
–	T11	TP2 TP27	F-A, S-A	Category B.	See entry above.	3278 △
–	T7	TP1 TP28	F-A, S-A	Category A.	See entry above.	3278 △
–	T14	TP2 TP13 TP27	F-E, S-D	Category B. Clear of living quarters.	A wide variety of toxic flammable liquids. Toxic if swallowed, by skin contact or by inhalation.	3279
–	T11	TP2 TP13 TP27	F-E, S-D	Category B. Clear of living quarters.	See entry above.	3279
–	T14	TP2 TP13 TP27	F-A, S-A	Category B.	Toxic if swallowed, by skin contact or by inhalation.	3280

UN No. (1)	Proper Shipping Name (PSN) (2) 3.1.2	Class or division (3) 2.0	Subsidiary risk(s) (4) 2.0	Packing group (5) 2.0.1.3	Special provisions (6) 3.3	Limited and excepted quantity provisions: Limited quantities (7a) 3.4	Excepted quantities (7b) 3.5	Packing: Instructions (8) 4.1.4	Packing: Provisions (9) 4.1.4	IBC: Instructions (10) 4.1.4	IBC: Provisions (11) 4.1.4
3280	ORGANOARSENIC COMPOUND, LIQUID, N.O.S.	6.1	–	II	274	100 mL	E4	P001	–	IBC02	–
3280	ORGANOARSENIC COMPOUND, LIQUID, N.O.S.	6.1	–	III	223 274	5 L	E1	P001 LP01	–	IBC03	–
3281	METAL CARBONYLS, LIQUID, N.O.S.	6.1	–	I	274 315	0	E5	P601	–	–	–
3281	METAL CARBONYLS, LIQUID, N.O.S.	6.1	–	II	274	100 mL	E4	P001	–	IBC02	–
3281	METAL CARBONYLS, LIQUID, N.O.S.	6.1	–	III	223 274	5 L	E1	P001 LP01	–	IBC03	–
△ 3282	ORGANOMETALLIC COMPOUND, LIQUID, TOXIC, N.O.S.	6.1	–	I	274	0	E5	P001	–	–	–
△ 3282	ORGANOMETALLIC COMPOUND, LIQUID, TOXIC, N.O.S.	6.1	–	II	274	100 mL	E4	P001	–	IBC02	–
△ 3282	ORGANOMETALLIC COMPOUND, LIQUID, TOXIC, N.O.S.	6.1	–	III	223 274	5 L	E1	P001 LP01	–	IBC03	–
3283	SELENIUM COMPOUND, SOLID, N.O.S.	6.1	–	I	274	0	E5	P002	–	IBC07	B1
3283	SELENIUM COMPOUND, SOLID, N.O.S.	6.1	–	II	274	500 g	E4	P002	–	IBC08	B2 B4
3283	SELENIUM COMPOUND, SOLID, N.O.S.	6.1	–	III	223 274	5 kg	E1	P002 LP02	–	IBC08	B3
3284	TELLURIUM COMPOUND, N.O.S.	6.1	–	I	274	0	E5	P002	–	IBC07	B1
3284	TELLURIUM COMPOUND, N.O.S.	6.1	–	II	274	500 g	E4	P002	–	IBC08	B2 B4
3284	TELLURIUM COMPOUND, N.O.S.	6.1	–	III	223 274	5 kg	E1	P002 LP02	–	IBC08	B3
3285	VANADIUM COMPOUND, N.O.S.	6.1	–	I	274	0	E5	P002	–	IBC07	B1
3285	VANADIUM COMPOUND, N.O.S.	6.1	–	II	274	500 g	E4	P002	–	IBC08	B2 B4
3285	VANADIUM COMPOUND, N.O.S.	6.1	–	III	223 274	5 kg	E1	P002 LP02	–	IBC08	B3
3286	FLAMMABLE LIQUID, TOXIC, CORROSIVE, N.O.S.	3	6.1/8	I	274	0	E0	P001	–	–	–
3286	FLAMMABLE LIQUID, TOXIC, CORROSIVE, N.O.S.	3	6.1/8	II	274	1 L	E2	P001	–	IBC99	–
3287	TOXIC LIQUID, INORGANIC, N.O.S.	6.1	–	I	274 315	0	E5	P001	–	–	–
3287	TOXIC LIQUID, INORGANIC, N.O.S.	6.1	–	II	274	100 mL	E4	P001	–	IBC02	–
3287	TOXIC LIQUID, INORGANIC, N.O.S.	6.1	–	III	223 274	5 L	E1	P001 LP01	–	IBC03	–
3288	TOXIC SOLID, INORGANIC, N.O.S.	6.1	–	I	274	0	E5	P002	–	IBC99	–
3288	TOXIC SOLID, INORGANIC, N.O.S.	6.1	–	II	274	500 g	E4	P002	–	IBC08	B2 B4

Portable tanks and bulk containers			EmS	Stowage and segregation	Properties and observations	UN No.
	Tank instructions	Provisions				
(12)	(13) 4.2.5 4.3	(14) 4.2.5	(15) 5.4.3.2 7.3	(16) 7.1 7.2	(17)	(18)
–	T11	TP2 TP27	F-A, S-A	Category B.	Toxic if swallowed, by skin contact or by inhalation.	3280
–	T7	TP1 TP28	F-A, S-A	Category A.	See entry above.	3280
–	T14	TP2 TP13 TP27	F-A, S-A	Category D. Clear of living quarters.	A range of metal carbonyls which, when heated, can give off carbon monoxide, a toxic gas. Immiscible with water. Toxic if swallowed, by skin contact or by inhalation.	3281
–	T11	TP2 TP27	F-A, S-A	Category B. Clear of living quarters.	See entry above.	3281
–	T7	TP1 TP28	F-A, S-A	Category B. Clear of living quarters.	See entry above.	3281
–	T14	TP2 TP13 TP27	F-A, S-A	Category B.	Toxic if swallowed, by skin contact or by inhalation.	3282 △
–	T11	TP2 TP27	F-A, S-A	Category B.	See entry above.	3282 △
–	T7	TP1 TP28	F-A, S-A	Category A.	See entry above.	3282 △
–	T6	TP33	F-A, S-A	Category B.	Toxic if swallowed, by skin contact or by inhalation.	3283
–	T3	TP33	F-A, S-A	Category B.	See entry above.	3283
–	T1	TP33	F-A, S-A	Category A.	See entry above.	3283
–	T6	TP33	F-A, S-A	Category B.	Toxic if swallowed, by skin contact or by inhalation.	3284
–	T3	TP33	F-A, S-A	Category B.	See entry above.	3284
–	T1	TP33	F-A, S-A	Category A.	See entry above.	3284
–	T6	TP33	F-A, S-A	Category B.	Toxic if swallowed, by skin contact or by inhalation.	3285
–	T3	TP33	F-A, S-A	Category B.	See entry above.	3285
–	T1	TP33	F-A, S-A	Category A.	See entry above.	3285
–	T14	TP2 TP13 TP27	F-E, S-C	Category E. Clear of living quarters. Segregation as for class 3 but "away from" class 4.1.	Flammable, toxic, corrosive liquid. Toxic if swallowed, by skin contact or by inhalation. Causes burns to skin, eyes and mucous membranes.	3286
–	T11	TP2 TP13 TP27	F-E, S-C	Category B. Clear of living quarters. Segregation as for class 3 but "away from" class 4.1.	See entry above.	3286
–	T14	TP2 TP13 TP27	F-A, S-A	Category B. Clear of living quarters.	Toxic if swallowed, by skin contact or by inhalation.	3287
–	T11	TP2 TP27	F-A, S-A	Category B. Clear of living quarters.	See entry above.	3287
–	T7	TP1 TP28	F-A, S-A	Category A. Clear of living quarters.	See entry above.	3287
–	T6	TP33	F-A, S-A	Category B.	Toxic if swallowed, by skin contact or by inhalation.	3288
–	T3	TP33	F-A, S-A	Category B.	See entry above.	3288

UN No.	Proper Shipping Name (PSN)	Class or division	Subsidiary risk(s)	Packing group	Special provisions	Limited and excepted quantity provisions		Packing		IBC	
						Limited quantities	Excepted quantities	Instructions	Provisions	Instructions	Provisions
(1)	(2) 3.1.2	(3) 2.0	(4) 2.0	(5) 2.0.1.3	(6) 3.3	(7a) 3.4	(7b) 3.5	(8) 4.1.4	(9) 4.1.4	(10) 4.1.4	(11) 4.1.4
3288	TOXIC SOLID, INORGANIC, N.O.S.	6.1	–	III	223 274	5 kg	E1	P002 LP02	–	IBC08	B3
3289	TOXIC LIQUID, CORROSIVE, INORGANIC, N.O.S.	6.1	8	I	274 315	0	E5	P001	–	–	–
3289	TOXIC LIQUID, CORROSIVE, INORGANIC, N.O.S.	6.1	8	II	274	100 mL	E4	P001	–	IBC02	–
3290	TOXIC SOLID, CORROSIVE, INORGANIC, N.O.S.	6.1	8	I	274	0	E5	P002	–	IBC99	–
3290	TOXIC SOLID, CORROSIVE, INORGANIC, N.O.S.	6.1	8	II	274	500 g	E4	P002	–	IBC06	B2
3291	CLINICAL WASTE, UNSPECIFIED, N.O.S. or (BIO) MEDICAL WASTE, N.O.S. or REGULATED MEDICAL WASTE, N.O.S.	6.2	–	II	–	0	E0	P621 LP621	–	IBC620	–
3292	BATTERIES, CONTAINING SODIUM or CELLS, CONTAINING SODIUM	4.3	–	II	239	0	E0	P408	–	–	–
3293	HYDRAZINE, AQUEOUS SOLUTION with not more than 37% hydrazine, by mass	6.1	–	III	223	5 L	E1	P001 LP01	–	IBC03	–
△ 3294	HYDROGEN CYANIDE, SOLUTION IN ALCOHOL with not more than 45% hydrogen cyanide	6.1	3 P	I	900	0	E5	P601	–	–	–
3295	HYDROCARBONS, LIQUID, N.O.S.	3	–	I	–	500 mL	E3	P001	–	–	–
3295	HYDROCARBONS, LIQUID, N.O.S.	3	–	II	–	1 L	E2	P001	–	IBC02	–
3295	HYDROCARBONS, LIQUID, N.O.S.	3	–	III	223	5 L	E1	P001 LP01	–	IBC03	–
3296	HEPTAFLUOROPROPANE (REFRIGERANT GAS R 227)	2.2	–	–	–	120 mL	E1	P200	–	–	–
3297	ETHYLENE OXIDE AND CHLOROTETRAFLUORO-ETHANE MIXTURE with not more than 8.8% ethylene oxide	2.2	–	–	–	120 mL	E1	P200	–	–	–
3298	ETHYLENE OXIDE AND PENTAFLUOROETHANE MIXTURE with not more than 7.9% ethylene oxide	2.2	–	–	–	120 mL	E1	P200	–	–	–
3299	ETHYLENE OXIDE AND TETRAFLUOROETHANE MIXTURE with not more than 5.6% ethylene oxide	2.2	–	–	–	120 mL	E1	P200	–	–	–

Portable tanks and bulk containers			EmS	Stowage and segregation	Properties and observations	UN No.
	Tank instructions	Provisions				
(12)	(13) 4.2.5 4.3	(14) 4.2.5	(15) 5.4.3.2 7.3	(16) 7.1 7.2	(17)	(18)
–	T1	TP33	F-A, S-A	Category A.	Toxic if swallowed, by skin contact or by inhalation.	3288
–	T14	TP2 TP13 TP27	F-A, S-B	Category B. Clear of living quarters.	Toxic if swallowed, by skin contact or by inhalation. Causes burns to skin, eyes and mucous membranes.	3289
–	T11	TP2 TP27	F-A, S-B	Category B. Clear of living quarters.	See entry above.	3289
–	T6	TP33	F-A, S-B	Category B. Clear of living quarters.	Toxic if swallowed, by skin contact or by inhalation. Causes burns to skin, eyes and mucous membranes.	3290
–	T3	TP33	F-A, S-B	Category B. Clear of living quarters.	See entry above.	3290
–	BK2	–	F-A, S-T	As approved by the competent authority of the country of origin.	Derived from the medical treatment of animals, humans or from bio-research.	3291
–	–	–	F-G, S-P	Category A.	Series of hermetically sealed metal cells containing sodium, electrically connected and secured within a metal casing. "Cold" batteries (batteries containing elemental sodium only in the solid state) are electrically inert. Batteries are activated by heating to between 300°C and 350°C before operating to produce electricity. Activated batteries (i.e., "hot" batteries containing liquid elemental sodium) may cause fire through short-circuit of the terminals. Batteries or cells should not be offered for transport at a temperature such that liquid elemental sodium is present in the battery or cell unless approved, and under conditions of transport established by the competent authority.	3292
–	T4	TP1	F-A, S-A	Category A. "Separated from" acids.	Colourless liquid. Reacts violently with acids. Toxic if swallowed, by skin contact or by inhalation.	3293
–	T14	TP2 TP13	F-E, S-D	Category D. Clear of living quarters.	Flammable solution, evolving extremely toxic flammable vapours. Miscible with water. Highly toxic if swallowed, by skin contact or by inhalation. Transport of HYDROGEN CYANIDE, SOLUTION IN ALCOHOL with more than 45% hydrogen cyanide is **prohibited**.	3294 △
–	T11	TP1 TP8 TP28	F-E, S-D	Category E.	Immiscible with water.	3295
–	T7	TP1 TP8 TP28	F-E, S-D	Category B.	See entry above.	3295
–	T4	TP1 TP29	F-E, S-D	Category A.	See entry above.	3295
–	T50	–	F-C, S-V	Category A.	Non-flammable compressed gas. Heavier than air (1.4).	3296
–	T50	–	F-C, S-V	Category A.	Liquefied, non-flammable, colourless gas with an ether-like odour. Much heavier than air.	3297
–	T50	–	F-C, S-V	Category A.	Liquefied, non-flammable, colourless gas with an ether-like odour. Much heavier than air.	3298
–	T50	–	F-C, S-V	Category A.	Liquefied, non-flammable, colourless gas with an ether-like odour. Much heavier than air.	3299

UN No.	Proper Shipping Name (PSN)	Class or division	Subsidiary risk(s)	Packing group	Special provisions	Limited and excepted quantity provisions		Packing		IBC	
						Limited quantities	Excepted quantities	Instructions	Provisions	Instructions	Provisions
(1)	(2) 3.1.2	(3) 2.0	(4) 2.0	(5) 2.0.1.3	(6) 3.3	(7a) 3.4	(7b) 3.5	(8) 4.1.4	(9) 4.1.4	(10) 4.1.4	(11) 4.1.4
3300	ETHYLENE OXIDE AND CARBON DIOXIDE MIXTURE with more than 87% ethylene oxide	2.3	2.1	–	–	0	E0	P200	–	–	–
3301	CORROSIVE LIQUID, SELF-HEATING, N.O.S.	8	4.2	I	274	0	E0	P001	–	–	–
3301	CORROSIVE LIQUID, SELF-HEATING, N.O.S.	8	4.2	II	274	0	E2	P001	–	–	–
△ 3302	2-DIMETHYLAMINOETHYL ACRYLATE	6.1	–	II	–	100 mL	E4	P001	–	IBC02	–
3303	COMPRESSED GAS, TOXIC, OXIDIZING, N.O.S.	2.3	5.1	–	274	0	E0	P200	–	–	–
3304	COMPRESSED GAS, TOXIC, CORROSIVE, N.O.S.	2.3	8	–	274	0	E0	P200	–	–	–
3305	COMPRESSED GAS, TOXIC, FLAMMABLE, CORROSIVE, N.O.S.	2.3	2.1/8	–	274	0	E0	P200	–	–	–
3306	COMPRESSED GAS, TOXIC, OXIDIZING, CORROSIVE, N.O.S.	2.3	5.1/8	–	274	0	E0	P200	–	–	–
3307	LIQUEFIED GAS, TOXIC, OXIDIZING, N.O.S.	2.3	5.1	–	274	0	E0	P200	–	–	–
3308	LIQUEFIED GAS, TOXIC, CORROSIVE, N.O.S.	2.3	8	–	274	0	E0	P200	–	–	–
3309	LIQUEFIED GAS, TOXIC, FLAMMABLE, CORROSIVE, N.O.S.	2.3	2.1/8	–	274	0	E0	P200	–	–	–
3310	LIQUEFIED GAS, TOXIC, OXIDIZING, CORROSIVE, N.O.S.	2.3	5.1/8	–	274	0	E0	P200	–	–	–
3311	GAS, REFRIGERATED LIQUID, OXIDIZING, N.O.S.	2.2	5.1	–	274	0	E0	P203	–	–	–
3312	GAS, REFRIGERATED LIQUID, FLAMMABLE, N.O.S.	2.1	–	–	274	0	E0	P203	–	–	–
3313	ORGANIC PIGMENTS, SELF-HEATING	4.2	–	II	–	0	E2	P002	–	IBC08	B2 B4
3313	ORGANIC PIGMENTS, SELF-HEATING	4.2	–	III	223	0	E1	P002 LP02	–	IBC08	B3
△ 3314	PLASTICS MOULDING COMPOUND in dough, sheet or extruded rope form, evolving flammable vapour	9	–	III	207 965	5 kg	E1	P002	PP14	IBC08	B3 B6

Portable tanks and bulk containers			EmS	Stowage and segregation	Properties and observations	UN No.
	Tank instructions	Provisions				
(12)	(13) 4.2.5 4.3	(14) 4.2.5	(15) 5.4.3.2 7.3	(16) 7.1 7.2	(17)	(18)
–	–	–	F-D, S-U	Category D. Clear of living quarters.	Liquefied, flammable, toxic colourless gas with an ether-like odour. Heavier than air (1.5).	3300
–	–	–	F-A, S-J	Category D.	Causes burns to skin, eyes and mucous membranes.	3301
–	–	–	F-A, S-J	Category D.	See entry above.	3301
–	T7	TP2	F-A, S-A	Category D. Protected from sources of heat.	Colourless to light yellow liquid. Acrid odour. Miscible with water. Causes tears. Stabilized with hydroquinone derivatives. Hydrolyses in water to give off acrylic acid and dimethylaminoethanol. Toxic if swallowed, by skin contact or by inhalation.	3302 △
–	–	–	F-C, S-W	Category D. Clear of living quarters.	–	3303
–	–	–	F-C, S-U	Category D. Clear of living quarters.	–	3304
–	–	–	F-D, S-U	Category D. Clear of living quarters. Segregation as for class 2.1 but "away from" class 4.3.	–	3305
–	–	–	F-C, S-W	Category D. Clear of living quarters. Segregation as for class 5.1 but "separated from" class 7.	–	3306
–	–	–	F-C, S-W	Category D. Clear of living quarters.	–	3307
–	–	–	F-C, S-U	Category D. Clear of living quarters.	–	3308
–	–	–	F-D, S-U	Category D. Clear of living quarters. Segregation as for class 2.1 but "away from" class 4.3.	–	3309
–	–	–	F-C, S-W	Category D. Clear of living quarters. Segregation as for class 5.1 but "separated from" class 7.	–	3310
–	T75	TP5 TP22	F-C, S-W	Category D.	–	3311
–	T75	TP5	F-D, S-U	Category D. Clear of living quarters.	–	3312
–	T3	TP33	F-A, S-J	Category C.	Self-heating coloured powder or granules. Odourless. Liable to self-heating or spontaneous combustion.	3313
–	T1	TP33	F-A, S-J	Category C.	See entry above.	3313
–	BK2	–	F-A, S-I	Category E. Protected from sources of heat. When stowed under-deck, mechanical ventilation shall be in accordance with SOLAS regulation II-2/19 (II-2/54) for flammable liquids with flashpoint below 23°C c.c. Segregation as for class 3 but "separated from" class 1 except division 1.4S.	A moulding material consisting predominantly of polystyrene, poly(methyl methacrylate) or other polymeric material and containing 5% to 8% of a volatile hydrocarbon which is predominantly pentane. During storage, a small proportion of this pentane is released to the atmosphere; this proportion increases at elevated temperatures.	3314 △

UN No.	Proper Shipping Name (PSN)	Class or division	Subsidiary risk(s)	Packing group	Special provisions	Limited and excepted quantity provisions		Packing		IBC	
						Limited quantities	Excepted quantities	Instructions	Provisions	Instructions	Provisions
(1)	(2) 3.1.2	(3) 2.0	(4) 2.0	(5) 2.0.1.3	(6) 3.3	(7a) 3.4	(7b) 3.5	(8) 4.1.4	(9) 4.1.4	(10) 4.1.4	(11) 4.1.4
3315	CHEMICAL SAMPLE, TOXIC	6.1	–	I	250	0	E5	P099	–	–	–
3316	CHEMICAL KIT or FIRST AID KIT	9	–	–	251 340	See SP251	See SP340	P901	–	–	–
3317	2-AMINO-4,6-DINITROPHENOL, WETTED with not less than 20% water, by mass	4.1	–	I	28	0	E0	P406	PP26 PP31	–	–
3318	AMMONIA SOLUTION relative density less than 0.880 at 15°C in water, with more than 50% ammonia	2.3	8	–	23	0	E0	P200	–	–	–
△ 3319	NITROGLYCERIN MIXTURE, DESENSITIZED, SOLID, N.O.S. with more than 2% but not more than 10% nitroglycerin, by mass	4.1	–	II	272 274	0	E0	P099	–	–	–
3320	SODIUM BOROHYDRIDE AND SODIUM HYDROXIDE SOLUTION with not more than 12% sodium borohydride and not more than 40% sodium hydroxide, by mass	8	–	II	–	1 L	E2	P001	–	IBC02	–
3320	SODIUM BOROHYDRIDE AND SODIUM HYDROXIDE SOLUTION with not more than 12% sodium borohydride and not more than 40% sodium hydroxide, by mass	8	–	III	223	5 L	E1	P001 LP01	–	IBC03	–
△ 3321	RADIOACTIVE MATERIAL, LOW SPECIFIC ACTIVITY (LSA-II), non fissile or fissile – excepted	7	See SP172	–	172 317 325	0	E0	See 4.1.9	See 4.1.9	See 4.1.9	See 4.1.9
△ 3322	RADIOACTIVE MATERIAL, LOW SPECIFIC ACTIVITY (LSA-III), non fissile or fissile – excepted	7	See SP172	–	172 317 325	0	E0	See 4.1.9	See 4.1.9	See 4.1.9	See 4.1.9
△ 3323	RADIOACTIVE MATERIAL, TYPE C PACKAGE, non fissile or fissile – excepted	7	See SP172	–	172 317 325	0	E0	See 4.1.9	See 4.1.9	See 4.1.9	See 4.1.9

Portable tanks and bulk containers			EmS	Stowage and segregation	Properties and observations	UN No.
	Tank instructions	Provisions				
(12)	(13) 4.2.5 4.3	(14) 4.2.5	(15) 5.4.3.2 7.3	(16) 7.1 7.2	(17)	(18)
–	–	–	F-A, S-A	Category D. Clear of living quarters.	This entry may only be used for samples of chemicals taken for analysis in connection with the implementation of the Convention on the Prohibition of the Development, Production, Stockpiling and Use of Chemical Weapons and on their Destruction. The transport of substances under this entry shall be in accordance with the chain of custody and security procedures specified by the Organization for the Prohibition of Chemical Weapons. The chemical sample may only be transported providing prior approval has been granted by the competent authority or the Director General of the Organization for the Prohibition of Chemical Weapons. During transport, the packaging shall be accompanied by a copy of the document of approval for transport, showing the quantity limitations and the packing requirements.	3315
–	–	–	F-A, S-P	Category A.	–	3316
–	–	–	F-B, S-J	Category D. "Away from" class 3 and heavy metals and their salts.	Desensitized explosive. Red crystals. Insoluble in water. Explosive in the dry state. May form extremely sensitive compounds with heavy metals or their salts. When involved in a fire, evolves toxic fumes; in closed compartments these fumes may form an explosive mixture with air. Harmful if swallowed or by skin contact.	3317
–	T50	–	F-C, S-U	Category D. Clear of living quarters. "Separated from" chlorine and acids.	Highly concentrated solution in water of a non-flammable, toxic and corrosive gas with a pungent odour. Even though this substance has a flammability hazard, it only exhibits such hazard under extreme fire conditions in confined areas. Reacts violently with acids. Highly irritating to skin, eyes and mucous membranes. Suffocating in low concentrations.	3318
–	–	–	F-B, S-J	Category E.	Desensitized explosive with lactose, glucose or cellulose. White solid. Soluble in water. When involved in a fire, the nitroglycerin may accumulate and may produce an explosion. Contact with water may dissolve the desensitizer (lactose or glucose), causing migration and accumulation of the nitroglycerin, which may explode. Nitroglycerin is more dense than water. When involved in a fire, evolves toxic fumes; in closed compartments these fumes may form an explosive mixture with air. Inhalation of vapours may cause headaches, dizziness and fainting.	3319 △
–	T7	TP2	F-A, S-B	Category A. "Separated from" acids.	Off-white clear liquid with a slight hydrocarbon odour. Reacts violently with acids. In contact with acids or if diluted with large amount of water, evolves hydrogen gas and heat. Causes burns to skin, eyes and mucous membranes.	3320
–	T4	TP2	F-A, S-B	Category A. "Separated from" acids.	See entry above.	3320
–	T5	TP4	F-I, S-S	Category A, except for uranyl nitrate hexahydrate solution for which category D applies.	See 1.5.1.	3321 △
–	T5	TP4	F-I, S-S	Category A, except for uranyl nitrate hexahydrate solution for which category D applies.	See 1.5.1.	3322 △
–	–	–	F-I, S-S	Category A, taking account of any supplementary requirements specified in the transport documents.	See 1.5.1. For ships transporting an INF cargo as defined in regulation VII/14 of the SOLAS Convention, 1974, as amended, refer also to the INF Code.	3323 △

UN No.	Proper Shipping Name (PSN)	Class or division	Subsidiary risk(s)	Packing group	Special provisions	Limited and excepted quantity provisions		Packing		IBC	
						Limited quantities	Excepted quantities	Instructions	Provisions	Instructions	Provisions
(1)	(2) 3.1.2	(3) 2.0	(4) 2.0	(5) 2.0.1.3	(6) 3.3	(7a) 3.4	(7b) 3.5	(8) 4.1.4	(9) 4.1.4	(10) 4.1.4	(11) 4.1.4
△ 3324	RADIOACTIVE MATERIAL, LOW SPECIFIC ACTIVITY (LSA-II), FISSILE	7	See SP172	–	172 326	0	E0	See 4.1.9	See 4.1.9	See 4.1.9	See 4.1.9
△ 3325	RADIOACTIVE MATERIAL, LOW SPECIFIC ACTIVITY, (LSA-III), FISSILE	7	See SP172	–	172 326	0	E0	See 4.1.9	See 4.1.9	See 4.1.9	See 4.1.9
△ 3326	RADIOACTIVE MATERIAL, SURFACE CONTAMINATED OBJECTS (SCO-I or SCO-II), FISSILE	7	See SP172	–	172	0	E0	See 4.1.9	See 4.1.9	See 4.1.9	See 4.1.9
△ 3327	RADIOACTIVE MATERIAL, TYPE A PACKAGE, FISSILE, non-special form	7	See SP172	–	172 326	0	E0	See 4.1.9	See 4.1.9	See 4.1.9	See 4.1.9
△ 3328	RADIOACTIVE MATERIAL, TYPE B(U) PACKAGE, FISSILE	7	See SP172	–	172 326	0	E0	See 4.1.9	See 4.1.9	See 4.1.9	See 4.1.9
△ 3329	RADIOACTIVE MATERIAL, TYPE B(M) PACKAGE, FISSILE	7	See SP172	–	172 326	0	E0	See 4.1.9	See 4.1.9	See 4.1.9	See 4.1.9
△ 3330	RADIOACTIVE MATERIAL, TYPE C PACKAGE, FISSILE	7	See SP172	–	172 326	0	E0	See 4.1.9	See 4.1.9	See 4.1.9	See 4.1.9
△ 3331	RADIOACTIVE MATERIAL, TRANSPORTED UNDER SPECIAL ARRANGEMENT, FISSILE	7	See SP172	–	172 326	0	E0	See 4.1.9	See 4.1.9	See 4.1.9	See 4.1.9
△ 3332	RADIOACTIVE MATERIAL, TYPE A PACKAGE, SPECIAL FORM, non fissile or fissile – excepted	7	See SP172	–	172 326	0	E0	See 4.1.9	See 4.1.9	See 4.1.9	See 4.1.9
△ 3333	RADIOACTIVE MATERIAL, TYPE A PACKAGE, SPECIAL FORM, FISSILE	7	See SP172	–	172	0	E0	See 4.1.9	See 4.1.9	See 4.1.9	See 4.1.9
3334	AVIATION REGULATED LIQUID, N.O.S.	9	–	–	960	–	–	–	–	–	–
3335	AVIATION REGULATED SOLID, N.O.S.	9	–	–	960	–	–	–	–	–	–

Portable tanks and bulk containers			EmS	Stowage and segregation	Properties and observations	UN No.
	Tank instructions	Provisions				
(12)	(13) 4.2.5 4.3	(14) 4.2.5	(15) 5.4.3.2 7.3	(16) 7.1 7.2	(17)	(18)
–	–	–	F-I, S-S	Category A, except for uranyl nitrate hexahydrate solution for which category D applies, taking account of any supplementary requirements specified in the transport documents.	See 1.5.1.	3324 △
–	–	–	F-I, S-S	Category A, taking account of any supplementary requirements specified in the transport documents.	See 1.5.1.	3325 △
–	–	–	F-I, S-S	Category A, taking account of any supplementary requirements specified in the transport documents.	See 1.5.1.	3326 △
–	–	–	F-I, S-S	Category A, except for uranyl nitrate hexahydrate solution, uranium metal pyrophoric and thorium metal pyrophoric for which category D applies, taking account of any supplementary requirements specified in the transport documents.	See 1.5.1.	3327 △
–	–	–	F-I, S-S	Category A, taking account of any supplementary requirements specified in the transport documents.	See 1.5.1. For ships transporting an INF cargo as defined in regulation VII/14 of the SOLAS Convention, 1974, as amended, refer also to the INF Code.	3328 △
–	–	–	F-I, S-S	Category A, taking account of any supplementary requirements specified in the transport documents.	See 1.5.1. For ships transporting an INF cargo as defined in regulation VII/14 of the SOLAS Convention, 1974, as amended, refer also to the INF Code.	3329 △
–	–	–	F-I, S-S	Category A, taking account of any supplementary requirements specified in the transport documents.	See 1.5.1. For ships transporting an INF cargo as defined in regulation VII/14 of the SOLAS Convention, 1974, as amended, refer also to the INF Code.	3330 △
–	–	–	F-I, S-S	Category A, taking account of any supplementary requirements specified in the competent authority approval certificate(s).	See 1.5.1. For ships transporting an INF cargo as defined in regulation VII/14 of the SOLAS Convention, 1974, as amended, refer also to the INF Code.	3331 △
–	–	–	F-I, S-S	Category A.	See 1.5.1.	3332 △
–	–	–	F-I, S-S	Category A, taking account of any supplementary requirements specified in the transport documents.	See 1.5.1.	3333 △
–	–	–	–	–	Not subject to the provisions of this Code but may be subject to provisions governing the transport of dangerous goods by other modes.	3334
–	–	–	–	–	Not subject to the provisions of this Code but may be subject to provisions governing the transport of dangerous goods by other modes.	3335

UN No.	Proper Shipping Name (PSN)	Class or division	Subsidiary risk(s)	Packing group	Special provisions	Limited and excepted quantity provisions		Packing		IBC	
						Limited quantities	Excepted quantities	Instructions	Provisions	Instructions	Provisions
(1)	(2) 3.1.2	(3) 2.0	(4) 2.0	(5) 2.0.1.3	(6) 3.3	(7a) 3.4	(7b) 3.5	(8) 4.1.4	(9) 4.1.4	(10) 4.1.4	(11) 4.1.4
△ 3336	MERCAPTANS, LIQUID, FLAMMABLE, N.O.S. or MERCAPTAN MIXTURE, LIQUID, FLAMMABLE, N.O.S.	3	–	I	274	0	E3	P001	–	–	–
△ 3336	MERCAPTANS, LIQUID, FLAMMABLE, N.O.S. or MERCAPTAN MIXTURE, LIQUID, FLAMMABLE, N.O.S.	3	–	II	274	1 L	E2	P001	–	IBC02	–
△ 3336	MERCAPTANS, LIQUID, FLAMMABLE, N.O.S. or MERCAPTAN MIXTURE, LIQUID, FLAMMABLE, N.O.S.	3	–	III	223 274	5 L	E1	P001 LP01	–	IBC03	–
3337	REFRIGERANT GAS R 404A	2.2	–	–	–	120 mL	E1	P200	–	–	–
3338	REFRIGERANT GAS R 407A	2.2	–	–	–	120 mL	E1	P200	–	–	–
3339	REFRIGERANT GAS R 407B	2.2	–	–	–	120 mL	E1	P200	–	–	–
3340	REFRIGERANT GAS R 407C	2.2	–	–	–	120 mL	E1	P200	–	–	–
3341	THIOUREA DIOXIDE	4.2	–	II	–	0	E2	P002	PP31	IBC06	B2
3341	THIOUREA DIOXIDE	4.2	–	III	223	0	E1	P002 LP02	PP31	IBC08	B3
3342	XANTHATES	4.2	–	II	–	0	E2	P002	PP31	IBC06	B2
3342	XANTHATES	4.2	–	III	223	0	E1	P002 LP02	PP31	IBC08	B3
3343	NITROGLYCERIN MIXTURE, DESENSITIZED, LIQUID, FLAMMABLE, N.O.S. with not more than 30% nitroglycerin, by mass	3	–	–	274 278	0	E0	P099	–	–	–
△ 3344	PENTAERYTHRITE TETRANITRATE (PENTAERYTHRITOL TETRANITRATE; PETN) MIXTURE, DESENSITIZED, SOLID, N.O.S. with more than 10% but not more than 20% PETN, by mass	4.1	–	II	272 274	0	E0	P406	PP26 PP80	–	–

Portable tanks and bulk containers			EmS	Stowage and segregation	Properties and observations	UN No.	
	Tank instructions	Provisions					
(12)	(13) 4.2.5 4.3	(14) 4.2.5	(15) 5.4.3.2 7.3	(16) 7.1 7.2	(17)	(18)	
–	T11	TP2	F-E, S-D	Category E. Segregation from foodstuffs as in 7.3.4.2.1, 7.6.3.1.2 or 7.7.3.6. "Separated from" odour-absorbing cargoes.	Colourless to yellow liquids with a garlic odour. Immiscible with water.	3336	△
–	T7	TP1 TP3 TP28	F-E, S-D	Category B. Segregation from foodstuffs as in 7.3.4.2.1, 7.6.3.1.2 or 7.7.3.6. "Separated from" odour-absorbing cargoes.	See entry above.	3336	△
–	T4	TP1 TP29	F-E, S-D	Category B. Segregation from foodstuffs as in 7.3.4.2.1, 7.6.3.1.2 or 7.7.3.6. "Separated from" odour-absorbing cargoes.	See entry above.	3336	△
–	T50	–	F-C, S-V	Category A.	Liquefied, non-flammable, colourless gas with a faint ether-like odour. Heavier than air (1.06). Very high exposures may cause anaesthetic effects and asphyxiation.	3337	
–	T50	–	F-C, S-V	Category A.	Liquefied, non-flammable, colourless gas with a faint ether-like odour. Heavier than air (1.17). Very high exposures may cause anaesthetic effects and asphyxiation.	3338	
–	T50	–	F-C, S-V	Category A.	Liquefied, non-flammable, colourless gas with a faint ether-like odour. Heavier than air (1.19). Very high exposures may cause anaesthetic effects and asphyxiation.	3339	
–	T50	–	F-C, S-V	Category A.	Liquefied, non-flammable, colourless gas with a faint ether-like odour. Heavier than air (1.16). Very high exposures may cause anaesthetic effects and asphyxiation.	3340	
–	T3	TP33	F-A, S-J	Category D.	White to yellow-white crystalline powder. Virtually odourless. Strong reducing agent. Violent exothermic decomposition above 100°C with emission of large amounts of sulphur oxides, ammonia, carbon monoxide, carbon dioxide, nitrogen oxides and hydrogen sulphide. Extended exposure to temperatures above 50°C and moisture may cause visible decomposition. Dust irritating to skin, eyes and mucous membranes.	3341	
–	T1	TP33	F-A, S-J	Category D.	See entry above.	3341	
–	T3	TP33	F-A, S-J	Category D. Clear of living quarters.	Hygroscopic yellow powder with an unpleasant odour. On contact with moisture, evolves highly flammable vapours such as carbon disulphide (UN 1131, which has a flashpoint of −30°C c.c. and a very low ignition temperature of 100°C). When confined, can cause an explosion due to the wide explosive limits of the vapours. Finely divided dust forms explosive mixtures in air. Care should be taken when opening cargo transport units in case carbon disulphide vapours are present.	3342	
–	T1	TP33	F-A, S-J	Category D. Clear of living quarters.	See entry above.	3342	
–	–	–	F-E, S-Y	Category D.	–	3343	
–	–	–	F-B, S-J	Category E.	–	3344	△

UN No.	Proper Shipping Name (PSN)	Class or division	Subsidiary risk(s)	Packing group	Special provisions	Limited and excepted quantity provisions		Packing		IBC	
						Limited quantities	Excepted quantities	Instructions	Provisions	Instructions	Provisions
(1)	(2) 3.1.2	(3) 2.0	(4) 2.0	(5) 2.0.1.3	(6) 3.3	(7a) 3.4	(7b) 3.5	(8) 4.1.4	(9) 4.1.4	(10) 4.1.4	(11) 4.1.4
3345	PHENOXYACETIC ACID DERIVATIVE PESTICIDE, SOLID, TOXIC	6.1	–	I	61 274	0	E5	P002	–	IBC07	B1
3345	PHENOXYACETIC ACID DERIVATIVE PESTICIDE, SOLID, TOXIC	6.1	–	II	61 274	500 g	E4	P002	–	IBC08	B2 B4
3345	PHENOXYACETIC ACID DERIVATIVE PESTICIDE, SOLID, TOXIC	6.1	–	III	61 223 274	5 kg	E1	P002 LP02	–	IBC08	B3
3346	PHENOXYACETIC ACID DERIVATIVE PESTICIDE, LIQUID, FLAMMABLE, TOXIC flashpoint less than 23°C	3	6.1	I	61 274	0	E0	P001	–	–	–
3346	PHENOXYACETIC ACID DERIVATIVE PESTICIDE, LIQUID, FLAMMABLE, TOXIC flashpoint less than 23°C	3	6.1	II	61 274	1 L	E2	P001	–	IBC02	–
3347	PHENOXYACETIC ACID DERIVATIVE PESTICIDE, LIQUID, TOXIC, FLAMMABLE, flashpoint not less than 23°C	6.1	3	I	61 274	0	E5	P001	–	–	–
3347	PHENOXYACETIC ACID DERIVATIVE PESTICIDE, LIQUID, TOXIC, FLAMMABLE, flashpoint not less than 23°C	6.1	3	II	61 274	100 mL	E4	P001	–	IBC02	–
3347	PHENOXYACETIC ACID DERIVATIVE PESTICIDE, LIQUID, TOXIC, FLAMMABLE, flashpoint not less than 23°C	6.1	3	III	61 223 274	5 L	E1	P001	–	IBC03	–
3348	PHENOXYACETIC ACID DERIVATIVE PESTICIDE, LIQUID, TOXIC	6.1	–	I	61 274	0	E5	P001	–	–	–
3348	PHENOXYACETIC ACID DERIVATIVE PESTICIDE, LIQUID, TOXIC	6.1	–	II	61 274	100 mL	E4	P001	–	IBC02	–
3348	PHENOXYACETIC ACID DERIVATIVE PESTICIDE, LIQUID, TOXIC	6.1	–	III	61 223 274	5 L	E1	P001 LP01	–	IBC03	–
3349	PYRETHROID PESTICIDE, SOLID, TOXIC	6.1	–	I	61 274	0	E5	P002	–	IBC07	B1
3349	PYRETHROID PESTICIDE, SOLID, TOXIC	6.1	–	II	61 274	500 g	E4	P002	–	IBC08	B2 B4
3349	PYRETHROID PESTICIDE, SOLID, TOXIC	6.1	–	III	61 223 274	5 kg	E1	P002 LP02	–	IBC08	B3
3350	PYRETHROID PESTICIDE, LIQUID, FLAMMABLE, TOXIC, flashpoint less than 23°C	3	6.1	I	61 274	0	E0	P001	–	–	–
3350	PYRETHROID PESTICIDE, LIQUID, FLAMMABLE, TOXIC, flashpoint less than 23°C	3	6.1	II	61 274	1 L	E2	P001	–	IBC02	–
3351	PYRETHROID PESTICIDE, LIQUID, TOXIC, FLAMMABLE, flashpoint not less than 23°C	6.1	3	I	61 274	0	E5	P001	–	–	–
3351	PYRETHROID PESTICIDE, LIQUID, TOXIC, FLAMMABLE, flashpoint not less than 23°C	6.1	3	II	61 274	100 mL	E4	P001	–	IBC02	–

Portable tanks and bulk containers			EmS	Stowage and segregation	Properties and observations	UN No.
	Tank instructions	Provisions				
(12)	(13) 4.2.5 4.3	(14) 4.2.5	(15) 5.4.3.2 7.3	(16) 7.1 7.2	(17)	(18)
–	T6	TP33	F-A, S-A	Category A. Clear of living quarters.	Solid pesticides present a very wide range of toxic hazard. Toxic if swallowed, by skin contact or by inhalation.	3345
–	T3	TP33	F-A, S-A	Category A. Clear of living quarters.	See entry above.	3345
–	T1	TP33	F-A, S-A	Category A. Clear of living quarters.	See entry above.	3345
–	T14	TP2 TP13 TP27	F-E, S-D	Category B. Clear of living quarters.	Pesticides frequently containing petroleum or coal tar distillates, or other flammable liquids. Miscibility with water depends upon the composition. Toxic if swallowed, by skin contact or by inhalation.	3346
–	T11	TP2 TP13 TP27	F-E, S-D	Category B. Clear of living quarters.	See entry above.	3346
–	T14	TP2 TP13 TP27	F-E, S-D	Category B. Clear of living quarters.	They frequently contain petroleum or coal tar distillates, or other flammable liquids. Flashpoint and miscibility with water depend upon the composition. Toxic if swallowed, by skin contact or by inhalation.	3347
–	T11	TP2 TP13 TP27	F-E, S-D	Category B. Clear of living quarters.	See entry above.	3347
–	T7	TP2 TP28	F-E, S-D	Category A. Clear of living quarters.	See entry above.	3347
–	T14	TP2 TP13 TP27	F-A, S-A	Category B. Clear of living quarters.	Liquid pesticides which present a very wide range of toxic hazard. Miscibility with water depends upon the composition. Toxic if swallowed, by skin contact or by inhalation.	3348
–	T11	TP2 TP27	F-A, S-A	Category B. Clear of living quarters.	See entry above.	3348
–	T7	TP2 TP28	F-A, S-A	Category A. Clear of living quarters.	See entry above.	3348
–	T6	TP33	F-A, S-A	Category A. Clear of living quarters.	Solid pesticides present a very wide range of toxic hazard. Toxic if swallowed, by skin contact or by inhalation.	3349
–	T3	TP33	F-A, S-A	Category A. Clear of living quarters.	See entry above.	3349
–	T1	TP33	F-A, S-A	Category A. Clear of living quarters.	See entry above.	3349
–	T14	TP2 TP13 TP27	F-E, S-D	Category B. Clear of living quarters.	Miscibility with water depends upon the composition. Toxic if swallowed, by skin contact or by inhalation.	3350
–	T11	TP2 TP13 TP27	F-E, S-D	Category B. Clear of living quarters.	See entry above.	3350
–	T14	TP2 TP13 TP27	F-E, S-D	Category B. Clear of living quarters.	They frequently contain petroleum or coal tar distillates, or other flammable liquids. Flashpoint and miscibility with water depend upon the composition. Toxic if swallowed, by skin contact or by inhalation.	3351
–	T11	TP2 TP13 TP27	F-E, S-D	Category B. Clear of living quarters.	See entry above.	3351

UN No.	Proper Shipping Name (PSN)	Class or division	Subsidiary risk(s)	Packing group	Special provisions	Limited and excepted quantity provisions		Packing		IBC	
						Limited quantities	Excepted quantities	Instructions	Provisions	Instructions	Provisions
(1)	(2) 3.1.2	(3) 2.0	(4) 2.0	(5) 2.0.1.3	(6) 3.3	(7a) 3.4	(7b) 3.5	(8) 4.1.4	(9) 4.1.4	(10) 4.1.4	(11) 4.1.4
3351	PYRETHROID PESTICIDE, LIQUID, TOXIC, FLAMMABLE, flashpoint not less than 23°C	6.1	3	III	61 223 274	5 L	E1	P001	–	IBC03	–
3352	PYRETHROID PESTICIDE, LIQUID, TOXIC	6.1	–	I	61 274	0	E5	P001	–	–	–
3352	PYRETHROID PESTICIDE, LIQUID, TOXIC	6.1	–	II	61 274	100 mL	E4	P001	–	IBC02	–
3352	PYRETHROID PESTICIDE, LIQUID, TOXIC	6.1	–	III	61 223 274	5 L	E1	P001 LP01	–	IBC03	–
3354	INSECTICIDE GAS, FLAMMABLE, N.O.S.	2.1	–	–	274	0	E0	P200	–	–	–
3355	INSECTICIDE GAS, TOXIC, FLAMMABLE, N.O.S.	2.3	2.1	–	274	0	E0	P200	–	–	–
3356	OXYGEN GENERATOR, CHEMICAL	5.1	–	II	284	0	E0	P500	–	–	–
3357	NITROGLYCERIN MIXTURE, DESENSITIZED, LIQUID, N.O.S. with not more than 30% nitroglycerin, by mass	3	–	II	274 288	0	E0	P099	–	–	–
3358	REFRIGERATING MACHINES containing flammable, non-toxic, liquefied gas	2.1	–	–	291	0	E0	P003	PP32	–	–
3359	FUMIGATED CARGO TRANSPORT UNIT	9	–	–	302	0	E0	–	–	–	–
3360	FIBRES, VEGETABLE, DRY	4.1	–	–	29 117 299	0	E0	P003	PP19	–	–
△ 3361	CHLOROSILANES, TOXIC, CORROSIVE, N.O.S.	6.1	8	II	274	0	E0	P010	–	–	–

Portable tanks and bulk containers			EmS	Stowage and segregation	Properties and observations	UN No.
	Tank instructions	Provisions				
(12)	(13) 4.2.5 4.3	(14) 4.2.5	(15) 5.4.3.2 7.3	(16) 7.1 7.2	(17)	(18)
–	T7	TP2 TP23	F-E, S-D	Category A. Clear of living quarters.	They frequently contain petroleum or coal tar distillates, or other flammable liquids. Flashpoint and miscibility with water depend upon the composition. Toxic if swallowed, by skin contact or by inhalation.	3351
–	T14	TP2 TP13 TP27	F-A, S-A	Category B. Clear of living quarters.	Liquid pesticides which present a very wide range of toxic hazard. Miscibility with water depends upon the composition. Toxic if swallowed, by skin contact or by inhalation.	3352
–	T11	TP2 TP27	F-A, S-A	Category B. Clear of living quarters.	See entry above.	3352
–	T7	TP2 TP23	F-A, S-A	Category A. Clear of living quarters.	See entry above.	3352
–	–	–	F-D, S-U	Category D.	Flammable mixtures of insecticides with liquefied gases.	3354
–	–	–	F-D, S-U	Category D. Clear of living quarters.	Toxic, flammable mixtures of insecticides with liquefied gases.	3355
–	–	–	F-H, S-Q	Category D.	Oxygen generators, chemical are devices containing chemicals which, upon activation, release oxygen as a product of chemical reaction. Chemical oxygen generators are used for the generation of oxygen for respiratory support, e.g., in aircraft, submarines, spacecraft, bomb shelters and breathing apparatus. Oxidizing salts such as chlorates and perchlorates of lithium, sodium and potassium, which are used in chemical oxygen generators, evolve oxygen when heated. These salts are mixed (compounded) with a fuel, usually iron powder, to form a chlorate candle, which produces oxygen by continuous reaction. The fuel is used to generate heat by oxidation. Once the reaction begins, oxygen is released from the hot salt by thermal decomposition (a thermal shield is used around the generator). A portion of the oxygen reacts with the fuel to produce more heat, which produces more oxygen, and so on. Initiation of the reaction can be achieved by a percussion device, friction device or electric wire.	3356
–	–	–	F-E, S-Y	Category D.	–	3357
–	–	–	F-D, S-U	Category D.	–	3358
–	–	–	F-A, S-D	Category B. Clear of living quarters.	A "FUMIGATED CARGO TRANSPORT UNIT" is a closed cargo transport unit containing goods or materials that either are or have been fumigated within the unit. The fumigant gases used are either poisonous or asphyxiant. The gases are usually evolved from solid or liquid preparations distributed within the unit. See also 5.5.2.	3359
–	–	–	F-A, S-I	Category A.	Ignite readily. Consignments of cotton, dry having a density not less than 360 kg/m^3, flax, dry having a density not less than 400 kg/m^3, sisal, dry having a density not less than 360 kg/m^3 (ISO Standard 8115 (1986)) and tampico fibre, dry having a density not less than 360 kg/m^3 are not subject to the provisions of this Code when carried in closed cargo transport units.	3360
–	T14	TP2 TP7 TP13 TP27	F-A, S-B	Category C. Clear of living quarters.	Colourless to yellow liquids with a pungent odour. Immiscible with water. React violently with water or steam, evolving hydrogen chloride, an irritating and corrosive gas apparent as white fumes. When involved in a fire, evolve toxic gas. In the presence of moisture, highly corrosive to most metals. Toxic if swallowed, by skin contact or by inhalation. Cause burns to skin, eyes and mucous membranes.	3361 △

3

DGL

	UN No.	Proper Shipping Name (PSN)	Class or division	Subsidiary risk(s)	Packing group	Special provisions	Limited and excepted quantity provisions		Packing		IBC	
							Limited quantities	Excepted quantities	Instructions	Provisions	Instructions	Provisions
	(1)	(2) 3.1.2	(3) 2.0	(4) 2.0	(5) 2.0.1.3	(6) 3.3	(7a) 3.4	(7b) 3.5	(8) 4.1.4	(9) 4.1.4	(10) 4.1.4	(11) 4.1.4
△	3362	CHLOROSILANES, TOXIC, CORROSIVE, FLAMMABLE, N.O.S.	6.1	3/8	II	274	0	E0	P010	–	–	–
	3363	DANGEROUS GOODS IN MACHINERY or DANGEROUS GOODS IN APPARATUS	9	–	–	301	See SP301	E0	P907	–	–	–
	3364	TRINITROPHENOL (PICRIC ACID), WETTED with not less than 10% water, by mass	4.1	–	I	28	0	E0	P406	PP24 PP31	–	–
	3365	TRINITROCHLOROBENZENE (PICRYL CHLORIDE), WETTED with not less than 10% water, by mass	4.1	–	I	28	0	E0	P406	PP24 PP31	–	–
	3366	TRINITROTOLUENE (TNT), WETTED with not less than 10% water, by mass	4.1	–	I	28	0	E0	P406	PP24 PP31	–	–
	3367	TRINITROBENZENE, WETTED with not less than 10% water, by mass	4.1	–	I	28	0	E0	P406	PP24 PP31	–	–
	3368	TRINITROBENZOIC ACID, WETTED with not less than 10% water, by mass	4.1	–	I	28	0	E0	P406	PP24 PP31	–	–
	3369	SODIUM DINITRO-*o*-CRESOLATE, WETTED with not less than 10% water, by mass	4.1	6.1 P	I	28	0	E0	P406	PP24 PP31	–	–
	3370	UREA NITRATE, WETTED with not less than 10% water, by mass	4.1	–	I	28	0	E0	P406	PP31 PP78	–	–
	3371	2-METHYLBUTANAL	3	–	II	–	1 L	E2	P001	–	IBC02	–
△	3373	BIOLOGICAL SUBSTANCE, CATEGORY B	6.2	–	–	319 341	0	E0	P650	–	–	–
△	3374	ACETYLENE, SOLVENT FREE	2.1	–	–	–	0	E0	P200	–	–	–

Portable tanks and bulk containers			EmS	Stowage and segregation	Properties and observations	UN No.	
	Tank instructions	Provisions					
(12)	(13) 4.2.5 4.3	(14) 4.2.5	(15) 5.4.3.2 7.3	(16) 7.1 7.2	(17)	(18)	
–	T14	TP2 TP7 TP13 TP27	F-E, S-C	Category C. Clear of living quarters. Segregation as for class 3 but "away from" class 4.1.	Colourless to yellow flammable liquids with a pungent odour. Immiscible with water. React violently with water or steam, evolving hydrogen chloride, an irritating and corrosive gas apparent as white fumes. When involved in a fire, evolve toxic gas. In the presence of moisture, highly corrosive to most metals. Toxic if swallowed, by skin contact or by inhalation. Cause burns to skin, eyes and mucous membranes.	3362	△
–	–	–	F-A, S-P	Category A.	Types of articles transported under this entry contain only limited quantities of dangerous goods.	3363	
–	–	–	F-B, S-J	Category E. "Away from" class 3 and heavy metals and their salts.	Desensitized explosive. Substance in pure form consists of yellow crystals. Soluble in water. Explosive and sensitive to friction in the dry state. May form extremely sensitive compounds with heavy metals or their salts. Harmful if swallowed or by skin contact.	3364	
–	–	–	F-B, S-J	Category E. "Away from" class 3 and heavy metals and their salts.	Desensitized explosive. Explosive and sensitive to shock and heat in the dry state. Reacts violently with heavy metals and their salts.	3365	
–	–	–	F-B, S-J	Category E. "Away from" class 3 and heavy metals and their salts.	Desensitized explosive. Substance in pure form consists of yellow crystals. When involved in a fire, evolves toxic fumes; in closed compartments, these fumes may form an explosive mixture with air. Explosive and sensitive to shock and heat in the dry state. Reacts violently with heavy metals and their salts.	3366	
–	–	–	F-B, S-J	Category E. "Away from" class 3 and heavy metals and their salts.	Desensitized explosive. Substance in pure form consists of odourless yellow crystals. When involved in a fire, evolves toxic fumes; in closed compartments, these fumes may form an explosive mixture with air. Explosive and sensitive to shock and heat in the dry state. Harmful if swallowed or by skin contact. Reacts violently with heavy metals and their salts.	3367	
–	–	–	F-B, S-J	Category E. "Away from" class 3 and heavy metals and their salts.	Desensitized explosive. Substance in pure form consists of yellow crystals. Soluble in water. When involved in a fire, evolves toxic fumes; in closed compartments, these fumes may form an explosive mixture with air. Explosive and sensitive to shock and heat in the dry state. Harmful if swallowed or by skin contact. Reacts violently with heavy metals and their salts.	3368	
–	–	–	F-B, S-J	Category E. "Away from" class 3 and heavy metals and their salts.	Desensitized explosive. Substance in pure form consists of yellow powder. May form extremely sensitive compounds with heavy metals or their salts. When involved in a fire, evolves toxic fumes; in closed compartments, these fumes may form an explosive mixture with air. Explosive and sensitive to friction in the dry state. Toxic if swallowed, by skin contact or by inhalation.	3369	
–	–	–	F-B, S-J	Category E. "Away from" class 3 and heavy metals and their salts.	Desensitized explosive. May form extremely sensitive compounds with heavy metals or their salts. Explosive and sensitive to friction in the dry state. Harmful if swallowed or by skin contact.	3370	
–	T4	TP1	F-E, S-D	Category B.	Colourless liquid. Flashpoint: −3.5°C. Explosive limits: 1.3 to 13.9%. Slightly miscible with water.	3371	
–	T1 BK2	TP1	F-A, S-T	Category A, when transported in accordance with P650. Otherwise Category C. Clear of living quarters.	Substances which are known or are reasonably expected to contain pathogens, transported in a form that, when exposure to it occurs, are not capable of causing permanent disability, life-threatening or fatal disease to humans or animals. Human or animal specimens for which there is minimal likelihood that pathogens are present are not subject to the provisions of this Code (see 2.6.3.2.3.6). Other exemptions are stated in 2.6.3.2.3.	3373	△
–	–	–	F-D, S-U	Category D. Protected from sources of heat. Clear of living quarters. "Separated from" chlorine.	Flammable gas with slight odour. Explosive limits: 2.1 to 80%. Lighter than air (0.907). Acetylene without solvent. Rough handling and exposure to local heating should be avoided, since these conditions may result in delayed explosion. Empty cylinders should be carried with the same precautions as filled cylinders.	3374	△

UN No.	Proper Shipping Name (PSN)	Class or division	Subsidiary risk(s)	Packing group	Special provisions	Limited and excepted quantity provisions		Packing		IBC	
						Limited quantities	Excepted quantities	Instructions	Provisions	Instructions	Provisions
(1)	(2) 3.1.2	(3) 2.0	(4) 2.0	(5) 2.0.1.3	(6) 3.3	(7a) 3.4	(7b) 3.5	(8) 4.1.4	(9) 4.1.4	(10) 4.1.4	(11) 4.1.4
△ 3375	**AMMONIUM NITRATE EMULSION or SUSPENSION or GEL intermediate for blasting explosives**	5.1	–	II	309	0	E2	P099	–	IBC99	–
3376	4-NITROPHENYLHYDRAZINE, with not less than 30% water, by mass	4.1	–	I	28	0	E0	P406	PP26 PP31	–	–
△ 3377	SODIUM PERBORATE MONOHYDRATE	5.1	–	III	967	5 kg	E1	P002 LP02	–	IBC08	B3
△ 3378	SODIUM CARBONATE PEROXYHYDRATE	5.1	–	II	967	1 kg	E2	P002	–	IBC08	B2 B4
△ 3378	SODIUM CARBONATE PEROXYHYDRATE	5.1	–	III	967	5 kg	E1	P002 LP02	–	IBC08	B3
3379	DESENSITIZED EXPLOSIVE, LIQUID, N.O.S.	3	–	I	274 311	0	E0	P099	–	–	–
3380	DESENSITIZED EXPLOSIVE, SOLID, N.O.S.	4.1	–	I	274 311	0	E0	P099	–	–	–
△ 3381	TOXIC BY INHALATION LIQUID, N.O.S. with an LC_{50} lower than or equal to 200 mL/m^3 and saturated vapour concentration greater than or equal to $500LC_{50}$	6.1	–	I	274	0	E0	P601	–	–	–
△ 3382	TOXIC BY INHALATION LIQUID, N.O.S. with an LC_{50} lower than or equal to 1000 mL/m^3 and saturated vapour concentration greater than or equal to $10LC_{50}$	6.1	–	I	274	0	E0	P602	–	–	–
△ 3383	TOXIC BY INHALATION LIQUID, FLAMMABLE, N.O.S. with an LC_{50} lower than or equal to 200 mL/m^3 and saturated vapour concentration greater than or equal to $500LC_{50}$	6.1	3	I	274	0	E0	P601	–	–	–

Portable tanks and bulk containers			EmS	Stowage and segregation	Properties and observations	UN No.	
	Tank instructions	Provisions					
(12)	(13) 4.2.5 4.3	(14) 4.2.5	(15) 5.4.3.2 7.3	(16) 7.1 7.2	(17)	(18)	
–	T1	TP1 TP9 TP17 TP32	F-H, S-Q	Category D. Protected from sources of heat. "Separated from" class 4.1, combustible material (particularly liquids), bromates, chlorates, chlorites, hypochlorites, nitrites, perchlorates, permanganates and powdered metals.	Non-sensitized emulsions, suspensions and gels consisting primarily of a mixture of ammonium nitrate and fuel, intended to produce a Type E blasting explosive only after further processing prior to use. Substances shall satisfactorily pass Test Series 8 of the United Nations *Manual of Tests and Criteria*, Part I, Section 18 and be approved by the competent authority.	3375	△
–	–	–	F-B, S-J	Category E. "Away from" class 3 and heavy metals and their salts.	Desensitized explosive. Dark orange solid. Explosive and sensitive to friction in the dry state. May form extremely sensitive compounds with heavy metals or their salts. Harmful if swallowed or by skin contact.	3376	
–	T1 BK2 BK3	TP33	F-A, S-Q	Category A. Keep as dry as reasonably practicable. Protected from sources of heat. "Separated from" permanganates. When transported in BK3 bulk container, see 7.6.2.12 and 7.7.3.9.	White crystals or powder. Partially soluble in water. Mixtures with combustible material are readily ignited and may burn fiercely. Risk of decomposition when exposed to continuous heat (exothermic decomposition ≥ 60°C). When involved in a fire or exposed to high temperatures, it may decompose, yielding oxygen and steam. Harmful if swallowed.	3377	△
–	T3 BK2	TP33	F-A, S-Q	Category A. Keep as dry as reasonably practicable. Protected from sources of heat. "Separated from" permanganates.	White crystals or powder. Soluble in water. Mixtures with combustible material are readily ignited. Decomposes in contact with water and acids, forming hydrogen peroxide. Risk of decomposition when exposed to continuous heat (exothermic decomposition ≥ 60°C). When involved in a fire or exposed to high temperatures, it may decompose, yielding oxygen and steam. Irritating to eyes, skin and mucous membranes. Harmful if swallowed.	3378	△
–	T1 BK2 BK3	TP33	F-A, S-Q	Category A. Keep as dry as reasonably practicable. Protected from sources of heat. "Separated from" permanganates. When transported in BK3 bulk container, see 7.6.2.12 and 7.7.3.9.	See entry above.	3378	△
–	–	–	F-E, S-Y	Category D. "Away from" heavy metals and their salts.	Desensitized explosive. Explosive and sensitive to friction in the dry state. May form extremely sensitive compounds with heavy metals and their salts.	3379	
–	–	–	F-B, S-J	Category D. "Away from" class 3 and heavy metals and their salts.	Desensitized explosive. Explosive and sensitive to friction in the dry state. May form extremely sensitive compounds with heavy metals and their salts.	3380	
–	T22	TP2 TP13	F-A, S-A	Category D. Clear of living quarters.	A variety of toxic liquids which present a highly toxic inhalation hazard. Highly toxic if swallowed, by skin contact or by inhalation.	3381	△
–	T20	TP2 TP13	F-A, S-A	Category D. Clear of living quarters.	A variety of toxic liquids which present a highly toxic inhalation hazard. Highly toxic if swallowed, by skin contact or by inhalation.	3382	△
–	T22	TP2 TP13	F-E, S-D	Category D. Clear of living quarters.	A variety of toxic liquids which present a highly toxic inhalation hazard as well as being flammable. Highly toxic if swallowed, by skin contact or by inhalation.	3383	△

UN No.	Proper Shipping Name (PSN)	Class or division	Subsidiary risk(s)	Packing group	Special provisions	Limited and excepted quantity provisions		Packing		IBC	
						Limited quantities	Excepted quantities	Instructions	Provisions	Instructions	Provisions
(1)	(2) 3.1.2	(3) 2.0	(4) 2.0	(5) 2.0.1.3	(6) 3.3	(7a) 3.4	(7b) 3.5	(8) 4.1.4	(9) 4.1.4	(10) 4.1.4	(11) 4.1.4
△ 3384	TOXIC BY INHALATION LIQUID, FLAMMABLE, N.O.S. with an LC_{50} lower than or equal to 1000 mL/m^3 and saturated vapour concentration greater than or equal to $10LC_{50}$	6.1	3	I	274	0	E0	P602	–	–	–
△ 3385	TOXIC BY INHALATION LIQUID, WATER-REACTIVE, N.O.S. with an LC_{50} lower than or equal to 200 mL/m^3 and saturated vapour concentration greater than or equal to $500LC_{50}$	6.1	4.3	I	274	0	E0	P601	–	–	–
△ 3386	TOXIC BY INHALATION LIQUID, WATER-REACTIVE, N.O.S. with an LC_{50} lower than or equal to 1000 mL/m^3 and saturated vapour concentration greater than or equal to $10LC_{50}$	6.1	4.3	I	274	0	E0	P602	–	–	–
△ 3387	TOXIC BY INHALATION LIQUID, OXIDIZING, N.O.S. with an LC_{50} lower than or equal to 200 mL/m^3 and saturated vapour concentration greater than or equal to $500LC_{50}$	6.1	5.1	I	274	0	E0	P601	–	–	–
△ 3388	TOXIC BY INHALATION LIQUID, OXIDIZING, N.O.S. with an LC_{50} lower than or equal to 1000 mL/m^3 and saturated vapour concentration greater than or equal to $10LC_{50}$	6.1	5.1	I	274	0	E0	P602	–	–	–
△ 3389	TOXIC BY INHALATION LIQUID, CORROSIVE, N.O.S. with an LC_{50} lower than or equal to 200 mL/m^3 and saturated vapour concentration greater than or equal to $500LC_{50}$	6.1	8	I	274	0	E0	P601	–	–	–
△ 3390	TOXIC BY INHALATION LIQUID, CORROSIVE, N.O.S. with an LC_{50} lower than or equal to 1000 mL/m^3 and saturated vapour concentration greater than or equal to $10LC_{50}$	6.1	8	I	274	0	E0	P602	–	–	–
3391	ORGANOMETALLIC SUBSTANCE, SOLID, PYROPHORIC	4.2	–	I	274	0	E0	P404	PP86	–	–
3392	ORGANOMETALLIC SUBSTANCE, LIQUID, PYROPHORIC	4.2	–	I	274	0	E0	P400	PP86	–	–
3393	ORGANOMETALLIC SUBSTANCE, SOLID, PYROPHORIC, WATER-REACTIVE	4.2	4.3	I	274	0	E0	P404	PP86	–	–
3394	ORGANOMETALLIC SUBSTANCE, LIQUID, PYROPHORIC, WATER-REACTIVE	4.2	4.3	I	274	0	E0	P400	PP86	–	–

Portable tanks and bulk containers			EmS	Stowage and segregation	Properties and observations	UN No.	
	Tank instructions	Provisions					
(12)	(13) 4.2.5 4.3	(14) 4.2.5	(15) 5.4.3.2 7.3	(16) 7.1 7.2	(17)	(18)	
–	T20	TP2 TP13	F-E, S-D	Category D. Clear of living quarters.	A variety of toxic liquids which present a highly toxic inhalation hazard as well as being flammable. Highly toxic if swallowed, by skin contact or by inhalation.	3384	△
–	T22	TP2 TP13	F-G, S-N	Category D. Clear of living quarters.	A variety of toxic liquids which present a highly toxic inhalation hazard as well as being water-reactive. Highly toxic if swallowed, by skin contact or by inhalation.	3385	△
–	T20	TP2 TP13	F-G, S-N	Category D. Clear of living quarters.	A variety of toxic liquids which present a highly toxic inhalation hazard as well as being water-reactive. Highly toxic if swallowed, by skin contact or by inhalation.	3386	△
–	T22	TP2 TP13	F-A, S-Q	Category D. Clear of living quarters.	A variety of toxic liquids which present a highly toxic inhalation hazard as well as being an oxidizer. Highly toxic if swallowed, by skin contact or by inhalation.	3387	△
–	T20	TP2 TP13	F-A, S-Q	Category D. Clear of living quarters.	A variety of toxic liquids which present a highly toxic inhalation hazard as well as being an oxidizer. Highly toxic if swallowed, by skin contact or by inhalation.	3388	△
–	T22	TP2 TP13	F-A, S-B	Category D. Clear of living quarters.	A variety of toxic liquids which present a highly toxic inhalation hazard as well as being corrosive. Highly toxic if swallowed, by skin contact or by inhalation.	3389	△
–	T20	TP2 TP13	F-A, S-B	Category D. Clear of living quarters.	A variety of toxic liquids which present a highly toxic inhalation hazard as well as being corrosive. Highly toxic if swallowed, by skin contact or by inhalation.	3390	△
–	T21	TP7 TP33 TP36	F-G, S-M	Category D.	Liable to ignite spontaneously in air. If shaken, may produce sparks.	3391	
–	T21	TP2 TP7 TP36	F-G, S-M	Category D. "Separated longitudinally by an intervening complete compartment or hold from" class 1.	Highly flammable liquid. Liable to ignite spontaneously in air. In contact with air, evolve irritating and slightly toxic fumes.	3392	
–	T21	TP7 TP33 TP36	F-G, S-M	Category D. "Separated from" acids.	Liable to ignite spontaneously in air. If shaken, may produce sparks. React violently with moisture, water and acids, evolving flammable gas.	3393	
–	T21	TP2 TP7 TP36	F-G, S-M	Category D. "Separated longitudinally by an intervening complete compartment or hold from" class 1. "Separated from" acids.	Highly flammable liquid. Liable to ignite spontaneously in air. In contact with air, evolve irritating and slightly toxic fumes. React violently with moisture, water and acids, evolving flammable gas.	3394	

UN No.	Proper Shipping Name (PSN)	Class or division	Subsidiary risk(s)	Packing group	Special provisions	Limited and excepted quantity provisions		Packing		IBC	
						Limited quantities	Excepted quantities	Instructions	Provisions	Instructions	Provisions
(1)	(2) 3.1.2	(3) 2.0	(4) 2.0	(5) 2.0.1.3	(6) 3.3	(7a) 3.4	(7b) 3.5	(8) 4.1.4	(9) 4.1.4	(10) 4.1.4	(11) 4.1.4
3395	ORGANOMETALLIC SUBSTANCE, SOLID, WATER-REACTIVE	4.3	–	I	274	0	E0	P403	–	–	–
3395	ORGANOMETALLIC SUBSTANCE, SOLID, WATER-REACTIVE	4.3	–	II	274	500 g	E2	P410	–	IBC04	–
3395	ORGANOMETALLIC SUBSTANCE, SOLID, WATER-REACTIVE	4.3	–	III	223 274	1 kg	E1	P410	–	IBC06	–
3396	ORGANOMETALLIC SUBSTANCE, SOLID, WATER-REACTIVE, FLAMMABLE	4.3	4.1	I	274	0	E0	P403	–	–	–
3396	ORGANOMETALLIC SUBSTANCE, SOLID, WATER-REACTIVE, FLAMMABLE	4.3	4.1	II	274	500 g	E2	P410	–	IBC04	–
3396	ORGANOMETALLIC SUBSTANCE, SOLID, WATER-REACTIVE, FLAMMABLE	4.3	4.1	III	223 274	1 kg	E1	P410	–	IBC06	–
3397	ORGANOMETALLIC SUBSTANCE, SOLID, WATER-REACTIVE, SELF-HEATING	4.3	4.2	I	274	0	E0	P403	–	–	–
3397	ORGANOMETALLIC SUBSTANCE, SOLID, WATER-REACTIVE, SELF-HEATING	4.3	4.2	II	274	500 g	E2	P410	–	IBC04	–
3397	ORGANOMETALLIC SUBSTANCE, SOLID, WATER-REACTIVE, SELF-HEATING	4.3	4.2	III	223 274	1 kg	E1	P410	–	IBC06	–
3398	ORGANOMETALLIC SUBSTANCE, LIQUID, WATER-REACTIVE	4.3	–	I	274	0	E0	P402	PP31	–	–
3398	ORGANOMETALLIC SUBSTANCE, LIQUID, WATER-REACTIVE	4.3	–	II	274	500 mL	E2	P001	PP31	IBC01	–
3398	ORGANOMETALLIC SUBSTANCE, LIQUID, WATER-REACTIVE	4.3	–	III	223 274	1 L	E1	P001	PP31	IBC02	–
3399	ORGANOMETALLIC SUBSTANCE, LIQUID, WATER-REACTIVE, FLAMMABLE	4.3	3	I	274	0	E0	P402	PP31	–	–
3399	ORGANOMETALLIC SUBSTANCE, LIQUID, WATER-REACTIVE, FLAMMABLE	4.3	3	II	274	500 mL	E2	P001	PP31	IBC01	–
3399	ORGANOMETALLIC SUBSTANCE, LIQUID, WATER-REACTIVE, FLAMMABLE	4.3	3	III	223 274	1 L	E1	P001	PP31	IBC02	–
3400	ORGANOMETALLIC SUBSTANCE, SOLID, SELF-HEATING	4.2	–	II	274	500 g	E2	P410	–	IBC06	–
3400	ORGANOMETALLIC SUBSTANCE, SOLID, SELF-HEATING	4.2	–	III	223 274	1 kg	E1	P002	–	IBC08	–
3401	ALKALI METAL AMALGAM, SOLID	4.3	–	I	182	0	E0	P403	PP31	–	–

Portable tanks and bulk containers			EmS	Stowage and segregation	Properties and observations	UN No.
	Tank instructions	Provisions				
(12)	(13) 4.2.5 4.3	(14) 4.2.5	(15) 5.4.3.2 7.3	(16) 7.1 7.2	(17)	(18)
–	T9	TP7 TP33 TP36	F-G, S-N	Category E. Clear of living quarters. "Separated from" acids.	Reacts violently with moisture, water and acids, evolving flammable gas.	3395
–	T3	TP33 TP36	F-G, S-N	Category E. Clear of living quarters. "Separated from" acids.	See entry above.	3395
–	T1	TP33 TP36	F-G, S-N	Category E. Clear of living quarters. "Separated from" acids.	See entry above.	3395
–	T9	TP7 TP33 TP36	F-G, S-N	Category E. Clear of living quarters. "Separated from" acids.	Flammable solid. Reacts violently with moisture, water and acids, evolving flammable gas.	3396
–	T3	TP33 TP36	F-G, S-N	Category E. Clear of living quarters. "Separated from" acids.	See entry above.	3396
–	T1	TP33 TP36	F-G, S-N	Category E. Clear of living quarters. "Separated from" acids.	See entry above.	3396
–	T9	TP7 TP33 TP36	F-G, S-N	Category E. Clear of living quarters. "Separated from" acids.	Liable to self-heating or spontaneous combustion. Reacts violently with moisture, water and acids, evolving flammable gas.	3397
–	T3	TP33 TP36	F-G, S-N	Category E. Clear of living quarters. "Separated from" acids.	See entry above.	3397
–	T1	TP33 TP36	F-G, S-N	Category E. Clear of living quarters. "Separated from" acids.	See entry above.	3397
–	T13	TP2 TP7 TP36	F-G, S-N	Category E. Clear of living quarters. "Separated from" acids.	Reacts violently with moisture, water and acids, evolving flammable gas.	3398
–	T7	TP2 TP7 TP36	F-G, S-N	Category E. Clear of living quarters. "Separated from" acids.	See entry above.	3398
–	T7	TP2 TP7 TP36	F-G, S-N	Category E. Clear of living quarters. "Separated from" acids.	See entry above.	3398
–	T13	TP2 TP7 TP36	F-G, S-N	Category D. Clear of living quarters. "Separated from" acids.	Flammable liquid. Reacts violently with moisture, water and acids, evolving flammable gas.	3399
–	T7	TP2 TP7 TP36	F-G, S-N	Category D. Clear of living quarters. "Separated from" acids.	See entry above.	3399
–	T7	TP2 TP7 TP36	F-G, S-N	Category E. Clear of living quarters. "Separated from" acids.	See entry above.	3399
–	T3	TP33 TP36	F-A, S-J	Category C.	Liable to self-heating or spontaneous combustion.	3400
–	T1	TP33 TP36	F-A, S-J	Category C.	See entry above.	3400
–	T9	TP7 TP33	F-G, S-N	Category D. "Separated from" acids.	Silvery solid, consisting of metal alloyed with mercury. Reacts with moisture, water or acids, evolving hydrogen, a flammable gas. When heated, evolves toxic vapours.	3401

UN No.	Proper Shipping Name (PSN)	Class or division	Subsidiary risk(s)	Packing group	Special provisions	Limited and excepted quantity provisions		Packing		IBC	
						Limited quantities	Excepted quantities	Instructions	Provisions	Instructions	Provisions
(1)	(2) 3.1.2	(3) 2.0	(4) 2.0	(5) 2.0.1.3	(6) 3.3	(7a) 3.4	(7b) 3.5	(8) 4.1.4	(9) 4.1.4	(10) 4.1.4	(11) 4.1.4
3402	ALKALINE EARTH METAL AMALGAM, SOLID	4.3	–	I	183	0	E0	P403	PP31	–	–
3403	POTASSIUM METAL ALLOYS, SOLID	4.3	–	I	–	0	E0	P403	PP31	–	–
3404	POTASSIUM SODIUM ALLOYS, SOLID	4.3	–	I	–	0	E0	P403	PP31	–	–
3405	BARIUM CHLORATE SOLUTION	5.1	6.1	II	–	1 L	E2	P504	–	IBC02	–
3405	BARIUM CHLORATE SOLUTION	5.1	6.1	III	223	5 L	E1	P001	–	IBC02	–
3406	BARIUM PERCHLORATE SOLUTION	5.1	6.1	II	–	1 L	E2	P504	–	IBC02	–
3406	BARIUM PERCHLORATE SOLUTION	5.1	6.1	III	223	5 L	E1	P001	–	IBC02	–
3407	CHLORATE AND MAGNESIUM CHLORIDE MIXTURE SOLUTION	5.1	–	II	–	1 L	E2	P504	–	IBC02	–
3407	CHLORATE AND MAGNESIUM CHLORIDE MIXTURE SOLUTION	5.1	–	III	223	5 L	E1	P504	–	IBC02	–
3408	LEAD PERCHLORATE SOLUTION	5.1	6.1 P	II	–	1 L	E2	P504	–	IBC02	–

Portable tanks and bulk containers			EmS	Stowage and segregation	Properties and observations	UN No.
	Tank instructions	Provisions				
(12)	(13) 4.2.5 4.3	(14) 4.2.5	(15) 5.4.3.2 7.3	(16) 7.1 7.2	(17)	(18)
–	T9	TP7 TP33	F-G, S-N	Category D. "Separated from" acids.	Consists of metal alloyed with mercury. Contains 2% to 10% alkaline earth metals and may contain up to 98% mercury. Reacts with moisture, water or acids, evolving hydrogen, a flammable gas. When heated, evolves toxic vapours.	3402
–	T9	TP7 TP33	F-G, S-L	Category D. "Separated from" acids.	Soft, silvery metal. Floats on water. Reacts violently with moisture, water or acids, evolving hydrogen, which may be ignited by the heat of the reaction. Highly reactive, sometimes with explosive effect.	3403
–	T9	TP7 TP33	F-G, S-L	Category D. "Separated from" acids.	Soft, silvery metal. Floats on water. Reacts violently with moisture, water or acids, evolving hydrogen, which may be ignited by the heat of the reaction. Highly reactive, sometimes with explosive effect.	3404
–	T4	TP1	F-H, S-Q	Category A. "Separated from" ammonium compounds, cyanides and sulphur.	Colourless aqueous solution. Reacts vigorously with sulphuric acid. Reacts fiercely with cyanides when heated. May form explosive mixtures with combustible material, powdered metals or ammonium compounds. These mixtures are liable to ignite. When involved in a fire, may cause an explosion. Toxic if swallowed, by skin contact or by inhalation. Leakage and subsequent evaporation of the water from the solutions may present increased dangers as follows: 1. in contact with combustible material (particularly fibrous material such as jute, cotton or sisal) or sulphur, danger of spontaneous combustion, 2. in contact with ammonium compounds, powdered metals or oils, danger of explosion.	3405
–	T4	TP1	F-H, S-Q	Category A. "Separated from" ammonium compounds, cyanides and sulphur.	See entry above.	3405
–	T4	TP1	F-H, S-Q	Category A. "Separated from" ammonium compounds, cyanides and sulphur.	Reacts vigorously with sulphuric acid. Reacts fiercely with cyanides when heated. May form explosive mixtures with combustible material, powdered metals or ammonium compounds. These mixtures are liable to ignite. When involved in a fire, may cause an explosion. Toxic if swallowed, by skin contact or by inhalation. Leakage and subsequent evaporation of the water from the solutions may present increased dangers as follows: 1. in contact with combustible material (particularly fibrous material such as jute, cotton or sisal) or sulphur, danger of spontaneous combustion, 2. in contact with ammonium compounds, powdered metals or oils, danger of explosion.	3406
–	T4	TP1	F-H, S-Q	Category A. "Separated from" ammonium compounds, cyanides and sulphur.	See entry above.	3406
–	T4	TP1	F-H, S-Q	Category A. "Separated from" ammonium compounds, cyanides and sulphur.	Reacts vigorously with sulphuric acid. Reacts fiercely with cyanides when heated. May form explosive mixtures with combustible material, powdered metals or ammonium compounds. These mixtures are liable to ignite. When involved in a fire, may cause an explosion. Leakage and subsequent evaporation of the water from the solutions may present increased dangers as follows: 1. in contact with combustible material (particularly fibrous material such as jute, cotton or sisal) or sulphur, danger of spontaneous combustion, 2. in contact with ammonium compounds, powdered metals or oils, danger of explosion.	3407
–	T4	TP1	F-H, S-Q	Category A. "Separated from" ammonium compounds, cyanides and sulphur.	See entry above.	3407
–	T4	TP1	F-H, S-Q	Category A. "Separated from" ammonium compounds and cyanides.	Reacts vigorously with sulphuric acid. Reacts fiercely with cyanides when heated. May form explosive mixtures with combustible material, powdered metals or ammonium compounds. These mixtures are liable to ignite. When involved in a fire, may cause an explosion.	3408

UN No. (1)	Proper Shipping Name (PSN) (2) 3.1.2	Class or division (3) 2.0	Subsidiary risk(s) (4) 2.0	Packing group (5) 2.0.1.3	Special provisions (6) 3.3	Limited and excepted quantity provisions: Limited quantities (7a) 3.4	Limited and excepted quantity provisions: Excepted quantities (7b) 3.5	Packing: Instructions (8) 4.1.4	Packing: Provisions (9) 4.1.4	IBC: Instructions (10) 4.1.4	IBC: Provisions (11) 4.1.4
3408	LEAD PERCHLORATE SOLUTION	5.1	6.1 P	III	223	5 L	E1	P001	–	IBC02	–
3409	CHLORONITROBENZENES, LIQUID	6.1	–	II	279	100 mL	E4	P001	–	IBC02	–
3410	4-CHLORO-*o*-TOLUIDINE HYDROCHLORIDE SOLUTION	6.1	–	III	223	5 L	E1	P001	–	IBC03	–
3411	*beta*-NAPHTHYLAMINE SOLUTION	6.1	–	II	–	100 mL	E4	P001	–	IBC02	–
3411	*beta*-NAPHTHYLAMINE SOLUTION	6.1	–	III	223	5 L	E1	P001	–	IBC02	–
3412	FORMIC ACID with not less than 10% but not more than 85% acid by mass	8	–	II	–	1 L	E2	P001	–	IBC02	–
3412	FORMIC ACID with not less than 5% but less than 10% acid by mass	8	–	III	–	5 L	E1	P001 LP01	–	IBC03	–
3413	POTASSIUM CYANIDE SOLUTION	6.1	– P	I	–	0	E5	P001	PP31	–	–
3413	POTASSIUM CYANIDE SOLUTION	6.1	– P	II	–	100 mL	E4	P001	PP31	IBC02	–
3413	POTASSIUM CYANIDE SOLUTION	6.1	– P	III	223	5 L	E1	P001 LP01	PP31	IBC03	–
3414	SODIUM CYANIDE SOLUTION	6.1	– P	I	–	0	E5	P001	PP31	–	–
3414	SODIUM CYANIDE SOLUTION	6.1	– P	II	–	100 mL	E4	P001	PP31	IBC02	–
3414	SODIUM CYANIDE SOLUTION	6.1	– P	III	223	5 L	E1	P001 LP01	PP31	IBC03	–
3415	SODIUM FLUORIDE SOLUTION	6.1	–	III	223	5 L	E1	P001 LP01	–	IBC03	–
3416	CHLOROACETOPHENONE, LIQUID	6.1	–	II	–	0	E4	P001	–	IBC02	–
3417	XYLYL BROMIDE, SOLID	6.1	–	II	–	0	E4	P002	–	IBC08	B2 B4
3418	2,4-TOLUYLENEDIAMINE SOLUTION	6.1	–	III	223	5 L	E1	P001 LP01	–	IBC03	–
3419	BORON TRIFLUORIDE ACETIC ACID COMPLEX, SOLID	8	–	II	–	1 kg	E2	P002	–	IBC08	B2 B4
3420	BORON TRIFLUORIDE PROPIONIC ACID COMPLEX, SOLID	8	–	II	–	1 kg	E2	P002	–	IBC08	B2 B4

Portable tanks and bulk containers			EmS	Stowage and segregation	Properties and observations	UN No.
	Tank instructions	Provisions				
(12)	(13) 4.2.5 4.3	(14) 4.2.5	(15) 5.4.3.2 7.3	(16) 7.1 7.2	(17)	(18)
–	T4	TP1	F-H, S-Q	Category A. "Separated from" ammonium compounds and cyanides.	Reacts vigorously with sulphuric acid. Reacts fiercely with cyanides when heated. May form explosive mixtures with combustible material, powdered metals or ammonium compounds. These mixtures are liable to ignite. When involved in a fire, may cause an explosion.	3408
–	T7	TP2	F-A, S-A	Category A.	Yellow liquid. Toxic if swallowed, by skin contact or by inhalation.	3409
–	T4	TP1	F-A, S-A	Category A.	Toxic if swallowed, by skin contact or by inhalation.	3410
–	T7	TP2	F-A, S-A	Category A.	Toxic if swallowed, by skin contact or by inhalation.	3411
–	T7	TP2	F-A, S-A	Category A.	See entry above.	3411
–	T7	TP2	F-A, S-B	Category A. Clear of living quarters.	Colourless liquid with a pungent odour. Corrosive to most metals. Causes burns to skin, eyes and mucous membranes.	3412
–	T4	TP1	F-A, S-B	Category A. Clear of living quarters.	See entry above.	3412
–	T14	TP2 TP13	F-A, S-A	Category B. "Separated from" acids.	Reacts with acids or acid fumes, evolving hydrogen cyanide, a highly toxic and flammable gas. Highly toxic if swallowed or by skin contact.	3413
–	T11	TP2 TP13 TP27	F-A, S-A	Category B. "Separated from" acids.	See entry above.	3413
–	T7	TP2 TP13 TP28	F-A, S-A	Category A. "Separated from" acids.	See entry above.	3413
–	T14	TP2 TP13	F-A, S-A	Category B. "Separated from" acids.	Reacts with acids or acid fumes, evolving hydrogen cyanide, a highly toxic and flammable gas. Highly toxic if swallowed or by skin contact.	3414
–	T11	TP2 TP13 TP27	F-A, S-A	Category B. "Separated from" acids.	See entry above.	3414
–	T7	TP2 TP13 TP28	F-A, S-A	Category A. "Separated from" acids.	See entry above.	3414
–	T4	TP1	F-A, S-A	Category A. "Separated from" acids.	Colourless liquid. Reacts with acids, evolving hydrogen fluoride, a toxic, irritating and corrosive gas, apparent as white fumes. Toxic if swallowed, by skin contact or by inhalation.	3415
–	T7	TP2 TP13	F-A, S-A	Category D. Keep as cool as reasonably practicable. Clear of living quarters.	Liquid evolving irritating vapour ("Tear Gas"). Toxic if swallowed, by skin contact or by inhalation.	3416
–	T3	TP33	F-A, S-G	Category D. Clear of living quarters.	Crystals or powder, evolving irritating vapour ("Tear Gas"). Toxic if swallowed, by skin contact or by inhalation.	3417
–	T4	TP1	F-A, S-A	Category A.	Toxic if swallowed, by skin contact or by inhalation.	3418
–	T3	TP33	F-A, S-B	Category A.	White crystalline solid. Melting point: 23°C. Highly corrosive to most metals. Causes burns to skin, eyes and mucous membranes.	3419
–	T3	TP33	F-A, S-B	Category A.	White crystalline solid. Melting point: 28°C. Highly corrosive to most metals. Causes burns to skin, eyes and mucous membranes.	3420

UN No.	Proper Shipping Name (PSN)	Class or division	Subsidiary risk(s)	Packing group	Special provisions	Limited and excepted quantity provisions		Packing		IBC	
						Limited quantities	Excepted quantities	Instructions	Provisions	Instructions	Provisions
(1)	(2) 3.1.2	(3) 2.0	(4) 2.0	(5) 2.0.1.3	(6) 3.3	(7a) 3.4	(7b) 3.5	(8) 4.1.4	(9) 4.1.4	(10) 4.1.4	(11) 4.1.4
△ 3421	POTASSIUM HYDROGEN DIFLUORIDE SOLUTION	8	6.1	II	–	1 L	E2	P001	–	IBC02	–
△ 3421	POTASSIUM HYDROGEN DIFLUORIDE SOLUTION	8	6.1	III	223	5 L	E1	P001	–	IBC03	–
3422	POTASSIUM FLUORIDE SOLUTION	6.1	–	III	223	5 L	E1	P001 LP01	–	IBC03	–
3423	TETRAMETHYLAMMONIUM HYDROXIDE, SOLID	8	–	II	–	1 kg	E2	P002	–	IBC08	B2 B4
3424	AMMONIUM DINITRO-*o*-CRESOLATE SOLUTION	6.1	– P	II	–	100 mL	E4	P001	–	IBC02	–
3424	AMMONIUM DINITRO-*o*-CRESOLATE SOLUTION	6.1	– P	III	223	5 L	E1	P001	–	IBC02	–
3425	BROMOACETIC ACID, SOLID	8	–	II	–	1 kg	E2	P002	–	IBC08	B2 B4
3426	ACRYLAMIDE SOLUTION	6.1	–	III	223	5 L	E1	P001 LP01	–	IBC03	–
3427	CHLOROBENZYL CHLORIDES, SOLID	6.1	– P	III	–	5 kg	E1	P002 LP02	–	IBC08	B3
3428	3-CHLORO-4-METHYLPHENYL ISOCYANATE, SOLID	6.1	–	II	–	500 g	E4	P002	–	IBC08	B2 B4
3429	CHLOROTOLUIDINES, LIQUID	6.1	–	III	–	5 L	E1	P001 LP01	–	IBC03	–
3430	XYLENOLS, LIQUID	6.1	–	II	–	100 mL	E4	P001	–	IBC02	–
3431	NITROBENZOTRIFLUORIDES, SOLID	6.1	– P	II	–	500 g	E4	P002	–	IBC08	B2 B4
△ 3432	POLYCHLORINATED BIPHENYLS, SOLID	9	– P	II	305 958	1 kg	E2	P906	–	IBC08	B2 B4
3434	NITROCRESOLS, LIQUID	6.1	–	III	–	5 L	E1	P001 LP01	–	IBC03	–
3436	HEXAFLUOROACETONE HYDRATE, SOLID	6.1	–	II	–	500 g	E4	P002	–	IBC08	B2 B4

Portable tanks and bulk containers			EmS	Stowage and segregation	Properties and observations	UN No.
	Tank instructions	Provisions				
(12)	(13) 4.2.5 4.3	(14) 4.2.5	(15) 5.4.3.2 7.3	(16) 7.1 7.2	(17)	(18)
–	T7	TP2	F-A, S-B	Category A. Protected from sources of heat. Clear of living quarters. "Separated from" acids.	Decomposed by heat or acids, evolving hydrogen fluoride, a toxic, extremely irritating and corrosive gas apparent as white fumes. In the presence of moisture, highly corrosive to glass, other siliceous materials and most metals. Toxic if swallowed, by skin contact or by inhalation. Causes burns to skin, eyes and mucous membranes.	3421 △
–	T4	TP1	F-A, S-B	Category A. Protected from sources of heat. Clear of living quarters. "Separated from" acids.	See entry above.	3421 △
–	T4	TP1	F-A, S-B	Category A. "Separated from" acids.	Decomposed by acids, evolving hydrogen fluoride, an irritating and corrosive gas. Toxic if swallowed, by skin contact or by inhalation.	3422
–	T3	TP33	F-A, S-B	Category A. "Separated from" acids.	Very soluble in water. Reacts violently with acids.	3423
–	T7	TP2	F-A, S-A	Category B. "Away from" heavy metals and their salts. "Separated from" classes 3 and 4.1. "Separated longitudinally by an intervening complete compartment or hold from" class 1.	The commercial product is a 50% suspension in water. May support combustion and burn without oxygen. When involved in a fire, evolves toxic fumes. Forms extremely sensitive explosive compounds with lead, silver or other heavy metals and their compounds. Toxic if swallowed, by skin contact or by inhalation.	3424
–	T7	TP2	F-A, S-A	Category A. "Away from" heavy metals and their salts. "Separated from" classes 3 and 4.1. "Separated longitudinally by an intervening complete compartment or hold from" class 1.	See entry above.	3424
–	T3	TP33	F-A, S-B	Category A.	Colourless, deliquescent crystals. Melting point: 51°C. Corrosive to most metals. Harmful if swallowed. Causes burns to eyes and skin.	3425
–	T4	TP1	F-A, S-A	Category A. Keep as cool as reasonably practicable.	Toxic if swallowed, by skin contact or by inhalation.	3426
–	T1	TP33	F-A, S-A	Category A.	Colourless crystalline solid. Melting point: 29°C. Immiscible with or insoluble in water. Toxic if swallowed, by skin contact or by inhalation.	3427
–	T3	TP33	F-A, S-A	Category B. Clear of living quarters.	Colourless solid with a pungent odour. Melting point: 23°C. Insoluble in water. Reacts with water, evolving carbon dioxide. Toxic if swallowed, by skin contact or by inhalation. Irritating to skin, eyes and mucous membranes.	3428
–	T4	TP1	F-A, S-A	Category A.	Brown liquid. Toxic if swallowed, by skin contact or by inhalation.	3429
–	T7	TP2	F-A, S-A	Category A.	The commercial products are liquids with a pungent tar odour. Toxic if swallowed, by skin contact or by inhalation.	3430
–	T3	TP33	F-A, S-A	Category A. Clear of living quarters.	Low melting point (31°C to 32°C) solids with an aromatic odour. Insoluble in water. Toxic if swallowed, by skin contact or by inhalation.	3431
–	T3	TP33	F-A, S-A	Category A. Segregation from foodstuffs as in 7.3.4.2.1, 7.6.3.1.2 or 7.7.3.6.	Solids with perceptible odour. Insoluble in water. Harmful by ingestion or by skin contact. If spilled, can be a persistent hazard to the environment. This entry covers articles, such as rags, cotton waste, clothing or sawdust, containing polychlorinated biphenyls where no free visible liquid is present.	3432 △
–	T4	TP1	F-A, S-A	Category A.	Slightly miscible in water. Toxic if swallowed, by skin contact or by inhalation.	3434
–	T3	TP33	F-A, S-A	Category B. Clear of living quarters.	This entry covers solid hydrate and hexafluoroacetone. Melting point of the pure substance: 23°C. Toxic if swallowed, by skin contact or by inhalation.	3436

UN No.	Proper Shipping Name (PSN)	Class or division	Subsidiary risk(s)	Packing group	Special provisions	Limited and excepted quantity provisions		Packing		IBC	
						Limited quantities	Excepted quantities	Instructions	Provisions	Instructions	Provisions
(1)	(2) 3.1.2	(3) 2.0	(4) 2.0	(5) 2.0.1.3	(6) 3.3	(7a) 3.4	(7b) 3.5	(8) 4.1.4	(9) 4.1.4	(10) 4.1.4	(11) 4.1.4
3437	CHLOROCRESOLS, SOLID	6.1	–	II	–	500 g	E4	P002	–	IBC08	B2 B4
3438	*alpha*-METHYLBENZYL ALCOHOL, SOLID	6.1	–	III	–	5 kg	E1	P002 LP02	–	IBC08	B3
△ 3439	NITRILES, SOLID, TOXIC, N.O.S.	6.1	–	I	274	0	E5	P002	–	IBC07	B1
△ 3439	NITRILES, SOLID, TOXIC, N.O.S.	6.1	–	II	274	500 g	E4	P002	–	IBC08	B2 B4
△ 3439	NITRILES, SOLID, TOXIC, N.O.S.	6.1	–	III	223 274	5 kg	E1	P002 LP02	–	IBC08	B3
3440	SELENIUM COMPOUND, LIQUID, N.O.S.	6.1	–	I	274	0	E5	P001	–	–	–
3440	SELENIUM COMPOUND, LIQUID, N.O.S.	6.1	–	II	274	100 mL	E4	P001	–	IBC02	–
3440	SELENIUM COMPOUND, LIQUID, N.O.S.	6.1	–	III	223 274	5 L	E1	P001	–	IBC03	–
3441	CHLORODINITROBENZENES, SOLID	6.1	– P	II	279	500 g	E4	P002	–	IBC08	B2 B4
3442	DICHLOROANILINES, SOLID	6.1	– P	II	279	500 g	E4	P002	–	IBC08	B2 B4
3443	DINITROBENZENES, SOLID	6.1	–	II	–	500 g	E4	P002	–	IBC08	B2 B4
3444	NICOTINE HYDROCHLORIDE, SOLID	6.1	–	II	43	500 g	E4	P002	–	IBC08	B2 B4
3445	NICOTINE SULPHATE, SOLID	6.1	–	II	–	500 g	E4	P002	–	IBC08	B2 B4
3446	NITROTOLUENES, SOLID	6.1	–	II	–	500 g	E4	P002	–	IBC08	B2 B4
3447	NITROXYLENES, SOLID	6.1	–	II	–	500 g	E4	P002	–	IBC08	B2 B4
3448	TEAR GAS SUBSTANCE, SOLID, N.O.S.	6.1	–	I	274	0	E5	P002	PP31	–	–
3448	TEAR GAS SUBSTANCE, SOLID, N.O.S.	6.1	–	II	274	0	E4	P002	PP31	IBC08	B2 B4
3449	BROMOBENZYL CYANIDES, SOLID	6.1	–	I	138	0	E5	P002	PP31	–	–
3450	DIPHENYLCHLOROARSINE, SOLID	6.1	– P	I	–	0	E5	P002	PP31	IBC07	B1
3451	TOLUIDINES, SOLID	6.1	–	II	279	500 g	E4	P002	–	IBC08	B2 B4
3452	XYLIDINES, SOLID	6.1	–	II	–	500 g	E4	P002	–	IBC08	B2 B4
3453	PHOSPHORIC ACID, SOLID	8	–	III	–	5 kg	E1	P002 LP02	–	IBC08	B3

Portable tanks and bulk containers			EmS	Stowage and segregation	Properties and observations	UN No.
	Tank instructions	Provisions				
(12)	(13) 4.2.5 4.3	(14) 4.2.5	(15) 5.4.3.2 7.3	(16) 7.1 7.2	(17)	(18)
–	T3	TP33	F-A, S-A	Category A. Keep as cool as reasonably practicable.	White or pink crystals with a phenol-like odour. Melting point: 45°C to 68°C. Slightly soluble in water. Decomposes when heated, evolving extremely toxic fumes (phosgene). Toxic if swallowed, by skin contact or by inhalation.	3437
–	T1	TP33	F-A, S-A	Category A.	Slightly soluble in water. Melting point: 21°C (pure substance). Toxic if swallowed, by skin contact or by inhalation.	3438
–	T6	TP33	F-A, S-A	Category B. "Separated from" acids.	Solid, evolving toxic vapours. Reacts with acids or acid fumes, evolving hydrogen cyanide, a highly toxic and flammable gas. Soluble in water. Toxic if swallowed, by skin contact or by inhalation.	3439 △
–	T3	TP33	F-A, S-A	Category B. "Separated from" acids.	See entry above.	3439 △
–	T1	TP33	F-A, S-A	Category A. "Separated from" acids.	See entry above.	3439 △
–	T14	TP2 TP27	F-A, S-A	Category B.	Toxic if swallowed, by skin contact or by inhalation.	3440
–	T11	TP2 TP27	F-A, S-A	Category B.	See entry above.	3440
–	T7	TP1 TP28	F-A, S-A	Category A.	See entry above.	3440
–	T3	TP33	F-A, S-A	Category A. "Separated from" class 3.	Crystals. Melting point: 27°C to 53°C. May explode if involved in a fire. Toxic if swallowed, by skin contact or by inhalation.	3441
–	T3	TP33	F-A, S-A	Category A. Clear of living quarters.	Solid with a penetrating odour. Liquid mixtures of various isomers of dichloroanilines, some of which in the pure state may be solid, with a melting point varying from 24°C to 72°C. Toxic if swallowed, by skin contact or by inhalation.	3442
–	T3	TP33	F-A, S-A	Category A. "Separated from" class 3.	May explode if involved in a fire. Toxic if swallowed, by skin contact or by inhalation.	3443
–	T3	TP33	F-A, S-A	Category A.	Deliquescent crystals or solids or pastes. Soluble in water. Toxic if swallowed, by skin contact or by inhalation.	3444
–	T3	TP33	F-A, S-A	Category A.	Solid or paste. Soluble in water. Toxic if swallowed, by skin contact or by inhalation.	3445
–	T3	TP33	F-A, S-A	Category A.	Yellow solid. Melting point: *para*-NITROTOLUENE: 52°C to 54°C. Toxic if swallowed, by skin contact or by inhalation.	3446
–	T3	TP33	F-A, S-A	Category A.	Yellow solid. Melting points: 4-NITRO-2-XYLENE: 29°C to 31°C, 5-NITRO-3-XYLENE: 72°C to 74°C. Insoluble in water. Toxic if swallowed, by skin contact or by inhalation.	3447
–	T6	TP33	F-A, S-A	Category D. Clear of living quarters.	"Tear gas substance" is a generic term for substances which, in minute quantities dispersed in air, cause extreme eye irritation and profuse tears. Toxic if swallowed, by skin contact or by inhalation.	3448
–	T3	TP33	F-A, S-A	Category D. Clear of living quarters.	See entry above.	3448
–	T6	TP33	F-A, S-A	Category D. Keep as cool as reasonably practicable. Clear of living quarters. "Separated from" acids.	Volatile, yellow crystals evolving irritating vapours ("Tear Gas"). Melting point: *meta*-BROMOBENZYL CYANIDE 25°C. Highly toxic if swallowed, by skin contact or by inhalation.	3449
–	T6	TP33	F-A, S-A	Category D. Clear of living quarters.	When pure, volatile, colourless crystals evolving an irritating vapour ("Tear Gas"). Melting point: 41°C. Highly toxic if swallowed, by skin contact or by inhalation.	3450
–	T3	TP33	F-A, S-A	Category A.	*para*-TOLUIDINE is solid in pure form, with a melting point of approximately 45°C. Toxic if swallowed, by skin contact or by inhalation.	3451
–	T3	TP33	F-A, S-A	Category A.	3,4-Dimethylaniline is a solid, which has a melting point of 47°C. Toxic if swallowed, by skin contact or by dust inhalation.	3452
–	T1	TP33	F-A, S-B	Category A.	Very deliquescent, crystalline solid. Melting point: 42°C. Soluble in water. Mildly corrosive to most metals.	3453

UN No. (1)	Proper Shipping Name (PSN) (2) 3.1.2	Class or division (3) 2.0	Subsidiary risk(s) (4) 2.0	Packing group (5) 2.0.1.3	Special provisions (6) 3.3	Limited and excepted quantity provisions: Limited quantities (7a) 3.4	Limited and excepted quantity provisions: Excepted quantities (7b) 3.5	Packing: Instructions (8) 4.1.4	Packing: Provisions (9) 4.1.4	IBC: Instructions (10) 4.1.4	IBC: Provisions (11) 4.1.4
3454	DINITROTOLUENES, SOLID	6.1	–	II	–	500 g	E4	P002	–	IBC08	B2 B4
3455	CRESOLS, SOLID	6.1	8	II	–	500 g	E4	P002	–	IBC08	B2 B4
3456	NITROSYLSULPHURIC ACID, SOLID	8	–	II	–	1 kg	E2	P002	–	IBC08	B2 B4
3457	CHLORONITROTOLUENES, SOLID	6.1	– P	III	–	5 kg	E1	P002 LP02	–	IBC08	B3
3458	NITROANISOLES, SOLID	6.1	–	III	279	5 kg	E1	P002 LP02	–	IBC08	B3
3459	NITROBROMOBENZENES, SOLID	6.1	–	III	–	5 kg	E1	P002 LP02	–	IBC08	B3
3460	*N*-ETHYLBENZYLTOLUIDINES, SOLID	6.1	–	III	–	5 kg	E1	P002 LP02	–	IBC08	B3
3462	TOXINS, EXTRACTED FROM LIVING SOURCES, SOLID, N.O.S.	6.1	–	I	210 274	0	E5	P002	–	IBC07	B1
3462	TOXINS, EXTRACTED FROM LIVING SOURCES, SOLID, N.O.S.	6.1	–	II	210 274	500 g	E4	P002	–	IBC08	B2 B4
3462	TOXINS, EXTRACTED FROM LIVING SOURCES, SOLID, N.O.S.	6.1	–	III	210 223 274	5 kg	E1	P002	–	IBC08	B3
3463	PROPIONIC ACID, with not less than 90% acid by mass	8	3	II	–	1 L	E2	P001	–	IBC02	–
△ 3464	ORGANOPHOSPHORUS COMPOUND, SOLID, TOXIC, N.O.S.	6.1	–	I	43 274	0	E5	P002	–	IBC07	B1
△ 3464	ORGANOPHOSPHORUS COMPOUND, SOLID, TOXIC, N.O.S.	6.1	–	II	43 274	500 g	E4	P002	–	IBC08	B2 B4
△ 3464	ORGANOPHOSPHORUS COMPOUND, SOLID, TOXIC, N.O.S.	6.1	–	III	43 223 274	5 kg	E1	P002 LP02	–	IBC08	B3
3465	ORGANOARSENIC COMPOUND, SOLID, N.O.S.	6.1	–	I	274	0	E5	P002	–	IBC07	B1
3465	ORGANOARSENIC COMPOUND, SOLID, N.O.S.	6.1	–	II	274	500 g	E4	P002	–	IBC08	B2 B4
3465	ORGANOARSENIC COMPOUND, SOLID, N.O.S.	6.1	–	III	223 274	5 kg	E1	P002 LP02	–	IBC08	B3
3466	METAL CARBONYLS, SOLID, N.O.S.	6.1	–	I	274	0	E5	P002	–	IBC07	B1
3466	METAL CARBONYLS, SOLID, N.O.S.	6.1	–	II	274	500 g	E4	P002	–	IBC08	B2 B4

Portable tanks and bulk containers			EmS	Stowage and segregation	Properties and observations	UN No.
	Tank instructions	Provisions				
(12)	(13) 4.2.5 4.3	(14) 4.2.5	(15) 5.4.3.2 7.3	(16) 7.1 7.2	(17)	(18)
–	T3	TP33	F-A, S-A	Category A.	Yellow crystals or flakes, insoluble in water. Toxic if swallowed, by skin contact or by inhalation.	3454
–	T3	TP33	F-A, S-B	Category B.	Light yellow solid. Soluble in water. Melting points of CRESOLS: *ortho*-CRESOL: 30°C, *para*-CRESOL: 35°C. Toxic if swallowed, by skin contact or by inhalation. Cause burns to skin, eyes and mucous membranes.	3455
–	T3	TP33	F-A, S-B	Category D. Clear of living quarters. Segregation as for class 5.1, but "separated from" classes 4.1, 5.1 and 7.	Crystalline solid. Oxidant which may cause fire with organic materials (such as wood, straw, etc.). When involved in a fire, evolves toxic gases. In the presence of moisture, highly corrosive to most metals. Causes burns to skin, eyes and mucous membranes.	3456
–	T1	TP33	F-A, S-A	Category A. Segregation as for class 5.1 but "away from" classes 4.1, 5.1 and 7.	Melting range 20°C to 40°C. Insoluble in water. Oxidizing substance which may explode or burn fiercely when in contact with organic materials. Toxic if swallowed, by skin contact or by inhalation.	3457
–	T1	TP33	F-A, S-A	Category A.	Light reddish or amber crystals. Melting points: 38°C to 54°C. Insoluble in water. Toxic if swallowed, by skin contact or by inhalation.	3458
–	T1	TP33	F-A, S-A	Category A.	Colourless to pale yellow crystals which may liquefy under transport conditions. Melting points: 1-BROMO-2-NITROBENZENE: 43°C. 1-BROMO-4-NITROBENZENE: 127°C. Insoluble in water. Toxic if swallowed, by skin contact or by inhalation.	3459
–	T1	TP33	F-A, S-A	Category A.	Solid which may liquefy under transport conditions. Strong odour. Insoluble in water. Toxic if swallowed, by skin contact or by inhalation.	3460
–	T6	TP33	F-A, S-A	Category B.	Toxins from plant, animal or bacterial sources which contain infectious substances or toxins that are contained in infectious substances should be classified in class 6.2. Toxic if swallowed, by skin contact or by inhalation.	3462
–	T3	TP33	F-A, S-A	Category B.	See entry above.	3462
–	T1	TP33	F-A, S-A	Category A.	See entry above.	3462
–	T7	TP2	F-E, S-C	Category A.	Colourless flammable liquid with a pungent odour. Miscible with water. Corrosive to lead and most other metals. Burns skin. Vapours irritate mucous membranes. Pure PROPIONIC ACID: flashpoint 50°C c.c.	3463
–	T6	TP33	F-A, S-A	Category B.	Toxic if swallowed, by skin contact or by inhalation.	3464 △
–	T3	TP33	F-A, S-A	Category B.	See entry above.	3464 △
–	T1	TP33	F-A, S-A	Category A.	See entry above.	3464 △
–	T6	TP33	F-A, S-A	Category B.	Toxic if swallowed, by skin contact or by inhalation.	3465
–	T3	TP33	F-A, S-A	Category B.	See entry above.	3465
–	T1	TP33	F-A, S-A	Category A.	See entry above.	3465
–	T6	TP33	F-A, S-A	Category D. Clear of living quarters.	Insoluble in water. Toxic if swallowed, by skin contact or by dust inhalation.	3466
–	T3	TP33	F-A, S-A	Category D. Clear of living quarters.	See entry above.	3466

3

DGL

UN No. (1)	Proper Shipping Name (PSN) (2) 3.1.2	Class or division (3) 2.0	Subsidiary risk(s) (4) 2.0	Packing group (5) 2.0.1.3	Special provisions (6) 3.3	Limited and excepted quantity provisions: Limited quantities (7a) 3.4	Excepted quantities (7b) 3.5	Packing: Instructions (8) 4.1.4	Provisions (9) 4.1.4	IBC: Instructions (10) 4.1.4	Provisions (11) 4.1.4
3466	METAL CARBONYLS, SOLID, N.O.S.	6.1	–	III	223 274	5 kg	E1	P002 LP02	–	IBC08	B3
△ 3467	ORGANOMETALLIC COMPOUND, SOLID, TOXIC, N.O.S.	6.1	–	I	274	0	E5	P002	–	IBC07	B1
△ 3467	ORGANOMETALLIC COMPOUND, SOLID, TOXIC, N.O.S.	6.1	–	II	274	500 g	E4	P002	–	IBC08	B2 B4
△ 3467	ORGANOMETALLIC COMPOUND, SOLID, TOXIC, N.O.S.	6.1	–	III	223 274	5 kg	E1	P002 LP02	–	IBC08	B3
3468	HYDROGEN IN A METAL HYDRIDE STORAGE SYSTEM or HYDROGEN IN A METAL HYDRIDE STORAGE SYSTEM CONTAINED IN EQUIPMENT or HYDROGEN IN A METAL HYDRIDE STORAGE SYSTEM PACKED WITH EQUIPMENT	2.1	–	–	321 356	0	E0	P205	–	–	–
3469	PAINT, FLAMMABLE, CORROSIVE (including paint, lacquer, enamel, stain, shellac, varnish, polish, liquid filler and liquid lacquer base) or PAINT RELATED MATERIAL, FLAMMABLE, CORROSIVE (including paint thinning or reducing compound)	3	8	I	163	0	E0	P001	–	–	–
3469	PAINT, FLAMMABLE, CORROSIVE (including paint, lacquer, enamel, stain, shellac, varnish, polish, liquid filler and liquid lacquer base) or PAINT RELATED MATERIAL, FLAMMABLE, CORROSIVE (including paint thinning or reducing compound)	3	8	II	163	1 L	E2	P001	–	IBC02	–
3469	PAINT, FLAMMABLE, CORROSIVE (including paint, lacquer, enamel, stain, shellac, varnish, polish, liquid filler and liquid lacquer base) or PAINT RELATED MATERIAL, FLAMMABLE, CORROSIVE (including paint thinning or reducing compound)	3	8	III	163 223	5 L	E1	P001	–	IBC03	–
3470	PAINT, CORROSIVE, FLAMMABLE (including paint, lacquer, enamel, stain, shellac, varnish, polish, liquid filler and liquid lacquer base) or PAINT RELATED MATERIAL CORROSIVE, FLAMMABLE (including paint thinning or reducing compound)	8	3	II	163	1 L	E2	P001	–	IBC02	–
△ 3471	HYDROGENDIFLUORIDES SOLUTION, N.O.S.	8	6.1	II	–	1 L	E2	P001	–	IBC02	–

Portable tanks and bulk containers			EmS	Stowage and segregation	Properties and observations	UN No.	
	Tank instructions	Provisions					
(12)	(13) 4.2.5 4.3	(14) 4.2.5	(15) 5.4.3.2 7.3	(16) 7.1 7.2	(17)	(18)	
–	T1	TP33	F-A, S-A	Category D. Clear of living quarters.	Insoluble in water. Toxic if swallowed, by skin contact or by dust inhalation.	3466	
–	T6	TP33	F-A, S-A	Category B.	Toxic if swallowed, by skin contact or by inhalation.	3467	△
–	T3	TP33	F-A, S-A	Category B.	See entry above.	3467	△
–	T1	TP33	F-A, S-A	Category A.	See entry above.	3467	△
–	–	–	F-D, S-U	Category D.	Article containing flammable odourless gas, which is much lighter than air.	3468	
–	T11	TP2 TP27	F-E, S-C	Category E. Clear of living quarters.	Miscibility with water depends upon the composition. Corrosive contents cause burns to skin, eyes and mucous membranes.	3469	
–	T7	TP2 TP8 TP28	F-E, S-C	Category B. Clear of living quarters.	See entry above.	3469	
–	T4	TP1 TP29	F-E, S-C	Category A. Clear of living quarters.	See entry above.	3469	
–	T7	TP2 TP8 TP28	F-E, S-C	Category B. Clear of living quarters.	Miscibility with water depends upon the composition. Corrosive contents cause burns to skin, eyes and mucous membranes.	3470	
–	T7	TP2	F-A, S-B	Category A. Protected from sources of heat. Clear of living quarters. "Separated from" acids.	When involved in a fire or in contact with acids, evolves hydrogen fluoride, an extremely irritating and corrosive gas. Corrosive to glass, other siliceous materials and most metals. Toxic if swallowed, by skin contact or by inhalation. Causes burns to skin, eyes and mucous membranes.	3471	△

UN No.	Proper Shipping Name (PSN)	Class or division	Subsidiary risk(s)	Packing group	Special provisions	Limited and excepted quantity provisions		Packing		IBC	
						Limited quantities	Excepted quantities	Instructions	Provisions	Instructions	Provisions
(1)	(2) 3.1.2	(3) 2.0	(4) 2.0	(5) 2.0.1.3	(6) 3.3	(7a) 3.4	(7b) 3.5	(8) 4.1.4	(9) 4.1.4	(10) 4.1.4	(11) 4.1.4
△ 3471	HYDROGENDIFLUORIDES SOLUTION, N.O.S.	8	6.1	III	223	5 L	E1	P001	–	IBC03	–
3472	CROTONIC ACID, LIQUID	8	–	III	–	5 L	E1	P001 LP01	–	IBC03	–
3473	FUEL CELL CARTRIDGES or FUEL CELL CARTRIDGES CONTAINED IN EQUIPMENT or FUEL CELL CARTRIDGES PACKED WITH EQUIPMENT containing flammable liquids	3	–	–	328	1 L	E0	P004	–	–	–
3474	1-HYDROXYBENZOTRIAZOLE MONOHYDRATE	4.1	–	I	–	0	E0	P406	PP48	–	–
△ 3475	ETHANOL AND GASOLINE MIXTURE or ETHANOL AND MOTOR SPIRIT MIXTURE or ETHANOL AND PETROL MIXTURE, with more than 10% ethanol	3	–	II	333 363	1 L	E2	P001	–	IBC02	–
3476	FUEL CELL CARTRIDGES or FUEL CELL CARTRIDGES CONTAINED IN EQUIPMENT or FUEL CELL CARTRIDGES PACKED WITH EQUIPMENT, containing water-reactive substances	4.3	–	–	328 334	500 mL or 500 g	E0	P004	–	–	–
3477	FUEL CELL CARTRIDGES or FUEL CELL CARTRIDGES CONTAINED IN EQUIPMENT or FUEL CELL CARTRIDGES PACKED WITH EQUIPMENT, containing corrosive substances	8	–	–	328 334	1 L or 1 kg	E0	P004	–	–	–
3478	FUEL CELL CARTRIDGES or FUEL CELL CARTRIDGES CONTAINED IN EQUIPMENT or FUEL CELL CARTRIDGES PACKED WITH EQUIPMENT, containing liquefied flammable gas	2.1	–	–	328 338	120 mL	E0	P004	–	–	–
3479	FUEL CELL CARTRIDGES or FUEL CELL CARTRIDGES CONTAINED IN EQUIPMENT or FUEL CELL CARTRIDGES PACKED WITH EQUIPMENT, containing hydrogen in metal hydride	2.1	–	–	328 339	120 mL	E0	P004	–	–	–
3480	LITHIUM ION BATTERIES (including lithium ion polymer batteries)	9	–	II	188 230 310 348 957	0	E0	P903	–	–	–

Portable tanks and bulk containers			EmS	Stowage and segregation	Properties and observations	UN No.
	Tank instructions	Provisions				
(12)	(13) 4.2.5 4.3	(14) 4.2.5	(15) 5.4.3.2 7.3	(16) 7.1 7.2	(17)	(18)
–	T4	TP1	F-A, S-B	Category A. Protected from sources of heat. Clear of living quarters. "Separated from" acids.	When involved in a fire or in contact with acids, evolves hydrogen fluoride, an extremely irritating and corrosive gas. Corrosive to glass, other siliceous materials and most metals. Toxic if swallowed, by skin contact or by inhalation. Causes burns to skin, eyes and mucous membranes.	3471 △
–	T4	TP1	F-A, S-B	Category A. Keep as cool as reasonably practicable.	Causes burns to skin, eyes and mucous membranes.	3472
–	–	–	F-E, S-D	Category A.	Fuel cell cartridges containing flammable liquids including methanol or methanol/water solutions. Fuel cell cartridges may also be shipped in, or packed with, equipment.	3473
–	–	–	F-B, S-J	Category D. "Away from" class 3 and heavy metals and their salts.	Desensitized explosive. White to light beige powder. Explosive and sensitive to friction in the dry state. When involved in a fire, evolves toxic fumes; in closed compartments these fumes may form an explosive mixture with air. May form extremely sensitive compounds with heavy metals or their salts.	3474
–	T4	TP1	F-E, S-E	Category E.	Colourless, volatile liquids. Miscibility with water depends on the composition.	3475 △
–	–	–	F-G, S-P	Category A.	Fuel cell cartridges containing water-reactive substances may also be shipped in, or packed with, equipment.	3476
–	–	–	F-A, S-B	Category A.	Fuel cell cartridges containing corrosive substances may also be shipped in, or packed with, equipment.	3477
–	–	–	F-D, S-U	Category B.	Fuel cell cartridges containing butane or other flammable liquefied gas may also be shipped in, or packed with, equipment.	3478
–	–	–	F-D, S-U	Category B.	Fuel cell cartridges containing hydrogen, butane or other flammable odourless gas, which is much lighter than air, may also be shipped in or packed with equipment.	3479
–	–	–	F-A, S-I	Category A.	Electrical batteries containing lithium ion encased in a rigid metallic body. Lithium ion batteries may also be shipped in, or packed with, equipment. Electrical lithium batteries may cause fire due to an explosive rupture of the body caused by improper construction or reaction with contaminants.	3480

3

DGL

UN No.	Proper Shipping Name (PSN)	Class or division	Subsidiary risk(s)	Packing group	Special provisions	Limited and excepted quantity provisions		Packing		IBC	
						Limited quantities	Excepted quantities	Instructions	Provisions	Instructions	Provisions
(1)	(2) 3.1.2	(3) 2.0	(4) 2.0	(5) 2.0.1.3	(6) 3.3	(7a) 3.4	(7b) 3.5	(8) 4.1.4	(9) 4.1.4	(10) 4.1.4	(11) 4.1.4
△ 3481	LITHIUM ION BATTERIES CONTAINED IN EQUIPMENT or LITHIUM ION BATTERIES PACKED WITH EQUIPMENT (including lithium ion polymer batteries)	9	–	II	188 230 348 360 957	0	E0	P903	–	–	–
3482	ALKALI METAL DISPERSION, FLAMMABLE or ALKALINE EARTH METAL DISPERSION, FLAMMABLE	4.3	3	I	182 183	0	E0	P402	PP31	–	–
△ 3483	MOTOR FUEL ANTI-KNOCK MIXTURE, FLAMMABLE	6.1	3 P	I	–	0	E5	P602	–	–	–
3484	HYDRAZINE AQUEOUS SOLUTION, FLAMMABLE with more than 37% hydrazine, by mass	8	3 6.1	I	–	0	E0	P001	–	–	–
△ 3485	CALCIUM HYPOCHLORITE, DRY, CORROSIVE or CALCIUM HYPOCHLORITE MIXTURE, DRY, CORROSIVE with more than 39% available chlorine (8.8% available oxygen)	5.1	8	II	314	1 kg	E2	P002	PP85	–	–
△ 3486	CALCIUM HYPOCHLORITE MIXTURE, DRY, CORROSIVE with more than 10% but not more than 39% available chlorine	5.1	8	III	314	5 kg	E1	P002	PP85	–	–
△ 3487	CALCIUM HYPOCHLORITE, HYDRATED, CORROSIVE or CALCIUM HYPOCHLORITE, HYDRATED MIXTURE, CORROSIVE with not less than 5.5% but not more than 16% water	5.1	8	II	314 322	1 kg	E2	P002	PP85	–	–

Portable tanks and bulk containers			EmS	Stowage and segregation	Properties and observations	UN No.	
	Tank instructions	Provisions					
(12)	(13) 4.2.5 4.3	(14) 4.2.5	(15) 5.4.3.2 7.3	(16) 7.1 7.2	(17)	(18)	
–	–	–	F-A, S-I	Category A.	Electrical batteries containing lithium ion encased in a rigid metallic body. Lithium ion batteries may also be shipped in, or packed with, equipment. Electrical lithium batteries may cause fire due to an explosive rupture of the body caused by improper construction or reaction with contaminants.	3481	△
–	–	–	F-G, S-N	Category D. "Separated from" acids.	Finely divided alkali or alkaline earth metal suspended in a flammable liquid. Reacts violently with moisture, water or acids, evolving hydrogen, which may be ignited by the heat of the reaction.	3482	
–	T14	TP2 TP13	F-E, S-D	Category D. Protected from sources of heat. Clear of living quarters.	Volatile flammable liquids evolving toxic vapour. Mixture of tetraethyllead or tetramethyllead with ethylene dibromide and ethylene dichloride. Insoluble in water. Highly toxic if swallowed, by skin contact or by inhalation.	3483	△
–	T10	TP2 TP13	F-E, S-C	Category D. Clear of living quarters. Segregation as for class 3, but "away from" class 4.1. "Separated from" acids.	Colourless flammable liquid. Powerful reducing agent, burns readily. Toxic if swallowed, by skin contact or by inhalation. Causes burns to skin, eyes and mucous membranes. Reacts violently with acids.	3484	
–	–	–	F-H, S-Q	Category D. Protected from sources of heat. Cargo transport units shall be shaded from direct sunlight. Packages in cargo transport units shall be stowed so as to allow for adequate air circulation throughout the cargo. "Separated from" ammonium compounds, acids, cyanides, hydrogen peroxides and liquid organic substances.	White or yellowish corrosive solid (powder, granules or tablets) with chlorine-like odour. May cause fire in contact with organic material or ammonium compounds. Substances are liable to exothermic decomposition at elevated temperatures. This condition may lead to fire or explosion. Decomposition can be initiated by heat or by impurities (e.g., powdered metals (iron, manganese, cobalt, magnesium) and their compounds). Liable to heat slowly. Reacts with acids, evolving chlorine, an irritating, corrosive and toxic gas. In the presence of moisture, corrosive to most metals. Causes burns to skin, eyes and mucous membranes.	3485	△
–	–	–	F-H, S-Q	Category D. Protected from sources of heat. Cargo transport units shall be shaded from direct sunlight. Packages in cargo transport units shall be stowed so as to allow for adequate air circulation throughout the cargo. "Separated from" ammonium compounds, acids, cyanides, hydrogen peroxides and liquid organic substances.	White or yellowish corrosive solid (powder, granules or tablets) with chlorine-like odour. Soluble in water. May cause fire in contact with organic material or ammonium compounds. Substances are liable to exothermic decomposition at elevated temperatures. This condition may lead to fire or explosion. Decomposition can be initiated by heat or by impurities (e.g., powdered metals (iron, manganese, cobalt, magnesium) and their compounds). Liable to heat slowly. Reacts with acids, evolving chlorine, an irritating, corrosive and toxic gas. In the presence of moisture, corrosive to most metals. Causes burns to skin, eyes and mucous membranes.	3486	△
–	–	–	F-H, S-Q	Category D. Protected from sources of heat. Cargo transport units shall be shaded from direct sunlight. Packages in cargo transport units shall be stowed so as to allow for adequate air circulation throughout the cargo. "Separated from" ammonium compounds, acids, cyanides, hydrogen peroxides and liquid organic substances.	White or yellowish corrosive solid (powder, granules or tablets) with chlorine-like odour. Soluble in water. May cause fire in contact with organic material or ammonium compounds. Substances are liable to exothermic decomposition at elevated temperatures. This condition may lead to fire or explosion. Decomposition can be initiated by heat or by impurities (e.g., powdered metals (iron, manganese, cobalt, magnesium) and their compounds). Liable to heat slowly. Reacts with acids, evolving chlorine, an irritating, corrosive and toxic gas. In the presence of moisture, corrosive to most metals. Causes burns to skin, eyes and mucous membranes.	3487	△

UN No. (1)	Proper Shipping Name (PSN) (2) 3.1.2	Class or division (3) 2.0	Subsidiary risk(s) (4) 2.0	Packing group (5) 2.0.1.3	Special provisions (6) 3.3	Limited and excepted quantity provisions		Packing		IBC	
						Limited quantities (7a) 3.4	Excepted quantities (7b) 3.5	Instructions (8) 4.1.4	Provisions (9) 4.1.4	Instructions (10) 4.1.4	Provisions (11) 4.1.4
△ 3487	CALCIUM HYPOCHLORITE, HYDRATED, CORROSIVE or CALCIUM HYPOCHLORITE, HYDRATED MIXTURE, CORROSIVE with not less than 5.5% but not more than 16% water	5.1	8	III	223 314	5 kg	E1	P002	PP85	–	–
△ 3488	TOXIC BY INHALATION LIQUID, FLAMMABLE, CORROSIVE, N.O.S. with an LC_{50} lower than or equal to 200 mL/m^3 and saturated vapour concentration greater than or equal to 500 LC_{50}	6.1	3 8	I	274	0	E0	P601	–	–	–
△ 3489	TOXIC BY INHALATION LIQUID, FLAMMABLE, CORROSIVE, N.O.S. with an LC_{50} lower than or equal to 1000 mL/m^3 and saturated vapour concentration greater than or equal to 10 LC_{50}	6.1	3 8	I	274	0	E0	P602	–	–	–
△ 3490	TOXIC BY INHALATION LIQUID, WATER-REACTIVE, FLAMMABLE, N.O.S. with an LC_{50} lower than or equal to 200 mL/m^3 and saturated vapour concentration greater than or equal to 500 LC_{50}	6.1	4.3 3	I	274	0	E0	P601	–	–	–
△ 3491	TOXIC BY INHALATION LIQUID, WATER-REACTIVE, FLAMMABLE, N.O.S. with an LC_{50} lower than or equal to 1000 mL/m^3 and saturated vapour concentration greater than or equal to 10 LC_{50}	6.1	4.3 3	I	274	0	E0	P602	–	–	–
⊗											
3494	PETROLEUM SOUR CRUDE OIL, FLAMMABLE, TOXIC	3	6.1	I	343	0	E0	P001	–	–	–
3494	PETROLEUM SOUR CRUDE OIL, FLAMMABLE, TOXIC	3	6.1	II	343	1 L	E2	P001	–	IBC02	–
3494	PETROLEUM SOUR CRUDE OIL, FLAMMABLE, TOXIC	3	6.1	III	343	5 L	E1	P001	–	IBC03	–
3495	IODINE	8	6.1	III	279	5 kg	E1	P002	–	IBC08	B3
△ 3496	BATTERIES, NICKEL-METAL HYDRIDE	9	–	–	117 963	0	E0	See SP963	–	IBC08	–
■ 3497	KRILL MEAL	4.2	–	II	300	0	E2	P410	–	IBC06	B2

Portable tanks and bulk containers			EmS	Stowage and segregation	Properties and observations	UN No.	
	Tank instructions	Provisions					
(12)	(13) 4.2.5 4.3	(14) 4.2.5	(15) 5.4.3.2 7.3	(16) 7.1 7.2	(17)	(18)	
–	–	–	F-H, S-Q	Category D. Protected from sources of heat. Cargo transport units shall be shaded from direct sunlight. Packages in cargo transport units shall be stowed so as to allow for adequate air circulation throughout the cargo. "Separated from" ammonium compounds, acids, cyanides, hydrogen peroxides and liquid organic substances.	White or yellowish corrosive solid (powder, granules or tablets) with chlorine-like odour. Soluble in water. May cause fire in contact with organic material or ammonium compounds. Substances are liable to exothermic decomposition at elevated temperatures. This condition may lead to fire or explosion. Decomposition can be initiated by heat or by impurities (e.g., powdered metals (iron, manganese, cobalt, magnesium) and their compounds). Liable to heat slowly. Reacts with acids, evolving chlorine, an irritating, corrosive and toxic gas. In the presence of moisture, corrosive to most metals. Causes burns to skin, eyes and mucous membranes.	3487	△
–	T22	TP2 TP13	F-E, S-D	Category D. Clear of living quarters. Segregation as for class 3 but "away from" class 4.1.	A variety of toxic liquids which present a highly toxic inhalation hazard as well as being flammable and corrosive. Highly toxic if swallowed, by skin contact or by inhalation. Causes burns to skin, eyes and mucous membranes.	3488	△
–	T20	TP2 TP13	F-E, S-D	Category D. Clear of living quarters. Segregation as for class 3 but "away from" class 4.1.	A variety of toxic liquids which present a highly toxic inhalation hazard as well as being flammable and corrosive. Highly toxic if swallowed, by skin contact or by inhalation. Causes burns to skin, eyes and mucous membranes.	3489	△
–	T22	TP2 TP13	F-G, S-N	Category D. Clear of living quarters. Segregation as for class 3 but "away from" classes 3 and 8.	A variety of toxic liquids which present a highly toxic inhalation hazard as well as being water-reactive and flammable. Highly toxic if swallowed, by skin contact or by inhalation.	3490	△
–	T20	TP2 TP13	F-G, S-N	Category D. Clear of living quarters. Segregation as for class 3 but "away from" classes 3 and 8.	A variety of toxic liquids which present a highly toxic inhalation hazard as well as being water-reactive and flammable. Highly toxic if swallowed, by skin contact or by inhalation.	3491	△
							⊗
–	T14	TP2 TP13	F-E, S-E	Category D. Clear of living quarters.	Immiscible with water. Evolves hydrogen sulphide, which is a flammable, toxic gas with a foul odour, heavier than air (1.2). Toxic if swallowed, by skin contact or by inhalation.	3494	
–	T7	TP2	F-E, S-E	Category D. Clear of living quarters.	See entry above.	3494	
–	T4	TP1	F-E, S-E	Category C. Clear of living quarters.	See entry above.	3494	
–	T1	TP33	F-A, S-B	Category B. Clear of living quarters. "Separated from" ammonia.	Bluish-black solid with a metallic lustre and a pungent odour. Melting point: 114°C. Below its melting point, may evolve vapours which are irritating to skin, eyes and mucous membranes. Slightly soluble in water but soluble in most organic solvents. Corrosive to most metals.	3495	
–	–	–	F-A, S-I	Category A. Protected from sources of heat.	Nickel-metal hydride button cells or nickel-metal hydride cells or batteries packed with or contained in equipment are not subject to the provisions of this Code.	3496	△
–	T3	TP33	F-A, S-J	Category B. "Separated by a complete compartment or hold from" class 1 except from division 1.4. For special stowage provisions, see 7.6.2.7.2.1.	Pink to red meal derived from Krill which is a shrimp-like marine organism. Medium odour, which may affect other sensitive cargo. Liable to self-heating. Naturally rich in anti-oxidants, which lessen the risk of spontaneous heating.	3497	■

UN No. (1)	Proper Shipping Name (PSN) (2) 3.1.2	Class or division (3) 2.0	Subsidiary risk(s) (4) 2.0	Packing group (5) 2.0.1.3	Special provisions (6) 3.3	Limited and excepted quantity provisions: Limited quantities (7a) 3.4	Limited and excepted quantity provisions: Excepted quantities (7b) 3.5	Packing: Instructions (8) 4.1.4	Packing: Provisions (9) 4.1.4	IBC: Instructions (10) 4.1.4	IBC: Provisions (11) 4.1.4
■ 3497	KRILL MEAL	4.2	–	III	223 300	0	E1	P002 LP02	–	IBC08	B3
■ 3498	IODINE MONOCHLORIDE, LIQUID	8	–	II	–	1 L	E2	P001	–	IBC02	–
■ 3499	CAPACITOR, electric double layer (with an energy storage capacity greater than 0.3 Wh)	9	–	–	361	0	E0	P003	–	–	–
■ 3500	CHEMICAL UNDER PRESSURE, N.O.S.	2.2	–	–	274 362	0	E0	P206	–	–	–
■ 3501	CHEMICAL UNDER PRESSURE, FLAMMABLE, N.O.S.	2.1	–	–	274 362	0	E0	P206	PP89	–	–
■ 3502	CHEMICAL UNDER PRESSURE, TOXIC, N.O.S.	2.2	6.1	–	274 362	0	E0	P206	PP89	–	–
■ 3503	CHEMICAL UNDER PRESSURE, CORROSIVE, N.O.S.	2.2	8	–	274 362	0	E0	P206	PP89	–	–
■ 3504	CHEMICAL UNDER PRESSURE, FLAMMABLE, TOXIC, N.O.S.	2.1	6.1	–	274 362	0	E0	P206	PP89	–	–
■ 3505	CHEMICAL UNDER PRESSURE, FLAMMABLE, CORROSIVE, N.O.S.	2.1	8	–	274 362	0	E0	P206	PP89	–	
■ 3506	MERCURY CONTAINED IN MANUFACTURED ARTICLES	8	6.1	III	366	5 kg	E0	P003	PP90		

Portable tanks and bulk containers			EmS	Stowage and segregation	Properties and observations	UN No.
	Tank instructions	Provisions				
(12)	(13) 4.2.5 4.3	(14) 4.2.5	(15) 5.4.3.2 7.3	(16) 7.1 7.2	(17)	(18)
–	T1	TP33	F-A, S-J	Category A	Pink to red meal derived from Krill which is a shrimp-like marine organism. Medium odour, which may affect other sensitive cargo. Liable to self-heating. Naturally rich in anti-oxidants, which lessen the risk of spontaneous heating.	3497 ■
–	T7	TP2	F-A, S-B	Category D. Clear of living quarters. Segregation as for class 5.1 but "separated from" classes 4.1, 5.1 and 7.	Red Liquid. Reacts violently with water, evolving irritating and corrosive gases apparent as white fumes. Powerful oxidant: may cause fire in contact with organic materials such as wood, cotton or straw. In the presence of moisture, highly corrosive to most metals. Vapour irritates mucous membranes.	3498 ■
–	–	–	F-A, S-I	Category A.	Articles intended to store energy containing a non-dangerous activated carbon and an electrolyte. Electric double layer capacitors installed in equipment may be transported in a charged state.	3499 ■
–	T50	TP4 TP40	F-C, S-V	Category B.	Liquids, pastes or powders, pressurized with a propellant which meets the definition of a gas.	3500 ■
–	T50	TP4 TP40	F-D, S-U	Category D. Clear of living quarters.	See entry above.	3501 ■
–	T50	TP4 TP40	F-C, S-V	Category D. Clear of living quarters.	See entry above.	3502 ■
–	T50	TP4 TP40	F-C, S-V	Category D. Clear of living quarters.	See entry above.	3503 ■
–	T50	TP4 TP40	F-D, S-U	Category D. Clear of living quarters.	See entry above.	3504 ■
–	T50	TP4 TP40	F-D, S-U	Category D. Clear of living quarters.	See entry above.	3505 ■
–	–	–	F-A, S-B	Category B. Clear of living quarters. "Away from" azides.	Articles containing mercury (UN 2809). Carriage should be prohibited in hovercraft and other ships constructed with aluminium.	3506 ■

3

DGL

Chapter 3.3

Special provisions applicable to certain substances, materials or articles

3.3.1 When column 6 of the Dangerous Goods List indicates that a special provision is relevant to a dangerous good, the meaning and requirement(s) of that special provision are as set out below:

16 Samples of new or existing explosive substances or articles may be transported as directed by the competent authority for purposes including: testing, classification, research and development, quality control, or as a commercial sample. Explosive samples which are not wetted or desensitized shall be limited to 10 kg in small packages as specified by the competent authority. Explosive samples which are wetted or desensitized shall be limited to 25 kg.

23 Even though this substance has a flammability hazard, it only exhibits such hazard under extreme fire conditions in confined areas.

26 This substance is not permitted for transport in portable tanks, or intermediate bulk containers with a capacity exceeding 450 L, due to the potential initiation of an explosion when transported in large volumes.

28 This substance may be transported under the provisions of class 4.1 only if it is so packaged that the percentage of diluent will not fall below that stated, at any time during transport (see 2.4.2.4).

29 The packages, including bales, are exempt from labelling provided that they are marked with the appropriate class (e.g., "class 4.2"). Packages, with the exception of bales, shall also display the Proper Shipping Name and the UN Number of the substance that they contain in accordance with 5.2.1. In any case, the packages, including bales, are exempt from class marking provided that they are loaded in a cargo transport unit and that they contain goods to which only one UN Number has been assigned. The cargo transport units in which the packages, including bales, are loaded shall display any relevant labels, placards and marks in accordance with chapter 5.3.

32 When in any other form, this substance is not subject to the provisions of this Code.

37 When coated, this substance is not subject to the provisions of this Code.

38 This substance, when it contains not more than 0.1% calcium carbide, is not subject to the provisions of this Code.

39 This substance, when it contains less than 30% or not less than 90% silicon, is not subject to the provisions of this Code.

43 When offered for transport as pesticides, these substances shall be transported under the relevant pesticide entry and in accordance with the relevant pesticide provisions (see 2.6.2.3 and 2.6.2.4).

45 Antimony sulphides and oxides which contain not more than 0.5% of arsenic, calculated on the total mass, are not subject to the provisions of this Code.

47 Ferricyanides and ferrocyanides are not subject to the provisions of this Code.

59 These substances, when they contain not more than 50% magnesium, are not subject to the provisions of this Code.

61 The technical name, which shall supplement the Proper Shipping Name, shall be the ISO common name, or other name listed in *The WHO Recommended Classification of Pesticides by Hazard and Guidelines to Classification* or the name of the active substance (see also 3.1.2.8.1.1).

62 This substance, when it contains not more than 4% sodium hydroxide, is not subject to the provisions of this Code.

63 The division of class 2 and the subsidiary risks depend on the nature of the contents of the aerosol dispenser. The following provisions shall apply:

.1 Class 2.1 applies if the contents include 85% by mass or more flammable components and the chemical heat of combustion is 30 kJ/g or more;

.2 Class 2.2 applies if the contents contain 1% by mass or less flammable components and the heat of combustion is less than 20 kJ/g.

.3 Otherwise the product shall be classified as tested by the tests described in the United Nations *Manual of Tests and Criteria*, Part III, section 31. Extremely flammable and flammable aerosols shall be classified in class 2.1; non-flammable in class 2.2;

.4 Gases of class 2.3 shall not be used as a propellant in an aerosol dispenser;

.5 Where the contents other than the propellant of aerosol dispensers to be ejected are classified as class 6.1 packing groups II or III or class 8 packing groups II or III, the aerosol shall have a subsidiary risk of class 6.1 or class 8;

.6 Aerosols with contents meeting the criteria for packing group I for toxicity or corrosivity shall be prohibited from transport;

.7 Except for consignments transported in limited quantities (see chapter 3.4), packages containing aerosols shall bear labels for the primary risk and for the subsidiary risk(s), if any.

Flammable components are flammable liquids, flammable solids or flammable gases and gas mixtures as defined in Notes 1 to 3 of sub-section 31.1.3 of Part III of the United Nations *Manual of Tests and Criteria.* This designation does not cover pyrophoric, self-heating or water-reactive substances. The chemical heat of combustion shall be determined by one of the following methods: ASTM D 240, ISO/FDIS 13943:1999 (E/F) 86.1 to 86.3 or NFPA 30B.

65 Hydrogen peroxide aqueous solutions with less than 8% hydrogen peroxide are not subject to the provisions of this Code.

66 Mercurous chloride shall be transported under UN 3077 and cinnabar is not subject to the provisions of this Code.

76 The transport of this substance shall be prohibited except with special authorization granted by the competent authority of the country concerned.

105 Nitrocellulose meeting the descriptions of UN 2556 or UN 2557 may be classified in class 4.1.

113 The transport of chemically unstable mixtures is prohibited.

117 Only regulated when transported by sea.

119 Refrigerating machines and refrigerating machinery components including machines or other appliances which have been designed for the specific purpose of keeping food or other items at a low temperature in an internal compartment, and air-conditioning units. Refrigerating machines and refrigerating machine components are not subject to the provisions of this Code if they contain less than 12 kg of gas in class 2.2 or less than 12 L of ammonia solution (UN 2672).

122 The subsidiary risk(s), the control and emergency temperatures, if any, and the generic entry number for each of the currently assigned organic peroxide formulations are given in 2.5.3.2.4.

127 Other inert material or inert material mixture may be used at the discretion of the competent authority, provided this inert material has identical phlegmatizing properties.

131 The phlegmatized substance shall be significantly less sensitive than dry PETN.

⊗

△ 133 If over-confined in packagings, this substance may exhibit explosive behaviour. Packagings authorized under packing instruction P409 are intended to prevent over-confinement. When a packaging other than those prescribed under packing instruction P409 is authorized by the competent authority of the country of origin in accordance with 4.1.3.7, the package shall bear an "EXPLOSIVE" subsidiary risk label (Model No. 1, see 5.2.2.2.2) unless the competent authority of the country of origin has permitted this label to be dispensed with for the specific packaging employed because test data have proved that the substance in this packaging does not exhibit explosive behaviour (see 5.4.1.5.5.1). The provisions of 7.2.3.3, 7.1.3.1 and 7.1.4.4 shall also be considered.

135 The dihydrated sodium salt of dichloroisocyanuric acid is not subject to the provisions of this Code.

138 *p*-Bromobenzyl cyanide is not subject to the provisions of this Code.

141 Products which have undergone sufficient heat treatment so that they present no hazard during transport are not subject to the provisions of this Code.

142 Solvent-extracted soya bean meal containing not more than 1.5% oil and 11% moisture, being substantially free from flammable solvents, which is accompanied by a certificate from the shipper stating that the substance, as offered for shipment, meets this requirement is not subject to the provisions of this Code.

144 An aqueous solution containing not more than 24% alcohol by volume is not subject to the provisions of this Code.

145 Alcoholic beverages of packing group III, when transported in receptacles of 250 L or less, are not subject to the provisions of this Code.

152 The classification of this substance will vary with particle size and packaging, but borderlines have not been experimentally determined. Appropriate classifications shall be made as required by 2.1.3.

153 This entry applies only if it is demonstrated, on the basis of tests, that the substance, when in contact with water, is not combustible nor shows a tendency to auto-ignition and that the mixture of gases evolved is not flammable.

163 A substance specifically listed by name in the Dangerous Goods List shall not be transported under this entry. Materials transported under this entry may contain 20% or less nitrocellulose provided the nitrocellulose contains not more than 12.6% nitrogen (by dry mass).

168 Asbestos which is immersed or fixed in a natural or artificial binder (such as cement, plastics, asphalt, resins or mineral ore) in such a way that no escape of hazardous quantities of respirable asbestos fibres can occur during transport is not subject to the provisions of this Code. Manufactured articles containing asbestos and not meeting this provision are nevertheless not subject to the provisions of this Code when packaged so that no escape of hazardous quantities of respirable asbestos fibres can occur during transport.

169 Phthalic anhydride in the solid state and tetrahydrophthalic anhydride, with not more than 0.05% maleic anhydride, are not subject to the provisions of this Code. Phthalic anhydride molten at a temperature above its flashpoint, with not more than 0.05% maleic anhydride, shall be classified under UN 3256.

172 Radioactive material with a subsidiary risk shall:

.1 be labelled with subsidiary risk labels corresponding to each subsidiary risk exhibited by the material; corresponding placards shall be affixed to transport units in accordance with the relevant provisions of 5.3.1;

.2 be allocated to packing groups I, II or III, as and if appropriate, by application of the grouping criteria provided in part 2 corresponding to the nature of the predominant subsidiary risk.

The description required in 5.4.1.5.7.1.2 shall include a description of these subsidiary risks (e.g., "Subsidiary risk: 3, 6.1"), the name of the constituents which most predominantly contribute to this (these) subsidiary risk(s), and, where applicable, the packing group.

177 Barium sulphate is not subject to the provisions of this Code.

178 This entry shall be used only when no other appropriate entry exists in the list, and only with the approval of the competent authority of the country of origin.

△ 181 Packages containing this type of substance shall bear the "EXPLOSIVE" subsidiary risk label (Model No. 1, see 5.2.2.2.2) unless the competent authority of the country of origin has permitted this label to be dispensed with for the specific packaging employed because test data have proved that the substance in this packaging does not exhibit explosive behaviour (see 5.4.1.5.5.1). The provisions of 7.2.3.3 shall also be considered.

182 The group of alkali metals includes lithium, sodium, potassium, rubidium and caesium.

183 The group of alkaline earth metals includes magnesium, calcium, strontium and barium.

186 In determining the ammonium nitrate content, all nitrate ions for which a molecular equivalent of ammonium ions is present in the mixture shall be calculated as ammonium nitrate.

188 Cells and batteries offered for transport are not subject to other provisions of this Code if they meet the following:

.1 For a lithium metal or lithium alloy cell, the lithium content is not more than 1 g, and for a lithium-ion cell, the watt-hour rating is not more than 20 W h;

.2 For a lithium metal or lithium alloy battery, the aggregate lithium content is not more than 2 g, and for a lithium-ion battery, the watt-hour rating is not more than 100 W h. Lithium-ion batteries subject to this provision shall be marked with the watt-hour rating on the outside case, except those manufactured before 1 January 2009;

△ .3 Each cell or battery meets the provisions of 2.9.4.1 and 2.9.4.5;

.4 Cells and batteries, except when installed in equipment, shall be packed in inner packagings that completely enclose the cell or battery. Cells and batteries shall be protected so as to prevent short circuits. This includes protection against contact with conductive materials within the same packaging that could lead to a short circuit. The inner packagings shall be packed in strong outer packagings which conform to the provisions of 4.1.1.1, 4.1.1.2, and 4.1.1.5.

△ .5 Cells and batteries when installed in equipment shall be protected from damage and short circuit, and the equipment shall be equipped with an effective means of preventing accidental activation. This requirement does not apply to devices which are intentionally active in transport (radio frequency identification (RFID) transmitters, watches, sensors, etc.) and which are not capable of generating a dangerous evolution of heat. When batteries are installed in equipment, the equipment shall be packed in strong outer packagings constructed of suitable material of adequate strength and design in relation to the packaging's capacity and its intended use unless the battery is afforded equivalent protection by the equipment in which it is contained.

.6 Except for packages containing button cell batteries installed in equipment (including circuit boards), or no more than four cells installed in equipment or no more than two batteries installed in equipment, each package shall be marked with the following:

(i) an indication that the package contains "lithium metal" or "lithium ion" cells or batteries, as appropriate;

(ii) an indication that the package shall be handled with care and that a flammability hazard exists if the package is damaged;

(iii) an indication that special procedures shall be followed in the event the package is damaged, to include inspection and repacking if necessary; and

(iv) a telephone number for additional information.

.7 Each consignment of one or more packages marked in accordance with paragraph .6 shall be accompanied with a document including the following:

(i) an indication that the package contains "lithium metal" or "lithium ion" cells or batteries, as appropriate;

(ii) an indication that the package shall be handled with care and that a flammability hazard exists if the package is damaged;

(iii) an indication that special procedures shall be followed in the event the package is damaged, to include inspection and repacking if necessary; and

(iv) a telephone number for additional information.

.8 Except when batteries are installed in equipment, each package shall be capable of withstanding a 1.2 m drop test in any orientation without damage to cells or batteries contained therein, without shifting of the contents so as to allow battery to battery (or cell to cell) contact and without release of contents; and

.9 Except when batteries are installed in or packed with equipment, packages shall not exceed 30 kg gross mass.

As used above and elsewhere in this Code, "lithium content" means the mass of lithium in the anode of a lithium metal or lithium alloy cell.

Separate entries exist for lithium metal batteries and lithium ion batteries to facilitate the transport of these batteries for specific modes of transport and to enable the application of different emergency response actions.

190 Aerosol dispensers shall be provided with protection against inadvertent discharge. Aerosols with a capacity not exceeding 50 mL containing only non-toxic constituents are not subject to the provisions of this Code.

191 Receptacles with a capacity not exceeding 50 mL containing only non-toxic constituents are not subject to the provisions of this Code.

193 This entry may only be used for uniform ammonium nitrate based fertilizer mixtures of the nitrogen, phosphate or potash type, containing not more than 70% ammonium nitrate and not more than 0.4% total combustible/organic material calculated as carbon or with not more than 45% ammonium nitrate and unrestricted combustible material. Fertilizers within these composition limits are not subject to the provisions of this Code when shown by a Trough Test (see United Nations *Manual of Tests and Criteria*, Part III, sub-section 38.2) that they are not liable to self-sustaining decomposition.

194 The control and emergency temperatures, if any, and the generic entry number for each of the currently assigned self-reactive substances are given in 2.4.2.3.2.3.

195 For certain organic peroxides types B or C, a smaller packaging than that allowed by packing methods OP5 or OP6 respectively has to be used (see 4.1.7 and 2.5.3.2.4).

196 Formulations which, in laboratory testing, neither detonate in the cavitated state nor deflagrate, which show no effect when heated under confinement and which exhibit no explosive power may be transported under this entry. The formulation must also be thermally stable (i.e., the SADT is 60°C or higher for a 50 kg package). Formulations not meeting these criteria shall be transported under the provisions of class 5.2 (see 2.5.3.2.4).

198 Nitrocellulose solutions containing not more than 20% nitrocellulose may be transported as paint, perfumery products or printing ink, as applicable. See UN Nos. 1210, 1263, 1266, 3066, 3469 and 3470.

199 Lead compounds which, when mixed in a ratio of 1:1000 with 0.07M hydrochloric acid and stirred for one hour at a temperature of 23°C ± 2°C, exhibit a solubility of 5% or less (see ISO 3711:1990 "Lead chromate pigments and lead chromate-molybdate pigments – Specifications and methods of test") are considered insoluble and are not subject to the provisions of this Code unless they meet the criteria for inclusion in another hazard class.

201 Lighters and lighter refills shall comply with the provisions of the country in which they were filled. They shall be provided with protection against inadvertent discharge. The liquid portion of the gas shall not exceed 85% of the capacity of the receptacle at 15°C. The receptacles, including the closures, shall be capable of withstanding an internal pressure of twice the pressure of the liquefied petroleum gas at 55°C. The valve mechanisms and ignition devices shall be securely sealed, taped or otherwise fastened or designed to prevent operation or leakage of the contents during transport. Lighters shall not contain more than 10 g of liquefied petroleum gas. Lighter refills shall not contain more than 65 g of liquefied petroleum gas.

203 This entry shall not be used for polychlorinated biphenyls, UN 2315.

204 Articles containing smoke-producing substance(s) corrosive according to the criteria for class 8 shall be labelled with a "CORROSIVE" subsidiary risk label (Model No. 8, see 5.2.2.2.2).

205 This entry shall not be used for PENTACHLOROPHENOL, UN 3155.

207 Polymeric beads and moulding compounds may be made from polystyrene, poly(methyl methacrylate) or other polymeric material.

208 The commercial grade of calcium nitrate fertilizer, when consisting mainly of a double salt (calcium nitrate and ammonium nitrate) containing not more than 10% ammonium nitrate and at least 12% water of crystallization, is not subject to the provisions of this Code.

209 The gas shall be at a pressure corresponding to ambient atmospheric pressure at the time the containment system is closed and this shall not exceed 105 kPa absolute.

210 Toxins from plant, animal or bacterial sources which contain infectious substances, or toxins that are contained in infectious substances, shall be classified under class 6.2.

215 This entry only applies to the technically pure substance or to formulations derived from it, having an SADT higher than 75°C, and, therefore, does not apply to formulations which are self-reactive substances (for self-reactive substances, see 2.4.2.3.2.3). Homogeneous mixtures containing not more than 35% by mass of azodicarbonamide and at least 65% of inert substance are not subject to this Code unless criteria of other classes are met.

216 Mixtures of solids which are not subject to the provisions of this Code and flammable liquids may be transported under this entry without first applying the classification criteria of class 4.1, provided there is no free liquid visible at the time the substance is loaded or at the time the packaging or cargo transport unit is closed. Each cargo transport unit shall be leakproof when used as a bulk container. Sealed packets and articles containing less than 10 mL of a packing group II or III flammable liquid absorbed into a solid material are not subject to the provisions of this Code provided there is no free liquid in the packet or article.

217 This entry shall only be used for mixtures of solids which are not subject to the provisions of this Code and toxic liquids may be transported under this entry without first applying the classification criteria of class 6.1, provided there is no free liquid visible at the time the substance is loaded or at the time the packaging or cargo transport unit is closed. Each cargo transport unit shall be leakproof when used as a bulk container. This entry shall not be used for solids containing a packing group I liquid.

218 This entry shall only be used for mixtures of solids which are not subject to the provisions of this Code and corrosive liquids may be transported under this entry without first applying the classification criteria of class 8, provided there is no free liquid visible at the time the substance is loaded or at the time the packaging or cargo transport unit is closed. Each cargo transport unit shall be leakproof when used as a bulk container. This entry shall not be used for solids containing a packing group I liquid.

219 Genetically modified microorganisms (GMMOs) and genetically modified organisms (GMOs) packed and marked in accordance with packing instruction P904 are not subject to any other provisions of this Code.

If GMMOs or GMOs meet the definition in chapter 2.6 of a toxic substance or an infectious substance and the criteria for inclusion in class 6.1 or 6.2, the provisions of this Code for transporting toxic substances or infectious substances apply.

220 The technical name of the flammable liquid component only of this solution or mixture shall be shown in parentheses immediately following the Proper Shipping Name.

221 Substances included under this entry shall not be of packing group I.

223 If the chemical or physical properties of a substance covered by this description are such that, when tested, it does not meet the established defining criteria for the class or division listed in column 3, or any other class or division, it is not subject to the provisions of this Code except in the case of a marine pollutant where 2.10.3 applies.

224 Unless it can be demonstrated by testing that the sensitivity of the substance in its frozen state is no greater than in its liquid state, the substance shall remain liquid during normal transport conditions. It shall not freeze at temperatures above −15°C.

225 Fire extinguishers under this entry may include installed actuating cartridges (cartridges, power device of division 1.4C or 1.4S) without changing the classification of class 2.2 provided the total quantity of deflagrating (propellant) explosives does not exceed 3.2 g per extinguishing unit.

226 Formulations of these substances containing not less than 30% non-volatile, non-flammable phlegmatizer are not subject to the provisions of this Code.

227 When phlegmatized with water and inorganic inert material, the content of urea nitrate may not exceed 75% by mass and the mixture shall not be capable of being detonated by the Series 1, type (a) test in the United Nations *Manual of Tests and Criteria*, Part I.

228 Mixtures not meeting the criteria for flammable gases (class 2.1) shall be transported under UN 3163.

△ 230 Lithium cells and batteries may be transported under this entry if they meet the provisions of 2.9.4.

232 This entry shall only be used when the substance does not meet the criteria of any other class. Transport in cargo transport units other than in tanks shall be in accordance with standards specified by the competent authority of the country of origin.

235 This entry applies to articles which contain class 1 explosive substances and which may also contain dangerous goods of other classes. These articles are used as life-saving vehicle air-bag inflators or air-bag modules or seat-belt pretensioners.

236 Polyester resin kits consist of two components: a base material (class 3, packing group II or III) and an activator (organic peroxide). The organic peroxide shall be type D, E or F, not requiring temperature control. Packing group shall be II or III, according to the criteria for class 3, applied to the base material. The quantity limit and the excepted quantity code shown in columns 7a and 7b of the Dangerous Goods List apply to the base material.

237 The membrane filters including paper separators, coating or backing materials, etc., that are present in transport, shall not be liable to propagate a detonation as tested by one of the tests described in the United Nations *Manual of Tests and Criteria*, Part I, Test series 1(a).

In addition, the competent authority may determine, on the basis of the results of suitable burning rate tests taking account of the standard tests in the United Nations *Manual of Tests and Criteria*, Part III, 33.2.1, that nitrocellulose membrane filters in the form in which they are to be transported are not subject to the provisions of this Code applicable to flammable solids in class 4.1.

238 .1 Batteries can be considered as non-spillable provided that they are capable of withstanding the vibration and pressure differential tests given below, without leakage of battery fluid:

Vibration test: The battery is rigidly clamped to the platform of a vibration machine and a simple harmonic motion having an amplitude of 0.8 mm (1.6 mm maximum total excursion) is applied. The frequency is varied at the rate of 1 Hz/min between the limits of 10 Hz and 55 Hz. The entire range of frequencies and return is traversed in 95 ± 5 minutes for each mounting position (direction of vibration) of the battery. The battery is tested in three mutually perpendicular positions (to include testing with fill openings and vents, if any, in an inverted position) for equal time periods.

Pressure differential test: Following the vibration test, the battery is stored for six hours at 24°C ± 4°C while subjected to a pressure differential of at least 88 kPa. The battery is tested in three mutually perpendicular positions (to include testing with fill openings and vents, if any, in an inverted position) for at least six hours in each position.

Non-spillable type batteries which are an integral part of and necessary for the operation of mechanical or electronic equipment shall be securely fastened in the battery holder on the equipment and protected in such a manner as to prevent damage and short circuits.

.2 Non-spillable batteries are not subject to the provisions of this Code if, at a temperature of 55°C, the electrolyte will not flow from a ruptured or cracked case and there is no free liquid to flow and if, when packaged for transport, the terminals are protected from short circuit.

239 Batteries or cells shall not contain dangerous goods other than sodium, sulphur or sodium compounds (e.g., sodium polysulphides and sodium tetrachloroaluminate). Batteries or cells shall not be offered for transport at a temperature such that liquid elemental sodium is present in the battery or cell, unless approved and under the conditions established by the competent authority.

Cells shall consist of hermetically sealed metal casings which fully enclose the dangerous goods and which are so constructed and closed as to prevent the release of the dangerous goods under normal conditions of transport.

Batteries shall consist of cells secured within and fully enclosed by a metal casing so constructed and closed as to prevent the release of the dangerous goods under normal conditions of transport.

Batteries installed in vehicles are not subject to the provisions of this Code.

△ 240 This entry only applies to vehicles powered by wet batteries, sodium batteries, lithium metal batteries or lithium ion batteries and equipment powered by wet batteries or sodium batteries transported with these batteries installed.

For the purpose of this special provision, vehicles are self-propelled apparatus designed to carry one or more persons or goods. Examples of such vehicles are electrically-powered cars, motorcycles, scooters, three- and four-wheeled vehicles or motorcycles, e-bikes, wheel-chairs, lawn tractors, boats and aircraft.

Examples of equipment are lawnmowers, cleaning machines or model boats and model aircraft. Equipment powered by lithium metal batteries or lithium ion batteries shall be consigned under the entries UN 3091 LITHIUM METAL BATTERIES CONTAINED IN EQUIPMENT or UN 3091 LITHIUM METAL BATTERIES PACKED WITH EQUIPMENT or UN 3481 LITHIUM ION BATTERIES CONTAINED IN EQUIPMENT or UN 3481 LITHIUM ION BATTERIES PACKED WITH EQUIPMENT, as appropriate.

Hybrid electric vehicles powered by both an internal combustion engine and wet batteries, sodium batteries, lithium metal batteries or lithium ion batteries, transported with the battery(ies) installed shall be consigned under the entries UN 3166 VEHICLE, FLAMMABLE GAS POWERED or UN 3166 VEHICLE, FLAMMABLE LIQUID POWERED, as appropriate. Vehicles which contain a fuel cell shall be consigned under the entries UN 3166 VEHICLE, FUEL CELL, FLAMMABLE GAS POWERED or UN 3166 VEHICLE, FUEL CELL, FLAMMABLE LIQUID POWERED, as appropriate.

241 The formulation shall be prepared so that it remains homogeneous and does not separate during transport. Formulations with low nitrocellulose contents and not showing dangerous properties when tested for their liability to detonate, deflagrate or explode when heated under defined confinement by tests of Test series 1(a), 2(b) and 2(c) respectively in the United Nations *Manual of Tests and Criteria*, Part I and not being a flammable solid when tested in accordance with test No. 1 in the United Nations *Manual of Tests and Criteria*, Part II , paragraph 33.2.1.4 (chips, if necessary, crushed and sieved to a particle size of less than 1.25 mm) are not subject to the provisions of this Code.

242 Sulphur is not subject to the provisions of this Code when it has been formed to a specific shape (such as prills, granules, pellets, pastilles or flakes).

243 Gasoline, motor spirit and petrol for use in spark-ignition engines (e.g., in automobiles, stationary engines and other engines) shall be assigned to this entry regardless of variations in volatility.

244 This entry includes materials and substances such as aluminium dross, aluminium skimmings, spent cathodes, spent potliner and aluminium salt slags

⊗

247 Alcoholic beverages containing more than 24% alcohol but not more than 70% by volume, when transported as part of the manufacturing process, may be transported in wooden barrels with a capacity of more than 250 L and not more than 500 L meeting the general requirements of 4.1.1, as appropriate, on the following conditions:

.1 the wooden barrels shall be checked and tightened before filling;

.2 sufficient ullage (not less than 3%) shall be left to allow for the expansion of the liquid;

.3 the wooden barrels shall be transported with the bungholes pointing upwards;

.4 the wooden barrels shall be transported in containers meeting the provisions of the International Convention for Safe Containers (CSC 1972), as amended, and each wooden barrel shall be secured in custom-made cradles and be wedged by appropriate means to prevent it from being displaced in any way during transport; and

.5 when carried on board ships, the containers shall be stowed in open cargo spaces or in enclosed cargo spaces complying with the requirements for class 3 flammable liquids with a flashpoint of 23°C c.c. or less in regulation II-2/19 of SOLAS, 74, as amended or regulation II-2/54 of SOLAS 74, as amended by the resolutions indicated in II-2/1.2.1, as applicable.

249 Ferrocerium, stabilized against corrosion, with a minimum iron content of 10% is not subject to the provisions of this Code.

250 This entry may only be used for samples of chemicals taken for analysis in connection with the implementation of the Convention on the Prohibition of the Development, Production, Stockpiling and Use of Chemical Weapons and on their Destruction. The transport of substances under this entry shall be in accordance with the chain of custody and security procedures specified by the Organization for the Prohibition of Chemical Weapons.

The chemical sample may only be transported provided prior approval has been granted by the competent authority or the Director General of the Organization for the Prohibition of Chemical Weapons and providing the sample complies with the following conditions:

.1 it shall be packaged according to packing instruction 623 in the International Civil Aviation Organization's Technical Instructions for the Safe Transport of Dangerous Goods by Air; and

.2 during transport, it shall be accompanied by a copy of the document of approval for transport, showing the quantity limitations and the packing provisions.

251 The entry CHEMICAL KIT or FIRST AID KIT is intended to apply to boxes, cases, etc., containing small quantities of various dangerous goods which are used, for example, for medical, analytical, testing or repair purposes. Such kits may not contain dangerous goods for which the quantity "0" has been indicated in column 7a of the Dangerous Goods List.

Components shall not react dangerously (see 4.1.1.6). The total quantity of dangerous goods in any one kit shall not exceed either 1 L or 1 kg. The packing group assigned to the kit as a whole shall be the most stringent packing group assigned to any individual substance in the kit.

Kits which are carried on board vehicles for first-aid or operating purposes are not subject to the provisions of this Code.

Chemical kits and first aid kits containing dangerous goods in inner packagings which do not exceed the quantity limits for limited quantities applicable to individual substances as specified in column 7a of the Dangerous Goods List may be transported in accordance with chapter 3.4.

252 Provided the ammonium nitrate remains in solution under all conditions of transport, aqueous solutions of ammonium nitrate, with not more than 0.2% combustible material, in a concentration not exceeding 80%, are not subject to the provisions of this Code.

266 This substance, when containing less alcohol, water or phlegmatizer than specified, shall not be transported, unless specifically authorized by the competent authority.

267 Any explosives, blasting, type C containing chlorates shall be segregated from explosives containing ammonium nitrate or other ammonium salts.

270 Aqueous solutions of class 5.1 inorganic solid nitrate substances are considered as not meeting the criteria of class 5.1 if the concentration of the substances in solution at the minimum temperature encountered in transport is not greater than 80% of the saturation limit.

271 Lactose or glucose or similar materials may be used as a phlegmatizer provided that the substance contains not less than 90%, by mass, of phlegmatizer. The competent authority may authorize these mixtures to be classified under class 4.1 on the basis of series 6(c) tests of Part I of the United Nations *Manual of Tests and Criteria* on at least three packages as prepared for transport. Mixtures containing at least 98%, by mass, of phlegmatizer are not subject to the provisions of this Code. Packages containing mixtures with not less than 90%, by mass, of phlegmatizer need not bear a "TOXIC" subsidiary risk label.

△ 272 This substance shall not be transported under the provisions of class 4.1 unless specifically authorized by the competent authority (see UN 0143 or UN 0150 as appropriate).

273 Maneb and maneb preparations stabilized against self-heating need not be classified in class 4.2 when it can be demonstrated by testing that a cubic volume of 1 m^3 of substance does not self-ignite and that the temperature at the centre of the sample does not exceed 200°C when the sample is maintained at a temperature of not less than 75°C $\pm$ 2°C for a period of 24 hours.

274 For the purposes of documentation and package marking, the Proper Shipping Name shall be supplemented with the technical name (see 3.1.2.8.1).

277 For aerosols or receptacles containing toxic substances, the limited quantity value is 120 mL. For all other aerosols or receptacles, the limited quantity value is 1000 mL.

278 These substances shall not be classified and transported unless authorized by the competent authority on the basis of results from series 2 tests and series 6(c) tests of Part I of the United Nations *Manual of Tests and Criteria* on packages as prepared for transport (see 2.1.3.1). The competent authority shall assign the packing group on the basis of the chapter 2.3 criteria and the package type used for the series 6(c) tests.

279 The substance is assigned to this classification or packing group based on human experience rather than the strict application of classification criteria set out in this Code.

△ 280 This entry applies to articles which are used as life-saving vehicle air bag inflators, or air bag modules or seat-belt pretensioners and which contain dangerous goods of class 1 or dangerous goods of other classes and when transported as component parts and when these articles as presented for transport have been tested in accordance with Test series 6(c) of Part I of the United Nations *Manual of Tests and Criteria*, with no explosion of the device, no fragmentation of device casing or pressure receptacle, and no projection hazard nor thermal effect which would significantly hinder fire-fighting or other emergency response efforts in the immediate vicinity.

281 Transport of hay, straw or bhusa when wet, damp or contaminated with oil is prohibited and when not wet or contaminated with oil is subject to the provisions of this Code.

283 Articles, containing gas, intended to function as shock absorbers, including impact-energy-absorbing devices or pneumatic springs, are not subject to the provisions of this Code provided:

.1 each article has a gas space capacity not exceeding 1.6 L and a charge pressure not exceeding 280 bar where the product of the capacity (litres) and charge pressure (bar) does not exceed 80 (i.e., 0.5 L gas space and 160 bar charge pressure, 1 L gas space and 80 bar charge pressure, 1.6 L gas space and 50 bar charge pressure, 0.28 L gas space and 280 bar charge pressure);

.2 each article has a minimum burst pressure of 4 times the charge pressure at 20°C for products not exceeding 0.5 L gas space capacity and 5 times charge pressure for products greater than 0.5 L gas space capacity;

.3 each article is manufactured from material which will not fragment upon rupture;

.4 each article is manufactured in accordance with a quality-assurance standard acceptable to the competent authority; and

.5 the design type has been subjected to a fire test demonstrating that pressure in the article is relieved by means of a fire-degradable seal or other pressure relief device, such that the article will not fragment and that the article does not rocket.

284 An oxygen generator, chemical, containing oxidizing substances shall meet the following conditions:

.1 the generator, when containing an explosive device, shall only be transported under this entry when excluded from class 1 in accordance with 2.1.3 of this Code;

.2 the generator, without its packaging, shall be capable of withstanding a 1.8 m drop test onto a rigid, non-resilient, flat and horizontal surface, in the position most likely to cause damage, without loss of its contents and without actuation; and

.3 when the generator is equipped with an actuating device, it shall have at least two positive means of preventing unintentional actuation.

286 Nitrocellulose membrane filters covered by this entry, each with a mass not exceeding 0.5 g, are not subject to the provisions of this Code when contained individually in an article or a sealed packet.

288 These substances shall not be classified and transported unless authorized by the competent authority on the basis of results from series 2 tests and series 6(c) tests of Part I of the United Nations *Manual of Tests and Criteria* on packages as prepared for transport (see 2.1.3).

△ 289 Air bag inflators, air bag modules or seat-belt pretensioners installed in vehicles, vessels or aircrafts or in completed components such as steering columns, door panels, seats, etc., are not subject to the provisions of this Code.

290 When this radioactive material meets the definitions and criteria of other classes or divisions as defined in part 2, it shall be classified in accordance with the following:

.1 where the substance meets the criteria for dangerous goods in excepted quantities as set out in chapter 3.5, the packagings shall be in accordance with 3.5.2 and meet the testing requirements of 3.5.3. All other requirements applicable to radioactive material, excepted packages as set out in 1.5.1.5 shall apply without reference to the other class or division;

.2 where the quantity exceeds the limits specified in 3.5.1.2, the substance shall be classified in accordance with the predominant subsidiary risk. The dangerous goods transport document shall describe the substance with the UN Number and Proper Shipping Name applicable to the other class supplemented with the name applicable to the radioactive excepted package according to column 2 in the Dangerous Goods List of chapter 3.2, and the substance shall be transported in accordance with the provisions applicable to that UN Number. An example of the information shown on the dangerous goods transport document is:

UN 1993, Flammable liquid, N.O.S. (ethanol and toluene mixture), Radioactive material, excepted package – limited quantity of material, class 3, PG II.

In addition, the provisions of 2.7.2.4.1 shall apply;

.3 the provisions of chapter 3.4 for the transport of dangerous goods packed in limited quantities shall not apply to substances classified in accordance with subparagraph .2;

.4 when the substance meets a special provision that exempts this substance from all dangerous goods provisions of the other classes, it shall be classified in accordance with the applicable UN Number of class 7 and all requirements specified in 1.5.1.5 shall apply.

291 Flammable liquefied gases shall be contained within refrigerating-machine components. These components shall be designed and tested to at least three times the working pressure of the machinery. The refrigerating machines and refrigerating machinery components shall be designed and constructed to contain the liquefied gas and preclude the risk of bursting or cracking of the pressure-retaining components during normal conditions of transport. Refrigerating machines and refrigerating-machine components are not subject to the provisions of this Code if they contain less than 12 kg of gas.

293 The following definitions apply to matches:

(a) *Fusee matches* are matches the heads of which are prepared with a friction-sensitive igniter composition and a pyrotechnic composition which burns with little or no flame, but with intense heat;

(b) *Safety matches* are combined with or attached to the box, book or card that can be ignited by friction only on a prepared surface;

(c) *"Strike anywhere" matches* are matches that can be ignited by friction on a solid surface;

(d) *Wax 'Vesta' matches* are matches that can be ignited by friction either on a prepared surface or on a solid surface.

294 Safety matches and wax 'Vesta' matches in an outer packaging not exceeding 25 kg net mass are not subject to any other provision (except marking) of this Code when packaged in accordance with packing instruction P407.

295 Batteries need not be individually marked and labelled if the pallet bears the appropriate mark and label.

296 These entries apply to life-saving appliances such as liferafts, personal flotation devices and self-inflating slides. UN 2990 applies to self-inflating appliances. UN 3072 applies to life-saving appliances that are not self-inflating. Life-saving appliances may contain:

.1 signal devices (class 1) which may include smoke and illumination signal flares packed in packagings that prevent them from being inadvertently activated;

.2 for UN 2990 only, cartridges, power device of division 1.4, compatibility group S, may be contained for purposes of the self-inflating mechanism and provided that the quantity of explosives per appliance does not exceed 3.2 g;

△ .3 class 2.2 compressed or liquefied gases;

.4 electric storage batteries (class 8) and lithium batteries (class 9);

.5 first aid kits or repair kits containing small quantities of dangerous goods (e.g., classes 3, 4.1, 5.2, 8 or 9 substances); or

.6 "Strike anywhere" matches packed in packagings that prevent them from being inadvertently activated.

■ Life-saving appliances packed in strong rigid outer packagings with a total maximum gross mass of 40 kg, containing no dangerous goods other than class 2.2 compressed or liquefied gases with no subsidiary risk in receptacles with a capacity not exceeding 120 mL, installed solely for the purpose of the activation of the appliance, are not subject to the provision of this Code.

⊗

299 Consignments of:

(i) Cotton, dry having a density not less than 360 kg/m^3;

(ii) Flax, dry having a density not less than 400 kg/m^3;

(iii) Sisal, dry having a density not less than 360 kg/m^3; and

(iv) Tampico fibre, dry having a density not less than 360 kg/m^3,

according to ISO 8115:1986, are not subject to the provisions of this Code when transported in closed cargo transport units.

△ 300 Fish meal, fish scrap and krill meal shall not be transported if the temperature at the time of loading exceeds 35°C or 5°C above the ambient temperature, whichever is higher.

△ 301 This entry only applies to machinery or apparatus containing dangerous substances as a residue or an integral element of the machinery or apparatus. It shall not be used for machinery or apparatus for which a Proper Shipping Name already exists in the Dangerous Goods List. Machinery and apparatus transported under this entry shall only contain dangerous goods which are authorized to be transported in accordance with the provisions in chapter 3.4 (Limited quantities). The quantity of dangerous goods in machinery or apparatus shall not exceed the quantity specified in column 7a of the Dangerous Goods List for each item of dangerous goods contained. If the machinery or apparatus contains more than one item of dangerous goods, the individual substances shall not be capable of reacting dangerously with one another (see 4.1.1.6). When it is required to ensure liquid dangerous goods remain in their intended orientation, package orientation labels meeting the specifications of ISO 780:1985 shall be affixed on at least two opposite vertical sides with the arrows pointing in the correct direction. The transport of dangerous goods in machinery or apparatus where the quantity of dangerous goods exceeds the quantity specified in column 7a of the Dangerous Goods List is authorized when approved by the competent authority, except where special provision 363 applies.

302 Fumigated cargo transport units containing no other dangerous goods are only subject to the provisions of 5.5.2.

303 Receptacles shall be assigned to the class and, if any, subsidiary hazard of the gas or mixture of gases contained therein determined in accordance with the provisions of chapter 2.2.

304 This entry may only be used for the transport of non-activated batteries which contain dry potassium hydroxide and which are intended to be activated prior to use by the addition of an appropriate amount of water to the individual cells.

305 These substances are not subject to the provisions of this Code when in concentrations of not more than 50 mg/kg.

306 This entry may only be used for substances that do not exhibit explosive properties of class 1 when tested in accordance to Test Series 1 and 2 of class 1 (see United Nations *Manual of Tests and Criteria*, Part I).

307 This entry shall be used for uniform mixtures containing ammonium nitrate as the main ingredient within the following composition limits:

.1 not less than 90% ammonium nitrate with not more than 0.2% total combustible/organic material calculated as carbon and with added matter, if any, which is inorganic and inert towards ammonium nitrate; or

.2 less than 90% but more than 70% ammonium nitrate with other inorganic materials or more than 80% but less than 90% ammonium nitrate mixed with calcium carbonate and/or dolomite and/or mineral calcium sulphate and not more than 0.4% total combustible/organic material calculated as carbon; or

.3 nitrogen type ammonium nitrate based fertilizers containing mixtures of ammonium nitrate and ammonium sulphate with more than 45% but less than 70% ammonium nitrate and not more than 0.4% total combustible/organic material calculated as carbon such that the sum of the percentage compositions of ammonium nitrate and ammonium sulphate exceeds 70%.

308 Fish scrap or fish meal shall contain at least 100 ppm of antioxidant (ethoxyquin) at the time of consignment.

309 This entry applies to non-sensitized emulsions, suspensions and gels consisting primarily of a mixture of ammonium nitrate and fuel, intended to produce a Type E blasting explosive only after further processing prior to use.

The mixture for emulsions typically has the following composition: 60–85% ammonium nitrate, 5–30% water, 2–8% fuel, 0.5–4% emulsifier agent, 0–10% soluble flame suppressants, and trace additives. Other inorganic nitrate salts may replace part of the ammonium nitrate.

The mixture for suspensions and gels typically has the following composition: 60–85% ammonium nitrate, 0–5% sodium or potassium perchlorate, 0–17% hexamine nitrate or monomethylamine nitrate, 5–30% water, 2–15% fuel, 0.5–4% thickening agent, 0–10% soluble flame suppressants, and trace additives. Other inorganic nitrate salts may replace part of the ammonium nitrate.

Substances shall satisfactorily pass Test Series 8 of the United Nations *Manual of Tests and Criteria*, Part I, section 18 and be approved by the competent authority.

310 The testing requirements in chapter 38.3 of the United Nations *Manual of Tests and Criteria* do not apply to production runs consisting of not more than 100 cells and batteries, or to pre-production prototypes of cells and batteries when these prototypes are transported for testing, if:

.1 the cells and batteries are transported in an outer packaging that is a metal, plastics or plywood drum or a metal, plastics or wooden box and that meets the criteria for packing group I packagings; and

.2 each cell and battery is individually packed in an inner packaging inside an outer packaging and is surrounded by cushioning material that is non-combustible, and non-conductive.

311 Substances shall not be transported under this entry unless approved by the competent authority on the basis of the results of appropriate tests according to Part I of the United Nations *Manual of Tests and Criteria*. Packaging shall ensure that the percentage of diluent does not fall below that stated in the competent authority approval at any time during transport.

△ 312 Vehicles or machinery powered by a fuel cell engine shall be consigned under the entries UN 3166 VEHICLE, FUEL CELL, FLAMMABLE GAS POWERED or UN 3166 VEHICLE, FUEL CELL, FLAMMABLE LIQUID POWERED, or UN 3166 ENGINE, FUEL CELL, FLAMMABLE GAS POWERED or UN 3166 ENGINE, FUEL CELL, FLAMMABLE LIQUID POWERED as appropriate. These entries include hybrid electric vehicles powered by both a fuel cell and an internal combustion engine with wet batteries, sodium batteries, lithium metal batteries or lithium ion batteries, transported with the battery(ies) installed.

Other vehicles which contain an internal combustion engine shall be consigned under the entries UN 3166 VEHICLE, FLAMMABLE GAS POWERED or UN 3166 VEHICLE, FLAMMABLE LIQUID POWERED, as appropriate. These entries include hybrid electric vehicles powered by both an internal combustion engine and wet batteries, sodium batteries, lithium metal batteries or lithium ion batteries, transported with the batteries installed.

314 (a) These substances are liable to exothermic decomposition at elevated temperatures. Decomposition can be initiated by heat or by impurities (e.g., powdered metals (iron, manganese, cobalt, magnesium) and their compounds).

(b) During the course of transport, these substances shall be shaded from direct sunlight and all sources of heat and be placed in adequately ventilated areas.

315 This entry shall not be used for class 6.1 substances which meet the inhalation toxicity criteria for packing group I described in 2.6.2.2.4.3.

316 This entry applies only to calcium hypochlorite, dry, when transported in non-friable tablet form.

317 "Fissile – excepted" applies only to those packages complying with 6.4.11.2.

318 For the purposes of documentation, the Proper Shipping Name shall be supplemented with the technical name (see 3.1.2.8). Technical names need not be shown on the package. When the infectious substances to be transported are unknown, but suspected of meeting the criteria for inclusion in category A and assignment to UN 2814 or UN 2900, the words "suspected category A infectious substance" shall be shown, in parentheses, following the Proper Shipping Name on the transport document, but not on the outer packagings.

319 Substances packed and packages marked in accordance with packing instruction P650 are not subject to any other provisions of this Code.

321 These storage systems shall always be considered as containing hydrogen.

322 When transported in non-friable tablet form, these goods are assigned to packing group III.

⊗

324 This substance needs to be stabilized when in concentrations of not more than 99%.

325 In the case of non-fissile or fissile excepted uranium hexafluoride, the material shall be classified under UN 2978.

326 In the case of fissile uranium hexafluoride, the material shall be classified under UN 2977.

△ 327 Waste aerosols consigned in accordance with 5.4.1.4.3.3 may be transported under this entry for the purposes of reprocessing or disposal. They need not be protected against inadvertent discharge provided that measures to prevent dangerous build-up of pressure and dangerous atmospheres are addressed. Waste aerosols, other than those leaking or severely deformed, shall be packed in accordance with packing instruction P207 and special provision PP87, or packing instruction LP02 and special packing provision L2. Leaking or severely deformed aerosols shall be transported in salvage packagings provided appropriate measures are taken to ensure there is no dangerous build-up of pressure. Waste aerosols shall not be transported in closed freight containers.

△ 328 This entry applies to fuel cell cartridges, including when contained in equipment or packed with equipment. Fuel cell cartridges installed in or integral to a fuel cell system are regarded as contained in equipment. "Fuel cell cartridge" means an article that stores fuel for discharge into the fuel cell through a valve(s) that controls the discharge of fuel into the fuel cell. Fuel cell cartridges, including when contained in equipment, shall be designed and constructed to prevent fuel leakage under normal conditions of transport.

Fuel cell cartridge design types using liquids as fuels shall pass an internal pressure test at a pressure of 100 kPa (gauge) without leakage.

Except for fuel cell cartridges containing hydrogen in metal hydride, which shall be in compliance with special provision 339, each fuel cell cartridge design type shall be shown to pass a 1.2 m drop test onto an unyielding surface, in the orientation most likely to result in failure of the containment system, with no loss of contents.

When lithium metal or lithium ion batteries are contained in the fuel cell system, the consignment shall be consigned under this entry and under the appropriate entries for UN 3091 LITHIUM METAL BATTERIES CONTAINED IN EQUIPMENT or UN 3481 LITHIUM ION BATTERIES CONTAINED IN EQUIPMENT.

332 Magnesium nitrate hexahydrate is not subject to the provisions of this Code.

333 Ethanol and gasoline, motor spirit or petrol mixtures for use in spark-ignition engines (e.g., in automobiles, stationary engines and other engines) shall be assigned to this entry regardless of variations in volatility.

334 A fuel cell cartridge may contain an activator provided it is fitted with two independent means of preventing unintended mixing with the fuel during transport.

335 Mixtures of solids which are not subject to the provisions of this Code and environmentally hazardous liquids assigned to UN 3082 may be classified and transported as UN 3077, provided there is no free liquid visible at the time the substance is loaded or at the time the packaging or cargo transport unit is closed. If free liquid is visible at the time the mixture is loaded or at the time the packaging or cargo transport unit is closed, the mixture shall be classified as UN 3082. Each cargo transport unit shall be leakproof when used as a bulk container. Sealed packets and articles containing less than 10 mL of an environmentally hazardous liquid assigned to UN 3082, absorbed into a solid material but with no free liquid in the packet or article, or containing less than 10 g of an environmentally hazardous solid assigned to UN 3077, are not subject to the provisions of this Code.

△ 338 Each fuel cell cartridge transported under this entry and designed to contain a liquefied flammable gas shall:

.1 be capable of withstanding, without leakage or bursting, a pressure of at least two times the equilibrium pressure of the contents at 55°C;

.2 not contain more than 200 mL liquefied flammable gas, the vapour pressure of which shall not exceed 1 000 kPa at 55°C; and

.3 pass the hot water bath test prescribed in 6.2.4.1 of chapter 6.2.

339 Fuel cell cartridges containing hydrogen in a metal hydride transported under this entry shall have a water capacity less than or equal to 120 mL. The pressure in the fuel cell cartridge shall not exceed 5 MPa at 55°C. The design type shall withstand, without leaking or bursting, a pressure of two (2) times the design pressure of the cartridge at 55°C or 200 kPa more than the design pressure of the cartridge at 55°C, whichever is greater. The pressure at which this test is conducted is referred to in the Drop Test and the Hydrogen Cycling Test as the "minimum shell burst pressure".

Fuel cell cartridges shall be filled in accordance with procedures provided by the manufacturer. The manufacturer shall provide the following information with each fuel cell cartridge:

.1 Inspection procedures to be carried out before initial filling and before refilling of the fuel cell cartridge;

.2 Safety precautions and potential hazards to be aware of;

.3 Method for determining when the rated capacity has been achieved;

.4 Minimum and maximum pressure range;

.5 Minimum and maximum temperature range; and

.6 Any other requirements to be met for initial filling and refilling, including the type of equipment to be used for initial filling and refilling.

The fuel cell cartridges shall be designed and constructed to prevent fuel leakage under normal conditions of transport. Each cartridge design type, including cartridges integral to a fuel cell, shall be subjected to and shall pass the following tests:

Drop test

A 1.8 m drop test onto an unyielding surface in four different orientations:

.1 Vertically, on the end containing the shut-off valve assembly;

.2 Vertically, on the end opposite to the shut-off valve assembly;

.3 Horizontally, onto a steel apex with a diameter of 38 mm, with the steel apex in the upward position; and

.4 At a 45° angle on the end containing the shut-off valve assembly.

There shall be no leakage, determined by using a soap bubble solution or other equivalent means on all possible leak locations, when the cartridge is charged to its rated charging pressure. The fuel cell cartridge shall then be hydrostatically pressurized to destruction. The recorded burst pressure shall exceed 85% of the minimum shell burst pressure.

Fire test

A fuel cell cartridge filled to rated capacity with hydrogen shall be subjected to a fire engulfment test. The cartridge design, which may include a vent feature integral to it, is deemed to have passed the fire test if:

.1 The internal pressure vents to zero gauge pressure without rupture of the cartridge; or

.2 The cartridge withstands the fire for a minimum of 20 minutes without rupture.

Hydrogen cycling test

This test is intended to ensure that a fuel cell cartridge design stress limits are not exceeded during use.

The fuel cell cartridge shall be cycled from not more than 5% rated hydrogen capacity to not less than 95% rated hydrogen capacity and back to not more than 5% rated hydrogen capacity. The rated charging pressure shall be used for charging and temperatures shall be held within the operating temperature range. The cycling shall be continued for at least 100 cycles.

Following the cycling test, the fuel cell cartridge shall be charged and the water volume displaced by the cartridge shall be measured. The cartridge design is deemed to have passed the hydrogen cycling test if the water volume displaced by the cycled cartridge does not exceed the water volume displaced by an uncycled cartridge charged to 95% rated capacity and pressurized to 75% of its minimum shell burst pressure.

Production leak test

Each fuel cell cartridge shall be tested for leaks at 15°C ± 5°C, while pressurized to its rated charging pressure. There shall be no leakage, determined by using a soap bubble solution or other equivalent means on all possible leak locations.

Each fuel cell cartridge shall be permanently marked with the following information:

.1 The rated charging pressure in megapascals (MPa);

.2 The manufacturer's serial number of the fuel cell cartridges or unique identification number; and

.3 The date of expiry based on the maximum service life (year in four digits; month in two digits).

340 Chemical kits, first aid kits and polyester resin kits containing dangerous substances in inner packagings which do not exceed the quantity limits for excepted quantities applicable to individual substances as specified in column 7b of the Dangerous Goods List may be transported in accordance with chapter 3.5. Class 5.2 substances, although not individually authorized as excepted quantities in the Dangerous Goods List, are authorized in such kits and are assigned code E2 (see 3.5.1.2).

341 Bulk transport of infectious substances in BK2 bulk containers is only permitted for infectious substances contained in animal material as defined in 1.2.1 (see 4.3.2.4.1).

342 Glass inner receptacles (such as ampoules or capsules) intended only for use in sterilization devices, when containing less than 30 mL of ethylene oxide per inner packaging with not more than 300 mL per outer packaging, may be transported in accordance with the provisions in chapter 3.5, irrespective of the indication of "E0" in column 7b of the Dangerous Goods List provided that:

.1 After filling, each glass inner receptacle has been determined to be leak tight by placing the glass inner receptacle in a hot water bath at a temperature, and for a period of time, sufficient to ensure that an internal pressure equal to the vapour pressure of ethylene oxide at 55°C is achieved. Any glass inner receptacle showing evidence of leakage, distortion or other defect under this test shall not be transported under the terms of this special provision;

.2 In addition to the packaging required by 3.5.2, each glass inner receptacle is placed in a sealed plastics bag compatible with ethylene oxide and capable of containing the contents in the event of breakage or leakage of the glass inner receptacle; and

.3 Each glass inner receptacle is protected by a means of preventing puncture of the plastics bag (e.g., sleeves or cushioning) in the event of damage to the packaging (e.g., by crushing).

343 This entry applies to crude oil containing hydrogen sulphide in sufficient concentration that vapours evolved from the crude oil can present an inhalation hazard. The packing group assigned shall be determined by the flammability hazard and inhalation hazard, in accordance with the degree of danger presented.

344 The provisions of 6.2.4 shall be met.

345 This gas contained in open cryogenic receptacles with a maximum capacity of 1 litre constructed with glass double walls having the space between the inner and outer wall evacuated (vacuum insulated) is not subject to the provisions of this Code provided each receptacle is transported in an outer packaging with suitable cushioning or absorbent materials to protect it from impact damage.

346 Open cryogenic receptacles conforming to the requirements of packing instruction P203 and containing no dangerous goods except for UN 1977, nitrogen, refrigerated liquid, which is fully absorbed in a porous material, are not subject to any other provisions of this Code.

347 This entry shall only be used if the results of Test series 6 (d) of part I of the United Nations *Manual of Tests and Criteria* have demonstrated that any hazardous effects arising from functioning are confined within the package.

348 Batteries manufactured after 31 December 2011 shall be marked with the Watt hour rating on the outside case.

349 Mixtures of a hypochlorite with an ammonium salt are not to be accepted for transport. UN 1791 hypochlorite solution is a substance of class 8.

350 Ammonium bromate and its aqueous solutions and mixtures of a bromate with an ammonium salt are not to be accepted for transport.

351 Ammonium chlorate and its aqueous solutions and mixtures of a chlorate with an ammonium salt are not to be accepted for transport.

352 Ammonium chlorite and its aqueous solutions and mixtures of a chlorite with an ammonium salt are not to be accepted for transport.

353 Ammonium permanganate and its aqueous solutions and mixtures of a permanganate with an ammonium salt are not to be accepted for transport.

354 This substance is toxic by inhalation.

355 Oxygen cylinders for emergency use transported under this entry may include installed actuating cartridges (cartridges, power device of class 1.4, compatibility group C or S), without changing the classification of class 2.2 provided the total quantity of deflagrating (propellant) explosives does not exceed 3.2 g per oxygen cylinder. The cylinders with the installed actuating cartridges as prepared for transport shall have an effective means of preventing inadvertent activation.

△ 356 Metal hydride storage systems installed in vehicles, vessels or aircrafts or in completed components or intended to be installed in vehicles, vessels or aircrafts shall be approved by the competent authority before acceptance for transport. The transport document shall include an indication that the package was approved by the competent authority or a copy of the competent authority approval shall accompany each consignment.

357 Petroleum crude oil containing hydrogen sulphide in sufficient concentration that vapours evolved from the crude oil can present an inhalation hazard shall be consigned under the entry UN 3494 PETROLEUM SOUR CRUDE OIL, FLAMMABLE, TOXIC.

■ 358 Nitroglycerin solution in alcohol with more than 1% but not more than 5% nitroglycerin may be classified in class 3 and assigned to UN 3064 provided all the requirements of packing instruction P300 are complied with.

■ 359 Nitroglycerin solution in alcohol with more than 1% but not more than 5% nitroglycerin shall be classified in class 1 and assigned to UN 0144 if not all the requirements of packing instruction P300 are complied with.

■ 360 Vehicles only powered by lithium metal batteries or lithium ion batteries shall be consigned under the entry UN 3171 BATTERY POWERED VEHICLE.

■ 361 This entry applies to electric double layer capacitors with an energy storage capacity greater than 0.3 Wh. Capacitors with an energy storage capacity of 0.3 Wh or less are not subject to the provisions of this Code. Energy storage capacity means the energy held by a capacitor, as calculated using the nominal voltage and capacitance. All capacitors to which this entry applies, including capacitors containing an electrolyte that does not meet the classification criteria of any class or division of dangerous goods, shall meet the following conditions:

.1 Capacitors not installed in equipment shall be transported in an uncharged state. Capacitors installed in equipment shall be transported either in an uncharged state or protected against short circuit;

.2 Each capacitor shall be protected against a potential short circuit hazard in transport as follows:

(i) when a capacitor's energy storage capacity is less than or equal to 10 Wh or when the energy storage capacity of each capacitor in a module is less than or equal to 10 Wh , the capacitor or module shall be protected against short circuit or be fitted with a metal strap connecting the terminals; and

(ii) when the energy storage capacity of a capacitor or a capacitor in a module is more than 10 Wh, the capacitor or module shall be fitted with a metal strap connecting the terminals;

.3 Capacitors containing dangerous goods shall be designed to withstand a 95 kPa pressure differential;

.4 Capacitors shall be designed and constructed to safely relieve pressure that may build up in use, through a vent or a weak point in the capacitor casing. Any liquid which is released upon venting shall be contained by the packaging or by the equipment in which a capacitor is installed; and

.5 Capacitors shall be marked with the energy storage capacity in Wh.

Capacitors containing an electrolyte not meeting the classification criteria of any class or division of dangerous goods, including when installed in equipment, are not subject to other provisions of this Code.

Capacitors containing an electrolyte meeting the classification criteria of any class or division of dangerous goods, with an energy storage capacity of 10 Wh or less are not subject to other provisions of this Code when they are capable of withstanding a 1.2 m drop test unpackaged on an unyielding surface without loss of contents.

Capacitors containing an electrolyte meeting the classification criteria of any class or division of dangerous goods that are not installed in equipment and with an energy storage capacity of more than 10 Wh are subject to the provisions of this Code.

Capacitors installed in the equipment and containing an electrolyte meeting the classification criteria of any class or division of dangerous goods, are not subject to other provisions of this Code provided the equipment is packaged in a strong outer packaging constructed of suitable material and of adequate strength and design, in relation to the packaging's intended use and in such a manner as to prevent accidental functioning of capacitors during transport. Large robust equipment containing capacitors may be offered for transport unpackaged or on pallets when capacitors are afforded equivalent protection by the equipment in which they are contained.

Note: Capacitors which by design maintain a terminal voltage (e.g., asymmetrical capacitors) do not belong to this entry.

■ 362 This entry applies to liquids, pastes or powders, pressurized with a propellant which meets the definition of a gas in 2.2.1.2.1 or 2.2.1.2.2.

Note: A chemical under pressure in an aerosol dispenser shall be transported under UN 1950.

The following provisions shall apply:

.1 the chemical under pressure shall be classified based on the hazard characteristics of the components in the different states:

– the propellant;

– the liquid; or

– the solid.

If one of these components, which can be a pure substance or a mixture, needs to be classified as flammable, the chemical under pressure shall be classified as flammable in class 2.1. Flammable components are flammable liquids and liquid mixtures, flammable solids and solid mixtures or flammable gases and gas mixtures meeting the following criteria:

(i) a flammable liquid is a liquid having a flashpoint of not more than 93°C;

(ii) a flammable solid is a solid which meets the criteria in 2.4.2.2 of this Code;

(iii) a flammable gas is a gas which meets the criteria in 2.2.2.1 of this Code;

.2 gases of class 2.3 and gases with a subsidiary risk of 5.1 shall not be used as a propellant in a chemical under pressure;

.3 where the liquid or solid components are classified as dangerous goods of class 6.1, packing groups II or III, or class 8, packing groups II or III, the chemical under pressure shall be assigned a subsidiary risk of class 6.1 or class 8 and the appropriate UN number shall be assigned. Components classified in class 6.1, packing group I, or class 8, packing group I, shall not be used for transport under this Proper Shipping Name;

.4 in addition, chemicals under pressure with components meeting the properties of: class 1, explosives; class 3, liquid desensitized explosives; class 4.1, self-reactive substances and solid desensitized explosives; class 4.2, substances liable to spontaneous combustion; class 4.3, substances which, in contact with water, emit flammable gases; class 5.1 oxidizing substances; class 5.2, organic peroxides; class 6.2, Infectious substances or class 7, Radioactive material, shall not be used for transport under this Proper Shipping Name;

.5 substances to which PP86 or TP7 are assigned in column 9 and column 14 of the Dangerous Goods List in chapter 3.2 and therefore require air to be eliminated from the vapour space, shall not be used for transport under this UN number but shall be transported under their respective UN numbers as listed in the Dangerous Goods List of chapter 3.2.

■ 363 This entry also applies to dangerous goods above the quantity specified in column 7a of the Dangerous Goods List of chapter 3.2 in means of containment (other than vehicles or means of containment defined in Part 6 of this Code subject to special provision 301) integral to equipment or machinery (e.g., generators, compressors, heating units, etc.) as part of their original design type. They are not subject to any other provisions of this Code if the following provisions are met:

.1 the means of containment shall be in compliance with the construction requirements of the competent authority;

.2 any valves or openings (e.g., venting devices) in the means of containment containing dangerous goods shall be closed during transport;

.3 the machinery or equipment shall be loaded in an orientation to prevent inadvertent leakage of dangerous goods and secured by means capable of restraining the machinery or equipment to prevent any movement during transport which would change the orientation or cause it to be damaged;

.4 where the means of containment has a capacity of not more than 450 L, the labelling requirements of 5.2.2 and, when necessary, the marking requirements of 5.2.1.6 shall apply and, where the capacity is greater than 450 L but not more than 1,500 L, the machinery or equipment shall be labelled on all four external sides in accordance with 5.2.2 and, when necessary, shall be marked in accordance with 5.2.1.6;

.5 where the means of containment has a capacity greater than 1,500 L, the machinery or equipment shall be placarded on all four external sides in accordance with 5.3.1.1.2 and, when necessary, shall be marked in accordance with 5.2.1.6;

.6 the provisions of part 2, part 3 except columns 8 to 14 of the Dangerous Goods List in 3.2, 5.1.1 except 5.1.1.4, 5.3 except 5.3.2.0 and 5.3.2.1, 5.4 and part 7 shall apply;

.7 the transport of machinery or equipment where the quantity of substance exceeds 1500 L is authorized when approved by the competent authority.

■ 364 This article may only be transported under the provisions of chapter 3.4 if, as presented for transport, the package is capable of passing the test in accordance with Test Series 6(d) of Part I of the *Manual of Tests and Criteria* as determined by the competent authority.

■ 365 For manufactured instruments and articles containing mercury, see UN 3506.

■ 366 Manufactured instruments and articles containing not more than 1 kg of mercury are not subject to the provisions of this Code.

△ 900 The transport of the following substances is prohibited:

AMMONIUM HYPOCHLORITE

AMMONIUM NITRATE liable to self-heating sufficient to initiate decomposition

AMMONIUM NITRITES and mixtures of an inorganic nitrite with an ammonium salt

CHLORIC ACID, AQUEOUS SOLUTION with more than 10% chloric acid

ETHYL NITRITE pure

HYDROCYANIC ACID, AQUEOUS SOLUTION (HYDROGEN CYANIDE, AQUEOUS SOLUTION) with more than 20% hydrogen cyanide

HYDROGEN CHLORIDE, REFRIGERATED LIQUID

HYDROGEN CYANIDE SOLUTION, IN ALCOHOL with more than 45% hydrogen cyanide

MERCURY OXYCYANIDE pure

METHYL NITRITE

PERCHLORIC ACID with more than 72% acid, by mass

SILVER PICRATE, dry or wetted with less than 30% water by mass

ZINC AMMONIUM NITRITE

See also special provisions 349, 350, 351, 352 and 353.

903 HYPOCHLORITE MIXTURES with 10% or less available CHLORINE are not subject to the provisions of this Code.

904 The provisions of this Code, except for the marine pollution aspects, do not apply to these substances if they are completely miscible with water, except when transported in receptacles with a capacity greater than 250 L and in tanks.

905 May only be shipped as an 80% solution in TOLUENE. The pure product is shock-sensitive and decomposes with explosive violence and the possibility of detonation when heated under confinement. Can be ignited by impact.

907 The consignment shall be accompanied by a certificate from a recognized authority stating:

- moisture content;
- fat content;
- details of anti-oxidant treatment for meals older than 6 months (for UN 2216 only);
- anti-oxidant concentration at the time of shipment, which must exceed 100 mg/kg (for UN 2216 only);
- packing, number of bags and total mass of the consignment;
- temperature of fishmeal at the time of despatch from the factory;
- date of production.

No weathering/curing is required prior to loading. Fishmeal under UN 1374 shall have been weathered for not less than 28 days before shipment.

When fishmeal is packed into containers, the containers shall be packed in such a way that the free air space has been restricted to the minimum.

912 This entry also covers solutions in water with concentrations above 70%.

⊗

915 This entry shall not be used for wetted explosives, self-reactive substances or metal powders.

916 The provisions of this Code do not apply to this substance when:

- mechanically produced, with a particle size of 53 microns or greater; or
- chemically produced, with a particle size of 840 microns or greater.

917 Scrap with rubber content below 45% or exceeding 840 microns and fully vulcanized hard rubber are not subject to the provisions of this Code.

919 UREA NITRATE, WETTED with not less than 10% water, by mass, may be transported under the provisions of this class, provided it is packaged in accordance with packing instruction P002.

920 Bars, ingots or sticks are not subject to the provisions of this Code.

921 Zirconium, dry, 254 microns or thicker is not subject to the provisions of this Code.

922 LEAD PHOSPHITE, DIBASIC which is accompanied by the certificate from the shipper stating that the substance, as offered for shipment, has been stabilized in such a way that it does not possess the properties of class 4.1 is not subject to the provisions of this Code.

923 The temperature shall be checked regularly.

⊗

925 The provisions of this Code do not apply to:

- non-activated carbon blacks of mineral origin;
- a consignment of carbon if it passes the tests for self-heating substances as reflected in the United Nations *Manual of Tests and Criteria* (see 33.3.1.3.3), and is accompanied by a certificate from a laboratory accredited by the competent authority, stating that the product to be loaded has been correctly sampled by trained staff from that laboratory and that the sample was correctly tested and has passed the test; and
- carbons made by a steam activation process.

926 This substance shall preferably have been weathered for not less than one month before shipment unless a certificate from a person recognized by the competent authority of the country of shipment states a maximum moisture content of 5%.

927 *p*-Nitrosodimethylaniline, wetted with more than 50% water is not subject to the provisions of this Code.

928 The provisions of this Code shall not apply to:

- fishmeal when acidified and wetted with more than 40% water, by mass, irrespective of other factors;
- consignments of fishmeal which are accompanied by a certificate issued by a recognized competent authority of the country of shipment or other recognized authority stating that the product has no self-heating properties when transported in packaged form; or
- fishmeal manufactured from "white" fish with a moisture content of not more than 12% and a fat content of not more than 5% by mass.

929 If satisfied, as a result of tests, that such relaxation is justified, the competent authority may permit:

- the seed cakes described as "SEED CAKE, containing vegetable oil (a) mechanically expelled seeds, containing more than 10% of oil or more than 20% of oil and moisture combined" to be transported under conditions governing "SEED CAKE, containing vegetable oil (b) solvent extractions and expelled seeds, containing not more than 10% of oil and, when the amount of moisture is higher than 10% not more than 20% of oil and moisture combined", and
- the seed cakes described as "SEED CAKE, containing vegetable oil (b) solvent extractions and expelled seeds, containing not more than 10% of oil and, when the amount of moisture is higher than 10% not more than 20% of oil and moisture combined" to be transported under conditions governing SEED CAKE, UN 2217.

Certificates from the shipper shall state oil content and moisture content and shall accompany the shipment.

930 All pesticides can only be carried under the provisions of this class if accompanied by a certificate supplied by the shipper stating that, when in contact with water, it is not combustible and does not show tendency to autoignition, and that the mixture of gases evolved is not flammable. Otherwise, the provisions of class 4.3 shall be applicable.

931 A consignment of this substance which is accompanied by a declaration from the shipper stating that it has no self-heating properties is not subject to the provisions of this Code.

932 Requires a certificate from the maker or shipper, stating that the shipment was stored under cover, but in the open air, in the size in which it was packaged, for not less than 3 days prior to shipment.

934 Requires the percentage range of calcium carbide impurity to be shown on the shipping documents.

935 Substances which do not evolve flammable gases when wet, which are accompanied by a certificate from the shipper stating that the substance, as offered for shipment, does not evolve flammable gases when wet, are not subject to the provisions of this Code.

937 The solid hydrated form of this substance is not subject to the provisions of this Code.

939 A consignment of this substance that is accompanied by a shipper's certificate stating that it does not contain more than 0.05% maleic anhydride is not subject to the provisions of this Code.

⊗

942 The concentration and temperature of the solution at the time of loading, its percentage of combustible material and of chlorides as well as the contents of free acid shall be certified.

943 Water-activated articles shall bear a subsidiary risk of class 4.3.

945 Stabilization of fishmeal shall be achieved to prevent spontaneous combustion by effective application of between 400 and 1000 mg/kg (ppm) ethoxyquin, or liquid BHT (butylated hydroxytoluene) or between 1000 and 4000 mg/kg (ppm) BHT in powder form at the time of production. The said application shall occur no longer than twelve months prior to shipment.

946 Requires certification from the shipper that the substance is not of class 4.2.

948 These substances may be transported in bulk in cargo transport units only if their melting point is 75°C or above.

951 Bulk container shall be hermetically sealed and under a nitrogen blanket.

952 UN 1942 may be transported in bulk container if approved by the competent authority.

954 The provisions of this Code shall not apply to consignments of compressed baled hay with a moisture content of less than 14% shipped in closed cargo transport units and accompanied by a certificate from the shipper stating that the product does not present any class 4.1, UN 1327, hazard in transport and that its moisture content is less than 14%.

955 If a viscous substance and its packaging fulfils the provisions of 2.3.2.5, the packing provisions of chapter 4.1, the marking and labelling provisions of chapter 5.2 and the package testing provisions of chapter 6.1 are not applicable.

⊗

957 Lithium cells and batteries manufactured before 1 January 2003 that have not been tested in accordance with the requirements in chapter 38.3 of the United Nations *Manual of Tests and Criteria*, as well as articles which contain such lithium cells or batteries, may be transported until 31 December 2013 if all applicable provisions of this Code are met.

958 This entry covers articles, such as rags, cotton waste, clothing or sawdust, containing polychlorinated biphenyls, polyhalogenated biphenyls or polyhalogenated terphenyls where no free visible liquid is present.

959 Waste aerosols authorized for transport under special provision 327 shall only be transported on short international voyages. Long international voyages are authorized only with the approval of the competent authority. Packagings shall be marked and labelled and cargo transport units shall be marked and placarded for appropriate sub-division of class 2 and, if applicable, the subsidiary risk(s).

960 Not subject to the provisions of this Code but may be subject to provisions governing the transport of dangerous goods by other modes.

961 Vehicles and equipment are not subject to the provisions of this Code if they are stowed on a roll-on/roll-off ship or in another cargo space designated by the Administration (flag State) as specifically designed and approved for the carriage of vehicles and equipment and there are no signs of leakage from the battery, engine, fuel cell, compressed gas cylinder or accumulator, or fuel tank when applicable.

In addition, vehicles and equipment are not subject to the provisions of this Code if any of the following conditions are met:

.1 The fuel tank(s) of the vehicle or equipment powered by a flammable liquid fuel is empty and installed batteries are protected from short circuit;

.2 The fuel tank(s) of the vehicle or equipment powered by a flammable gas is emptied of liquefied or compressed gas, the positive pressure in the tank does not exceed 2 bar, the fuel shut-off or isolation valve is closed and secured, and installed batteries are protected from short circuit; or

.3 The vehicle or equipment is solely powered by a wet or dry electric storage battery or a sodium battery, and the battery is protected from short circuit.

962 Vehicles or equipment powered by internal combustion engines, fuel cells or batteries not meeting the conditions of special provision 961 shall be assigned to class 9 and shall meet the following requirements:

.1 vehicles and equipment shall not show signs of leakage from batteries, engines, fuel cells, compressed gas cylinders or accumulators, or fuel tank(s) when applicable;

.2 for flammable liquid powered vehicles and equipment, the fuel tank(s) containing the flammable liquid shall not be more than one fourth full and in any case the flammable liquid shall not exceed 250 L;

.3 for flammable gas powered vehicles and equipment, the fuel shut-off valve of the fuel tank(s) shall be securely closed;

.4 installed batteries shall be protected from damage, short circuit, and accidental activation during transport. Lithium ion or lithium metal batteries shall meet the requirements of the United Nations *Manual of Tests and Criteria*, part III, subsection 38.3, unless otherwise approved by the competent authority; and

.5 dangerous goods required for the operation of the vehicle or equipment such as fire extinguishers, compressed gas accumulators, airbag inflators, etc., shall be securely mounted in the vehicle or equipment.

The marking, labelling and placarding provisions of this Code shall not apply.

963 Nickel-metal hydride button cells or nickel-metal hydride cells or batteries packed with or contained in equipment are not subject to the provisions of this Code.

All other nickel-metal hydride cells or batteries shall be securely packed and protected from short circuit. They are not subject to other provisions of this Code provided that they are loaded in a cargo transport unit in a total quantity of less than 100 kg gross mass. When loaded in a cargo transport unit in a total quantity of 100 kg gross mass or more, they are not subject to other provisions of this Code except those of 5.4.1, 5.4.3 and column 16 of the Dangerous Goods List in chapter 3.2.

964 This substance is not subject to the provisions of this Code when transported in non-friable prills or granules form and if it passes the test for oxidizing solid substances as reflected in the United Nations *Manual of Tests and Criteria* (see 34.4.1) and is accompanied by a certificate from a laboratory accredited by a competent authority, stating that the product has been correctly sampled by trained staff from the laboratory and that the sample was correctly tested and has passed the test.

■ 965 .1 When transported in cargo transport units, the cargo transport units shall provide an adequate exchange of air in the unit (e.g., by using a ventilated container, open-top container or container in one door off operation) to prevent the build-up of an explosive atmosphere. Alternatively, these entries shall be transported under temperature control in refrigerated cargo transport units that comply with the provisions of 7.3.7.6. When cargo transport units with venting devices are used, these devices shall be kept clear and operable. When mechanical devices are used for ventilation, they shall be explosion-proof to prevent ignition of flammable vapours from the substances.

.2 The provisions of .1 do not apply if:

(a) the substance is packed in hermetically sealed packagings or IBCs, which conform to packing group II performance level for liquid dangerous goods according to the provisions of 6.1 or 6.5, respectively; and

(b) the marked hydraulic test pressure exceeds 1.5 times the total gauge pressure in the packagings or IBCs determined at 55°C for the respective filling goods according to 4.1.1.10.1.

.3 Where the substance is loaded in closed cargo transport units, the provisions of 7.3.6.1 shall be met.

.4 Cargo transport units shall be marked with a warning mark including the words "CAUTION – MAY CONTAIN FLAMMABLE VAPOUR" with lettering not less than 25 mm high. This mark shall be affixed at each access point in a location where it will be easily seen by persons prior to opening or entering the cargo transport unit and shall remain on the cargo transport unit until the following provisions are met:

(a) the cargo transport unit has been completely ventilated to remove any hazardous concentration of vapour or gas;

(b) the immediate vicinity of the cargo transport unit is clear of any source of ignition; and

(c) the goods have been unloaded.

■ 966 Sheeted bulk containers (BK1) are only permitted in accordance with 4.3.3.

■ 967 Flexible bulk containers (BK3) are only permitted in accordance with 4.3.4.

△ Chapter 3.4

Dangerous goods packed in limited quantities

3.4.1 General

3.4.1.1 This chapter provides the provisions applicable to the transport of dangerous goods of certain classes packed in limited quantities. The applicable quantity limit for the inner packaging or article is specified for each substance in column 7a of the Dangerous Goods List of chapter 3.2. In addition, the quantity "0" has been indicated in this column for each entry not permitted to be transported in accordance with this chapter.

3.4.1.2 Limited quantities of dangerous goods packed in such limited quantities, meeting the provisions of this chapter, are not subject to any other provisions of this Code except the relevant provisions of:

.1 Part 1, chapters 1.1, 1.2 and 1.3;

.2 Part 2;

.3 Part 3, chapters 3.1, 3.2, 3.3;

.4 Part 4, 4.1.1.1, 4.1.1.2 and 4.1.1.4 to 4.1.1.8;

.5 Part 5, 5.1.1 except 5.1.1.4, 5.1.2.3, 5.2.1.7, 5.2.1.9, 5.3.2.3, 5.3.2.4, and chapter 5.4;

.6 Part 6, construction requirements of 6.1.4, 6.2.1.2 and 6.2.4;

.7 Part 7, 7.1.3.2, 7.6.3.1 and 7.3 except 7.3.3.15 and 7.3.4.1.

3.4.2 Packing

3.4.2.1 Dangerous goods shall be packed only in inner packagings placed in suitable outer packagings. Intermediate packagings may be used. In addition, for articles of Division 1.4, Compatibility Group S, the provisions of section 4.1.5 shall be fully complied with. The use of inner packagings is not necessary for the transport of articles such as aerosols or "receptacles, small, containing gas". The total gross mass of the package shall not exceed 30 kg.

3.4.2.2 Except for articles of Division 1.4, Compatibility Group S, shrink-wrapped or stretch-wrapped trays meeting the conditions of 4.1.1.1, 4.1.1.2 and 4.1.1.4 to 4.1.1.8 are acceptable as outer packagings for articles or inner packagings containing dangerous goods transported in accordance with this chapter. Inner packagings that are liable to break or be easily punctured, such as those made of glass, porcelain, stoneware or certain plastics, shall be placed in suitable intermediate packagings meeting the provisions of 4.1.1.1, 4.1.1.2 and 4.1.1.4 to 4.1.1.8, and be so designed that they meet the construction requirements of 6.1.4. The total gross mass of the package shall not exceed 20 kg

3.4.2.3 Liquid goods of Class 8, packing group II in glass, porcelain or stoneware inner packagings shall be enclosed in a compatible and rigid intermediate packaging.

3.4.3 Stowage

Dangerous goods packed in limited quantity are allocated stowage category A as defined in 7.1.3.2. The other stowage provisions indicated in column 16 of the Dangerous Goods List are not applicable.

3.4.4 Segregation

3.4.4.1 Different dangerous substances in limited quantities may be packed in the same outer packaging, provided:

.1 the substances comply with the provisions of 7.2.6.1; and

.2 the segregation provisions of chapter 7.2, including the segregation provisions in column 16 of the Dangerous Goods List, are taken into account. However, notwithstanding the individual provisions specified in the Dangerous Goods List, substances in packing group III within the same class may be packed together subject to compliance with 3.4.4.1.1 of the IMDG Code. The following statement shall be included in the transport document: "Transport in accordance with 3.4.4.1.2 of the IMDG Code" (see 5.4.1.5.2.2).

3.4.4.2 The segregation provisions of chapter 7.2 are not applicable for packagings containing dangerous goods in limited quantities or in relation to other dangerous goods.

3.4.5 Marking and placarding

3.4.5.1 Except for air transport, packages containing dangerous goods in limited quantities shall bear the marking shown below:

Marking for packages containing limited quantities

The marking shall be readily visible, legible and able to withstand open weather exposure without a substantial reduction in effectiveness. The top and bottom portions and the surrounding line shall be black. The centre area shall be white or a suitable contrasting background. The minimum dimensions shall be 100 mm × 100 mm and the minimum width of line forming the diamond shall be 2 mm. If the size of the package so requires, the dimension may be reduced, to be not less than 50 mm × 50 mm provided the marking remains clearly visible.

3.4.5.2 Packages containing dangerous goods consigned for air transport in conformity with the provisions of Part 3, Chapter 4 of the *ICAO Technical Instructions for the Safe Transport of Dangerous Goods by Air* shall bear the marking shown below:

Marking for packages containing limited quantities conforming to part 3, chapter 4 of the ICAO *Technical Instructions for the Safe Transport of Dangerous Goods by Air*

The marking shall be readily visible, legible and able to withstand open weather exposure without a substantial reduction in effectiveness. The top and bottom portions and the surrounding line shall be black. The centre area shall be white or a suitable contrasting background. The minimum dimensions shall be 100 mm × 100 mm. The minimum width of line forming diamond shall be 2 mm. The symbol "Y" shall be placed in the centre of the mark and shall be clearly visible. If the size of the package so requires, the dimension may be reduced, to be not less than 50 mm × 50 mm provided the marking remains clearly visible.

3.4.5.3 Packages containing dangerous goods bearing the marking shown in 3.4.5.2 shall be deemed to meet the provisions of sections 3.4.1 and 3.4.2 of this chapter and need not bear the marking shown in 3.4.5.1.

3.4.5.4 When packages containing dangerous goods packed in limited quantities are placed in an overpack or in a unit load, the overpack or the unit load shall be marked with the marking required by this chapter unless the markings representative of all dangerous goods in the overpack or the unit load are visible. In addition, an overpack shall be marked with the word "OVERPACK" unless markings representative of all dangerous goods, as required by this chapter, in the overpack are visible. The other provisions of 5.1.2.1 apply only if other dangerous goods which are not packed in limited quantities are contained in the overpack or in a unit load and only in relation to these other dangerous goods.

3.4.5.5 Placarding and marking of cargo transport units

3.4.5.5.1 Cargo transport units containing dangerous goods packed in limited quantities with no other dangerous goods shall not be placarded nor marked according to 5.3.2.0 and 5.3.2.1. However, they shall be suitably marked on the exterior with the mark in 3.4.5.5.4.

3.4.5.5.2 Cargo transport units containing dangerous goods and dangerous goods packed in limited quantities shall be placarded and marked according to the provisions applicable to the dangerous goods which are not packed in limited quantities. However, if no placard or mark is required for the dangerous goods not packed in limited quantities, the cargo transport units shall be marked with the mark in 3.4.5.5.4.

3.4.5.5.3 In all cases, if the dangerous goods packed in limited quantities are marine pollutants, the cargo transport units shall bear the marine pollutant mark as indicated in 5.3.2.3.

3.4.5.5.4 When required in 3.4.5.5.1 or 3.4.5.5.2, the following mark shall be affixed on cargo transport units:

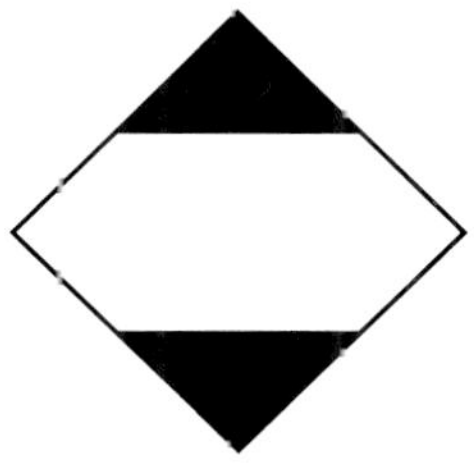

The marking shall be readily visible, legible and be such that this information will still be identifiable on cargo transport units surviving at least three months' immersion in the sea. In considering suitable marking methods, account shall be taken of ease with which the surface of the cargo transport unit can be marked. The top and bottom portions and the surrounding line shall be black. The centre area shall be white or a suitable contrasting background. The minimum dimensions shall be of 250 mm × 250 mm in locations indicated in 5.3.1.1.4.1.

3.4.6 Documentation

3.4.6.1 In addition to the provisions for documentation specified in chapter 5.4, the words "limited quantity" or "LTD QTY" shall be included on the dangerous goods declaration together with the description of the shipment.

3

Chapter 3.5

Dangerous goods packed in excepted quantities

3.5.1 Excepted quantities

3.5.1.1 Excepted quantities of dangerous goods of certain classes, other than articles, meeting the provisions of this chapter, are not subject to any other provisions of this Code except for:

.1 The training provisions in chapter 1.3;

.2 The classification procedures and packing group criteria in Part 2, Classification;

.3 The packaging provisions of 4.1.1.1, 4.1.1.2, 4.1.1.4, 4.1.1.4.1 and 4.1.1.6 in Part 4; and

.4 The provisions for documentation specified in chapter 5.4.

Note: In the case of radioactive material, the provisions for radioactive material in excepted packages in 1.5.1.5 apply.

3.5.1.2 Dangerous goods which may be carried as excepted quantities in accordance with the provisions of this chapter are shown in column 7b of the Dangerous Goods List by means of an alphanumeric code as follows:

Code	**Maximum net quantity per inner packaging** (in grams for solids and mL for liquids and gases)	**Maximum net quantity per outer packaging** (in grams for solids and mL for liquids and gases, or sum of grams and mL in the case of mixed packaging)
E0	Not permitted as Excepted Quantity	
E1	30	1,000
E2	30	500
E3	30	300
E4	1	500
E5	1	300

For gases, the volume indicated for inner packagings refers to the water capacity of the inner receptacle and the volume indicated for outer packagings refers to the combined water capacity of all inner packagings within a single outer packaging.

3.5.1.3 Where dangerous goods in excepted quantities for which different codes are assigned are packaged together, the total quantity per outer packaging shall be limited to that corresponding to the most restrictive code.

■ 3.5.1.4 Excepted quantities of dangerous goods assigned to codes E1, E2, E4 and E5 are not subject to the provisions of this Code provided that:

.1 The maximum net quantity of material per inner packaging is limited to 1 mL for liquids and gases and 1 g for solids;

.2 The provisions of 3.5.2 are met, except that an intermediate packaging is not required if the inner packagings are securely packed in an outer packaging with cushioning material in such a way that, under normal conditions of transport, they cannot break, be punctured, or leak their contents; and for liquid dangerous goods, the outer packaging contains sufficient absorbent material to absorb the entire contents of the inner packagings;

.3 The provisions of 3.5.3 are complied with; and

.4 The maximum net quantity of dangerous goods per outer packaging does not exceed 100 g for solids or 100 mL for liquids and gases.

3.5.2 Packagings

3.5.2.1 Packagings used for the transport of dangerous goods in excepted quantities shall be in compliance with the following:

.1 There shall be an inner packaging and each inner packaging shall be constructed of plastic (when used for liquid dangerous goods it shall have a thickness of not less than 0.2 mm), or of glass, porcelain, stoneware, earthenware or metal (see also 4.1.1.2) and the closure of each inner packaging shall be held securely in place with wire, tape or other positive means; any receptacle having a neck with moulded screw threads shall have a leakproof threaded-type cap. The closure shall be resistant to the contents;

.2 Each inner packaging shall be securely packed in an intermediate packaging with cushioning material in such a way that, under normal conditions of transport, they cannot break, be punctured or leak their contents. The intermediate packaging shall completely contain the contents in case of breakage or leakage, regardless of package orientation. For liquid dangerous goods, the intermediate packaging shall contain sufficient absorbent material to absorb the entire contents of the inner packaging. In such cases, the absorbent material may be the cushioning material. Dangerous goods shall not react dangerously with cushioning, absorbent material and packaging material or reduce the integrity or function of the materials;

.3 The intermediate packaging shall be securely packed in a strong, rigid outer packaging (wooden, fibreboard or other equally strong material);

.4 Each package type shall be in compliance with the provisions in 3.5.3;

.5 Each package shall be of such a size that there is adequate space to apply all necessary markings; and

.6 Overpacks may be used and may also contain packages of dangerous goods or goods not subject to the provisions of this Code.

3.5.3 Tests for packages

3.5.3.1 The complete package as prepared for transport, with inner packagings filled to not less than 95% of their capacity for solids or 98% for liquids, shall be capable of withstanding, as demonstrated by testing which is appropriately documented, without breakage or leakage of any inner packaging and without significant reduction in effectiveness:

.1 Drops onto a rigid, non-resilient flat and horizontal surface from a height of 1.8 m:

(i) Where the sample is in the shape of a box, it shall be dropped in each of the following orientations:

- flat on the base;
- flat on the top;
- flat on the longest side;
- flat on the shortest side;
- on a corner

(ii) Where the sample is in the shape of a drum, it shall be dropped in each of the following orientations:

- diagonally on the top chime, with the centre of gravity directly above the point of impact;
- diagonally on the base chime;
- flat on the side.

Note: Each of the above drops may be performed on different but identical packages.

.2 A force applied to the top surface for a duration of 24 h, equivalent to the total weight of identical packages if stacked to a height of 3 m (including the sample).

3.5.3.2 For the purposes of testing, the substances to be transported in the packaging may be replaced by other substances except where this would invalidate the results of the tests. For solids, when another substance is used, it shall have the same physical characteristics (mass, grain size, etc.) as the substance to be carried. In the drop tests for liquids, when another substance is used, its relative density (specific gravity) and viscosity shall be similar to those of the substance to be transported.

3.5.4 Marking of packages

3.5.4.1 Packages containing excepted quantities of dangerous goods prepared in accordance with this chapter shall be durably and legibly marked with the mark shown below. The primary hazard class of each of the dangerous goods contained in the package shall be shown in the mark. Where the name of the consignor or consignee is not shown elsewhere on the package, this information shall be included within the mark.

Excepted quantities mark

Hatching and symbol of the same colour, black or red,
on white or suitable contrasting background

* The class, or, when assigned, the division number(s), shall be shown in this location.
** The name of the consignor or of the consignee shall be shown in this location if not shown elsewhere on the package.

3.5.4.2 The dimensions of the mark shall be a minimum of 100 mm × 100 mm.

3.5.4.3 An overpack containing dangerous goods in excepted quantities shall display the markings required by 3.5.4.1, unless such markings on packages within the overpack are clearly visible.

3.5.5 Maximum number of packages in any cargo transport unit

3.5.5.1 The number of packages containing dangerous goods packed in excepted quantities in any cargo transport unit shall not exceed 1000.

3.5.6 Documentation

3.5.6.1 In addition to the provisions for documentation specified in chapter 5.4, the words "dangerous goods in excepted quantities" and the number of packages shall be included on the dangerous goods declaration together with the description of the shipment.

3.5.7 Stowage

△ 3.5.7.1 Dangerous goods packed in excepted quantity are allocated stowage category A as defined in 7.1.3.2. The other stowage provisions indicated in column 16 of the Dangerous Goods List are not applicable.

△ 3.5.8 Segregation

3.5.8.1 The segregation provisions of chapters 7.2 to 7.7, including the segregation provisions in column 16 of the Dangerous Goods List, are not applicable for packagings containing dangerous goods packed in excepted quantities or in relation to other dangerous goods.

3.5.8.2 The segregation provisions of chapters 7.2 to 7.7, including the segregation provisions in column 16 of the Dangerous Goods List, are not applicable for different dangerous goods in excepted quantities in the same outer packaging provided that they do not react dangerously with each other (see 4.1.1.6).

APPENDICES

Appendix A

List of generic and N.O.S. Proper Shipping Names

△ Substances or articles not mentioned specifically by name in the Dangerous Goods List in chapter 3.2 shall be classified in accordance with 3.1.1.2. Thus the name in the Dangerous Goods List which most appropriately describes the substance or article shall be used as the Proper Shipping Name. The main generic entries and all the N.O.S. entries given in the Dangerous Goods List are listed below. This Proper Shipping Name shall be supplemented by the technical name when special provision 274 or 318 has been assigned to the entry in column 6 of the Dangerous Goods List. For marine pollutants, see also 3.1.2.9.

In this list, general and N.O.S. names are grouped according to their hazard class or division. Within each hazard class or division, the names have been placed into three groups as follows:

- specific entries covering a group of substances or articles of a particular chemical or technical nature;
- pesticide entries, for class 3 and class 6.1;
- general entries covering a group of substances or articles having one or more general dangerous properties.

THE MOST SPECIFIC APPLICABLE NAME SHALL ALWAYS BE USED.

Class or division	Subsidiary risk	UN Number	Proper Shipping Name
			CLASS 1
1		0190	SAMPLES, EXPLOSIVE, other than initiating explosive
			Division 1.1
1.1A		0473	SUBSTANCES, EXPLOSIVE, N.O.S.
1.1B		0461	COMPONENTS, EXPLOSIVE TRAIN, N.O.S.
1.1C		0462	ARTICLES, EXPLOSIVE, N.O.S.
1.1C		0474	SUBSTANCES, EXPLOSIVE, N.O.S.
1.1C		0497	PROPELLANT, LIQUID
1.1C		0498	PROPELLANT, SOLID
1.1D		0463	ARTICLES, EXPLOSIVE, N.O.S.
1.1D		0475	SUBSTANCES, EXPLOSIVE, N.O.S.
1.1E		0464	ARTICLES, EXPLOSIVE, N.O.S.
1.1F		0465	ARTICLES, EXPLOSIVE, N.O.S.
1.1G		0476	SUBSTANCES, EXPLOSIVE, N.O.S.
1.1L		0354	ARTICLES, EXPLOSIVE, N.O.S.
1.1L		0357	SUBSTANCES, EXPLOSIVE, N.O.S.
			Division 1.2
1.2B		0382	COMPONENTS, EXPLOSIVE TRAIN, N.O.S.
1.2C		0466	ARTICLES, EXPLOSIVE, N.O.S.
1.2D		0467	ARTICLES, EXPLOSIVE, N.O.S.
1.2E		0468	ARTICLES, EXPLOSIVE, N.O.S.
1.2F		0469	ARTICLES, EXPLOSIVE, N.O.S.
1.2K	6.1	0020	AMMUNITION, TOXIC with burster, expelling charge or propelling charge
1.2L	4.3	0248	CONTRIVANCES, WATER-ACTIVATED with burster, expelling charge or propelling charge
1.2L		0355	ARTICLES, EXPLOSIVE, N.O.S.
1.2L		0358	SUBSTANCES, EXPLOSIVE, N.O.S.
			Division 1.3
1.3C		0132	DEFLAGRATING METAL SALTS OF AROMATIC NITRO-DERIVATIVES, N.O.S.
1.3C		0470	ARTICLES, EXPLOSIVE, N.O.S.
1.3C		0477	SUBSTANCES, EXPLOSIVE, N.O.S.
1.3C		0495	PROPELLANT, LIQUID
1.3C		0499	PROPELLANT, SOLID
1.3G		0478	SUBSTANCES, EXPLOSIVE, N.O.S.
1.3K	6.1	0021	AMMUNITION, TOXIC with burster, expelling charge or propelling charge
1.3L	4.3	0249	CONTRIVANCES, WATER-ACTIVATED with burster, expelling charge or propelling charge
1.3L		0356	ARTICLES, EXPLOSIVE, N.O.S.
1.3L		0359	SUBSTANCES, EXPLOSIVE, N.O.S.
			Division 1.4
1.4B		0350	ARTICLES, EXPLOSIVE, N.O.S.
1.4B		0383	COMPONENTS, EXPLOSIVE TRAIN, N.O.S.
1.4C		0351	ARTICLES, EXPLOSIVE, N.O.S.
1.4C		0479	SUBSTANCES, EXPLOSIVE, N.O.S.
1.4C		0501	PROPELLANT, SOLID
1.4D		0352	ARTICLES, EXPLOSIVE, N.O.S.
1.4D		0480	SUBSTANCES, EXPLOSIVE, N.O.S.
1.4E		0471	ARTICLES, EXPLOSIVE, N.O.S.
1.4F		0472	ARTICLES, EXPLOSIVE, N.O.S.
1.4G		0353	ARTICLES, EXPLOSIVE, N.O.S.
1.4G		0485	SUBSTANCES, EXPLOSIVE, N.O.S.
1.4S		0349	ARTICLES, EXPLOSIVE, N.O.S.
1.4S		0384	COMPONENTS, EXPLOSIVE TRAIN, N.O.S.
1.4S		0481	SUBSTANCES, EXPLOSIVE, N.O.S.
			Division 1.5
1.5D		0482	SUBSTANCES, EXPLOSIVE, VERY INSENSITIVE (SUBSTANCES, EVI), N.O.S.
			Division 1.6
1.6N		0486	ARTICLES, EXPLOSIVE, EXTREMELY INSENSITIVE (ARTICLES, EEI)

APPENDIX A

	Class or division	Subsidiary risk	UN Number	Proper Shipping Name
				CLASS 2
				Class 2.1
				Specific entries
	2.1		1964	HYDROCARBON GAS MIXTURE, COMPRESSED, N.O.S.
	2.1		1965	HYDROCARBON GAS MIXTURE, LIQUEFIED, N.O.S.
	2.1		3354	INSECTICIDE GAS, FLAMMABLE, N.O.S.
				General entries
	2.1		1954	COMPRESSED GAS, FLAMMABLE, N.O.S.
	2.1		3161	LIQUEFIED GAS, FLAMMABLE, N.O.S.
	2.1		3167	GAS SAMPLE, NON-PRESSURIZED, FLAMMABLE, N.O.S., not refrigerated liquid
	2.1		3312	GAS, REFRIGERATED LIQUID, FLAMMABLE, N.O.S.
■	2.1		3501	CHEMICAL UNDER PRESSURE, FLAMMABLE, N.O.S.
■	2.1	6.1	3504	CHEMICAL UNDER PRESSURE, FLAMMABLE, TOXIC, N.O.S.
■	2.1	8	3505	CHEMICAL UNDER PRESSURE, FLAMMABLE, CORROSIVE, N.O.S.
				Class 2.2
				Specific entries
	2.2		1078	REFRIGERANT GAS, N.O.S.
	2.2		1968	INSECTICIDE GAS, N.O.S.
				General entries
	2.2		1956	COMPRESSED GAS, N.O.S.
	2.2		3163	LIQUEFIED GAS, N.O.S.
	2.2		3158	GAS, REFRIGERATED LIQUID, N.O.S.
■	2.2		3500	CHEMICAL UNDER PRESSURE, N.O.S.
	2.2	5.1	3156	COMPRESSED GAS, OXIDIZING, N.O.S.
	2.2	5.1	3157	LIQUEFIED GAS, OXIDIZING, N.O.S.
	2.2	5.1	3311	GAS, REFRIGERATED LIQUID, OXIDIZING, N.O.S.
■	2.2	6.1	3502	CHEMICAL UNDER PRESSURE, TOXIC, N.O.S.
■	2.2	8	3503	CHEMICAL UNDER PRESSURE, CORROSIVE, N.O.S.
				Class 2.3
				Specific entries
	2.3		1967	INSECTICIDE GAS, TOXIC, N.O.S.
	2.3	2.1	3355	INSECTICIDE GAS, TOXIC, FLAMMABLE, N.O.S.
				General entries
	2.3		1955	COMPRESSED GAS, TOXIC, N.O.S.
	2.3		3162	LIQUEFIED GAS, TOXIC, N.O.S.
	2.3		3169	GAS SAMPLE, NON-PRESSURIZED, TOXIC, N.O.S., not refrigerated liquid
	2.3	2.1	1953	COMPRESSED GAS, TOXIC, FLAMMABLE, N.O.S.
	2.3	2.1	3160	LIQUEFIED GAS, TOXIC, FLAMMABLE, N.O.S.
	2.3	2.1	3168	GAS SAMPLE, NON-PRESSURIZED, TOXIC, FLAMMABLE, N.O.S., not refrigerated liquid
	2.3	2.1 + 8	3305	COMPRESSED GAS, TOXIC, FLAMMABLE, CORROSIVE, N.O.S.
	2.3	2.1 + 8	3309	LIQUEFIED GAS, TOXIC, FLAMMABLE, CORROSIVE, N.O.S.
	2.3	5.1	3303	COMPRESSED GAS, TOXIC, OXIDIZING, N.O.S.
	2.3	5.1	3307	LIQUEFIED GAS, TOXIC, OXIDIZING, N.O.S.
	2.3	5.1 + 8	3306	COMPRESSED GAS, TOXIC, OXIDIZING, CORROSIVE, N.O.S.
	2.3	5.1 + 8	3310	LIQUEFIED GAS, TOXIC, OXIDIZING, CORROSIVE, N.O.S.
	2.3	8	3304	COMPRESSED GAS, TOXIC, CORROSIVE, N.O.S.
	2.3	8	3308	LIQUEFIED GAS, TOXIC, CORROSIVE, N.O.S.

Class or division	Subsidiary risk	UN Number	Proper Shipping Name
			CLASS 3
			Specific entries
3		1224	KETONES, LIQUID, N.O.S.
3		1268	PETROLEUM DISTILLATES, N.O.S. or PETROLEUM PRODUCTS, N.O.S.
3		1987	ALCOHOLS, N.O.S.
3		1989	ALDEHYDES, N.O.S.
3		2319	TERPENE HYDROCARBONS, N.O.S.
3		3271	ETHERS, N.O.S.
3		3272	ESTERS, N.O.S.
3		3295	HYDROCARBONS, LIQUID, N.O.S.
3		3336	MERCAPTANS, LIQUID, FLAMMABLE, N.O.S. or MERCAPTAN MIXTURE, LIQUID, FLAMMABLE, N.O.S.
3		3343	NITROGLYCERIN MIXTURE, DESENSITIZED, LIQUID, FLAMMABLE, N.O.S. with not more than 30% nitroglycerin, by mass
3		3357	NITROGLYCERIN MIXTURE, DESENSITIZED, LIQUID, N.O.S. with not more than 30% nitroglycerin, by mass
3		3379	DESENSITIZED EXPLOSIVE, LIQUID, N.O.S.
3	6.1	1228	MERCAPTANS, LIQUID, FLAMMABLE, TOXIC, N.O.S. or MERCAPTAN MIXTURE, LIQUID, FLAMMABLE, TOXIC, N.O.S.
3	6.1	1986	ALCOHOLS, FLAMMABLE, TOXIC, N.O.S.
3	6.1	1988	ALDEHYDES, FLAMMABLE, TOXIC, N.O.S.
3	6.1	2478	ISOCYANATES, FLAMMABLE, TOXIC, N.O.S. or ISOCYANATE SOLUTION, FLAMMABLE, TOXIC, N.O.S.
3	6.1	3248	MEDICINE, LIQUID, FLAMMABLE, TOXIC, N.O.S.
3	6.1	3273	NITRILES, FLAMMABLE, TOXIC, N.O.S.
3	8	2733	AMINES, FLAMMABLE, CORROSIVE, N.O.S. or POLYAMINES, FLAMMABLE, CORROSIVE, N.O.S.
3	8	2985	CHLOROSILANES, FLAMMABLE, CORROSIVE, N.O.S.
3	8	3274	ALCOHOLATES SOLUTION, N.O.S. in alcohol
			Pesticides
3	6.1	2758	CARBAMATE PESTICIDE, LIQUID, FLAMMABLE, TOXIC flashpoint < 23°C
3	6.1	2760	ARSENICAL PESTICIDE, LIQUID, FLAMMABLE, TOXIC flashpoint < 23°C
3	6.1	2762	ORGANOCHLORINE PESTICIDE, LIQUID, FLAMMABLE, TOXIC flashpoint < 23°C
3	6.1	2764	TRIAZINE PESTICIDE, LIQUID, FLAMMABLE, TOXIC flashpoint < 23°C
3	6.1	2772	THIOCARBAMATE PESTICIDE, LIQUID, FLAMMABLE, TOXIC flashpoint < 23°C
3	6.1	2776	COPPER BASED PESTICIDE, LIQUID, FLAMMABLE, TOXIC flashpoint < 23°C
3	6.1	2778	MERCURY BASED PESTICIDE, LIQUID, FLAMMABLE, TOXIC flashpoint < 23°C
3	6.1	2780	SUBSTITUTED NITROPHENOL PESTICIDE, LIQUID, FLAMMABLE, TOXIC flashpoint < 23°C
3	6.1	2782	BIPYRIDILIUM PESTICIDE, LIQUID, FLAMMABLE, TOXIC flashpoint < 23°C
3	6.1	2784	ORGANOPHOSPHORUS PESTICIDE, LIQUID, FLAMMABLE, TOXIC flashpoint < 23°C
3	6.1	2787	ORGANOTIN PESTICIDE, LIQUID, FLAMMABLE, TOXIC flashpoint < 23°C
3	6.1	3021	PESTICIDE, LIQUID, FLAMMABLE, TOXIC, N.O.S. flashpoint < 23°C
3	6.1	3024	COUMARIN DERIVATIVE PESTICIDE, LIQUID, FLAMMABLE, TOXIC flashpoint < 23°C
3	6.1	3346	PHENOXYACETIC ACID DERIVATIVE PESTICIDE, LIQUID, FLAMMABLE, TOXIC flashpoint < 23°C
3	6.1	3350	PYRETHROID PESTICIDE, LIQUID, FLAMMABLE, TOXIC flashpoint < 23°C
			General entries
3		1993	FLAMMABLE LIQUID, N.O.S.
3		3256	ELEVATED TEMPERATURE LIQUID, FLAMMABLE, N.O.S. with flashpoint above 60°C, at or above its flashpoint
3	6.1	1992	FLAMMABLE LIQUID, TOXIC, N.O.S.
3	6.1 + 8	3286	FLAMMABLE LIQUID, TOXIC, CORROSIVE, N.O.S.
3	8	2924	FLAMMABLE LIQUID, CORROSIVE, N.O.S.

APPENDIX A

Class or division	Subsidiary risk	UN Number	Proper Shipping Name
			CLASS 4
			Class 4.1
			Specific entries
4.1		1353	FIBRES or FABRICS IMPREGNATED WITH WEAKLY NITRATED NITROCELLULOSE, N.O.S.
4.1		3089	METAL POWDER, FLAMMABLE, N.O.S.
4.1		3182	METAL HYDRIDES, FLAMMABLE, N.O.S.
4.1		3221	SELF-REACTIVE LIQUID TYPE B
4.1		3222	SELF-REACTIVE SOLID TYPE B
4.1		3223	SELF-REACTIVE LIQUID TYPE C
4.1		3224	SELF-REACTIVE SOLID TYPE C
4.1		3225	SELF-REACTIVE LIQUID TYPE D
4.1		3226	SELF-REACTIVE SOLID TYPE D
4.1		3227	SELF-REACTIVE LIQUID TYPE E
4.1		3228	SELF-REACTIVE SOLID TYPE E
4.1		3229	SELF-REACTIVE LIQUID TYPE F
4.1		3230	SELF-REACTIVE SOLID TYPE F
4.1		3231	SELF-REACTIVE LIQUID TYPE B, TEMPERATURE CONTROLLED
4.1		3232	SELF-REACTIVE SOLID TYPE B, TEMPERATURE CONTROLLED
4.1		3233	SELF-REACTIVE LIQUID TYPE C, TEMPERATURE CONTROLLED
4.1		3234	SELF-REACTIVE SOLID TYPE C, TEMPERATURE CONTROLLED
4.1		3235	SELF-REACTIVE LIQUID TYPE D, TEMPERATURE CONTROLLED
4.1		3236	SELF-REACTIVE SOLID TYPE D, TEMPERATURE CONTROLLED
4.1		3237	SELF-REACTIVE LIQUID TYPE E, TEMPERATURE CONTROLLED
4.1		3238	SELF-REACTIVE SOLID TYPE E, TEMPERATURE CONTROLLED
4.1		3239	SELF-REACTIVE LIQUID TYPE F, TEMPERATURE CONTROLLED
4.1		3240	SELF-REACTIVE SOLID TYPE F, TEMPERATURE CONTROLLED
4.1		3319	NITROGLYCERIN MIXTURE, DESENSITIZED, SOLID, N.O.S. with more than 2% but not more than 10% nitroglycerin, by mass
4.1		3344	PENTAERYTHRITE TETRANITRATE (PENTAERYTHRITOL TETRANITRATE; PETN) MIXTURE, DESENSITIZED, SOLID, N.O.S. with more than 10% but not more than 20% PETN, by mass
4.1		3380	DESENSITIZED EXPLOSIVE, SOLID, N.O.S.
			General entries
4.1		1325	FLAMMABLE SOLID, ORGANIC, N.O.S.
4.1		3175	SOLIDS CONTAINING FLAMMABLE LIQUID, N.O.S.
4.1		3176	FLAMMABLE SOLID, ORGANIC, MOLTEN, N.O.S.
4.1		3178	FLAMMABLE SOLID, INORGANIC, N.O.S.
4.1		3181	METAL SALTS OF ORGANIC COMPOUNDS, FLAMMABLE, N.O.S.
4.1	5.1	3097	FLAMMABLE SOLID, OXIDIZING, N.O.S.
4.1	6.1	2926	FLAMMABLE SOLID, TOXIC, ORGANIC, N.O.S.
4.1	6.1	3179	FLAMMABLE SOLID, TOXIC, INORGANIC, N.O.S.
4.1	8	2925	FLAMMABLE SOLID, CORROSIVE, ORGANIC, N.O.S.
4.1	8	3180	FLAMMABLE SOLID, CORROSIVE, INORGANIC, N.O.S.
			Class 4.2
			Specific entries
4.2		1373	FIBRES or FABRICS, ANIMAL or VEGETABLE or SYNTHETIC, N.O.S. with oil
4.2		1378	METAL CATALYST, WETTED with a visible excess of liquid
4.2		1383	PYROPHORIC METAL, N.O.S. or PYROPHORIC ALLOY, N.O.S.
4.2		2006	PLASTICS, NITROCELLULOSE-BASED, SELF-HEATING, N.O.S.
4.2		2881	METAL CATALYST, DRY
4.2		3189	METAL POWDER, SELF-HEATING, N.O.S.
4.2		3205	ALKALINE EARTH METAL ALCOHOLATES, N.O.S.
4.2		3313	ORGANIC PIGMENTS, SELF-HEATING
4.2		3342	XANTHATES
4.2		3391	ORGANOMETALLIC SUBSTANCE, SOLID, PYROPHORIC
4.2		3392	ORGANOMETALLIC SUBSTANCE, LIQUID, PYROPHORIC

Class or division	Subsidiary risk	UN Number	Proper Shipping Name
			CLASS 4 *(continued)*
			Class 4.2 *(continued)*
			Specific entries *(continued)*
4.2		3400	ORGANOMETALLIC SUBSTANCE, SOLID, SELF-HEATING
4.2	4.3	3393	ORGANOMETALLIC SUBSTANCE, SOLID, PYROPHORIC, WATER-REACTIVE
4.2	4.3	3394	ORGANOMETALLIC SUBSTANCE, LIQUID, PYROPHORIC, WATER-REACTIVE
4.2	8	3206	ALKALI METAL ALCOHOLATES, SELF-HEATING, CORROSIVE, N.O.S.
			General entries
4.2		2845	PYROPHORIC LIQUID, ORGANIC, N.O.S.
4.2		2846	PYROPHORIC SOLID, ORGANIC, N.O.S.
4.2		3088	SELF-HEATING SOLID, ORGANIC, N.O.S.
4.2		3183	SELF-HEATING LIQUID, ORGANIC, N.O.S.
4.2		3186	SELF-HEATING LIQUID, INORGANIC, N.O.S.
4.2		3190	SELF-HEATING SOLID, INORGANIC, N.O.S.
4.2		3194	PYROPHORIC LIQUID, INORGANIC, N.O.S.
4.2		3200	PYROPHORIC SOLID, INORGANIC, N.O.S.
4.2	5.1	3127	SELF-HEATING SOLID, OXIDIZING, N.O.S.
4.2	6.1	3128	SELF-HEATING SOLID, TOXIC, ORGANIC, N.O.S.
4.2	6.1	3184	SELF-HEATING LIQUID, TOXIC, ORGANIC, N.O.S.
4.2	6.1	3187	SELF-HEATING LIQUID, TOXIC, INORGANIC, N.O.S.
4.2	6.1	3191	SELF-HEATING SOLID, TOXIC, INORGANIC, N.O.S.
4.2	8	3126	SELF-HEATING SOLID, CORROSIVE, ORGANIC, N.O.S.
4.2	8	3185	SELF-HEATING LIQUID, CORROSIVE, ORGANIC, N.O.S.
4.2	8	3188	SELF-HEATING LIQUID, CORROSIVE, INORGANIC, N.O.S.
4.2	8	3192	SELF-HEATING SOLID, CORROSIVE, INORGANIC, N.O.S.
			Class 4.3
			Specific entries
4.3		1389	ALKALI METAL AMALGAM, LIQUID
4.3		1390	ALKALI METAL AMIDES
4.3		1391	ALKALI METAL DISPERSION or ALKALINE EARTH METAL DISPERSION
4.3		1392	ALKALINE EARTH METAL AMALGAM, LIQUID
4.3		1393	ALKALINE EARTH METAL ALLOY, N.O.S.
4.3		1409	METAL HYDRIDES, WATER-REACTIVE, N.O.S.
4.3		1421	ALKALI METAL ALLOY, LIQUID, N.O.S.
4.3		3208	METALLIC SUBSTANCE, WATER-REACTIVE, N.O.S.
4.3		3395	ORGANOMETALLIC SUBSTANCE, SOLID, WATER-REACTIVE
4.3		3398	ORGANOMETALLIC SUBSTANCE, LIQUID, WATER-REACTIVE
4.3		3401	ALKALI METAL AMALGAM, SOLID
4.3		3402	ALKALINE EARTH METAL AMALGAM, SOLID
4.3	3	3399	ORGANOMETALLIC SUBSTANCE, LIQUID, WATER-REACTIVE, FLAMMABLE
4.3	3	3482	ALKALI METAL DISPERSION, FLAMMABLE or ALKALINE EARTH METAL DISPERSION, FLAMMABLE
4.3	3 + 8	2988	CHLOROSILANES, WATER-REACTIVE, FLAMMABLE, CORROSIVE, N.O.S.
4.3	4.1	3396	ORGANOMETALLIC SUBSTANCE, SOLID, WATER-REACTIVE, FLAMMABLE
4.3	4.2	3209	METALLIC SUBSTANCE, WATER-REACTIVE, SELF-HEATING, N.O.S.
4.3	4.2	3397	ORGANOMETALLIC SUBSTANCE, SOLID, WATER-REACTIVE, SELF-HEATING
			General entries
4.3		3148	WATER-REACTIVE LIQUID, N.O.S.
4.3		2813	WATER-REACTIVE SOLID, N.O.S.
4.3	4.1	3132	WATER-REACTIVE SOLID, FLAMMABLE, N.O.S.
4.3	4.2	3135	WATER-REACTIVE SOLID, SELF-HEATING, N.O.S.
4.3	5.1	3133	WATER-REACTIVE SOLID, OXIDIZING, N.O.S.
4.3	6.1	3130	WATER-REACTIVE LIQUID, TOXIC, N.O.S.
4.3	6.1	3134	WATER-REACTIVE SOLID, TOXIC, N.O.S.
4.3	8	3129	WATER-REACTIVE LIQUID, CORROSIVE, N.O.S.
4.3	8	3131	WATER-REACTIVE SOLID, CORROSIVE, N.O.S.

Class or division	Subsidiary risk	UN Number	Proper Shipping Name
			CLASS 5
			Class 5.1
			Specific entries
5.1		1450	BROMATES, INORGANIC, N.O.S.
5.1		1461	CHLORATES, INORGANIC, N.O.S.
5.1		1462	CHLORITES, INORGANIC, N.O.S.
5.1		1477	NITRATES, INORGANIC, N.O.S.
5.1		1481	PERCHLORATES, INORGANIC, N.O.S.
5.1		1482	PERMANGANATES, INORGANIC, N.O.S.
5.1		1483	PEROXIDES, INORGANIC, N.O.S.
5.1		2627	NITRITES, INORGANIC, N.O.S.
5.1		3210	CHLORATES, INORGANIC, AQUEOUS SOLUTION, N.O.S.
5.1		3211	PERCHLORATES, INORGANIC, AQUEOUS SOLUTION, N.O.S.
5.1		3212	HYPOCHLORITES, INORGANIC, N.O.S.
5.1		3213	BROMATES, INORGANIC, AQUEOUS SOLUTION, N.O.S.
5.1		3214	PERMANGANATES, INORGANIC, AQUEOUS SOLUTION, N.O.S.
5.1		3215	PERSULPHATES, INORGANIC, N.O.S.
5.1		3216	PERSULPHATES, INORGANIC, AQUEOUS SOLUTION, N.O.S.
5.1		3218	NITRATES, INORGANIC, AQUEOUS SOLUTION, N.O.S.
5.1		3219	NITRITES, INORGANIC, AQUEOUS SOLUTION, N.O.S.
			General entries
5.1		1479	OXIDIZING SOLID, N.O.S.
5.1		3139	OXIDIZING LIQUID, N.O.S.
5.1	4.1	3137	OXIDIZING SOLID, FLAMMABLE, N.O.S.
5.1	4.2	3100	OXIDIZING SOLID, SELF-HEATING, N.O.S.
5.1	4.3	3121	OXIDIZING SOLID, WATER-REACTIVE, N.O.S.
5.1	6.1	3087	OXIDIZING SOLID, TOXIC, N.O.S.
5.1	6.1	3099	OXIDIZING LIQUID, TOXIC, N.O.S.
5.1	8	3085	OXIDIZING SOLID, CORROSIVE, N.O.S.
5.1	8	3098	OXIDIZING LIQUID, CORROSIVE, N.O.S.
			Class 5.2
			Specific entries
5.2		3101	ORGANIC PEROXIDE TYPE B, LIQUID
5.2		3102	ORGANIC PEROXIDE TYPE B, SOLID
5.2		3103	ORGANIC PEROXIDE TYPE C, LIQUID
5.2		3104	ORGANIC PEROXIDE TYPE C, SOLID
5.2		3105	ORGANIC PEROXIDE TYPE D, LIQUID
5.2		3106	ORGANIC PEROXIDE TYPE D, SOLID
5.2		3107	ORGANIC PEROXIDE TYPE E, LIQUID
5.2		3108	ORGANIC PEROXIDE TYPE E, SOLID
5.2		3109	ORGANIC PEROXIDE TYPE F, LIQUID
5.2		3110	ORGANIC PEROXIDE TYPE F, SOLID
5.2		3111	ORGANIC PEROXIDE TYPE B, LIQUID, TEMPERATURE CONTROLLED
5.2		3112	ORGANIC PEROXIDE TYPE B, SOLID, TEMPERATURE CONTROLLED
5.2		3113	ORGANIC PEROXIDE TYPE C, LIQUID, TEMPERATURE CONTROLLED
5.2		3114	ORGANIC PEROXIDE TYPE C, SOLID, TEMPERATURE CONTROLLED
5.2		3115	ORGANIC PEROXIDE TYPE D, LIQUID, TEMPERATURE CONTROLLED
5.2		3116	ORGANIC PEROXIDE TYPE D, SOLID, TEMPERATURE CONTROLLED
5.2		3117	ORGANIC PEROXIDE TYPE E, LIQUID, TEMPERATURE CONTROLLED
5.2		3118	ORGANIC PEROXIDE TYPE E, SOLID, TEMPERATURE CONTROLLED
5.2		3119	ORGANIC PEROXIDE TYPE F, LIQUID, TEMPERATURE CONTROLLED
5.2		3120	ORGANIC PEROXIDE TYPE F, SOLID, TEMPERATURE CONTROLLED

Class or division	Subsidiary risk	UN Number	Proper Shipping Name
			CLASS 6
			Class 6.1
			Specific entries
6.1		1544	ALKALOIDS, SOLID, N.O.S. or ALKALOID SALTS, SOLID, N.O.S.
6.1		1549	ANTIMONY COMPOUND, INORGANIC, SOLID, N.O.S.
6.1		1556	ARSENIC COMPOUND, LIQUID, N.O.S. inorganic, including: Arsenates, n.o.s., Arsenites, n.o.s., and Arsenic sulphides, n.o.s.
6.1		1557	ARSENIC COMPOUND, SOLID, N.O.S. inorganic, including: Arsenates, n.o.s., Arsenites, n.o.s., and Arsenic sulphides, n.o.s.
6.1		1564	BARIUM COMPOUND, N.O.S.
6.1		1566	BERYLLIUM COMPOUND, N.O.S.
6.1		1583	CHLOROPICRIN MIXTURE, N.O.S.
6.1		1588	CYANIDES, INORGANIC, SOLID, N.O.S.
6.1		1601	DISINFECTANT, SOLID, TOXIC, N.O.S.
6.1		1602	DYE, LIQUID, TOXIC, N.O.S. or DYE INTERMEDIATE, LIQUID, TOXIC, N.O.S.
6.1		1655	NICOTINE COMPOUND, SOLID, N.O.S. or NICOTINE PREPARATION, SOLID, N.O.S.
6.1		1693	TEAR GAS SUBSTANCE, LIQUID, N.O.S.
6.1		1707	THALLIUM COMPOUND, N.O.S.
6.1		1851	MEDICINE, LIQUID, TOXIC, N.O.S.
6.1		1935	CYANIDE SOLUTION, N.O.S.
6.1		2024	MERCURY COMPOUND, LIQUID, N.O.S.
6.1		2025	MERCURY COMPOUND, SOLID, N.O.S.
6.1		2026	PHENYLMERCURIC COMPOUND, N.O.S.
6.1		2206	ISOCYANATES, TOXIC, N.O.S. or ISOCYANATE SOLUTION, TOXIC, N.O.S.
6.1		2291	LEAD COMPOUND, SOLUBLE, N.O.S.
6.1		2570	CADMIUM COMPOUND
6.1		2788	ORGANOTIN COMPOUND, LIQUID, N.O.S.
6.1		2856	FLUOROSILICATES, N.O.S.
6.1		3140	ALKALOIDS, LIQUID, N.O.S or ALKALOIDS SALTS, LIQUID, N.O.S.
6.1		3141	ANTIMONY COMPOUND, INORGANIC, LIQUID, N.O.S.
6.1		3142	DISINFECTANT, LIQUID, TOXIC, N.O.S.
6.1		3143	DYE, SOLID, TOXIC, N.O.S. or DYE INTERMEDIATE, SOLID, TOXIC, N.O.S.
6.1		3144	NICOTINE COMPOUND, LIQUID, N.O.S. or NICOTINE PREPARATION, LIQUID, N.O.S.
6.1		3146	ORGANOTIN COMPOUND, SOLID, N.O.S.
6.1		3249	MEDICINE, SOLID, TOXIC, N.O.S.
6.1		3276	NITRILES, TOXIC, LIQUID, N.O.S.
6.1		3278	ORGANOPHOSPHORUS COMPOUND, TOXIC, LIQUID, N.O.S.
6.1		3280	ORGANOARSENIC COMPOUND, LIQUID, N.O.S.
6.1		3281	METAL CARBONYLS, LIQUID, N.O.S. with LC_{50}
6.1		3282	ORGANOMETALLIC COMPOUND, TOXIC, LIQUID, N.O.S. with LC_{50}
6.1		3283	SELENIUM COMPOUND, SOLID, N.O.S. with LC_{50}
6.1		3284	TELLURIUM COMPOUND, N.O.S. with LC_{50}
6.1		3285	VANADIUM COMPOUND, N.O.S.
6.1		3439	NITRILES, TOXIC, SOLID, N.O.S.
6.1		3440	SELENIUM COMPOUND, LIQUID, N.O.S.
6.1		3448	TEAR GAS SUBSTANCE, SOLID, N.O.S.
6.1		3462	TOXINS EXTRACTED FROM LIVING SOURCES, SOLID, N.O.S.
6.1		3464	ORGANOPHOSPHORUS COMPOUND, TOXIC, SOLID, N.O.S.
6.1		3465	ORGANOARSENIC COMPOUND, SOLID, N.O.S.
6.1		3466	METAL CARBONYLS, SOLID, N.O.S.
6.1		3467	ORGANOMETALLIC COMPOUND, TOXIC, SOLID, N.O.S.
6.1	3	3071	MERCAPTANS, LIQUID, TOXIC, FLAMMABLE, N.O.S. or MERCAPTAN MIXTURE, LIQUID, TOXIC, FLAMMABLE, N.O.S.
6.1	3	3080	ISOCYANATES, TOXIC, FLAMMABLE, N.O.S or ISOCYANATE SOLUTION, TOXIC, FLAMMABLE, N.O.S.
6.1	3	3275	NITRILES, TOXIC, FLAMMABLE, N.O.S.
6.1	3	3279	ORGANOPHOSPHORUS COMPOUND, TOXIC, FLAMMABLE, N.O.S.
6.1	3 + 8	2742	CHLOROFORMATES, TOXIC, CORROSIVE, FLAMMABLE, N.O.S.

APPENDIX A

Class or division	Subsidiary risk	UN Number	Proper Shipping Name
			CLASS 6 *(continued)*
			Class 6.1 *(continued)*
			Specific entries *(continued)*
6.1	3 + 8	3362	CHLOROSILANES, TOXIC, CORROSIVE, FLAMMABLE, N.O.S.
6.1	8	3277	CHLOROFORMATES, TOXIC, CORROSIVE, N.O.S.
6.1	8	3361	CHLOROSILANES, TOXIC, CORROSIVE, N.O.S.
			Pesticides
			(a) Solid
6.1		2588	PESTICIDE, SOLID, TOXIC, N.O.S.
6.1		2757	CARBAMATE PESTICIDE, SOLID, TOXIC
6.1		2759	ARSENICAL PESTICIDE, SOLID, TOXIC
6.1		2761	ORGANOCHLORINE PESTICIDE, SOLID, TOXIC
6.1		2763	TRIAZINE PESTICIDE, SOLID, TOXIC
6.1		2771	THIOCARBAMATE PESTICIDE, SOLID, TOXIC
6.1		2775	COPPER BASED PESTICIDE, SOLID, TOXIC
6.1		2777	MERCURY BASED PESTICIDE, SOLID, TOXIC
6.1		2779	SUBSTITUTED NITROPHENOL PESTICIDE, SOLID, TOXIC
6.1		2781	BIPYRIDILIUM PESTICIDE, SOLID, TOXIC
6.1		2783	ORGANOPHOSPHORUS PESTICIDE, SOLID, TOXIC
6.1		2786	ORGANOTIN PESTICIDE, SOLID, TOXIC
6.1		3027	COUMARIN DERIVATIVE PESTICIDE, SOLID, TOXIC
6.1		3345	PHENOXYACETIC ACID DERIVATIVE PESTICIDE, SOLID, TOXIC
6.1		3349	PYRETHROID PESTICIDE, SOLID, TOXIC
			(b) Liquid
6.1		2902	PESTICIDE, LIQUID, TOXIC, N.O.S.
6.1		2992	CARBAMATE PESTICIDE, LIQUID, TOXIC
6.1		2994	ARSENICAL PESTICIDE, LIQUID, TOXIC
6.1		2996	ORGANOCHLORINE PESTICIDE, LIQUID, TOXIC
6.1		2998	TRIAZINE PESTICIDE, LIQUID, TOXIC
6.1		3006	THIOCARBAMATE PESTICIDE, LIQUID, TOXIC
6.1		3010	COPPER BASED PESTICIDE, LIQUID, TOXIC
6.1		3012	MERCURY BASED PESTICIDE, LIQUID, TOXIC
6.1		3014	SUBSTITUTED NITROPHENOL PESTICIDE, LIQUID, TOXIC
6.1		3016	BIPYRIDILIUM PESTICIDE, LIQUID, TOXIC
6.1		3018	ORGANOPHOSPHORUS PESTICIDE, LIQUID, TOXIC
6.1		3020	ORGANOTIN PESTICIDE, LIQUID, TOXIC
6.1		3026	COUMARIN DERIVATIVE PESTICIDE, LIQUID, TOXIC
6.1		3348	PHENOXYACETIC ACID DERIVATIVE PESTICIDE, LIQUID, TOXIC
6.1		3352	PYRETHROID PESTICIDE, LIQUID, TOXIC
6.1	3	2903	PESTICIDE, LIQUID, TOXIC, FLAMMABLE, N.O.S., flashpoint $\geq$ 23°C
6.1	3	2991	CARBAMATE PESTICIDE, LIQUID, TOXIC, FLAMMABLE, flashpoint $\geq$ 23°C
6.1	3	2993	ARSENICAL PESTICIDE, LIQUID, TOXIC, FLAMMABLE flashpoint $\geq$ 23°C
6.1	3	2995	ORGANOCHLORINE PESTICIDE, LIQUID, TOXIC, FLAMMABLE flashpoint $\geq$ 23°C
6.1	3	2997	TRIAZINE PESTICIDE, LIQUID, TOXIC, FLAMMABLE, flashpoint $\geq$ 23°C
6.1	3	3005	THIOCARBAMATE PESTICIDE, LIQUID, TOXIC, FLAMMABLE flashpoint $\geq$ 23°C
6.1	3	3009	COPPER BASED PESTICIDE, LIQUID, TOXIC, FLAMMABLE flashpoint $\geq$ 23°C
6.1	3	3011	MERCURY BASED PESTICIDE, LIQUID, TOXIC, FLAMMABLE flashpoint $\geq$ 23°C
6.1	3	3013	SUBSTITUTED NITROPHENOL PESTICIDE, LIQUID, TOXIC, FLAMMABLE flashpoint $\geq$ 23°C
6.1	3	3015	BIPYRIDILIUM PESTICIDE, LIQUID, TOXIC, FLAMMABLE flashpoint $\geq$ 23°C
6.1	3	3017	ORGANOPHOSPHORUS PESTICIDE, LIQUID, TOXIC, FLAMMABLE flashpoint $\geq$ 23°C
6.1	3	3019	ORGANOTIN PESTICIDE, LIQUID, TOXIC, FLAMMABLE flashpoint $\geq$ 23°C
6.1	3	3025	COUMARIN DERIVATIVE PESTICIDE, LIQUID, TOXIC, FLAMMABLE flashpoint $\geq$ 23°C
6.1	3	3347	PHENOXYACETIC ACID DERIVATIVE PESTICIDE, LIQUID, TOXIC, FLAMMABLE flashpoint $\geq$ 23°C
6.1	3	3351	PYRETHROID PESTICIDE, LIQUID, TOXIC, FLAMMABLE flashpoint $\geq$ 23°C

APPENDIX A

	Class or division	Subsidiary risk	UN Number	Proper Shipping Name
				CLASS 6 *(continued)*
				Class 6.1 *(continued)*
				General entries
	6.1		2810	TOXIC LIQUID, ORGANIC, N.O.S.
	6.1		2811	TOXIC SOLID, ORGANIC, N.O.S.
	6.1		3172	TOXINS, EXTRACTED FROM LIVING SOURCES, LIQUID, N.O.S.
	6.1		3243	SOLIDS CONTAINING TOXIC LIQUID, N.O.S.
	6.1		3287	TOXIC LIQUID, INORGANIC, N.O.S.
	6.1		3288	TOXIC SOLID, INORGANIC, N.O.S.
	6.1		3315	CHEMICAL SAMPLE, TOXIC
△	6.1		3381	TOXIC BY INHALATION LIQUID, N.O.S. with an LC_{50} lower than or equal to 200 mL/m^3 and saturated vapour concentration greater than or equal to 500 LC_{50}
△	6.1		3382	TOXIC BY INHALATION LIQUID, N.O.S. with an LC_{50} lower than or equal to 1000 mL/m^3 and saturated vapour concentration greater than or equal to 10 LC_{50}
■	6.1		3462	TOXINS, EXTRACTED FROM LIVING SOURCES, SOLID, N.O.S.
	6.1	3	2929	TOXIC LIQUID, FLAMMABLE, ORGANIC, N.O.S.
△	6.1	3	3383	TOXIC BY INHALATION LIQUID, FLAMMABLE, N.O.S. with an LC_{50} lower than or equal to 200 mL/m^3 and saturated vapour concentration greater than or equal to 500 LC_{50}
△	6.1	3	3384	TOXIC BY INHALATION LIQUID, FLAMMABLE, N.O.S. with an LC_{50} lower than or equal to 1000 mL/m^3 and saturated vapour concentration greater than or equal to 10 LC_{50}
△	6.1	3 + 8	3488	TOXIC BY INHALATION LIQUID, FLAMMABLE, CORROSIVE, N.O.S. with an LC_{50} lower than or equal to 200 mL/m^3 and saturated vapour concentration greater than or equal to 500 LC_{50}
△	6.1	3 + 8	3489	TOXIC BY INHALATION LIQUID, FLAMMABLE, CORROSIVE, N.O.S. with an LC_{50} lower than or equal to 1000 mL/m^3 and saturated vapour concentration greater than or equal to 10 LC_{50}
	6.1	4.1	2930	TOXIC SOLID, FLAMMABLE, ORGANIC, N.O.S.
	6.1	4.2	3124	TOXIC SOLID, SELF-HEATING, N.O.S.
	6.1	4.3	3123	TOXIC LIQUID, WATER-REACTIVE, N.O.S.
	6.1	4.3	3125	TOXIC SOLID, WATER-REACTIVE, N.O.S.
△	6.1	4.3	3385	TOXIC BY INHALATION LIQUID, WATER-REACTIVE, N.O.S. with an LC_{50} lower than or equal to 200 mL/m^3 and saturated vapour concentration greater than or equal to 500 LC_{50}
△	6.1	4.3	3386	TOXIC BY INHALATION LIQUID, WATER-REACTIVE, N.O.S. with an LC_{50} lower than or equal to 1000 mL/m^3 and saturated vapour concentration greater than or equal to 10 LC_{50}
△	6.1	4.3 + 3	3490	TOXIC BY INHALATION LIQUID, WATER-REACTIVE, FLAMMABLE, N.O.S. with an LC_{50} lower than or equal to 200 mL/m^3 and saturated vapour concentration greater than or equal to 500 LC_{50}
△	6.1	4.3 + 3	3491	TOXIC BY INHALATION LIQUID, WATER-REACTIVE, FLAMMABLE, N.O.S. with an LC_{50} lower than or equal to 1000 mL/m^3 and saturated vapour concentration greater than or equal to 10 LC_{50}
	6.1	5.1	3122	TOXIC LIQUID, OXIDIZING, N.O.S.
	6.1	5.1	3086	TOXIC SOLID, OXIDIZING, N.O.S.
△	6.1	5.1	3387	TOXIC BY INHALATION LIQUID, OXIDIZING, N.O.S. with an LC_{50} lower than or equal to 200 mL/m^3 and saturated vapour concentration greater than or equal to 500 LC_{50}
△	6.1	5.1	3388	TOXIC BY INHALATION LIQUID, OXIDIZING, N.O.S. with an LC_{50} lower than or equal to 1000 mL/m^3 and saturated vapour concentration greater than or equal to 10 LC_{50}
	6.1	8	2927	TOXIC LIQUID, CORROSIVE, ORGANIC, N.O.S.
	6.1	8	2928	TOXIC SOLID, CORROSIVE, ORGANIC, N.O.S.
	6.1	8	3289	TOXIC LIQUID, CORROSIVE, INORGANIC, N.O.S.
	6.1	8	3290	TOXIC SOLID, CORROSIVE, INORGANIC, N.O.S.
△	6.1	8	3389	TOXIC BY INHALATION LIQUID, CORROSIVE, N.O.S. with an LC_{50} lower than or equal to 200 mL/m^3 and saturated vapour concentration greater than or equal to 500 LC_{50}
△	6.1	8	3390	TOXIC BY INHALATION LIQUID, CORROSIVE, N.O.S. with an LC_{50} lower than or equal to 1000 mL/m^3 and saturated vapour concentration greater than or equal to 10 LC_{50}
⊗				

Class or division	Subsidiary risk	UN Number	Proper Shipping Name
			CLASS 6 *(continued)*
			Class 6.2
			Specific entries
6.2		3291	CLINICAL WASTE, UNSPECIFIED, N.O.S. or (BIO)MEDICAL WASTE, N.O.S. or REGULATED MEDICAL WASTE, N.O.S.
6.2		3373	BIOLOGICAL SUBSTANCE, CATEGORY B
			General entries
6.2		2814	INFECTIOUS SUBSTANCE, AFFECTING HUMANS
6.2		2900	INFECTIOUS SUBSTANCE, AFFECTING ANIMALS only

Class or division	Subsidiary risk	UN Number	Proper Shipping Name
			CLASS 7
			General entries
7		2908	RADIOACTIVE MATERIAL, EXCEPTED PACKAGE – EMPTY PACKAGING
7		2909	RADIOACTIVE MATERIAL, EXCEPTED PACKAGE – ARTICLES MANUFACTURED FROM NATURAL URANIUM or DEPLETED URANIUM or NATURAL THORIUM
7		2910	RADIOACTIVE MATERIAL, EXCEPTED PACKAGE – LIMITED QUANTITY OF MATERIAL
7		2911	RADIOACTIVE MATERIAL, EXCEPTED PACKAGE – INSTRUMENTS or ARTICLES
7		2912	RADIOACTIVE MATERIAL, LOW SPECIFIC ACTIVITY (LSA-I) non fissile or fissile – excepted
7		2913	RADIOACTIVE MATERIAL, SURFACE CONTAMINATED OBJECTS (SCO-I or SCO-II) non fissile or fissile – excepted
7		2915	RADIOACTIVE MATERIAL, TYPE A PACKAGE, non-special form, non fissile or fissile – excepted
7		2916	RADIOACTIVE MATERIAL, TYPE B(U) PACKAGE non fissile or fissile – excepted
7		2917	RADIOACTIVE MATERIAL, TYPE B(M) PACKAGE non fissile or fissile – excepted
7		2919	RADIOACTIVE MATERIAL, TRANSPORTED UNDER SPECIAL ARRANGEMENT non fissile or fissile – excepted
7		3321	RADIOACTIVE MATERIAL, LOW SPECIFIC ACTIVITY (LSA-II), non fissile or fissile – excepted
7		3322	RADIOACTIVE MATERIAL, LOW SPECIFIC ACTIVITY (LSA-III), non fissile or fissile – excepted
7		3323	RADIOACTIVE MATERIAL, TYPE C PACKAGE non fissile or fissile – excepted
7		3324	RADIOACTIVE MATERIAL, LOW SPECIFIC ACTIVITY (LSA-II), FISSILE
7		3325	RADIOACTIVE MATERIAL, LOW SPECIFIC ACTIVITY (LSA-III), FISSILE
7		3326	RADIOACTIVE MATERIAL, SURFACE CONTAMINATED OBJECTS (SCO-I or SCO-II), FISSILE
7		3327	RADIOACTIVE MATERIAL, TYPE A PACKAGE, FISSILE non-special form
7		3328	RADIOACTIVE MATERIAL, TYPE B(U) PACKAGE, FISSILE
7		3329	RADIOACTIVE MATERIAL, TYPE B(M) PACKAGE, FISSILE
7		3330	RADIOACTIVE MATERIAL, TYPE C PACKAGE, FISSILE
7		3331	RADIOACTIVE MATERIAL, TRANSPORTED UNDER SPECIAL ARRANGEMENT, FISSILE
7		3332	RADIOACTIVE MATERIAL, TYPE A PACKAGE, SPECIAL FORM non fissile or fissile – excepted
7		3333	RADIOACTIVE MATERIAL, TYPE A PACKAGE, SPECIAL FORM, FISSILE

Class or division	Subsidiary risk	UN Number	Proper Shipping Name
			CLASS 8
			Specific entries
8		1719	CAUSTIC ALKALI LIQUID, N.O.S.
8		1740	HYDROGENDIFLUORIDES, SOLID, N.O.S.
8		1903	DISINFECTANT, LIQUID, CORROSIVE, N.O.S.
8		2430	ALKYLPHENOLS, SOLID, N.O.S. (including C_2–C_{12} homologues)
8		2693	BISULPHITES, AQUEOUS SOLUTION, N.O.S.
8		2735	AMINES, LIQUID, CORROSIVE, N.O.S. or POLYAMINES, LIQUID, CORROSIVE, N.O.S.
8		2801	DYE, LIQUID, CORROSIVE, N.O.S. or DYE INTERMEDIATE, LIQUID, CORROSIVE, N.O.S.
8		2837	BISULPHATES, AQUEOUS SOLUTION
8		2987	CHLOROSILANES, CORROSIVE, N.O.S.
8		3145	ALKYLPHENOLS, LIQUID, N.O.S. (including C_2–C_{12} homologues)
8		3147	DYE, SOLID, CORROSIVE, N.O.S. or DYE INTERMEDIATE, SOLID, CORROSIVE, N.O.S.
8		3259	AMINES, SOLID, CORROSIVE, N.O.S. or POLYAMINES, SOLID, CORROSIVE, N.O.S.
8	3	2734	AMINES, LIQUID, CORROSIVE, FLAMMABLE, N.O.S. or POLYAMINES, LIQUID, CORROSIVE, FLAMMABLE, N.O.S.
8	3	2986	CHLOROSILANES, CORROSIVE, FLAMMABLE, N.O.S.
8	6.1	3471	HYDROGENDIFLUORIDES SOLUTION, N.O.S.
			General entries
8		1759	CORROSIVE SOLID, N.O.S.
8		1760	CORROSIVE LIQUID, N.O.S.
8		3244	SOLIDS CONTAINING CORROSIVE LIQUID, N.O.S.
8		3260	CORROSIVE SOLID, ACIDIC, INORGANIC, N.O.S.
8		3261	CORROSIVE SOLID, ACIDIC, ORGANIC, N.O.S.
8		3262	CORROSIVE SOLID, BASIC, INORGANIC, N.O.S.
8		3263	CORROSIVE SOLID, BASIC, ORGANIC, N.O.S.
8		3264	CORROSIVE LIQUID, ACIDIC, INORGANIC, N.O.S.
8		3265	CORROSIVE LIQUID, ACIDIC, ORGANIC, N.O.S.
8		3266	CORROSIVE LIQUID, BASIC, INORGANIC, N.O.S.
8		3267	CORROSIVE LIQUID, BASIC, ORGANIC, N.O.S.
8	3	2920	CORROSIVE LIQUID, FLAMMABLE, N.O.S.
8	4.1	2921	CORROSIVE SOLID, FLAMMABLE, N.O.S.
8	4.2	3095	CORROSIVE SOLID, SELF-HEATING, N.O.S.
8	4.2	3301	CORROSIVE LIQUID, SELF-HEATING, N.O.S.
8	4.3	3094	CORROSIVE LIQUID, WATER-REACTIVE, N.O.S.
8	4.3	3096	CORROSIVE SOLID, WATER-REACTIVE, N.O.S.
8	5.1	3084	CORROSIVE SOLID, OXIDIZING, N.O.S.
8	5.1	3093	CORROSIVE LIQUID, OXIDIZING, N.O.S.
8	6.1	2922	CORROSIVE LIQUID, TOXIC, N.O.S.
8	6.1	2923	CORROSIVE SOLID, TOXIC, N.O.S.

Class or division	Subsidiary risk	UN Number	Proper Shipping Name
			CLASS 9
			General entries
9		3077	ENVIRONMENTALLY HAZARDOUS SUBSTANCE, SOLID, N.O.S.
9		3082	ENVIRONMENTALLY HAZARDOUS SUBSTANCE, LIQUID, N.O.S.
9		3245	GENETICALLY MODIFIED MICROORGANISMS or GENETICALLY MODIFIED ORGANISMS
9		3257	ELEVATED TEMPERATURE LIQUID, N.O.S. at or above 100°C and below its flashpoint (including molten metals, molten salts, etc.)
9		3258	ELEVATED TEMPERATURE SOLID, N.O.S., at or above 240°C
see SP960		3334	AVIATION REGULATED LIQUID, N.O.S.
see SP960		3335	AVIATION REGULATED SOLID, N.O.S.

APPENDIX A

△ Appendix B

Glossary of terms

Note: The provisions of this appendix are not mandatory.

Caution: The explanations in this glossary are for information only and are not to be used for purposes of hazard classification.

AIR BAG INFLATORS, PYROTECHNIC or AIR BAG MODULES, PYROTECHNIC or SEAT-BELT PRETENSIONERS, PYROTECHNIC	Articles which contain pyrotechnic substances and are used as life-saving vehicle air bags or seat-belts.
Ammunition	Generic term related mainly to articles of military application consisting of all kind of bombs, grenades, rockets, mines, projectiles and other similar devices or contrivances.
AMMUNITION, ILLUMINATING with or without burster, expelling charge or propelling charge	Ammunition designed to produce a single source of intense light for lighting up an area. The term includes illuminating cartridges, grenades and projectiles; and illuminating and target identification bombs. The term excludes the following articles which are listed separately: CARTRIDGES, SIGNAL; SIGNAL DEVICES, HAND; SIGNALS, DISTRESS; FLARES, AERIAL and FLARES, SURFACE.
AMMUNITION, INCENDIARY	Ammunition containing incendiary substances which may be a solid, liquid or gel including white phosphorus. Except when the composition is an explosive *per se*, it also contains one or more of the following: a propelling charge with primer and igniter charge; a fuze with burster or expelling charge. The term includes: AMMUNITION, INCENDIARY, liquid or gel, with burster, expelling charge or propelling charge; AMMUNITION, INCENDIARY with or without burster, expelling charge or propelling charge; AMMUNITION, INCENDIARY, WHITE PHOSPHORUS with burster, expelling charge or propelling charge.
AMMUNITION, PRACTICE	Ammunition without a main bursting charge, containing a burster or expelling charge. Normally it also contains a fuze and a propelling charge. The term excludes the following articles which are listed separately: GRENADES, PRACTICE.
AMMUNITION, PROOF	Ammunition containing pyrotechnic substances, used to test the performance or strength of new ammunition, weapon component or assemblies.

AMMUNITION, SMOKE

Ammunition containing a smoke-producing substance such as chlorosulphonic acid mixture, titanium tetrachloride or white phosphorus; or smoke-producing pyrotechnic composition based on hexachloroethane or red phosphorus. Except when the substance is an explosive *per se*, the ammunition also contains one or more of the following: a propelling charge with primer and igniter charge; a fuze with burster or expelling charge. The term includes grenades, smoke but excludes SIGNALS, SMOKE which are listed separately. The term includes:

AMMUNITION, SMOKE with or without burster, expelling charge or propelling charge;

AMMUNITION, SMOKE, WHITE PHOSPHORUS with burster, expelling charge or propelling charge.

AMMUNITION, TEAR-PRODUCING with burster, expelling charge or propelling charge

Ammunition containing tear-producing substance. It also contains one or more of the following: a pyrotechnic substance; a propelling charge with primer and igniter charge; a fuze with burster or expelling charge.

AMMUNITION, TOXIC with burster, expelling charge or propelling charge

Ammunition containing toxic agent. It also contains one or more of the following: a pyrotechnic substance; a propelling charge with primer and igniter charge; a fuze with burster or expelling charge.

ARTICLES, EXPLOSIVE, EXTREMELY INSENSITIVE (ARTICLES, EEI)

Articles that contain only extremely insensitive substances and which demonstrate a negligible probability of accidental initiation or propagation (under normal conditions of transport) and which have passed Test Series 7.

ARTICLES, PYROPHORIC

Articles which contain a pyrophoric substance (capable of spontaneous ignition when exposed to air) and an explosive substance or component. The term excludes articles containing white phosphorus.

ARTICLES, PYROTECHNIC for technical purposes

Articles which contain pyrotechnic substances and are used for technical purposes such as heat generation, gas generation, theatrical effects, etc. The term excludes the following articles which are listed separately: all ammunition; CARTRIDGES, SIGNAL; CUTTERS, CABLE, EXPLOSIVE; FIREWORKS; FLARES, AERIAL; FLARES, SURFACE; RELEASE DEVICES, EXPLOSIVE; RIVETS, EXPLOSIVE; SIGNAL DEVICES, HAND; SIGNALS, DISTRESS; SIGNALS, RAILWAY TRACK, EXPLOSIVE; SIGNALS, SMOKE.

Auxiliary explosive component, isolated

An "isolated auxiliary explosive component" is a small device that explosively performs an operation related to the article's functioning, other than its main explosive load's performance. Functioning of the component does not cause any reaction of the main explosive loads contained within the article.

BLACK POWDER (GUNPOWDER)

Substance consisting of an intimate mixture of charcoal or other carbon and either potassium nitrate or sodium nitrate, with or without sulphur. It may be meal, granular, compressed or pelletized.

Bombs

Explosive articles which are dropped from aircraft. They may contain a flammable liquid with bursting charge, a photo-flash composition or a bursting charge. The term excludes torpedoes (aerial) and includes:

BOMBS, PHOTO-FLASH;

BOMBS with bursting charge;

BOMBS WITH FLAMMABLE LIQUID with bursting charge.

BOOSTERS	Articles consisting of a charge of detonating explosive with or without means of initiation. They are used to increase the initiating power of detonators or detonating cord.
BURSTERS, explosive	Articles consisting of a small charge of explosive used to open projectiles or other ammunition in order to disperse their contents.
Cartridges, blank	Articles which consist of a cartridge case with a centre or rim fire primer and a confined charge of smokeless or black powder but no projectile. Used for training, saluting or in starter pistols, tools, etc.
CARTRIDGES, FLASH	Articles consisting of a casing, a primer and flash powder, all assembled in one piece ready for firing.
Cartridges for Weapons	(1) Fixed (assembled) or semi-fixed (partially-assembled) ammunition designed to be fired from weapons. Each cartridge includes all the components necessary to function the weapon once. The name and description shall be used for small arms cartridges that cannot be described as "cartridges, small arms". Separate loading ammunition is included under this name and description when the propelling charge and projectile are packed together (see also "Cartridges, blank"). (2) Incendiary, smoke, toxic and tear-producing cartridges are described in this Glossary under AMMUNITION, INCENDIARY etc.
CARTRIDGES FOR WEAPONS, INERT PROJECTILE	Ammunition consisting of a projectile without bursting charge but with a propelling charge. The presence of a tracer can be disregarded for classification purposes provided that the predominant hazard is that of the propelling charge.
CARTRIDGES, OIL WELL	Articles consisting of a casing of thin fibre, metal or other material containing only propellant which projects a hardened projectile. The term excludes the following articles which are listed separately: CHARGES, SHAPED.
CARTRIDGES, POWER DEVICE	Articles designed to accomplish mechanical actions. They consist of a casing with a charge of deflagrating explosive and a means of ignition. The gaseous products of the deflagration produce inflation, or linear or rotary motion, or activate diaphragms, valves or switches or project fastening devices or extinguishing agents.
CARTRIDGES, SIGNAL	Articles designed to fire coloured flares or other signals from signal pistols, etc.
CARTRIDGES, SMALL ARMS	Ammunition consisting of a cartridge case fitted with a centre or rim fire primer and containing both a propelling charge and a solid projectile. They are designed to be fired in weapons of calibre not larger than 19.1 mm. Shotgun cartridges of any calibre are included in this description. The term excludes: CARTRIDGES, SMALL ARMS, BLANK listed separately in the Dangerous Goods List; and some small arms cartridges which are listed under CARTRIDGES FOR WEAPONS, INERT PROJECTILE.
CASES, CARTRIDGE, EMPTY, WITH PRIMER	Articles consisting of a cartridge case made from metal, plastics or other non-flammable material, in which the only explosive component is the primer.
CASES, COMBUSTIBLE, EMPTY, WITHOUT PRIMER	Articles consisting of cartridge cases made partly or entirely from nitrocellulose.
Charges, bursting	Articles consisting of a charge of detonating explosive such as hexolite, octolite or plastics bonded explosive designed to produce effect by blast or fragmentation.

APPENDIX B

CHARGES, DEMOLITION	Articles containing a charge of a detonating explosive in a casing of fibreboard, plastics, metal or other material. The term excludes the following articles which are listed separately: bombs, mines, etc.
CHARGES, DEPTH	Articles consisting of a charge of detonating explosive contained in a drum or projectile. They are designed to detonate under water.
Charges, expelling	A charge of deflagrating explosive designed to eject the payload from the parent articles without damage.
CHARGES, EXPLOSIVE, COMMERCIAL without detonator	Articles consisting of a charge of detonating explosive without means of initiation, used for explosive welding, jointing, forming and other metallurgical processes.
CHARGES, PROPELLING	Articles consisting of a propellant charge in any physical form, with or without a casing, for use as a component of rocket motors or for reducing the drag of projectiles.
CHARGES, PROPELLING FOR CANNON	Articles consisting of a propellant charge in any physical form, with or without a casing, for use in a cannon.
CHARGES, SHAPED, without detonator	Articles consisting of a casing containing a charge of detonating explosive with a cavity lined with rigid material, without means of initiation. They are designed to produce a powerful, penetrating jet effect.
CHARGES, SHAPED, FLEXIBLE, LINEAR	Articles consisting of a V-shaped core of a detonating explosive clad by a flexible metal sheath.
CHARGES, SUPPLEMENTARY, EXPLOSIVE	Articles consisting of a small removable booster used in the cavity of a projectile between the fuze and the bursting charge.
COMPONENTS, EXPLOSIVE TRAIN, N.O.S.	Articles containing an explosive designed to transmit the detonation or deflagration within an explosive train.
CONTRIVANCES, WATER-ACTIVATED with burster, expelling charge or propelling charge	Articles whose functioning depends upon physico-chemical reaction of their contents with water.
CORD, DETONATING, flexible	Article consisting of a core of detonating explosive enclosed in spun fabric, with plastics or other covering unless the spun fabric is sift-proof.
CORD (FUSE), DETONATING, metal clad	Article consisting of a core of detonating explosive clad by a soft metal tube with or without protective covering. When the core contains a sufficiently small quantity of explosive, the words "MILD EFFECT" are added.
CORD, IGNITER	Article consisting of textile yarns covered with black powder or another fast-burning pyrotechnic composition and of a flexible protective covering; or it consists of a core of black powder surrounded by a flexible woven fabric. It burns progressively along its length with an external flame and is used to transmit ignition from a device to a charge or primer.
CUTTERS, CABLE, EXPLOSIVE	Articles consisting of a knife-edged device which is driven by a small charge of deflagrating explosive into an anvil.
DETONATOR ASSEMBLIES, NON-ELECTRIC for blasting	Non-electric detonators assembled with and activated by such means as safety fuse, shock tube, flash tube or detonating cord. They may be of instantaneous design or incorporate delay elements. Detonating relays incorporating detonating cord are included. Other detonating relays are included in "Detonators, non-electric".

Detonators

Articles consisting of a small metal or plastics tube containing explosives such as lead azide, PETN or combinations of explosives. They are designed to start a detonation train. They may be constructed to detonate instantaneously, or may contain a delay element. The term includes:

DETONATORS FOR AMMUNITION and

DETONATORS for blasting, both ELECTRIC and NON-ELECTRIC.

Detonating relays without flexible detonating cord are included.

Entire load and total contents

The phrases "entire load" and "total contents" mean such a substantial proportion that the practical hazard shall be assessed by assuming simultaneous explosion of the whole of the explosive content of the load or package.

Explode

The verb used to indicate those explosive effects capable of endangering life and property through blast, heat and projection of missiles. It encompasses both deflagration and detonation.

Explosion of the total contents

The phrase "explosion of the total contents" is used in testing a single article or package or a small stack of articles or packages.

Explosive, blasting

Detonating explosive substances used in mining, construction and similar tasks. Blasting explosives are assigned to one of five types. In addition to the ingredients listed, blasting explosives may also contain inert components such as kieselguhr, and minor ingredients such as colouring agents and stabilizers.

EXPLOSIVE, BLASTING, TYPE A

Substances consisting of liquid organic nitrates such as nitroglycerin or a mixture of such ingredients with one or more of the following: nitrocellulose; ammonium nitrate or other inorganic nitrates; aromatic nitro-derivatives, or combustible materials, such as wood-meal and aluminium powder. Such explosives shall be in powdery, gelatinous or elastic form.

The term includes dynamite gelatine, blasting and gelatine dynamites.

EXPLOSIVE, BLASTING, TYPE B

Substances consisting of (a) a mixture of ammonium nitrate or other inorganic nitrates with an explosive such as trinitrotoluene, with or without other substances such as wood-meal and aluminium powder, or (b) a mixture of ammonium nitrate or other inorganic nitrates with other combustible substances which are not explosive ingredients. Such explosives shall not contain nitroglycerin, similar liquid organic nitrates, or chlorates.

EXPLOSIVE, BLASTING, TYPE C

Substances consisting of a mixture of either potassium or sodium chlorate or potassium, sodium or ammonium perchlorate with organic nitro-derivatives or combustible materials such as wood-meal or aluminium powder or a hydrocarbon. Such explosives shall not contain nitroglycerin or similar liquid organic nitrates.

EXPLOSIVE, BLASTING, TYPE D

Substances consisting of a mixture of organic nitrated compounds and combustible materials such as hydrocarbons and aluminium powder. Such explosives shall not contain nitroglycerin, similar liquid organic nitrates, chlorates or ammonium nitrate. The term generally includes plastic explosives.

EXPLOSIVE, BLASTING, TYPE E	Substances consisting of water as an essential ingredient and high proportions of ammonium nitrate or other oxidizers, some or all of which are in solution. The other constituents may include nitro-derivatives such as trinitrotoluene, hydrocarbons or aluminium powder. The term includes explosives, emulsion; explosives slurry and explosives, water gel.
Explosive, deflagrating	A substance which reacts by detonation rather than deflagration when initiated and used in its normal manner.
Explosive, extremely insensitive substance (EIS)	A substance which has demonstrated through tests that it is so insensitive that there is very little probability of accidental initiation.
Explosive, primary	Explosive substance manufactured with a view to producing a practical effect by explosion which is very sensitive to heat, impact or friction and which, even in very small quantities, either detonates or burns very rapidly. It is able to transmit detonation (in the case of initiating explosive) or deflagration to secondary explosives close to it. The main primary explosives are mercury fulminate, lead azide and lead styphnate.
Explosive, secondary	Explosive substance which is relatively insensitive (when compared to primary explosives), which is usually initiated by primary explosives with or without the aid of boosters or supplementary charges. Such an explosive may react as a deflagrating or as a detonating explosive.
FIREWORKS	Pyrotechnic articles designed for entertainment.
Flares	Articles containing pyrotechnic substances which are designed for use to illuminate, identify, signal or warn. The term includes: FLARES, AERIAL; FLARES, SURFACE.
FLASH POWDER	Pyrotechnic substance which, when ignited, produces an intense light.
FRACTURING DEVICES, EXPLOSIVE for oil wells, without detonator	Articles consisting of a charge of detonating explosive contained in a casing without means of initiation. They are used to fracture the rock around a drill shaft to assist the flow of crude oil from the rock.
Fuse/Fuze	Although these two words have a common origin (French fusée, fusil) and are sometimes considered to be different spellings, it is useful to maintain the convention that fuse refers to a cord-like igniting device whereas fuze refers to a device used in ammunition which incorporates mechanical, electrical, chemical or hydrostatic components to initiate a train by deflagration or detonation.
FUSE, IGNITER, tubular, metal clad	Article consisting of a metal tube with a core of deflagrating explosive.
FUSE, INSTANTANEOUS, NON-DETONATING (QUICKMATCH)	Article consisting of cotton yarns impregnated with fine black powder (Quickmatch). It burns with an external flame and is used in ignition trains for fireworks, etc.
FUSE, SAFETY	Article consisting of a core of fine-grained black powder surrounded by a flexible woven fabric with one or more protective outer coverings. When ignited, it burns at a predetermined rate without any external explosive effect.

Fuzes	Articles designed to start a detonation or a deflagration in ammunition. They incorporate mechanical, electrical, chemical or hydrostatic components and generally protective features. The term includes: FUZES, DETONATING; FUZES, DETONATING with protective features; FUZES, IGNITING.
GRENADES, hand or rifle	Articles which are designed to be thrown by hand or to be projected by a rifle. The term includes: GRENADES, hand or rifle, with bursting charge; GRENADES, PRACTICE, hand or rifle. The term excludes grenades, smoke which are listed under AMMUNITION, SMOKE.
IGNITERS	Articles containing one or more explosive substances used to start deflagration in an explosive train. They may be actuated chemically, electrically or mechanically. This term excludes the following articles which are listed separately: CORD, IGNITER; FUSE, IGNITER; FUSE, NON-DETONATING; FUZES, IGNITING; LIGHTERS, FUSE; PRIMERS, CAP TYPE; PRIMERS, TUBULAR.
Ignition, means of	A general term used in connection with the method employed to ignite a deflagrating train of explosive or pyrotechnic substances (for example: a primer for a propelling charge; an igniter for a rocket motor; an igniting fuze).
Initiation, means of	(1) A device intended to cause the detonation of an explosive (for example: detonator; detonator for ammunition; detonating fuze). (2) The term "with its own means of initiation" means that the contrivance has its normal initiating device assembled to it and this device is considered to present a significant risk during transport but not one great enough to be unacceptable. The term does not apply, however, to a contrivance packed together with its means of initiation provided the device is packaged so as to eliminate the risk of causing detonation of the contrivance in the event of accidental functioning of the initiating device. The means of initiating can even be assembled to the contrivance provided there are protective features such that the device is very unlikely to cause detonation of the contrivance in conditions which are associated with transport. (3) For the purposes of classification any means of initiation without two effective protective features shall be regarded as Compatibility Group B; an article with its own means of initiation, without two effective protective features, would be Compatibility Group F. On the other hand a means of initiation which itself possesses two effective protective features would be Compatibility Group D; and an article with a means of initiation which possesses two effective protective features would be Compatibility Group D or E. Means of initiation adjudged as having two effective protective features shall have been approved by the competent national authority. A common and effective way of achieving the necessary degree of protection is to use a means of initiation which incorporates two or more independent safety features.

JET PERFORATING GUNS, CHARGED, oil well, without detonator	Articles consisting of a steel tube or metallic strip, into which are inserted shaped charges connected by detonating cord, without means of initiation.
LIGHTERS, FUSE	Articles of various design actuated by friction, percussion or electricity and used to ignite safety fuse.
Mass explosion	Explosion which affects almost the entire load virtually instantaneously.
MINES	Articles consisting normally of metal or composition receptacles and a bursting charge. They are designed to be operated by the passage of ships, vehicles or personnel. The term includes "Bangalore torpedoes".
OXYGEN GENERATORS, CHEMICAL	Oxygen generators, chemical, are devices containing chemicals which upon activation release oxygen as a product of chemical reaction. Chemical oxygen generators are used for the generation of oxygen for respiratory support, e.g., in aircraft, submarines, spacecraft, bomb shelters and breathing apparatus. Oxidizing salts such as chlorates and perchlorates of lithium, sodium and potassium, which are used in chemical oxygen generators, evolve oxygen when heated. These salts are mixed (compounded) with a fuel, usually iron powder, to form a chlorate candle, which produces oxygen by continuous reaction. The fuel is used to generate heat by oxidation. Once the reaction begins, oxygen is released from the hot salt by thermal decomposition (a thermal shield is used around the generator). A portion of the oxygen reacts with the fuel to produce more heat which produces more oxygen, and so on. Initiation of the reaction can be achieved by a percussion device, friction device or electric wire.
POWDER CAKE (POWDER PASTE), WETTED	Substance consisting of nitrocellulose impregnated with not more than 60% of nitroglycerin or other liquid organic nitrates or a mixture of these.
POWDER, SMOKELESS	Substance based on nitrocellulose used as propellant. The term includes propellants with a single base (nitrocellulose (NC) alone), those with a double base (such as NC and nitroglycerin (NG)) and those with a triple base (such as NC/NG/nitroguanidine). Cast, pressed or bag-charges of smokeless powder are listed under "CHARGES, PROPELLING" or "CHARGES, PROPELLING FOR CANNON".
PRIMERS, CAP TYPE	Articles consisting of a metal or plastics cap containing a small amount of primary explosive mixture that is readily ignited by impact. They serve as igniting elements in small arms cartridges, and in percussion primers for propelling charges.
PRIMERS, TUBULAR	Articles consisting of a primer for ignition and an auxiliary charge of deflagrating explosive such as black powder used to ignite the propelling charge in a cartridge case for cannon, etc.
PROJECTILES	Articles such as a shell or bullet which are projected from a cannon or other artillery gun, rifle or other small arm. They may be inert, with or without tracer, or may contain a burster or expelling charge or a bursting charge. The term includes: PROJECTILES, inert, with tracer; PROJECTILES with burster or expelling charge; PROJECTILES with bursting charge.

PROPELLANTS	Deflagrating explosive used for propulsion or for reducing the drag of projectiles.
PROPELLANTS, LIQUID	Substances consisting of a deflagrating liquid explosive, used for propulsion.
PROPELLANTS, SOLID	Substances consisting of a deflagrating solid explosive, used for propulsion.
RELEASE DEVICES, EXPLOSIVE	Articles consisting of a small charge of explosive with means of initiation. They sever rods or links to release equipment quickly.
ROCKET MOTORS	Articles consisting of a solid, liquid or hypergolic fuel contained in a cylinder fitted with one or more nozzles. They are designed to propel a rocket or a guided missile. The term includes: ROCKET MOTORS; ROCKET MOTORS WITH HYPERGOLIC LIQUIDS with or without expelling charge; ROCKET MOTORS, LIQUID FUELLED.
ROCKETS	Articles consisting of a rocket motor and a payload which may be an explosive warhead or other device. The term includes guided missiles and: ROCKETS, LINE-THROWING; ROCKETS, LIQUID FUELLED with bursting charge; ROCKETS with bursting charge; ROCKETS with expelling charge; ROCKETS with inert head.
SIGNALS	Articles containing pyrotechnic substances designed to produce signals by means of sound, flame or smoke or any combinations thereof. The term includes: SIGNAL DEVICES, HAND; SIGNALS, DISTRESS, ship; SIGNALS, RAILWAY TRACK, EXPLOSIVE; SIGNALS, SMOKE.
SOUNDING DEVICES, EXPLOSIVE	Articles consisting of a charge of detonating explosive. They are dropped from ships and function when they reach a predetermined depth or the sea bed.
STABILIZED	Stabilized means that the substance is in a condition that precludes uncontrolled reaction. This may be achieved by methods such as the addition of an inhibiting chemical, degassing the substance to remove dissolved oxygen and inerting the air space in the package, or maintaining the substance under temperature control.
SUBSTANCES, EXPLOSIVE, VERY INSENSITIVE (SUBSTANCES, EVI), N.O.S.	Substances which present a mass explosion hazard but which are so insensitive that there is very little probability of initiation, or of transition from burning to detonation (under normal conditions of transport) and which have passed Test Series 5.
TORPEDOES	Articles containing an explosive or non-explosive propulsion system and designed to be propelled through water. They may contain an inert head or a warhead. The term includes: TORPEDOES, LIQUID FUELLED with inert head; TORPEDOES, LIQUID FUELLED with or without bursting charge; TORPEDOES with bursting charge.

TRACERS FOR AMMUNITION	Sealed articles containing pyrotechnic substances, designed to reveal the trajectory of a projectile.
Warheads	Articles consisting of detonating explosives. They are designed to be fitted to a rocket, guided missile or torpedo. They may contain a burster or expelling charge or bursting charge. The term includes: WARHEADS, ROCKET with burster or expelling charge; WARHEADS, ROCKET with bursting charge; WARHEADS, TORPEDO with bursting charge.

INDEX

Index

In this Index, the word "*see*", after the name in the substance, material or article column, means that it is a synonym and for details regarding the transport provisions reference shall be made to the entry in the Dangerous Goods List (chapter 3.2) which is relevant to the UN Number/Proper Shipping Name stated against the synonym.

Method of indexing

Substances, materials and articles have been listed in the alphabetical order of their names. For the purpose of determining the alphabetical order, numbers and roman numerals (I), (II) etc. and the prefixes listed below have been disregarded, although they form an integral part of the name:

N-	*sym-*
n- or *normal-*	*uns-*
sec- or *secondary-*	*cis-*
tert- or *tertiary-*	*trans-*
o- or *ortho-*	*d-*
m- or *meta-*	α- or *alpha-*
p- or *para-*	β- or *beta-*
	γ- or *gamma-*

Note 1

Certain marine pollutants are identified only in the Index. These marine pollutants have not been assigned to an N.O.S. or generic entry. These marine pollutants may possess properties of classes 1 to 8 and should be classified accordingly. A substance which does not fall within the criteria of these classes should be offered for transport as an ENVIRONMENTALLY HAZARDOUS SUBSTANCE, SOLID, N.O.S., UN 3077, or as an ENVIRONMENTALLY HAZARDOUS SUBSTANCE, LIQUID, N.O.S., UN 3082, under these entries in class 9.

Substance, material or article	MP	Class	UN No.
Acid mixture, hydrofluoric and sulphuric, *see*	–	8	1786
Acid mixture, nitrating acid, *see*	–	8	1796
Acid mixture, spent, nitrating acid, *see*	–	8	1826
Acraldehyde, stabilized, *see*	P	6.1	1092
ACRIDINE	–	6.1	2713
Acroleic acid, stabilized, *see*	–	8	2218
Acrolein diethyl acetal, *see*	–	3	2374
ACROLEIN DIMER, STABILIZED	–	3	2607
ACROLEIN, STABILIZED	P	6.1	1092
ACRYLAMIDE, SOLID	–	6.1	2074
ACRYLAMIDE SOLUTION	–	6.1	3426
Acrylic acid isobutyl ester, stabilized, *see*	–	3	2527
ACRYLIC ACID, STABILIZED	–	8	2218
Acrylic aldehyde, stabilized, *see*	P	6.1	1092
ACRYLONITRILE, STABILIZED	–	3	1093
Actinolite, *see*	–	9	2590
Activated carbon, *see*	–	4.2	1362
Activated charcoal, *see*	–	4.2	1362
ADHESIVES containing flammable liquid	–	3	1133
ADIPONITRILE	–	6.1	2205
Aeroplane flares, *see* FLARES, AERIAL	–	–	–
AEROSOLS	–	2	1950
AGENT, BLASTING, TYPE B	–	1.5D	0331
AGENT, BLASTING, TYPE E	–	1.5D	0332
AIR BAG INFLATORS	–	1.4G	0503
AIR BAG INFLATORS	–	9	3268
AIR BAG MODULES	–	1.4G	0503
AIR BAG MODULES	–	9	3268
AIR, COMPRESSED	–	2.2	1002
AIRCRAFT HYDRAULIC POWER UNIT FUEL TANK (containing a mixture of anhydrous hydrazine and methylhydrazine)	–	3	3165
AIR, REFRIGERATED LIQUID	–	2.2	1003
⊗			
ALCOHOLATES SOLUTION, N.O.S. in alcohol	–	3	3274
Alcohol C_{12}–C_{16} poly(1–6)ethoxylate, *see*	P	9	3082
Alcohol C_6–C_{17} (secondary) poly(3–6)ethoxylate, *see*	P	9	3082
⊗			
ALCOHOLIC BEVERAGES, with more than 24% but not more than 70% alcohol by volume	–	3	3065
ALCOHOLIC BEVERAGES, with more than 70% alcohol by volume	–	3	3065
⊗			
ALCOHOLS, FLAMMABLE, TOXIC, N.O.S.	–	3	1986
ALCOHOLS, N.O.S.	–	3	1987
⊗			

INDEX

Substance, material or article	MP	Class	UN No.
Ammonium bifluoride solution, *see*	–	8	2817
Ammonium bisulphate, *see*	–	8	2506
Ammonium bisulphite solution, *see*	–	8	2693
Ammonium bromate **(transport prohibited)**	–	–	–
Ammonium bromate solution **(transport prohibited)**	–	–	–
Ammonium chlorate **(transport prohibited)**	–	–	–
Ammonium chlorate solution **(transport prohibited)**	–	–	–
Ammonium chlorite **(transport prohibited)**	–	–	–
AMMONIUM DICHROMATE	–	5.1	1439
AMMONIUM DINITRO-*o*-CRESOLATE, SOLID	P	6.1	1843
AMMONIUM DINITRO-*o*-CRESOLATE SOLUTION	P	6.1	3424
AMMONIUM FLUORIDE	–	6.1	2505
AMMONIUM FLUOROSILICATE	–	6.1	2854
Ammonium hexafluorosilicate, *see*	–	6.1	2854
AMMONIUM HYDROGENDIFLUORIDE, SOLID	–	8	1727
AMMONIUM HYDROGENDIFLUORIDE SOLUTION	–	8	2817
AMMONIUM HYDROGEN SULPHATE	–	8	2506
Ammonium hypochlorite **(transport prohibited)**	–	–	–
AMMONIUM METAVANADATE	–	6.1	2859
AMMONIUM NITRATE BASED FERTILIZER	–	5.1	2067
AMMONIUM NITRATE BASED FERTILIZER	–	9	2071
AMMONIUM NITRATE EMULSION intermediate for blasting explosives	–	5.1	3375
AMMONIUM NITRATE GEL intermediate for blasting explosives	–	5.1	3375
AMMONIUM NITRATE liable to self-heating sufficient to initiate decomposition **(transport prohibited)**	–	–	–
AMMONIUM NITRATE, LIQUID (hot concentrated solution)	–	5.1	2426
AMMONIUM NITRATE SUSPENSION intermediate for blasting explosives	–	5.1	3375
△ AMMONIUM NITRATE with more than 0.2% combustible substances, including any organic substance calculated as carbon to the exclusion of any other added substance	–	1.1D	0222
AMMONIUM NITRATE with not more than 0.2% total combustible material, including any organic substance calculated as carbon to the exclusion of any other added substance	–	5.1	1942
Ammonium nitrite **(transport prohibited)**	–	–	–
AMMONIUM PERCHLORATE	–	1.1D	0402
AMMONIUM PERCHLORATE	–	5.1	1442
Ammonium permanganate **(transport prohibited)**	–	–	–
Ammonium permanganate solution **(transport prohibited)**	–	–	–
AMMONIUM PERSULPHATE	–	5.1	1444
AMMONIUM PICRATE dry or wetted with less than 10% water, by mass	–	1.1D	0004
AMMONIUM PICRATE, WETTED with not less than 10% water, by mass	–	4.1	1310
AMMONIUM POLYSULPHIDE SOLUTION	–	8	2818
AMMONIUM POLYVANADATE	–	6.1	2861

INDEX

Substance, material or article	MP	Class	UN No.
ARTICLES, EXPLOSIVE, N.O.S.	–	1.3L	0356
ARTICLES, EXPLOSIVE, N.O.S.	–	1.4B	0350
ARTICLES, EXPLOSIVE, N.O.S.	–	1.4C	0351
ARTICLES, EXPLOSIVE, N.O.S.	–	1.4D	0352
ARTICLES, EXPLOSIVE, N.O.S.	–	1.4E	0471
ARTICLES, EXPLOSIVE, N.O.S.	–	1.4F	0472
ARTICLES, EXPLOSIVE, N.O.S.	–	1.4G	0353
ARTICLES, EXPLOSIVE, N.O.S.	–	1.4S	0349
ARTICLES, PRESSURIZED, HYDRAULIC (containing non-flammable gas)	–	2.2	3164
ARTICLES, PRESSURIZED, PNEUMATIC (containing non-flammable gas)	–	2.2	3164
ARTICLES, PYROPHORIC	–	1.2L	0380
ARTICLES, PYROTECHNIC for technical purposes	–	1.1G	0428
ARTICLES, PYROTECHNIC for technical purposes	–	1.2G	0429
ARTICLES, PYROTECHNIC for technical purposes	–	1.3G	0430
ARTICLES, PYROTECHNIC for technical purposes	–	1.4G	0431
ARTICLES, PYROTECHNIC for technical purposes	–	1.4S	0432
ARYLSULPHONIC ACIDS, LIQUID with more than 5% free sulphuric acid	–	8	2584
ARYLSULPHONIC ACIDS, LIQUID with not more than 5% free sulphuric acid	–	8	2586
ARYLSULPHONIC ACIDS, SOLID with more than 5% free sulphuric acid	–	8	2583
ARYLSULPHONIC ACIDS, SOLID with not more than 5% free sulphuric acid	–	8	2585
Asbestos, *see* BLUE ASBESTOS	–	9	2212
Asbestos, *see* BROWN ASBESTOS	–	9	2212
Asbestos, *see* WHITE ASBESTOS	–	9	2590
Asphalt, *see*	–	3	1999
AVIATION REGULATED LIQUID, N.O.S.	–	9	3334
AVIATION REGULATED SOLID, N.O.S.	–	9	3335
Azinphos-ethyl, *see* ORGANOPHOSPHORUS PESTICIDE	P	–	–
Azinphos-methyl, *see* ORGANOPHOSPHORUS PESTICIDE	P	–	–
Aziridine, stabilized, *see*	–	6.1	1185
AZODICARBONAMIDE	–	4.1	3242
△ Azodicarbonamide formulation Type B (concentration $<$ 100%, temperature controlled), *see*	–	4.1	3232
Azodicarbonamide formulation Type C (concentration $<$ 100%), *see*	–	4.1	3224
Azodicarbonamide formulation Type C (concentration $<$ 100%, temperature controlled), *see*	–	4.1	3234
Azodicarbonamide formulation Type D (concentration $<$ 100%), *see*	–	4.1	3226
Azodicarbonamide formulation Type D (concentration $<$ 100%, temperature controlled), *see*	–	4.1	3236
2,2′-Azodi(2,4-dimethyl-4-methoxyvaleronitrile) (concentration 100%), *see*	–	4.1	3236

INDEX

INDEX

INDEX

INDEX

INDEX

INDEX

Substance, material or article	MP	Class	UN No.
CHLOROTRIFLUOROMETHANE	–	2.2	1022
CHLOROTRIFLUOROMETHANE AND TRIFLUOROMETHANE AZEOTROPIC MIXTURE with approximately 60% chlorotrifluoromethane	–	2.2	2599
2-Chloro-5-trifluoromethylnitrobenzene, *see*	P	6.1	2307
Chlorovinyl acetate, *see*	–	6.1	2589
Chlorphacinone, *see* ORGANOCHLORINE PESTICIDE	–	–	–
Chlorpyriphos, *see* ORGANOPHOSPHORUS PESTICIDE	P	–	–
Chlorthiophos, *see* ORGANOPHOSPHORUS PESTICIDE	P	–	–
Chromic acid, solid, *see*	–	5.1	1463
CHROMIC ACID SOLUTION	–	8	1755
Chromic anhydride, *see*	–	5.1	1463
CHROMIC FLUORIDE, SOLID	–	8	1756
CHROMIC FLUORIDE SOLUTION	–	8	1757
Chromic nitrate, *see*	–	5.1	2720
Chromium(VI) dichloride dioxide, *see*	–	8	1758
Chromium(III) fluoride, solid, *see*	–	8	1756
Chromium fluoride, solid, *see*	–	8	1756
Chromium fluoride solution, *see*	–	8	1757
CHROMIUM NITRATE	–	5.1	2720
Chromium(III) nitrate, *see*	–	5.1	2720
CHROMIUM OXYCHLORIDE	–	8	1758
CHROMIUM TRIOXIDE, ANHYDROUS	–	5.1	1463
CHROMOSULPHURIC ACID	–	8	2240
Chromyl chloride, *see*	–	8	1758
Chrysotile, *see*	–	9	2590
Cinene, *see*	P	3	2052
Cinnamene, *see*	–	3	2055
Cinnamol, *see*	–	3	2055
CLINICAL WASTE, UNSPECIFIED, N.O.S.	–	6.2	3291
COAL GAS, COMPRESSED	–	2.3	1023
Coal tar, *see*	P	9	3082
COAL TAR DISTILLATES, FLAMMABLE	–	3	1136
Coal tar naphtha, *see*	–	3	1268
△ Coal tar oil, *see*	–	3	1136
COATING SOLUTION (includes surface treatments or coatings used for industrial purposes such as vehicle under-coating, drum or barrel lining)	–	3	1139
COBALT NAPHTHENATES, POWDER	–	4.1	2001
COBALT RESINATE, PRECIPITATED	–	4.1	1318
Cocculus, *see*	P	6.1	3172
Coconitrile, *see*	P	9	3082
Collodion cottons (class 1), *see* NITROCELLULOSE	–	–	–
Collodion cotton with alcohol, *see*	–	4.1	2556

INDEX

INDEX

INDEX

INDEX

INDEX

Substance, material or article	MP	Class	UN No.
Di-(*tert*-butylperoxyisopropyl)benzene(s) (concentration >42–100%, with inert solid), *see*	–	5.2	3106
Di-(*tert*-butylperoxy) phthalate (concentration ≤42%, with diluent Type A), *see*	–	5.2	3107
Di-(*tert*-butylperoxy) phthalate (concentration >42–52%, with diluent Type A), *see*	–	5.2	3105
Di-(*tert*-butylperoxy) phthalate (concentration ≤52%, as a paste with diluent Type A, with or without water), *see*	–	5.2	3106
△ 2,2-Di-(*tert*-butylperoxy)propane (concentration ≤42%, with diluent Type A, with inert solid), *see*	–	5.2	3106
2,2-Di-(*tert*-butylperoxy)propane (concentration ≤52% with diluent Type A), *see*	–	5.2	3105
1,1-Di-(*tert*-butylperoxy)-3,3,5-trimethylcyclohexane (concentration ≤32%, with diluents Type A and B), *see*	–	5.2	3107
1,1-Di-(*tert*-butylperoxy)-3,3,5-trimethylcyclohexane (concentration ≤57%, with diluent Type A), *see*	–	5.2	3107
1,1-Di-(*tert*-butylperoxy)-3,3,5-trimethylcyclohexane (concentration ≤57%, with inert solid), *see*	–	5.2	3110
1,1-Di-(*tert*-butylperoxy)-3,3,5-trimethylcyclohexane (concentration >57–90%, with diluent Type A), *see*	–	5.2	3103
1,1-Di-(*tert*-butylperoxy)-3,3,5-trimethylcyclohexane (concentration ≤77%, with diluent Type B), *see*	–	5.2	3103
1,1-Di-(*tert*-butylperoxy)-3,3,5-trimethylcyclohexane (concentration ≤90%, with diluent Type B)	–	5.2	3103
1,1-Di-(*tert*-butylperoxy)-3,3,5-trimethylcyclohexane (concentration >90–100%), *see*	–	5.2	3101
2,4-Di-*tert*-butylphenol, *see* **Note 1**	–	–	–
2,6-Di-*tert*-butylphenol, *see* **Note 1**	–	–	–
Di-*n*-butyl phthalate, *see*	P	9	3082
Dicetyl peroxydicarbonate (concentration ≤42% as a stable dispersion in water), *see*	–	5.2	3119
Dicetyl peroxydicarbonate (concentration ≤100%), *see*	–	5.2	3116
Dichlofenthion, *see* ORGANOPHOSPHORUS PESTICIDE	P	–	–
1,1-DICHLORO-1-NITROETHANE	–	6.1	2650
DICHLOROACETIC ACID	–	8	1764
1,3-DICHLOROACETONE	–	6.1	2649
DICHLOROACETYL CHLORIDE	–	8	1765
DICHLOROANILINES, LIQUID	P	6.1	1590
DICHLOROANILINES, SOLID	P	6.1	3442
1,2-Dichlorobenzene, *see*	–	6.1	1591
1,3-Dichlorobenzene, *see*	P	6.1	2810
1,4-Dichlorobenzene, *see*	P	9	3082
meta-Dichlorobenzene, *see*	P	6.1	2810
o-DICHLOROBENZENE	–	6.1	1591
para-Dichlorobenzene, *see*	P	9	3082
Di-(4-chlorobenzoyl) peroxide (concentration ≤32%, with inert solid) **(exempt)**	–	–	–

Substance, material or article	MP	Class	UN No.
N,N-DIETHYLANILINE	–	6.1	2432
DIETHYLBENZENES	–	3	2049
Diethyl carbinol, *see*	–	3	1105
DIETHYL CARBONATE	–	3	2366
DIETHYLDICHLOROSILANE	–	8	1767
Diethylenediamine, *see*	–	8	2579
Diethylenediamine, solid, *see*	–	8	2579
1,4-Diethylene dioxide, *see*	–	3	1165
Diethyleneglycol bis(allyl carbonate) + di-isopropyl peroxycicarbonate (concentration ≥88% + ≤12%), *see*	–	4.1	3237
DIETHYLENEGLYCOL DINITRATE, DESENSITIZED with not less than 25% non-volatile water-insoluble phlegmatizer, by mass	–	1.1D	0075
Diethylene oxide, *see*	–	3	1165
DIETHYLENETRIAMINE	–	8	2079
N,N-Diethylethanolamine, *see*	–	8	2686
DIETHYL ETHER	–	3	1155
N,N-DIETHYLETHYLENEDIAMINE	–	8	2685
Diethyl formal, *see*	–	3	2373
Di-(2-ethylhexyl) peroxydicarbonate (concentration ≤52%, as a stable dispersion in water (frozen)), *see*	–	5.2	3120
Di-(2-ethylhexyl) peroxydicarbonate (concentration ≤62%, as a stable dispersion in water), *see*	–	5.2	3119
Di-(2-ethylhexyl) peroxydicarbonate (concentration ≤77%, with diluent Type B), *see*	–	5.2	3115
Di-(2-ethylhexyl) peroxydicarbonate (concentration >77–100%), *see*	–	5.2	3113
Di-(2-ethylhexyl)phosphoric acid, *see*	–	8	1902
DIETHYL KETONE	–	3	1156
Diethyl oxalate, *see*	–	6.1	2525
N,N-Diethyl-1,3-propanediamine, *see*	–	3	2684
DIETHYL SULPHATE	–	6.1	1594
DIETHYL SULPHIDE	–	3	2375
DIETHYLTHIOPHOSPHORYL CHLORIDE	–	8	2751
Diethylzinc, *see*	–	4.2	3394
Difenacoum, *see* COUMARIN DERIVATIVE PESTICIDE	–	–	–
Difenzoquat, *see* PESTICIDE, N.O.S.	–	–	–
2,4-Difluoroaniline, *see*	–	6.1	2941
Difluorochloroethane, *see*	–	2.1	2517
Difluorodibromomethane, *see*	–	9	1941
1,1-DIFLUOROETHANE	–	2.1	1030
Difluoroethane and dichlorodifluoromethane, azeotropic mixture with approximately 74% dichlorodifluoromethane, *see* DICHLORODIFLUOROMETHANE AND DIFLUOROETHANE, AZEOTROPIC MIXTURE	–	–	–
1,1-DIFLUOROETHYLENE	–	2.1	1959
DIFLUOROMETHANE	–	2.1	3252

INDEX

Substance, material or article	MP	Class	UN No.
4-(Dimethylamino)benzenediazonium trichlorozincate(–1) (concentration 100%), *see*	–	4.1	3228
4-Dimethylamino-6-(2-dimethylaminoethoxy)toluene-2-diazonium zinc chloride (concentration 100%), *see*	–	4.1	3236
2-DIMETHYLAMINOETHANOL	–	8	2051
2-DIMETHYLAMINOETHYL ACRYLATE	–	6.1	3302
2-DIMETHYLAMINOETHYL METHACRYLATE	–	6.1	2522
N,N-DIMETHYLANILINE	–	6.1	2253
3,4-Dimethylaniline, *see*	–	6.1	1711
Dimethylarsinic acid, *see*	–	6.1	1572
Dimethylbenzenes, *see*	–	3	1307
Di-(2-methylbenzoyl) peroxide (concentration ≤87%, with water), *see*	–	5.2	3112
Di-(3-methylbenzoyl) peroxide (concentration ≤20%), with benzoyl (3-methylbenzoyl) peroxide (concentration ≤18%), with dibenzoyl peroxide (concentration ≤4%) and diluent Type B, *see*	–	5.2	3115
Di-(4-methylbenzoyl) peroxide (concentration ≤52%, as a paste with silicon oil), *see*	–	5.2	3106
Dimethylbenzylamine, *see*	–	8	2619
N,N-Dimethylbenzylamine, *see*	–	8	2619
2,3-DIMETHYLBUTANE	–	3	2457
1,3-DIMETHYLBUTYLAMINE	–	3	2379
DIMETHYLCARBAMOYL CHLORIDE	–	8	2262
Dimethyl carbinol, *see*	–	3	1219
DIMETHYL CARBONATE	–	3	1161
DIMETHYLCYCLOHEXANES	–	3	2263
N,N-DIMETHYLCYCLOHEXYLAMINE	–	8	2264
2,5-Dimethyl-2,5-di-(benzoylperoxy)hexane (concentration ≤82%, with inert solid), *see*	–	5.2	3106
2,5-Dimethyl-2,5-di-(benzoylperoxy)hexane (concentration ≤82%, with water), *see*	–	5.2	3104
2,5-Dimethyl-2,5-di-(benzoylperoxy)hexane (concentration >82–100%), *see*	–	5.2	3102
2,5-Dimethyl-2,5-di-(*tert*-butylperoxy)hexane (concentration ≤47%, as a paste), *see*	–	5.2	3108
2,5-Dimethyl-2,5-di-(*tert*-butylperoxy)hexane (concentration ≤52%, with diluent Type A), *see*	–	5.2	3109
2,5-Dimethyl-2,5-di-(*tert*-butylperoxy)hexane (concentration >52–90%), *see*	–	5.2	3105
2,5-Dimethyl-2,5-di-(*tert*-butylperoxy)hexane (concentration ≤77%, with inert solid), *see*	–	5.2	3108
2,5-Dimethyl-2,5-di-(*tert*-butylperoxy)hexane (concentration >90–100%), *see*	–	5.2	3103
2,5-Dimethyl-2,5-di-(*tert*-butylperoxy)hexyne-3 (concentration ≤52%, with inert solid), *see*	–	5.2	3106
2,5-Dimethyl-2,5-di-(*tert*-butylperoxy)hexyne-3 (concentration >52–86%, with diluent Type A), *see*	–	5.2	3103
2,5-Dimethyl-2,5-di-(*tert*-butylperoxy)hexyne-3 (concentration >86–100%), *see*	–	5.2	3101

INDEX

INDEX

Substance, material or article	MP	Class	UN No.
Di-(2-neodecanoylperoxyisopropyl)benzene (concentration ≤52%, with diluent Type A), *see*	–	5.2	3115
DINGU	–	1.1D	0489
DINITROANILINES	–	6.1	1596
DINITROBENZENES, LIQUID	–	6.1	1597
DINITROBENZENES, SOLID	–	6.1	3443
Dinitrochlorobenzenes, liquid, *see*	P	6.1	1577
Dinitrochlorobenzenes, solid, *see*	P	6.1	3441
DINITRO-*o*-CRESOL	P	6.1	1598
Dinitrogen oxide, *see*	–	2.2	1070
Dinitrogen oxide, refrigerated liquid, *see*	–	2.2	2201
DINITROGEN TETROXIDE	–	2.3	1067
Dinitrogen tetroxide and nitric oxide mixtures, *see* NITRIC OXIDE AND DINITROGEN TETROXIDE MIXTURE	–	–	–
Dinitrogen trioxide, *see*	–	2.3	2421
DINITROGLYCOLURIL	–	1.1D	0489
Dinitrophenates (class 1), *see*	P	1.3C	0077
Dinitrophenates, wetted, *see*	P	4.1	1321
DINITROPHENOLATES alkali metals, dry or wetted with less than 15% water, by mass	P	1.3C	0077
DINITROPHENOLATES, WETTED with not less than 15% water, by mass	P	4.1	1321
DINITROPHENOL dry or wetted with less than 15% water, by mass	P	1.1D	0076
DINITROPHENOL SOLUTION	P	6.1	1599
DINITROPHENOL, WETTED with not less than 15% water, by mass	P	4.1	1320
DINITRORESORCINOL dry or wetted with less than 15% water, by mass	–	1.1D	0078
DINITRORESORCINOL, WETTED with not less than 15% water, by mass	–	4.1	1322
DINITROSOBENZENE	–	1.3C	0406
N,N'-Dinitroso-*N,N'*-dimethylterephthalamide, as a paste (concentration 72%), *see*	–	4.1	3224
N,N'-Dinitrosopentamethylenetetramine (concentration 82%), *see*	–	4.1	3224
Dinitrotoluene mixed with sodium chlorate, *see*	–	1.1D	0083
DINITROTOLUENES, LIQUID	–	6.1	2038
DINITROTOLUENES, MOLTEN	–	6.1	1600
DINITROTOLUENES, SOLID	–	6.1	3454
Dinobuton, *see* SUBSTITUTED NITROPHENOL PESTICIDE	P	–	–
Di-*n*-nonanoyl peroxide (concentration ≤100%), *see*	–	5.2	3116
Dinoseb, *see* SUBSTITUTED NITROPHENOL PESTICIDE	P	–	–
Dinoseb acetate, *see* SUBSTITUTED NITROPHENOL PESTICIDE	P	–	–
Dinoterb, *see* SUBSTITUTED NITROPHENOL PESTICIDE	–	–	–
Dinoterb acetate, *see* SUBSTITUTED NITROPHENOL PESTICIDE	–	–	–
Di-*n*-octanoyl peroxide (concentration ≤100%), *see*	–	5.2	3114
Dioxacarb, *see* CARBAMATE PESTICIDE	P	–	–

INDEX

Substance, material or article	MP	Class	UN No.
Endrin, *see* ORGANOCHLORINE PESTICIDE	P	–	–
ENGINE, FUEL CELL, FLAMMABLE GAS POWERED	–	9	3166
ENGINE, FUEL CELL, FLAMMABLE LIQUID POWERED	–	9	3166
ENGINE, INTERNAL COMBUSTION	–	9	3166
Engines, rocket, *see* ROCKET MOTORS WITH HYPERGOLIC LIQUIDS	–	–	–
ENVIRONMENTALLY HAZARDOUS SUBSTANCE, LIQUID, N.O.S.	–	9	3082
ENVIRONMENTALLY HAZARDOUS SUBSTANCE, SOLID, N.O.S.	–	9	3077
EPIBROMOHYDRIN	P	6.1	2558
EPICHLOROHYDRIN	P	6.1	2023
EPN, *see* ORGANOPHOSPHORUS PESTICIDE	P	–	–
1,2-Epoxybutane, stabilized, *see*	–	3	3022
1,2-Epoxyethane, *see*	–	2.3	1040
1,2-Epoxyethane with nitrogen up to a total pressure of 1 MPa (10 bar) at 50°C, *see*	–	2.3	1040
1,2-EPOXY-3-ETHOXYPROPANE	–	3	2752
2,3-Epoxy-1-propanal, *see*	–	3	2622
1,2-Epoxypropane, *see*	–	3	1280
2,3-Epoxypropionaldehyde, *see*	–	3	2622
2,3-Epoxypropyl ethyl ether, *see*	–	3	2752
Esfenvalerate, *see* **Note 1**	P	–	–
ESTERS, N.O.S.	–	3	3272
Ethanal, *see*	–	3	1089
ETHANE	–	2.1	1035
ETHANE, REFRIGERATED LIQUID	–	2.1	1961
Ethanethiol, *see*	P	3	2363
Ethanoic anhydride, *see*	–	8	1715
ETHANOL	–	3	1170
ETHANOLAMINE	–	8	2491
ETHANOLAMINE SOLUTION	–	8	2491
ETHANOL AND GASOLINE MIXTURE, with more than 10% ethanol	–	3	3475
ETHANOL AND MOTOR SPIRIT MIXTURE, with more than 10% ethanol	–	3	3475
ETHANOL AND PETROL MIXTURE, with more than 10% ethanol	–	3	3475
ETHANOL SOLUTION	–	3	1170
Ethanoyl chloride, *see*	–	3	1717
Ether, *see*	–	3	1155
ETHERS, N.O.S.	–	3	3271
Ethion, *see* ORGANOPHOSPHORUS PESTICIDE	P	–	–
Ethoate-methyl, *see* ORGANOPHOSPHORUS PESTICIDE	–	–	–
Ethoprophos, *see* ORGANOPHOSPHORUS PESTICIDE	P	–	–
2-(*N*,*N*-Ethoxycarbonylphenylamino)-3-methoxy-4-(*N*-methyl-*N*-cyclohexylamino)benzenediazonium zinc chloride (concentration 62%), *see*	–	4.1	3236
2-(*N*,*N*-Ethoxycarbonylphenylamino)-3-methoxy-4-(*N*-methyl-*N*-cyclohexylamino)benzenediazonium zinc chloride (concentration 63–92%), *see*	–	4.1	3236

INDEX

INDEX

Substance, material or article	MP	Class	UN No.
Fire extinguisher charges, expelling, explosive, *see* CARTRIDGES, POWER DEVICE	–	–	–
FIRE EXTINGUISHERS with compressed or liquefied gas	–	2.2	1044
FIRELIGHTERS, SOLID with flammable liquid	–	4.1	2623
FIREWORKS	–	1.1G	0333
FIREWORKS	–	1.2G	0334
FIREWORKS	–	1.3G	0335
FIREWORKS	–	1.4G	0336
FIREWORKS	–	1.4S	0337
FIRST AID KIT	–	9	3316
FISHMEAL, STABILIZED anti-oxidant treated. Moisture content greater than 5% but not exceeding 12%, by mass. Fat content not more than 15%	–	4.2	2216
FISHMEAL, UNSTABILIZED. High hazard. Unrestricted moisture content. Unrestricted fat content in excess of 12%, by mass. Unrestricted fat content in excess of 15%, by mass, in the case of anti-oxidant treated fishmeal	–	4.2	1374
FISHMEAL, UNSTABILIZED not anti-oxidant treated. Moisture content: more than 5% but not more than 12%, by mass. Fat content: not more than 12%, by mass	–	4.2	1374
FISHSCRAP, STABILIZED anti-oxidant treated. Moisture content greater than 5% but not exceeding 12%, by mass. Fat content not more than 15%	–	9	2216
FISHSCRAP, UNSTABILIZED. High hazard. Unrestricted moisture content. Unrestricted fat content in excess of 12%, by mass. Unrestricted fat content in excess of 15% by mass, in the case of anti-oxidant treated fishscrap	–	4.2	1374
FISHSCRAP, UNSTABILIZED not anti-oxidant treated. Moisture content: more than 5% but not more than 12%, by mass. Fat content: not more than 12%, by mass	–	4.2	1374
Flammable gas in lighters, *see*	–	2.1	1057
FLAMMABLE LIQUID, CORROSIVE, N.O.S.	–	3	2924
FLAMMABLE LIQUID, N.O.S.	–	3	1993
FLAMMABLE LIQUID, TOXIC, CORROSIVE, N.O.S.	–	3	3286
FLAMMABLE LIQUID, TOXIC, N.O.S.	–	3	1992
FLAMMABLE SOLID, CORROSIVE, INORGANIC, N.O.S.	–	4.1	3180
FLAMMABLE SOLID, CORROSIVE, ORGANIC, N.O.S.	–	4.1	2925
FLAMMABLE SOLID, INORGANIC, N.O.S.	–	4.1	3178
FLAMMABLE SOLID, ORGANIC, MOLTEN, N.O.S.	–	4.1	3176
FLAMMABLE SOLID, ORGANIC, N.O.S.	–	4.1	1325
FLAMMABLE SOLID, OXIDIZING, N.O.S.	–	4.1	3097
FLAMMABLE SOLID, TOXIC, INORGANIC, N.O.S.	–	4.1	3179
FLAMMABLE SOLID, TOXIC, ORGANIC, N.O.S.	–	4.1	2926
FLARES, AERIAL	–	1.1G	0420
FLARES, AERIAL	–	1.2G	0421
FLARES, AERIAL	–	1.3G	0093
FLARES, AERIAL	–	1.4G	0403
FLARES, AERIAL	–	1.4S	0404

Substance, material or article	MP	Class	UN No.
Flares, distress, small, *see* SIGNAL DEVICES, HAND	–	–	–
Flares, highway or railway, *see* SIGNAL DEVICES, HAND	–	–	–
FLARES, SURFACE	–	1.1G	0418
FLARES, SURFACE	–	1.2G	0419
FLARES, SURFACE	–	1.3G	0092
Flares, water-activated, *see* CONTRIVANCES, WATER-ACTIVATED	–	–	–
FLASH POWDER	–	1.1G	0094
FLASH POWDER	–	1.3G	0305
Flax, dry, *see*	–	4.1	3360
Flowers of sulphur, *see*	–	4.1	1350
Flue dust, arsenical, *see*	–	6.1	1562
Fluoric acid, *see*	–	8	1790
Fluorine compounds (pesticides), *see* PESTICIDE, N.O.S.	–	–	–
FLUORINE, COMPRESSED	–	2.3	1045
Fluorine monoxide, compressed, *see*	–	2.3	2190
Fluoroacetamide, *see* PESTICIDE, N.O.S.	–	–	–
FLUOROACETIC ACID	–	6.1	2642
FLUOROANILINES	–	6.1	2941
FLUOROBENZENE	–	3	2387
FLUOROBORIC ACID	–	8	1775
Fluoroethane, *see*	–	2.1	2453
Fluoroethanoic acid, *see*	–	6.1	2642
Fluoroform, *see*	–	2.2	1984
Fluoroformyl fluoride, compressed, *see*	–	2.3	2417
Fluoromethane, *see*	–	2.1	2454
FLUOROPHOSPHORIC ACID, ANHYDROUS	–	8	1776
FLUOROSILICATES, N.O.S.	–	6.1	2856
FLUOROSILICIC ACID	–	8	1778
FLUOROSULPHONIC ACID	–	8	1777
FLUOROTOLUENES	–	3	2388
Fonofos, *see* ORGANOPHOSPHORUS PESTICIDE	P	–	–
Formal, *see*	–	3	1234
Formaldehyde dimethylacetal, *see*	–	3	1234
FORMALDEHYDE SOLUTION, FLAMMABLE	–	3	1198
FORMALDEHYDE SOLUTION with not less than 25% formaldehyde	–	8	2209
Formalin solution, flammable, *see*	–	3	1198
Formalin solution with not less than 25% formaldehyde, *see*	–	8	2209
Formamidine sulphinic acid, *see*	–	4.2	3341
Formetanate, *see* CARBAMATE PESTICIDE	P	–	–
Formic acid ethyl ester, *see*	–	3	1190
FORMIC ACID with more than 85% acid, by mass	–	8	1779
FORMIC ACID with not less than 5% but less than 10% acid, by mass	–	8	3412

INDEX

INDEX

INDEX

INDEX

Substance, material or article	MP	Class	UN No.
HYDROGEN AND METHANE MIXTURE, COMPRESSED	–	2.1	2034
Hydrogen antimonide, *see*	–	2.3	2676
Hydrogen arsenide, *see*	–	2.3	2188
Hydrogen bromide, *see*	–	8	1788
HYDROGEN BROMIDE, ANHYDROUS	–	2.3	1048
Hydrogen bromide solution, *see*	–	8	1788
Hydrogencarboxylic acid, *see*	–	8	1779
Hydrogen chloride, *see*	–	8	1789
HYDROGEN CHLORIDE, ANHYDROUS	–	2.3	1050
HYDROGEN CHLORIDE, REFRIGERATED LIQUID **(transport prohibited)**	–	2.3	2186
HYDROGEN, COMPRESSED	–	2.1	1049
HYDROGEN CYANIDE, AQUEOUS SOLUTION with not more than 20% hydrogen cyanide	P	6.1	1613
△ HYDROGEN CYANIDE, SOLUTION IN ALCOHOL with more than 45% hydrogen cyanide **(transport prohibited)**	–	–	–
△ HYDROGEN CYANIDE, SOLUTION IN ALCOHOL with not more than 45% hydrogen cyanide	P	6.1	3294
HYDROGEN CYANIDE, STABILIZED containing less than 3% water	P	6.1	1051
HYDROGEN CYANIDE, STABILIZED containing less than 3% water and absorbed in a porous inert material	P	6.1	1614
HYDROGENDIFLUORIDES, SOLID, N.O.S.	–	8	1740
HYDROGENDIFLUORIDES SOLUTION, N.O.S.	–	8	3471
Hydrogen fluoride, *see*	–	8	1790
HYDROGEN FLUORIDE, ANHYDROUS	–	8	1052
HYDROGEN IN A METAL HYDRIDE STORAGE SYSTEM	–	2.1	3468
HYDROGEN IN A METAL HYDRIDE STORAGE SYSTEM CONTAINED IN EQUIPMENT	–	2.1	3468
HYDROGEN IN A METAL HYDRIDE STORAGE SYSTEM PACKED WITH EQUIPMENT	–	2.1	3468
Hydrogen iodide, *see*	–	8	1787
HYDROGEN IODIDE, ANHYDROUS	–	2.3	2197
HYDROGEN PEROXIDE AND PEROXYACETIC ACID MIXTURE with acid(s), water and not more than 5% peroxyacetic acid, STABILIZED	–	5.1	3149
HYDROGEN PEROXIDE, AQUEOUS SOLUTION, STABILIZED with more than 60% hydrogen peroxide	–	5.1	2015
HYDROGEN PEROXIDE, AQUEOUS SOLUTION with not less than 8% but less than 20% hydrogen peroxide (stabilized as necessary)	–	5.1	2984
HYDROGEN PEROXIDE, AQUEOUS SOLUTION with not less than 20% but not more than 60% hydrogen peroxide (stabilized as necessary)	–	5.1	2014
△ Hydrogen peroxide carbamide, solid, *see*	–	5.1	1511
HYDROGEN PEROXIDE, STABILIZED	–	5.1	2015
Hydrogen phosphide, *see*	–	2.3	2199
HYDROGEN, REFRIGERATED LIQUID	–	2.1	1966
HYDROGEN SELENIDE, ANHYDROUS	–	2.3	2202
Hydrogen silicide, compressed, *see*	–	2.1	2203

INDEX

INDEX

Substance, material or article	MP	Class	UN No.
ISOPHORONE DIISOCYANATE	–	6.1	2290
ISOPRENE, STABILIZED	–	3	1218
Isoprocarb, *see* CARBAMATE PESTICIDE	P	–	–
ISOPROPANOL	–	3	1219
ISOPROPENYL ACETATE	–	3	2403
ISOPROPENYLBENZENE	–	3	2303
Isopropenyl carbinol, *see*	–	3	2614
Isopropenyl chloride, *see*	–	3	2456
2-Isopropoxypropane, *see*	–	3	1159
ISOPROPYL ACETATE	–	3	1220
ISOPROPYL ACID PHOSPHATE	–	8	1793
ISOPROPYL ALCOHOL	–	3	1219
ISOPROPYLAMINE	–	3	1221
ISOPROPYLBENZENE	–	3	1918
Isopropyl bromide, *see*	–	3	2344
Isopropyl *sec*-butyl peroxydicarbonate (concentration ≤32%) with di-*sec*-butyl peroxydicarbonate (concentration ≤15–18%) and di-isopropyl peroxydicarbonate (concentration ≤12–15%), with diluent Type A, *see*	–	5.2	3115
Isopropyl *sec*-butyl peroxydicarbonate (concentration ≤52%) with di-*sec*-butyl peroxydicarbonate (concentration ≤28%) and di-isopropyl peroxydicarbonate (concentration ≤22%), *see*	–	5.2	3111
ISOPROPYL BUTYRATE	–	3	2405
Isopropyl carbinol, *see*	–	3	1212
Isopropyl chloride, *see*	–	3	2356
ISOPROPYL CHLOROACETATE	–	3	2947
Isopropyl chlorocarbonate, *see*	–	6.1	2407
ISOPROPYL CHLOROFORMATE	–	6.1	2407
Isopropyl chloromethanoate, *see*	–	6.1	2407
ISOPROPYL 2-CHLOROPROPIONATE	–	3	2934
alpha-Isopropyl *alpha*-chloropropionate, *see*	–	3	2934
Isopropylcumyl hydroperoxide (concentration ≤72%, with diluent Type A), *see*	–	5.2	3109
Isopropyl cyanide, *see*	–	3	2284
Isopropyl ether, *see*	–	3	1159
Isopropylethylene, *see*	–	3	2561
Isopropyl formate, *see*	–	3	1281
Isopropylideneacetone, *see*	–	3	1229
ISOPROPYL ISOBUTYRATE	–	3	2406
△ ISOPROPYL ISOCYANATE	–	6.1	2483
Isopropyl mercaptan, *see*	–	3	2402
Isopropyl methanoate, *see*	–	3	1281
ISOPROPYL NITRATE	–	3	1222
ISOPROPYL PROPIONATE	–	3	2409
Isopropyltoluene, *see*	P	3	2046

INDEX

INDEX

INDEX

INDEX

INDEX

INDEX

Substance, material or article	MP	Class	UN No.
ORGANIC PIGMENTS, SELF-HEATING	–	4.2	3313
ORGANOARSENIC COMPOUND, LIQUID, N.O.S.	–	6.1	3280
ORGANOARSENIC COMPOUND, SOLID, N.O.S.	–	6.1	3465
ORGANOCHLORINE PESTICIDE, LIQUID, FLAMMABLE, TOXIC, flashpoint less than 23°C	–	3	2762
ORGANOCHLORINE PESTICIDE, LIQUID, TOXIC	–	6.1	2996
ORGANOCHLORINE PESTICIDE, LIQUID, TOXIC, FLAMMABLE, flashpoint not less than 23°C	–	6.1	2995
ORGANOCHLORINE PESTICIDE, SOLID, TOXIC	–	6.1	2761
Organometallic compound dispersion, water-reactive, flammable, *see*	–	4.3	3399
Organometallic compound solid, water-reactive, flammable, *see*	–	4.3	3396
Organometallic compound solution, water-reactive, flammable, *see*	–	4.3	3399
ORGANOMETALLIC COMPOUND, LIQUID, TOXIC, N.O.S.	–	6.1	3282
ORGANOMETALLIC COMPOUND, SOLID, TOXIC, N.O.S.	–	6.1	3467
ORGANOMETALLIC SUBSTANCE, LIQUID, PYROPHORIC	–	4.2	3392
ORGANOMETALLIC SUBSTANCE, LIQUID, PYROPHORIC, WATER-REACTIVE	–	4.2	3394
ORGANOMETALLIC SUBSTANCE, LIQUID, WATER-REACTIVE	–	4.3	3398
ORGANOMETALLIC SUBSTANCE, LIQUID, WATER-REACTIVE, FLAMMABLE	–	4.3	3399
ORGANOMETALLIC SUBSTANCE, SOLID, PYROPHORIC	–	4.2	3391
ORGANOMETALLIC SUBSTANCE, SOLID, PYROPHORIC, WATER-REACTIVE	–	4.2	3393
ORGANOMETALLIC SUBSTANCE, SOLID, SELF-HEATING	–	4.2	3400
ORGANOMETALLIC SUBSTANCE, SOLID, WATER-REACTIVE	–	4.3	3395
ORGANOMETALLIC SUBSTANCE, SOLID, WATER-REACTIVE, FLAMMABLE	–	4.3	3396
ORGANOMETALLIC SUBSTANCE, SOLID, WATER-REACTIVE, SELF-HEATING	–	4.3	3397
ORGANOPHOSPHORUS COMPOUND, TOXIC, FLAMMABLE, N.O.S.	–	6.1	3279
ORGANOPHOSPHORUS COMPOUND, LIQUID, TOXIC, N.O.S.	–	6.1	3278
ORGANOPHOSPHORUS COMPOUND, SOLID, TOXIC, N.O.S.	–	6.1	3464
ORGANOPHOSPHORUS PESTICIDE, LIQUID, FLAMMABLE, TOXIC, flashpoint less than 23°C	–	3	2784
ORGANOPHOSPHORUS PESTICIDE, LIQUID, TOXIC	–	6.1	3018
ORGANOPHOSPHORUS PESTICIDE, LIQUID, TOXIC, FLAMMABLE, flashpoint not less than 23°C	–	6.1	3017
ORGANOPHOSPHORUS PESTICIDE, SOLID, TOXIC	–	6.1	2783
ORGANOTIN COMPOUND, LIQUID, N.O.S.	P	6.1	2788
ORGANOTIN COMPOUND, SOLID, N.O.S.	P	6.1	3146
Organotin compounds (pesticides), *see* ORGANOTIN PESTICIDE	P	–	–
ORGANOTIN PESTICIDE, LIQUID, FLAMMABLE, TOXIC, flashpoint less than 23°C	P	3	2787
ORGANOTIN PESTICIDE, LIQUID, TOXIC	P	6.1	3020
ORGANOTIN PESTICIDE, LIQUID, TOXIC, FLAMMABLE, flashpoint not less than 23°C	P	6.1	3019

Substance, material or article	MP	Class	UN No.
ORGANOTIN PESTICIDE, SOLID, TOXIC	P	6.1	2786
Orthoarsenic acid, *see*	–	6.1	1553
Orthophosphoric acid, liquid, *see*	–	8	1805
Orthophosphoric acid, solid, *see*	–	8	3453
OSMIUM TETROXIDE	P	6.1	2471
Oxamyl, *see* PESTICIDE, N.O.S.	P	–	–
OXIDIZING LIQUID, CORROSIVE, N.O.S.	–	5.1	3098
OXIDIZING LIQUID, N.O.S.	–	5.1	3139
OXIDIZING LIQUID, TOXIC, N.O.S.	–	5.1	3099
OXIDIZING SOLID, CORROSIVE, N.O.S.	–	5.1	3085
OXIDIZING SOLID, FLAMMABLE, N.O.S.	–	5.1	3137
OXIDIZING SOLID, N.O.S.	–	5.1	1479
OXIDIZING SOLID, SELF-HEATING, N.O.S.	–	5.1	3100
OXIDIZING SOLID, TOXIC, N.O.S.	–	5.1	3087
OXIDIZING SOLID, WATER-REACTIVE, N.O.S.	–	5.1	3121
Oxirane, *see*	–	2.3	1040
Oxirane with nitrogen up to a total pressure of 1 MPa (10 bar) at 50°C	–	2.3	1040
Oxydemeton-methyl, *see* ORGANOPHOSPHORUS PESTICIDE	–	–	–
Oxydisulfoton, *see* ORGANOPHOSPHORUS PESTICIDE	P	–	–
OXYGEN, COMPRESSED	–	2.2	1072
OXYGEN DIFLUORIDE, COMPRESSED	–	2.3	2190
Oxygen fluoride, compressed, *see*	–	2.3	2190
OXYGEN GENERATOR, CHEMICAL	–	5.1	3356
OXYGEN, REFRIGERATED LIQUID	–	2.2	1073
1-Oxy-4-nitrobenzene, *see*	–	6.1	1662
△ PAINT (including paint, lacquer, enamel, stain, shellac, varnish, polish, liquid filler and liquid lacquer base)	–	3	1263
PAINT (including paint, lacquer, enamel, stain, shellac, varnish, polish, liquid filler and liquid lacquer base)	–	8	3066
PAINT, CORROSIVE, FLAMMABLE (including paint, lacquer, enamel, stain, shellac, varnish, polish, liquid filler and liquid lacquer base)	–	8	3470
PAINT, FLAMMABLE, CORROSIVE (including paint, lacquer, enamel, stain, shellac, varnish, polish, liquid filler and liquid lacquer base)	–	3	3469
PAINT RELATED MATERIAL (including paint thinning or reducing compound)	–	3	1263
PAINT RELATED MATERIAL (including paint thinning or reducing compound)	–	8	3066
PAPER, UNSATURATED OIL TREATED incompletely dried (including carbon paper)	–	4.2	1379
Para-acetaldehyde, *see*	–	3	1264
⊗			
PARAFORMALDEHYDE	–	4.1	2213
PARALDEHYDE	–	3	1264
Paraoxon, *see* ORGANOPHOSPHORUS PESTICIDE	P	–	–

INDEX

INDEX

INDEX

Substance, material or article	MP	Class	UN No.
POTASSIUM SODIUM ALLOYS, SOLID	–	4.3	3404
POTASSIUM SULPHIDE, ANHYDROUS	–	4.2	1382
POTASSIUM SULPHIDE, HYDRATED with not less than 30% water of crystallization	–	8	1847
POTASSIUM SULPHIDE with less than 30% water of crystallization	–	4.2	1382
POTASSIUM SUPEROXIDE	–	5.1	2466
Potassium tetracyanomercurate(II), *see*	–	6.1	1626
Potassium vanadate, *see*	–	6.1	2864
POWDER CAKE, WETTED with not less than 17% alcohol, by mass	–	1.1C	0433
POWDER CAKE, WETTED with not less than 25% water, by mass	–	1.3C	0159
POWDER PASTE, WETTED with not less than 17% alcohol, by mass	–	1.1C	0433
POWDER PASTE, WETTED with not less than 25% water, by mass	–	1.3C	0159
POWDER, SMOKELESS	–	1.1C	0160
POWDER, SMOKELESS	–	1.3C	0161
POWDER, SMOKELESS	–	1.4C	0509
Power devices, explosive, *see* CARTRIDGES, POWER DEVICE	–	–	–
Praseodymium nitrate and neodymium nitrate mixture, *see*	–	5.1	1465
PRIMERS, CAP TYPE	–	1.1B	0377
PRIMERS, CAP TYPE	–	1.4B	0378
PRIMERS, CAP TYPE	–	1.4S	0044
Primers, small arms, *see*	–	1.4S	0044
PRIMERS, TUBULAR	–	1.3G	0319
PRIMERS, TUBULAR	–	1.4G	0320
PRIMERS, TUBULAR	–	1.4S	0376
PRINTING INK flammable	–	3	1210
PRINTING INK RELATED MATERIAL (including printing ink thinning or reducing compound), flammable	–	3	1210
Projectiles, illuminating, *see* AMMUNITION, ILLUMINATING	–	–	–
PROJECTILES inert, with tracer	–	1.3G	0424
PROJECTILES inert, with tracer	–	1.4G	0425
PROJECTILES inert, with tracer	–	1.4S	0345
PROJECTILES with burster or expelling charge	–	1.2D	0346
PROJECTILES with burster or expelling charge	–	1.2F	0426
PROJECTILES with burster or expelling charge	–	1.2G	0434
PROJECTILES with burster or expelling charge	–	1.4D	0347
PROJECTILES with burster or expelling charge	–	1.4F	0427
PROJECTILES with burster or expelling charge	–	1.4G	0435
PROJECTILES with bursting charge	–	1.1D	0168
PROJECTILES with bursting charge	–	1.1F	0167
PROJECTILES with bursting charge	–	1.2D	0169
PROJECTILES with bursting charge	–	1.2F	0324
PROJECTILES with bursting charge	–	1.4D	0344
Promecarb, *see* CARBAMATE PESTICIDE	P	–	–
Promurit, *see* CARBAMATE PESTICIDE	–	–	–

INDEX

INDEX

INDEX

INDEX

INDEX

Substance, material or article	MP	Class	UN No.
Trichlorobutylene, *see*	P	6.1	2322
1,1,1-TRICHLOROETHANE	–	6.1	2831
1,1,2-Trichloroethane, *see*	–	9	3082
TRICHLOROETHYLENE	–	6.1	1710
TRICHLOROISOCYANURIC ACID, DRY	–	5.1	2468
Trichloromethane, *see*	–	6.1	1888
Trichloromethanesulphuryl chloride, *see*	P	6.1	1670
Trichloromethyl sulphochloride, *see*	P	6.1	1670
Trichloronat, *see* ORGANOPHOSPHORUS PESTICIDE	P	–	–
Trichloronitromethane, *see*	–	6.1	1580
TRICHLOROSILANE	–	4.3	1295
2,4,6-Trichloro-1,3,5-triazine, *see*	–	8	2670
1,3,5-Trichloro-*s*-triazine-2,4,6-trione, *see*	–	5.1	2468
Tricresyl phosphate, less than 1% *ortho*-isomer, *see*	P	9	3082
Tricresyl phosphate, not less than 1% but not more than 3% *ortho*-isomer, *see*	P	9	3082
TRICRESYL PHOSPHATE with more than 3% *ortho*-isomer	P	6.1	2574
Tricyanogen chloride, *see*	–	8	2670
Triethoxyboron, *see*	–	3	1176
Triethoxymethane, *see*	–	3	2524
TRIETHYLAMINE	–	3	1296
Triethylbenzene, *see*	P	9	3082
Triethyl borate, *see*	–	3	1176
Triethylenephosphoramide solution, *see*	–	6.1	2501
TRIETHYLENETETRAMINE	–	8	2259
■ 3,6,9-TRIETHYL-3,6,9-TRIMETHYL-1,4,7-TRIPEROXONATE (concentration ≤ 17%, with diluent Type A, with inert solid)	–	5.2	3110
Triethyl orthoformate, *see*	–	3	2524
TRIETHYL PHOSPHITE	–	3	2323
3,6,9-Triethyl-3,6,9-trimethyl-1,4,7-triperoxonane (concentration ≤ 42%, with diluent Type A, available oxygen ≤ 7.6%), *see*	–	5.2	3105
TRIFLUOROACETIC ACID	–	8	2699
TRIFLUOROACETYL CHLORIDE	–	2.3	3057
Trifluorobromomethane, *see*	–	2.2	1009
Trifluorochloroethane, *see*	–	2.2	1983
TRIFLUOROCHLOROETHYLENE, STABILIZED	–	2.3	1082
Trifluorochloromethane, *see*	–	2.2	1022
1,1,1-TRIFLUOROETHANE	–	2.1	2035
TRIFLUOROMETHANE	–	2.2	1984
Trifluoromethane and chlorotrifluoromethane azeotropic mixture, *see* CHLOROTRIFLUOROMETHANE AND TRIFLUOROMETHANE AZEOTROPIC MIXTURE	–	–	–
TRIFLUOROMETHANE, REFRIGERATED LIQUID	–	2.2	3136
Trifluoromethoxytrifluoroethylene, *see*	–	2.1	3153
2-TRIFLUOROMETHYLANILINE	–	6.1	2942

INDEX

Substance, material or article	MP	Class	UN No.
ZIRCONIUM POWDER, WETTED with not less than 25% water (a visible excess of water must be present) (b) chemically produced, particle size less than 840 microns	–	4.1	1358
ZIRCONIUM, SCRAP	–	4.2	1932
ZIRCONIUM, SUSPENDED IN A FLAMMABLE LIQUID	–	3	1308
ZIRCONIUM TETRACHLORIDE	–	8	2503

Notes

Notes

Notes

Notes

Notes

Notes

Notes

Notes

Notes

Notes

Notes

Notes